D0025248

DESIGN OF FEEDBACK CONTROL SYSTEMS

Gene H. Hostetter
University of California, Irvine

Clement J. Savant, Jr.
California State University, Long Beach

Raymond T. Stefani
California State University, Long Beach

Holt, Rinehart and Winston

New York Chicago San Francisco
Philadelphia Montreal Toronto London
Sydney Tokyo Mexico City
Rio de Janeiro Madrid

Library of Congress Cataloging in Publication Data

Hostetter, G. H., 1939–
Design of feedback control systems.

Includes bibliographies and index.
1. Feedback control systems. I. Savant, C. J.
II. Stefani, R. T. III. Title.
TJ216.H63 629.8′312 81-6371
ISBN 0-03-057593-1 AACR2

Address correspondence to:
383 Madison Avenue
New York, N.Y. 10017

Printed in the United States of America

2 3 4 038 9 8 7 6 5 4 3 2

CBS COLLEGE PUBLISHING
Holt, Rinehart and Winston
The Dryden Press
Saunders College Publishing

To
Donna, Kelly, and Kristen
Barbara
Val, Ted, and Rick

Preface

This is a design-oriented control systems text intended for use in an introductory academic course and for reference and self-study by electrical and mechanical engineers in industry. Laplace transforms and electrical and/or mechanical network analysis are the prerequisite subjects upon which this text builds. It is especially well suited for a one-term junior- or early senior-level first course in control systems.

The manuscript evolved over several years of combined effort to provide an interesting, relevant, and effective introductory control system design class at California State University, Long Beach. Much of the understanding and skill we once taught specifically as it applied to networks is here presented in a more general context, with a wider immediate applicability. When used as a prerequisite for a sequence of modern control courses, this material also greatly reduces the large amount of time and effort that would otherwise be expended in establishing (then parting from) classical concepts. And it serves to encourage broad interests, perspectives, and skills at an early stage.

The greatly increased availability of digital computers naturally poses questions as to the proper and best role of computers in design. As the creative aspect of system design continues to involve the *directed* use of analytical tools, the emphasis here is upon the understanding, practical experience, and judgment necessary to be a creative designer. The manner in which the analytic tools are employed (hand calculation, pocket calculator, or computer) is taken to be of secondary concern at this introductory stage.

This text is designed to guide the reader in gaining the following:

1. A review of the fundamentals of electrical, translational mechanical, rotational mechanical, and electromechanical networks
2. Confidence in the use of Laplace transform methods in system response calculation and an understanding of commonly used response components
3. Familiarity with the use of transfer functions for linear, time-invariant systems, including asymptotic stability concepts and multivariable relations
4. Capability with block diagram manipulations, signal flow graphs, and the use of Mason's gain rule
5. Thorough acquaintance with Routh-Hurwitz polynomial testing and the ability to determine root distributions, to test adjustable systems, and to axis-shift to find relative stability
6. Appreciation of the feedback concept and its importance to tracking and other systems; familiarity with steady state error concepts and calculation and an acquaintance with

system parameter sensitivity, susceptability to disturbances, and the use of performance indices

7. A thorough understanding of root locus methods, including those for adjustable systems other than the unity subtractive feedback type
8. Experience with classical compensation methods and real understanding of the principles guiding compensator design
9. Good comprehension and ability with frequency response methods, including those for systems involving delay elements and unstable components; appreciation of the powerful approach of incorporating experimental data
10. Competence with polar response plots and Nyquist methods
11. Familiarity with basic state variable concepts, including simulation diagrams, time-domain vector-matrix equations and solutions, transfer function matrices, stability, diagonalization, observability, and controllability; an introduction to state feedback design
12. Acquaintance with digital control concepts, sampling, and discrete-time system models at a level suitable for easy transition to study of computer and microprocessor-based

Quarter System Schedule

Week	*Chapters*	*Topics*
1	1	Introduction to the course System equations and terminology Review of Laplace transform Transfer functions
2	2–3	Block diagrams and signal flow graphs System response Stability and Routh-Hurwitz testing
3	4	Steady state errors Sensitivity and disturbance rejection Performance indices, optimality, and design
4, 5	5	Root locus construction and examples System compensation Design using root locus methods
		Midterm Examination
6, 7	6	Bode plot construction Frequency response examples Gain and phase margins Design using frequency response methods The Nyquist criterion
8, 9	7	State variable system models Controllability and observability Time-domain response Response computation
10	8	Digital control concepts Sampling Discrete-time system models Introduction to digital control system design

Semester System Schedule

Week	*Chapters*	*Topics*
1, 2	1	Introduction to the course System equations and terminology Review of Laplace transform Transfer functions
3	2	Block diagrams and signal flow graphs Response of first-, second-, and higher-order systems
4	3	Stability and Routh-Hurwitz testing
5, 6	4	Steady state errors Sensitivity and disturbance rejection Performance indices, optimality, and design
7, 8	5	Root locus construction and examples System compensation Design using root locus methods
		Midterm Examination
9, 10	6	Bode plot construction Frequency response examples Gain and phase margins Design using frequency response methods The Nyquist criterion
11, 12	7	State variable system models Controllability and observability Time-domain response Response computation
13, 14	8	Digital control concepts Sampling Discrete-time system models Introduction to digital control system design

real-time systems such as that given in B. C. Kuo, *Digital Control Systems* (New York: Holt, Rinehart and Winston, 1980), G. F. Franklin and J. D. Powell, *Digital Control of Dynamic Systems* (Reading, Mass.: Addison-Wesley, 1980), and similar texts

Along the way, it is hoped that the reader will learn much about the iterative process of engineering design. We have found the large number of example systems included here to be invaluable to this learning process.

Suggested class textbook schedules for quarters and semesters are given in the accompanying tables. For some classes—for example, for an introductory controls course early in an engineering program—the range of material available here is more than should be covered in a single term. The text is designed so that it is easy to abbreviate or delete topics to achieve a desired emphasis. In a course emphasizing transition to modern control theory, the compensation material of Secs. 5.7 through the end of Chap. 5 may be omitted with no penalty in understanding of the later topics. In a course emphasizing the classical viewpoint, Chap. 7 on state space methods and Sec. 8.8, which ties these ideas to the digital domain, may be omitted without disturbing the flow of topics. Chapter 8 may be easily omitted or abridged as desired.

We sincerely hope that you or your students will enjoy reading and using this text as much as we have enjoyed its development. Our students have enthusiastically contributed to it as have our colleagues, especially L. Bailin, G. H. Cain, E. N. Evans, H. J. Lane, C. S. Lindquist, M. Santina, R. Rountree, and S. Wolf. Special thanks are due to Cynthia Klepadlo, who supervised the manuscript typing and the drafting of many of the original figures and to Mohammed Santina, who tirelessly compiled the problem solutions.

Gene H. Hostetter
Clement J. Savant
Raymond T. Stefani

Contents

2 TRANSFER FUNCTIONS AND SYSTEM RESPONSE 63

3 CHARACTERISTIC POLYNOMIAL STABILITY TESTING 111

4 PERFORMANCE SPECIFICATIONS 149

5 ROOT LOCUS ANALYSIS AND DESIGN 199

DESIGN OF FEEDBACK CONTROL SYSTEMS

one

INTRODUCTORY CONCEPTS

1.1 PREVIEW

This study begins by introducing control system terminology. The feedback concept is very important to control system design, and examples are given to demonstrate associated ideas and properties in Sec. 1.3.

Control system components are typically electrical, electronic, mechanical, and electromechanical devices. Mathematical models for selected components are then introduced and the relations are summarized in several tables in Sec. 1.4.

The manipulation and solution of system equations is greatly aided by Laplace transform methods which are reviewed and summarized in Sec. 1.5. Transfer functions, both for single-input, single-output and for multivariable systems are then defined and explained. The decomposition of system response into zero-state/zero-input and into forced/natural components is discussed carefully, and stability for linear, time-invariant systems is defined.

A position servo system is then discussed and used to reinforce understanding of the feedback principle. DC and AC control motors are analyzed, and transfer functions, including the coupling of a disturbance input, are computed from the system equations.

It is expected that much of the material of this first chapter will involve subjects known to the reader from previous study and experience. Our initial purpose is to bring together these introductory topics from several areas, relating them to systems and control.

1.2 CONTROL SYSTEMS AND TERMINOLOGY

Control systems influence every facet of modern life. Automatic washers and dryers, microwave ovens, chemical process plants, navigation and guidance systems, space satellites, pollution control, mass transit, and economic regulation are a few examples. A control system is, in the broadest sense, any interconnection of components to provide a desired function.

The portion of a system which is to be controlled is called the *plant* or the *process.* It is affected by applied signals, called *inputs*, and produces signals of particular interest, called *outputs*, as indicated in Fig. 1-1.

A *controller* may be used to produce a desired behavior of the plant, as shown in Fig. 1-2. The controller generates plant input signals designed to produce desired outputs. Some of the plant inputs are accessible to the designer and some are generally not available. The inaccessible input signals are often disturbances to the plant. The double lines in the figure indicate that several signals of each type may be involved. This system is termed *open-loop* because the control inputs are not influenced by the plant outputs; that is, there is no *feedback* around the plant.

Such an open-loop control system has the advantage of simplicity, but its performance is highly dependent upon the properties of the plant, which may vary with time. The disturbances to the plant may also create unwanted response which it would be desirable to reduce.

As an example, suppose a gasoline engine is used to drive a large pump, as depicted in Fig. 1-3. The carburetor and engine comprise a common type of control system wherein large-power output is controlled with a small-power input. The carburetor is the controller in this case, and the engine is the plant. The fuel rate is the control input, and the pump load is a disturbance signal. The

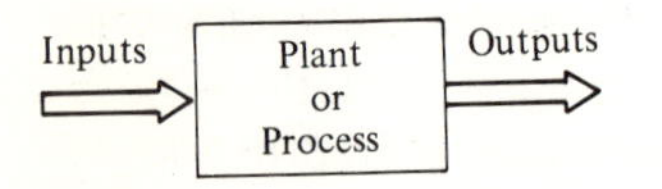

FIGURE 1-1
A plant or process to be controlled.

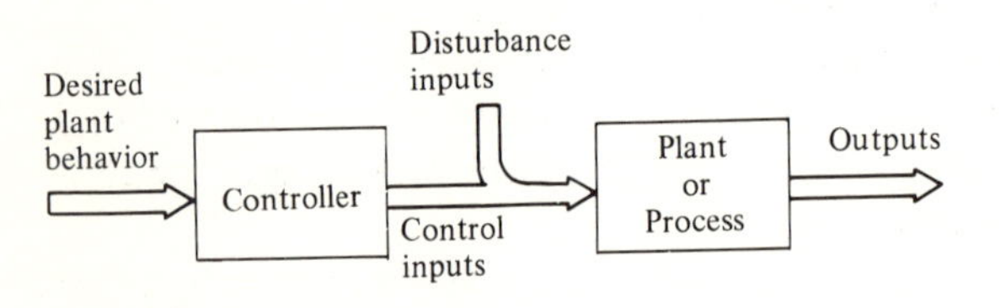

FIGURE 1-2
An open-loop control system.

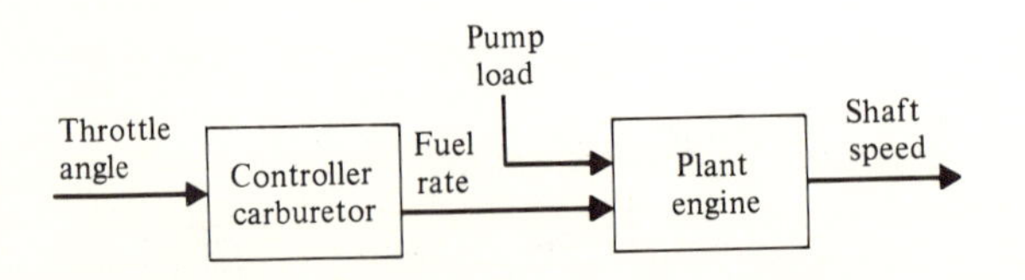

FIGURE 1-3
Example of an open-loop control system.

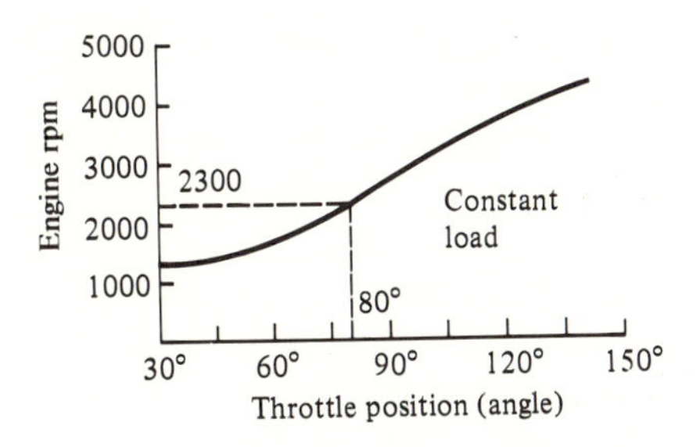

FIGURE 1-4
An engine speed vs. throttle angle curve.

desired plant output, a certain engine shaft speed, may be obtained by adjusting the throttle angle. The single lines in the figure indicate individual signals.

A representative plot of engine speed versus throttle angle is sketched in Fig. 1-4. This "calibration curve" gives the engine speed for a given throttle setting, at constant load on the engine. To produce an engine speed of 2300 rpm, for example, set the throttle angle to 80°. If the engine should become untuned (a change in the plant) or if the load should change (a disturbance), the calibration curve would change, and an 80° throttle angle would no longer produce a 2300-rpm engine speed.

In applications such as automatic washing machines and variable-speed hand drills, maintaining an accurate calibration curve is of little importance, within bounds. In other applications, such as laboratory instrumentation systems, it suffices to calibrate the system, reestablishing knowledge of the input-output relation often enough to obtain the desired accuracy. In systems such as the automobile, the human operator is capable of adjusting to changes and disturbances in the plant. In driving another's automobile for the first time, a new sense of "feel" must be established because no two automobiles produce exactly the same engine performance with the same accelerator setting.

1.3 THE FEEDBACK CONCEPT

If the requirements of the system cannot be satisfied with an open-loop control system, a *closed-loop* or *feedback* system is desirable. A path (or loop) is provided from the output back to the controller. Some or all of the system outputs are measured and used by the controller, as indicated in Fig. 1-5. The controller may then compare a desired plant output with the actual output and act to reduce the difference between the two.

Suppose that the system comprising a gasoline engine driving a pump is arranged in a closed-loop manner. One possible feedback control configuration is shown in Fig. 1-6. A tachometer produces a voltage proportional to the engine shaft speed. The input voltage, which is proportional to the desired speed, is set with a potentiometer. The tachometer voltage is subtracted from the input

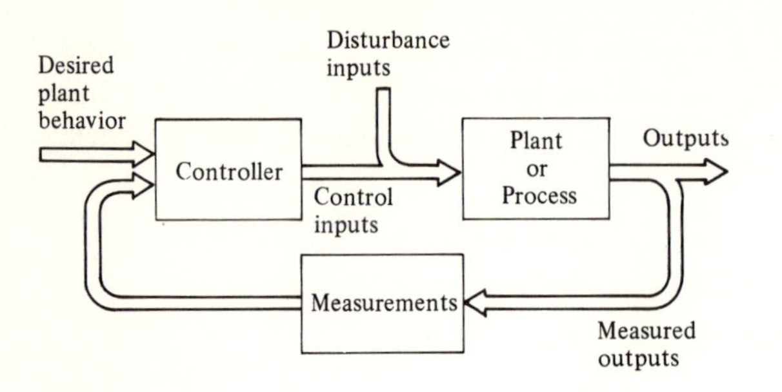

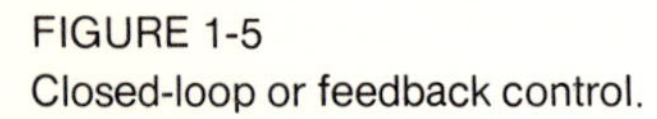
FIGURE 1-5
Closed-loop or feedback control.

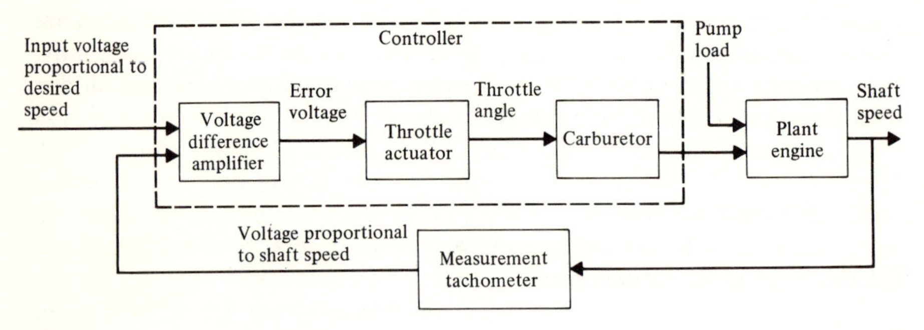

FIGURE 1-6
A closed-loop engine control system.

voltage, giving an error voltage which is proportional to the difference between the actual speed and the desired speed.

The error voltage is then amplified and used to position the throttle. The throttle actuator could be a reversible electric motor, geared to the throttle arm. When the engine shaft speed is equal to the desired speed (when the difference or *error* is zero), the throttle remains fixed. If a change in load or a change in the engine components should occur in the system, and the actual speed is no longer equal to the desired speed, the error voltage becomes nonzero, causing the throttle setting to change so that the actual speed approaches the desired speed. The controller here consists of the voltage difference amplifier, throttle actuator, and carburetor.

Some of the advantages which feedback control offer to the designer are

1. *Increased accuracy.* The closed-loop system may be designed to drive the error between desired and measured response to zero.
2. *Reduced sensitivity to changes in components.* As in the previous example, the system may be designed to seek zero error despite changes in the plant.
3. *Reduced effects of disturbances.* The effects of disturbances to the system may be greatly attenuated.
4. *Increased speed of response and bandwidth.* Feedback may be used to increase

the range of frequencies over which a system will respond and to make it respond more desirably. A satellite booster rocket, for example, has aerodynamics resembling those of a giant broomstick. It may, with feedback, behave with beauty and grace.

1.4 SYSTEM EQUATIONS AND MODELING

1.4.1 Control System Components

The first step in designing a control system is to obtain differential equations for those parts of the system which are fixed. The components of control systems commonly include electrical, electronic, mechanical, and electromechanical elements. This section gives a brief outline of the equations which characterize some common control system components and their connections. Many other less common types of elements, such as hydraulic, thermal, biological, and chemical ones, may also be involved in control system design.

1.4.2 Electrical Networks

Electrical networks are governed by the two Kirchhoff laws:

1. The algebraic sum of voltages around a closed loop equals zero.
2. The algebraic sum of currents flowing into a circuit node equals zero.

Network element models include resistors, capacitors, inductors, voltage sources, and current sources. The voltage-current relations for these are summarized in Table 1-1.

One systematic method of network analysis consists of defining a loop current in each mesh of a network, equating the algebraic sums of the voltages around each mesh to zero. It is assumed that current sources are absent. If current sources are present, the number of unknowns would include the voltages across such current sources. In return for each unknown current source voltage, a loop current or difference between loop currents is known.

As an example, consider the network of Fig. 1-7a, where the loop currents $i_1(t)$ and $i_2(t)$ have been defined. In Fig. 1-7b the resistor, inductor, and capacitor voltages have been expressed in terms of the loop currents. The application of Kirchhoff's voltage law around the i_1 loop gives

$$2i_1 + 3\frac{di_1}{dt} + 4(i_1 - i_2) + \frac{1}{5}\int_{-\infty}^{t} (i_1 - i_2)\, dt = 8 \cos 9t$$

Similarly, around the i_2 loop,

$$6\frac{di_2}{dt} + 7i_2 - 4(i_1 - i_2) - \frac{1}{5}\int_{-\infty}^{t} (i_1 - i_2)\, dt = 0$$

TABLE 1-1
Electrical network elements and voltage-current relations.

Independent Sources

Voltage source

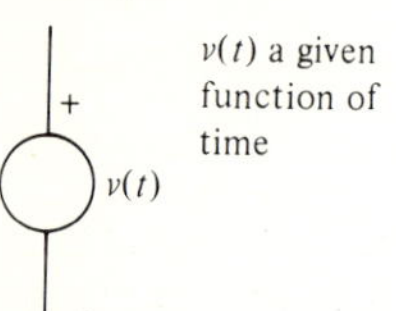

$v(t)$ a given function of time

Current source

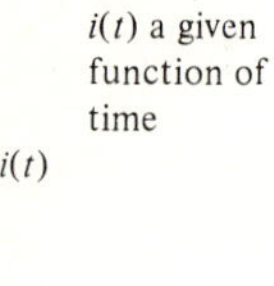

$i(t)$ a given function of time

Dependent Sources

Controlled voltage source

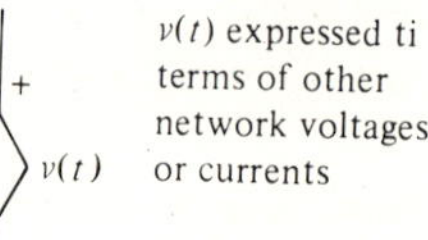

$v(t)$ expressed ti terms of other network voltages or currents

Controlled current source

$i(t)$ expressed in terms of other network voltages or currents

Element	*Differential Equation*
Resistor	$v(t) = Ri(t)$ $i(t) = \frac{1}{R} v(t)$
Inductor	$v(t) = L\frac{di}{dt}$ $i(t) = \frac{1}{L}\int_{-\infty}^{t} v(t)\,dt$
Capacitor	$v(t) = \frac{1}{C}\int_{-\infty}^{t} i(t)\,dt$ $i(t) = C\frac{dv}{dt}$

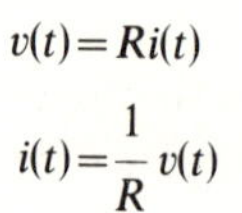

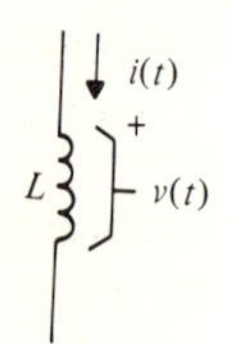

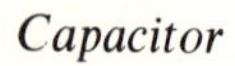

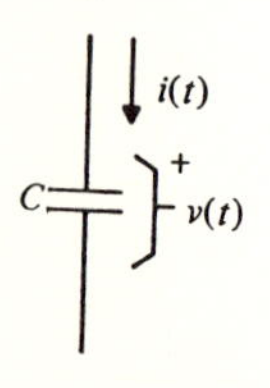

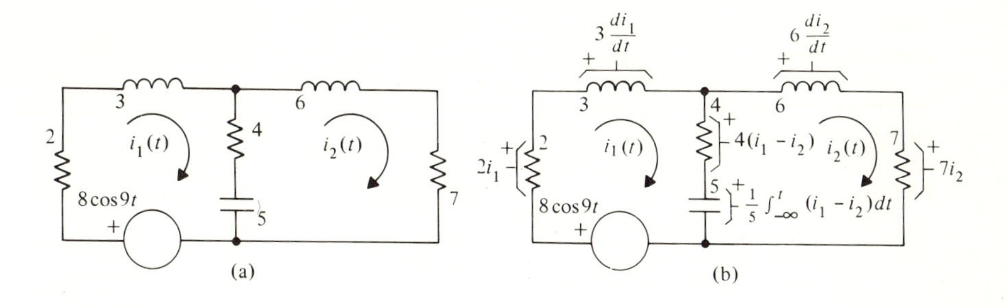

FIGURE 1-7
Writing systematic loop equations for an electrical network.
(a) Electrical network example for loop analysis. (b) Network with element voltages expressed in terms of the loop currents.

Collecting terms, the following simultaneous integrodifferential equations in i_1 and i_2 result:

$$\begin{cases} 3\dfrac{di_1}{dt} + 6i_1 + \dfrac{1}{5}\displaystyle\int_{-\infty}^{t} i_1\,dt - 4i_2 - \dfrac{1}{5}\displaystyle\int_{-\infty}^{t} i_2\,dt = 8\cos 9t \\ -4i_1 - \dfrac{1}{5}\displaystyle\int_{-\infty}^{t} i_1\,dt + 6\dfrac{di_2}{dt} + 11i_2 + \dfrac{1}{5}\displaystyle\int_{-\infty}^{t} i_2\,dt = 0 \end{cases}$$

In the nodal method of network analysis, one node in the network is chosen as the reference node and voltages between the reference node and each other node are defined. Expressing the element currents in terms of node voltages and applying Kirchhoff's current law at each node except the reference node gives the same number of independent simultaneous integrodifferential equations as there are node voltages. If voltage sources are present, the unknowns include the currents through the voltage sources, while the sources fix node-to-node voltages.

Nodal analysis is used to analyze the network of Fig. 1–8a, where the node voltages are labeled $v_1(t)$ and $v_2(t)$. In Fig. 1–8b, the branch currents are expressed in terms of these node voltages. Applying Kirchhoff's current law at node 1 gives

$$\frac{1}{2}\int_{-\infty}^{t} v_1\,dt + \frac{1}{3}v_1 + 4\frac{d}{dt}(v_1 - v_2) + \frac{1}{5}(v_1 - v_2) = 12$$

At node 2,

$$\frac{1}{6}v_2 - 4\frac{d}{dt}(v_1 - v_2) - \frac{1}{5}(v_1 - v_2) = -\sin t$$

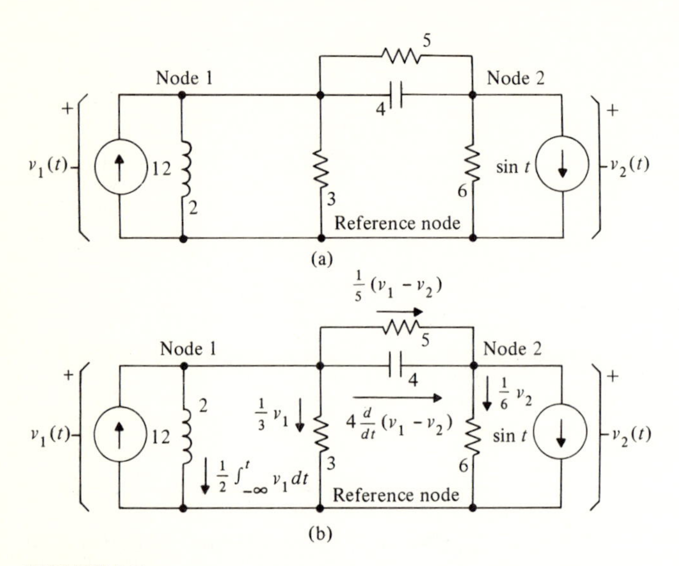

FIGURE 1-8
Writing systematic nodal equations for an electrical network. (a) Electrical network example for nodal analysis. (b) Network with element currents expressed in terms of the node voltages.

Collecting terms, the following two simultaneous integrodifferential equations in $v_1(t)$ and $v_2(t)$ result:

$$\begin{cases} 4\dfrac{dv_1}{dt}+\dfrac{8}{15}v_1+\dfrac{1}{2}\displaystyle\int_{-\infty}^{t} v_1\,dt-4\dfrac{dv_2}{dt}-\dfrac{1}{5}v_2=12 \\ -4\dfrac{dv_1}{dt}-\dfrac{1}{5}v_1+4\dfrac{dv_2}{dt}+\dfrac{11}{30}v_2=-\sin t \end{cases}$$

Simple models for other common electrical and electronic devices, expressed in terms of the basic R, L, C, and source elements, are given in Table 1-2.

TABLE 1-2
Simple models of some electrical and electronic devices.

1. *Transformer* with core in linear region

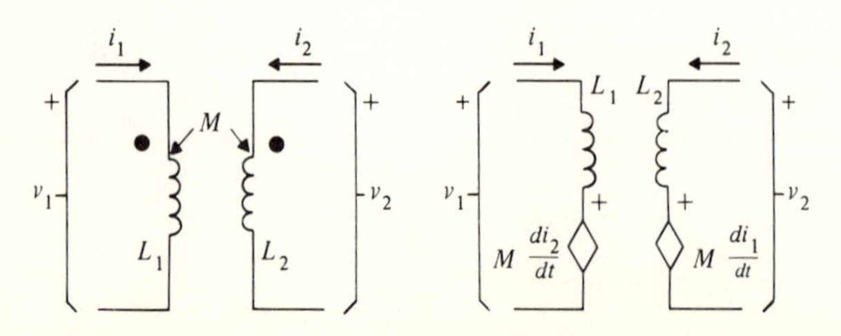

Resistance of the coil wires may be included as additional resistors at each port.

2. *Ideal transformer*

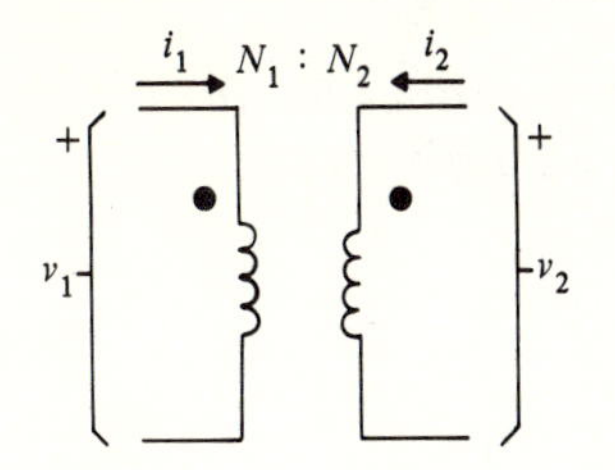

The ideal transformer models a transformer with perfect magnetic coupling, for which

$$M = \sqrt{L_1 L_2}$$

For the ideal transformer,

$$v_2 = \frac{N_2}{N_1} v_1$$

$$i_2 = -\frac{N_1}{N_2} i_1$$

where N_1 and N_2 are the number of turns of the L_1 and L_2 coils, respectively.

3. *Operational amplifier*

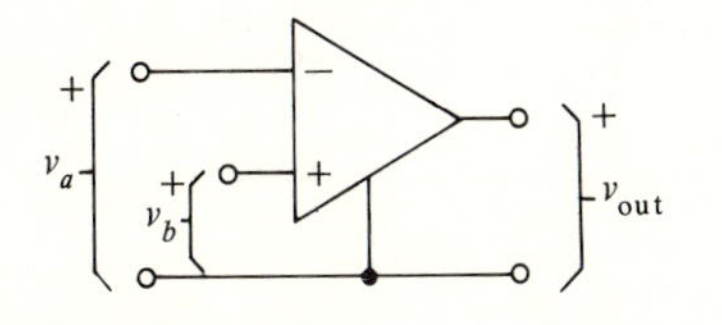

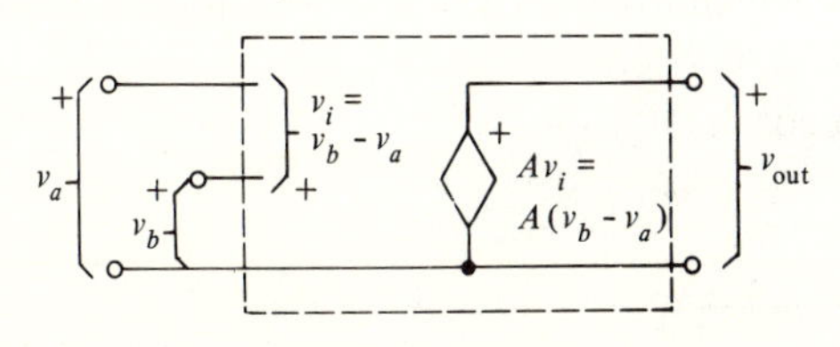

The idealized operational amplifier produces an output voltage which is proportional to the difference between two input voltages.

DRILL PROBLEMS

D1-1. Write systematic integrodifferential loop equations for the following electrical networks in terms of the indicated mesh currents:

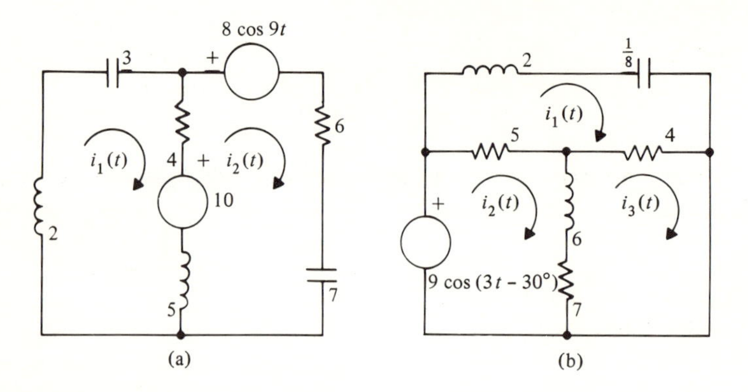

D1-2. Write systematic integrodifferential nodal equations for the following electrical networks in terms of the indicated node voltages. Node numbers are circled.

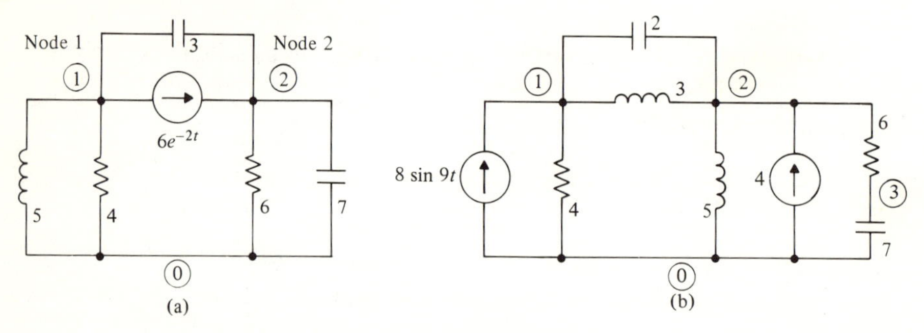

D1-3. For the following network which involves an ideal transformer model, find $i(t)$:

ans. 4.3 cos 3*t*

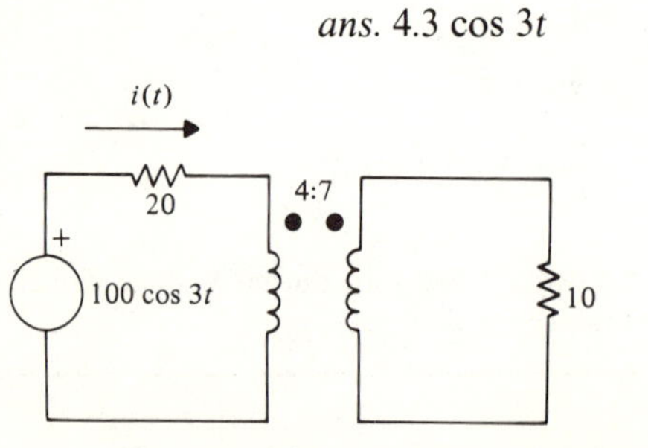

D1-4. For the following network, which involves an operational amplifier, find $v(t)$. Let the operational amplifier gain be $A = 100$.

ans. 49.6 cos 30*t*

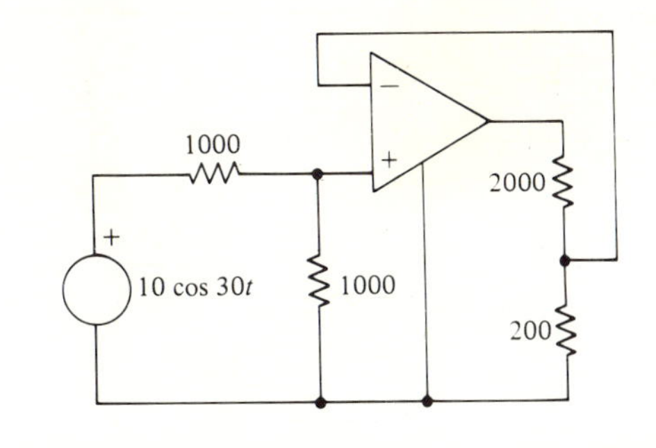

1.4.3 Translational Mechanical Networks

The force and position relations for the translational mechanical mass, spring, and damper elements are given in Table 1-3. A method of analysis for translational mechanical systems involving these elements is as follows:

1. Define positions with directional senses for each mass in the system.
2. Draw free-body diagrams for each of the masses, expressing the forces on them in terms of mass positions.
3. Write an equation for each mass, equating the algebraic sum of forces on it to the inertial force.

This procedure is applied to the system of Fig. 1-9a, where mass positions

TABLE 1-3
Translational mechanical element force-position relations.

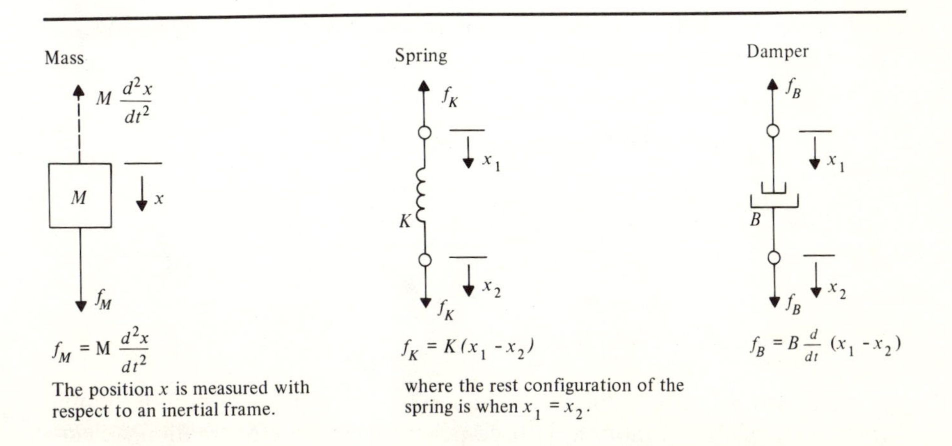

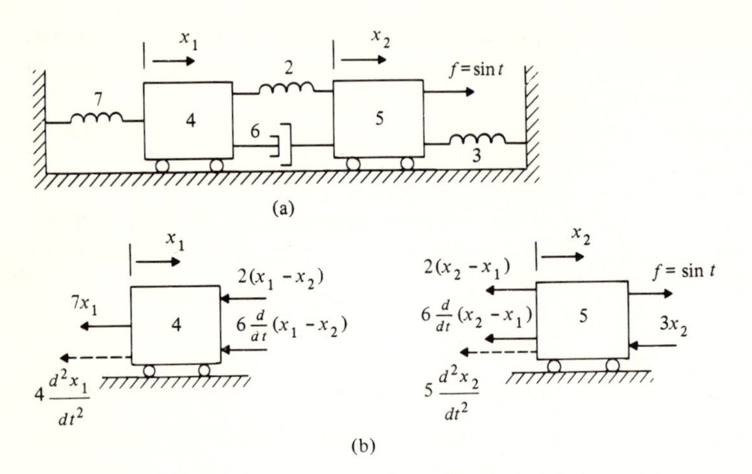

FIGURE 1-9
Writing systematic translational mechanical network equations. (a) A translational mechanical network. (b) Free-body diagrams for the masses.

have been defined. In Fig. 1-9b, free-body diagrams for the two masses are shown. Equating forces for the first mass gives

$$4\frac{d^2x_1}{dt^2}=-7x_1-2(x_1-x_2)-6\frac{d}{dt}(x_1-x_2)$$

The applied force to the second mass, $f=\sin t$, is in the same sense as x_2, so it carries a positive sign in the equation of forces for that mass:

$$5\frac{d^2x_2}{dt^2}=f-2(x_2-x_1)-6\frac{d}{dt}(x_2-x_1)-3x_2$$

Collecting terms, the two simultaneous differential equations, in x_1 and x_2, for the translational system are

$$\begin{cases} 4\dfrac{d^2x_1}{dt^2}+6\dfrac{dx_1}{dt}+9x_1-6\dfrac{dx_2}{dt}-2x_2=0 \\ -6\dfrac{dx_1}{dt}-2x_1+5\dfrac{d^2x_2}{dt^2}+6\dfrac{dx_2}{dt}+5x_2=\sin t \end{cases}$$

1.4.4 Rotational Mechanical Networks

A method of obtaining differential equations for angular motion is similar to that for translational motion. Torque-position relations for rotational elements are summarized in Table 1-4. An analysis procedure is as follows:

1. Define angular positions with directional senses for each rotational mass.
2. Draw free-body diagrams for each of the rotational masses, expressing each torque in terms of the angular positions of the masses.
3. Write an equation for each rotational mass, equating the algebraic sum of torques on it to the inertial torque.

This procedure is applied to the rotational system of Fig. 1-10a. Free-body diagrams for this system are drawn in Fig. 1-10b. For the first rotational mass, equating torques gives the equation

$$3\frac{d^2\theta_1}{dt^2} = -2\theta_1 - 5(\theta_1 - \theta_2) - 6\frac{d\theta_1}{dt}$$

The applied torque, $\tau = 20$, to the second rotational mass is in the sense of increasing θ_2, so the torque equation involving it is as follows:

$$4\frac{d^2\theta_2}{dt^2} = -5(\theta_2 - \theta_1) - 8\frac{d\theta_2}{dt} + 20$$

Collecting terms, the two simultaneous differential equations in θ_1 and θ_2 are

$$\begin{cases} 3\dfrac{d^2\theta_1}{dt^2} + 6\dfrac{d\theta_1}{dt} + 7\theta_1 - 5\theta_2 = 0 \\ -5\theta_1 + 4\dfrac{d^2\theta_2}{dt^2} + 8\dfrac{d\theta_2}{dt} + 5\theta_2 = 20 \end{cases}$$

TABLE 1-4
Rotational mechanical element torque-angle relations.

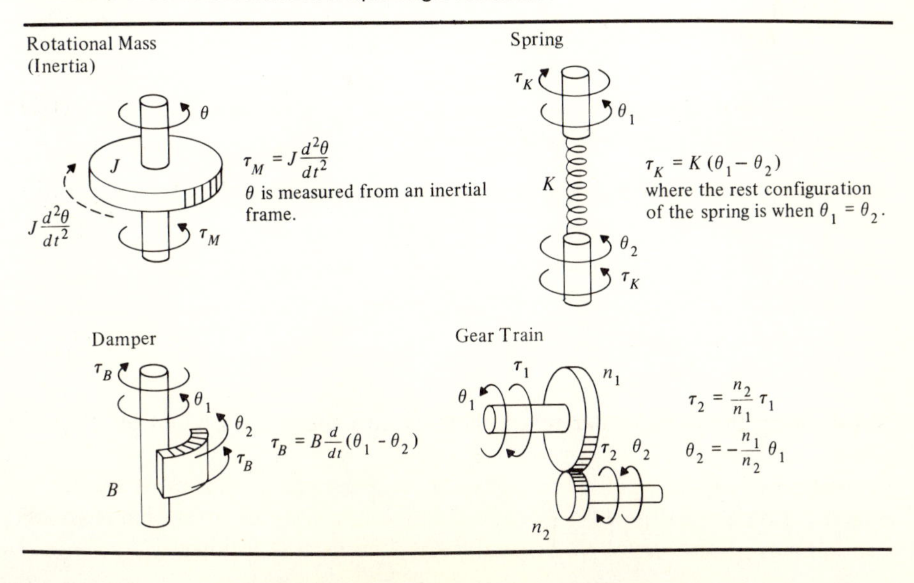

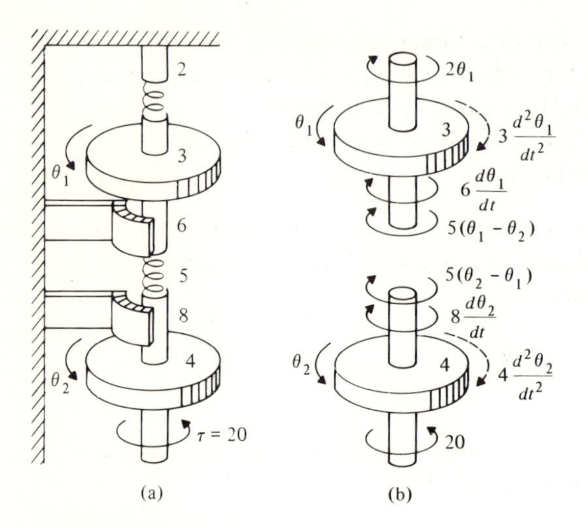

FIGURE 1-10
Writing systematic rotational mechanical network equations. (a) A rotational mechanical network. (b) Free-body diagrams for the inertias.

1.4.5 Mechanical-Electrical Analog

The electrical network of Fig. 1-11a involves an ideal transformer. The transformer windings are labeled "primary" and "secondary," with the primary winding closest to the voltage source $v(t)$. Systematic simultaneous loop equations for the network are

$$\begin{cases} L_1 \dfrac{di_1}{dt} + R_1 i_1 + v_1 = v \\ L_2 \dfrac{di_2}{dt} + R_2 i_2 + v_2 = 0 \end{cases} \tag{1-1}$$

where the ideal transformer voltages and currents are related by the turns ratio

$$\begin{cases} v_2 = \dfrac{N_2}{N_1} v_1 \\ i_2 = -\dfrac{N_1}{N_2} i_1 \end{cases} \tag{1-2}$$

Substituting the transformer relations (1-2) into the second equation of (1-1) gives

$$v_1 = \left(\frac{N_1}{N_2}\right)^2 L_2 \frac{di_1}{dt} + \left(\frac{N_1}{N_2}\right)^2 R_2 i_1$$

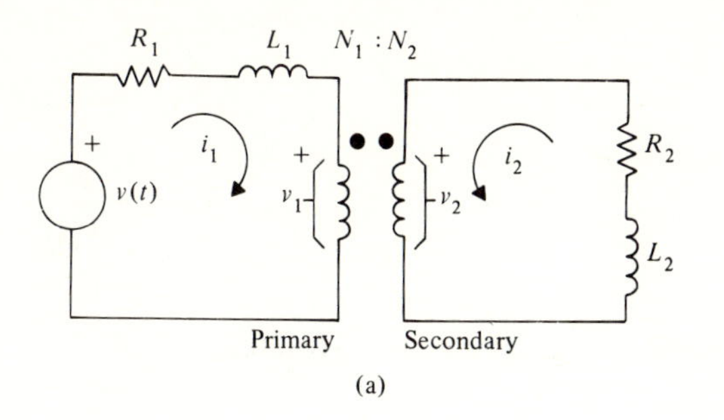

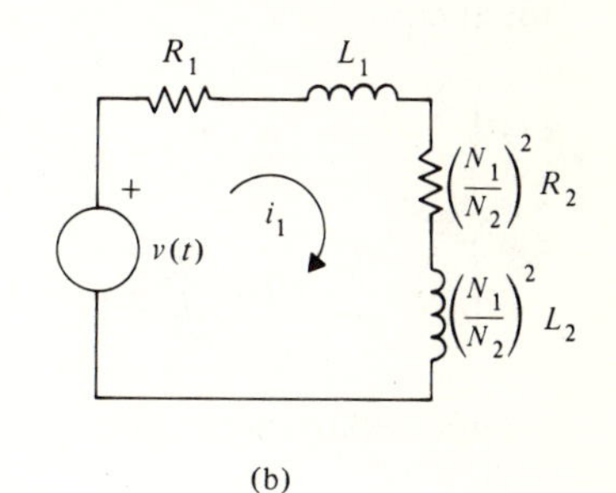

τ θ1 J1 B1 θ2 J2 B2

(c)

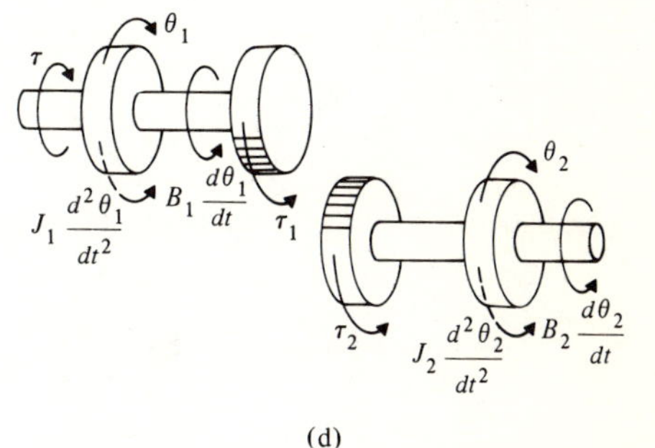

τ θ1 J1 B1 $\left(\frac{n_1}{n_2}\right)^2 J_2$ $\left(\frac{n_1}{n_2}\right)^2 B_2$

(e)

FIGURE 1-11
Analogy between electrical transformers and gear trains.
(a) Transformer-coupled electrical network. (b) Equivalent network with secondary load reflected to the primary circuit. (c) Mechanical network coupled by gears. (d) Free-body diagrams. (e) Equivalent mechanical network with load reflected through gears.

Eliminating v_1 in the first equation of (1-1), there results

$$L_1 \frac{di_1}{dt} + R_1 i_1 + \left(\frac{N_1}{N_2}\right)^2 L_2 \frac{di_1}{dt} + \left(\frac{N_1}{N_2}\right)^2 R_2 i_1 = v$$

The secondary load, consisting of L_2 and R_2, is said to be reflected to the primary side of the transformer, through the square of the transformer turns ratio, as in the equivalent circuit of Fig. 1-11b.

The mechanical network of Fig. 1-11c involves a gear train and has rotational equations

$$\begin{cases} J_1 \dfrac{d^2\theta_1}{dt^2} + B_1 \dfrac{d\theta_1}{dt} + \tau_1 = \tau \\ J_2 \dfrac{d^2\theta_2}{dt^2} + B_2 \dfrac{d\theta_2}{dt} + \tau_2 = 0 \end{cases} \tag{1-3}$$

as indicated in the free-body diagrams of Fig. 1-11d. The gear relations are

$$\begin{cases} \tau_2 = \dfrac{n_2}{n_1} \tau_1 \\ \theta_2 = -\dfrac{n_1}{n_2} \theta_1 \end{cases} \tag{1-4}$$

Substituting the gear relations (1-4) into the second equation of (1-3) gives

$$\tau_1 = \left(\frac{n_1}{n_2}\right)^2 \left(J_2 \frac{d^2\theta_2}{dt^2} + B_2 \frac{d\theta_2}{dt}\right)$$

Eliminating τ_1 in the first equation of (1-3), there results

$$J_1 \frac{d^2\theta_1}{dt^2} + B_1 \frac{d\theta_1}{dt} + \left(\frac{n_1}{n_2}\right)^2 J_2 \frac{d^2\theta_1}{dt^2} + \left(\frac{n_1}{n_2}\right)^2 B_2 \frac{d\theta_1}{dt} = \tau$$

in which the gear load has been reflected to the left side of the gear train. An equivalent mechanical network is given in Fig. 1-11e.

The example electrical and mechanical systems of Fig. 1-11 are analogous to one another, with the following equivalent quantities:

$$i_1 \leftrightarrow \frac{d\theta_1}{dt}$$

$$i_2 \leftrightarrow \frac{d\theta_2}{dt}$$

$$v \leftrightarrow \tau$$

$$L_1 \leftrightarrow J_1$$

$$R_1 \leftrightarrow B_1$$

$$L_2 \leftrightarrow J_2$$

$$R_2 \leftrightarrow B_2$$

$$\frac{N_1}{N_2} \leftrightarrow \frac{n_1}{n_2}$$

DRILL PROBLEMS

D1-5. Write systematic differential equations for the following translational mechanical networks in terms of the indicated mass positions. The mass positions are defined so that when all positions are zero, the spring forces are zero.

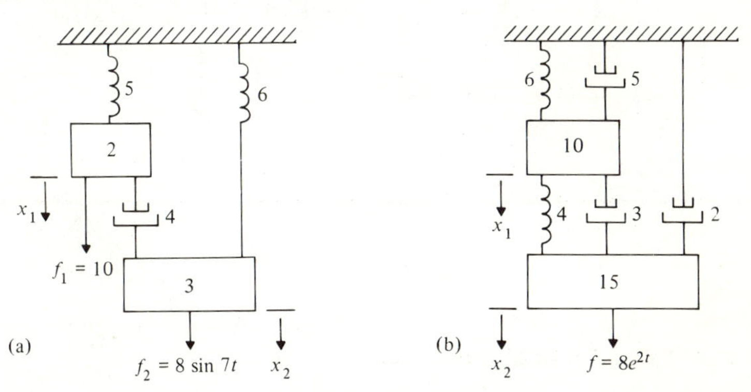

D1-6. Write systematic differential equations for the following rotational mechanical networks in terms of the indicated rotational mass angles. The angles are defined so that when all angles are zero, the spring torques are zero.

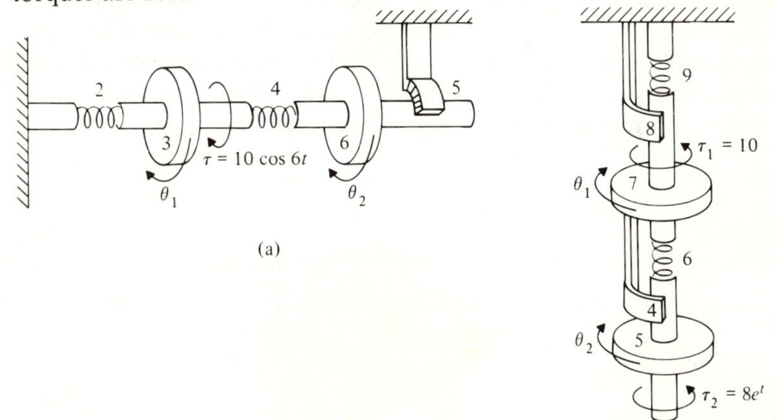

D1-7. Find $\theta(t)$. Mass and damping effects in this system are negligible.

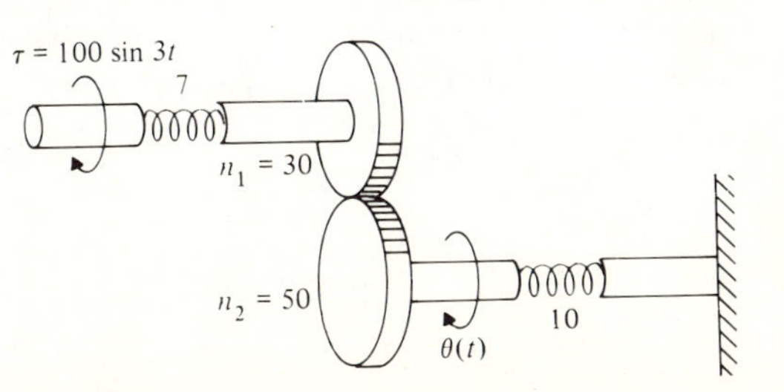

D1-8. Draw an electrical network which is analogous to the following translational mechanical network:

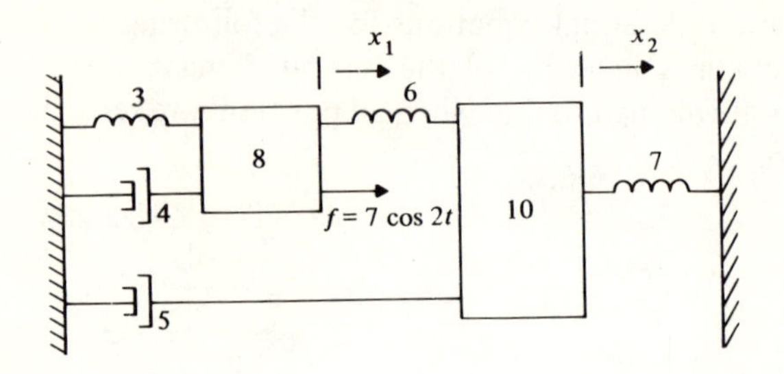

1.4.6 Electromechanical Networks

Numerous electromechanical devices are encountered in engineering and scientific applications. Solenoids, actuators, motors, generators, gyroscopes, accelerometers, and loudspeakers are just a few. For many control systems it is necessary to deal with equations for a combination of electrical and mechanical components. Table 1-5 gives idealized equations for several common electromechanical devices. As with all physical elements, electromechanical devices are ultimately nonlinear. The models given in the table are linearized, over a suitable and sufficiently small range of the variables.

TABLE 1-5
Simple models for common electromechanical elements.

1. *Voltage-Driven Potentiometer*

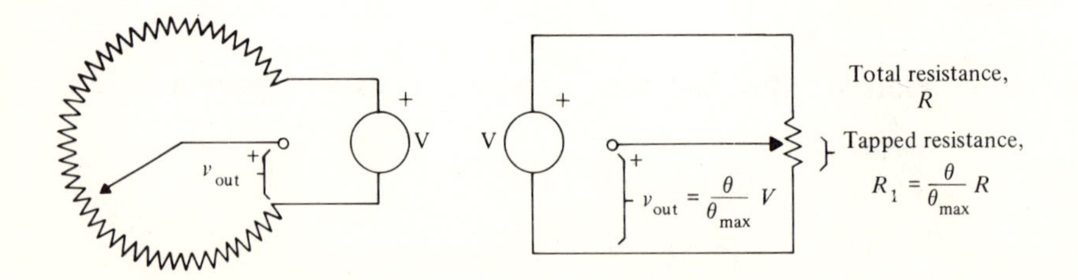

θ_{max} is the maximum angular position θ. For $\theta = \theta_{max}$, $R_1 = R$.

2. *DC Motor-Generator*

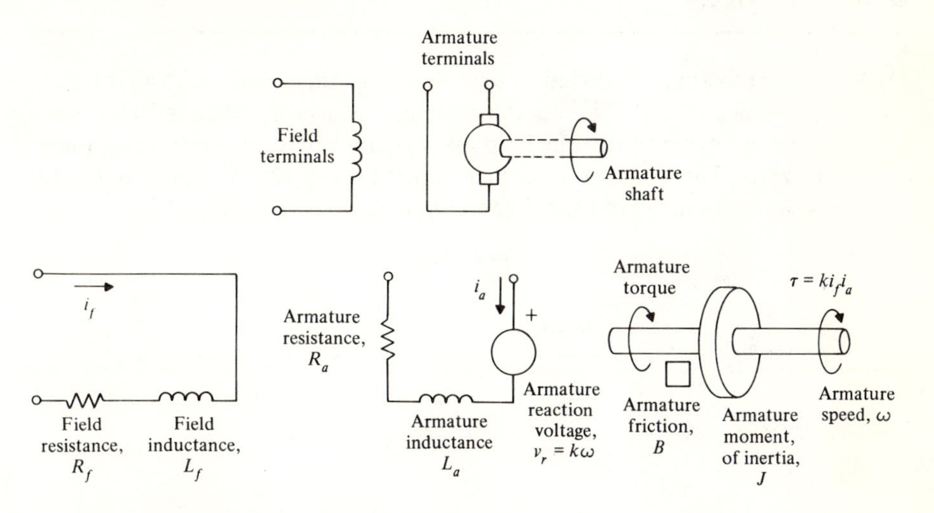

The constants k_r and k_τ depends upon the degrees of electromechanical coupling.

3. *Tachometer*

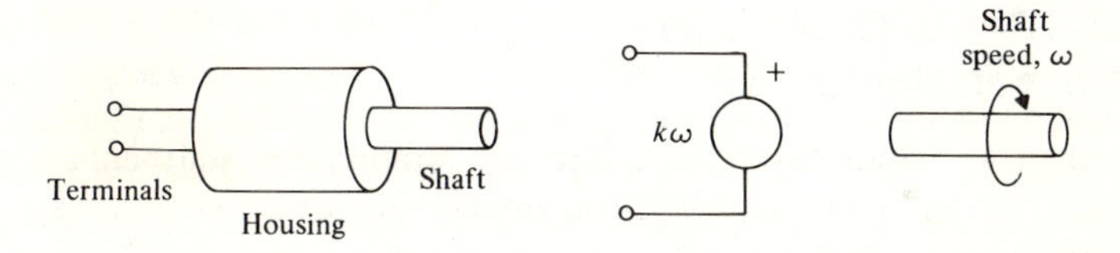

The tachometer is a special case of a DC generator in which the field is replaced by a permanent magnet, which is equivalent to having a constant (and relatively small) field current. It is normally used with a very small electrical load, so that i_a is nearly zero. Friction and inertia are typically made as small as is practical.

4. *Linear Actuator* (*Solenoid*)

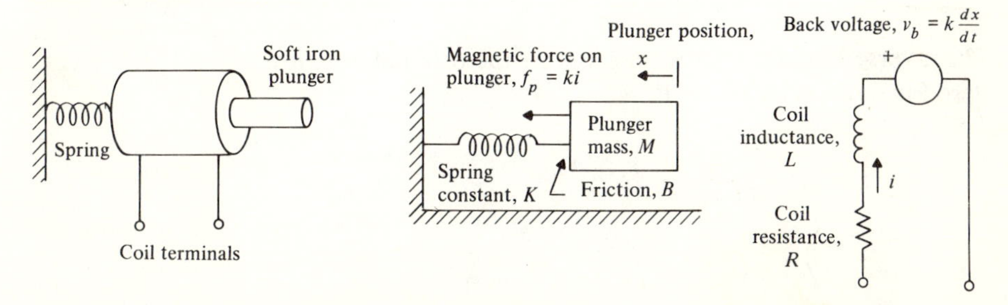

The constant k depends upon the amount of electromechanical coupling and is the same number in both relations when they are expressed in consistent units.

DRILL PROBLEM

D1-9. A linear actuator is driven by a sinusoidal voltage source with an internal resistance of 1 Ω. The actuator plunger is connected to a 20-kg load mass as shown. When the load mass position x is zero, the spring force is zero. The solenoid's electromagnetic coupling constant is $k=40$ V-sec/coulomb. Find the differential equations describing $x(t)$.

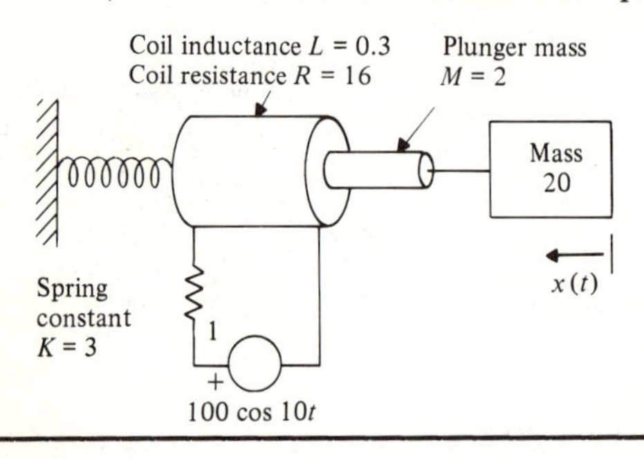

1.5 LAPLACE TRANSFORMATION

1.5.1 Transforms and Properties

The translation of a system description into the language of mathematics is termed *modeling*. Modeling involves idealization whereby the important aspects of a problem are isolated and the minor ones ignored, so that simplicity with sufficient accuracy is obtained.

When the system models consist of linear, constant-coefficient integro-differential equations, Laplace transform methods may be used to advantage. The Laplace transform* of a function is

$$\mathscr{L}[f(t)]=F(s)=\int_{0^-}^{\infty} f(t)e^{-st}\,dt$$

The inverse Laplace transform recovers the original function for $t\geqslant 0$ and gives zero for times prior to $t=0$:

$$\begin{aligned}\mathscr{L}^{-1}[F(s)]&=\frac{1}{2\pi j}\int_{\sigma-j\infty}^{\sigma+j\infty} F(s)e^{st}\,ds\\&=\begin{cases}f(t) & t\geqslant 0\\ 0 & t<0\end{cases}\\&=f(t)u(t)\end{aligned}$$

*When functions with discontinuities at $t=0$ are involved, $t=0$ is an especially inconvenient lower limit for the Laplace transform integral. We use the most common and useful definition which includes $t=0$ in the transform. The choice $t=0^+$ is also workable but has the disadvantages of giving $\mathscr{L}[\delta(t)]=0$ and of requiring translation of initial conditions to conditions at $t=0^+$.

TABLE 1-6
Some Laplace transform properties.

$\mathscr{L}[kf(t)] = kF(s)$, k a constant

$\mathscr{L}[f_1(t)+f_2(t)] = F_1(s)+F_2(s)$

$\mathscr{L}[f_1(t)f_2(t)]$ does *not* equal $F_1(s)F_2(s)$

$\mathscr{L}[f(t-T)] = e^{-sT}F(s)$, T a constant, provided $f(t)$ and $f(t-T)$ are both zero prior to $t=0$

$\mathscr{L}[f(at)] = \dfrac{1}{a}F\left(\dfrac{s}{a}\right)$, a is a positive constant

$\mathscr{L}[e^{-at}f(t)] = F(s+a)$

$$\mathscr{L}\left[\frac{df}{dt}\right] = sF(s) - f(0^-)$$

$$\mathscr{L}\left[\frac{d^2f}{dt^2}\right] = s^2F(s) - sf(0^-) - f'(0^-)$$

$$\mathscr{L}\left[\frac{d^nf}{dt^n}\right] = s^nF(s) - s^{n-1}f(0^-) - s^{n-2}f'(0^-) - \cdots - sf^{[n-2]}(0^-) - f^{[n-1]}(0^-)$$

$$\mathscr{L}\left[\int_{0^-}^{t} f(t)\,dt\right] = \frac{F(s)}{s}$$

$$\mathscr{L}\left[\int_{-\infty}^{t} f(t)\,dt\right] = \frac{F(s)}{s} + \frac{1}{s}\int_{-\infty}^{0^-} f(t)\,dt$$

$$\mathscr{L}[tf(t)] = -\frac{dF(s)}{ds}$$

$$\mathscr{L}[t^2f(t)] = \frac{d^2F(s)}{ds^2}$$

$$\mathscr{L}[t^nf(t)] = (-1)^n\frac{d^nF(s)}{ds^n}$$

Some important properties of the Laplace transform are given in Table 1-6, and important transform pairs are given in Table 1-7.

1.5.2 Solving Differential Equations Using Laplace Transforms

By Laplace-transforming linear constant-coefficient integrodifferential equations, one obtains linear algebraic equations, which can then be solved for the transforms of the solutions. Initial conditions may be included when the equations are transformed. For example, transforming

$$\frac{d^2y}{dt^2} + 9\frac{dy}{dt} + 2y = 6e^{-4t}$$

with

$$y(0^-) = 2$$
$$y'(0^-) = -4$$

TABLE 1-7
Some Laplace transform pairs.

$f(t)$	$F(s)$
$\delta(t)$, unit impulse	1
$u(t)$, unit step	$\frac{1}{s}$
$tu(t)$	$\frac{1}{s^2}$
$t^n u(t)$	$\frac{n!}{s^{n+1}}$
$e^{-at}u(t)$, a a constant	$\frac{1}{s+a}$
$te^{-at}u(t)$	$\frac{1}{(s+a)^2}$
$t^n e^{-at}u(t)$	$\frac{n!}{(s+a)^{n+1}}$
$(\sin bt)u(t)$, b a constant	$\frac{b}{s^2+b^2}$
$(\cos bt)u(t)$	$\frac{s}{s^2+b^2}$
$(t \sin bt)u(t)$	$\frac{2bs}{(s^2+b^2)^2}$
$(t \cos bt)u(t)$	$\frac{s^2-b^2}{(s^2+b^2)^2}$
$(e^{-at} \sin bt)u(t)$	$\frac{b}{(s+a)^2+b^2}$
$(e^{-at} \cos bt)u(t)$	$\frac{(s+a)}{(s+a)^2+b^2}$
$2Me^{-at} \cos(bt+\theta)u(t)$	$\frac{Me^{j\theta}}{s+a-jb}+\frac{Me^{-j\theta}}{s+a+jb}=\frac{\text{numerator}}{(s+a)^2+b^2}$

gives

$$s^2 Y(s)-2s+4+9[sY(s)-2]+2Y(s)=\frac{6}{s+4}$$

$$Y(s)=\frac{6}{(s+4)(s^2+9s+2)}+\frac{2s+14}{s^2+9s+2}$$

The Laplace transform method is also easily applied to systems described by simultaneous integrodifferential equations. Transforming the simultaneous equations

$$\begin{cases} \dfrac{dy_1}{dt} + 2y_1 - \dfrac{3dy_2}{dt} = r(t) \\[2ex] y_1 + \dfrac{dy_2}{dt} + 4\displaystyle\int_0^t y_2\, dt = 0 \end{cases}$$

gives

$$\begin{cases} sY_1(s) - y_1(0^-) + 2Y_1(s) - 3sY_2(s) + 3y_2(0^-) = R(s) \\[2ex] Y_1(s) + sY_2(s) - y_2(0^-) + \dfrac{4}{s}\, Y_2(s) = 0 \end{cases}$$

or

$$\begin{cases} (s+2)Y_1(s) - 3sY_2(s) = R(s) + y_1(0^-) - 3y_2(0^-) \\[2ex] Y_1(s) + \left(s + \dfrac{4}{s}\right) Y_2(s) = y_2(0^-) \end{cases}$$

which may be solved for $Y_1(s)$ and $Y_2(s)$, given the input $R(s)$ and the initial conditions $y_1(0^-)$ and $y_2(0^-)$.

1.5.3 Partial Fraction Expansion

Finding an inverse Laplace transform frequently requires expanding a rational function of s in partial fractions. If the numerator polynomial is of lower order than the denominator polynomial and if the denominator polynomial has no repeated roots, then constants $K_1, K_2, \ldots,$ termed the *residues*, may be found such that

$$Y(s) = \frac{\text{numerator polynomial}}{(s+a)(s+b)(s+c)\cdots} = \frac{K_1}{s+a} + \frac{K_2}{s+b} + \frac{K_3}{s+c} + \cdots$$

The individual terms in the expansion represent exponential functions of time after $t = 0$:

$$y(t) = K_1 e^{-at} + K_2 e^{-bt} + K_3 e^{-ct} + \cdots \qquad t \geq 0$$

If the numerator polynomial is not of lower order than the denominator polynomial, the denominator may be divided into the numerator until a remainder polynomial of lower order than the denominator is obtained, yielding an expansion of the form

$$\frac{\text{Numerator polynomial}}{(s+a)(s+b)(s+c)\cdots} = \frac{\text{dividend}}{\text{polynomial}} + \frac{\text{remainder polynomial}}{(s+a)(s+b)(s+c)\cdots}$$

$$= \frac{\text{dividend}}{\text{polynomial}} + \frac{K_1}{s+a} + \frac{K_2}{s+b} + \frac{K_3}{s+c} + \cdots$$

For example,

$$\frac{2s^2+3}{s^2+3s+2} = 2 + \frac{-6s-1}{s^2+3s+2}$$

A constant term in the Laplace transform corresponds to an impulse time function.

If denominator roots are repeated, the corresponding terms in the partial fraction expansion are as follows:

$$\frac{\text{Numerator}}{(s+a)^n} + \frac{K_1}{s+a} + \frac{K_2}{(s+a)^2} + \cdots + \frac{K_n}{(s+a)^n}$$

The inverse Laplace transform of a repeated root is of the form

$$\mathcal{L}^{-1}\left[\frac{K_n}{(s+a)^n}\right] = \frac{K_n}{(n-1)!} t^{n-1} e^{-at} u(t)$$

1.5.4 Residue Calculation

One method of finding the residues of a partial fraction expansion is to rationalize the expansion and equate coefficients. For example,

$$Y(s) = \frac{-3s+1}{(s+2)(s+4)} = \frac{K_1}{s+2} + \frac{K_2}{s+4}$$

$$= \frac{(K_1+K_2)s + (4K_1+2K_2)}{(s+2)(s+4)}$$

$$\begin{cases} K_1 + K_2 = -3 \\ 4K_1 + 2K_2 = 1 \end{cases}$$

$$K_1 = \frac{\begin{vmatrix} -3 & 1 \\ 1 & 2 \end{vmatrix}}{\begin{vmatrix} 1 & 1 \\ 4 & 2 \end{vmatrix}} = \frac{-7}{-2} = \frac{7}{2}$$

$$K_2 = \frac{\begin{vmatrix} 1 & -3 \\ 4 & 1 \end{vmatrix}}{-2} = \frac{13}{-2} = -\frac{13}{2}$$

A faster method for finding a residue other than for a repeated root is to multiply each side of the expansion by the denominator term, then evaluate the resulting expansion at the value of s which makes that denominator term zero (that is, the root location). To obtain the coefficient K_1 associated with root at $s=-2$, one proceeds as follows:

$$\frac{\text{Numerator}}{(s+a)(s+b)\cdots} = \frac{K_1}{s+a} + \frac{K_2}{s+b} + \cdots$$

$$\frac{(s+a)(\text{numerator})}{(s+a)(s+b)} = \frac{(s+a)K_1}{s+a} + (s+a)\left[\frac{K_2}{s+b} + \cdots\right]$$

$$\left.\frac{\text{Numerator}}{(s+b)\cdots}\right|_{s=-a} = K_1 + 0 \cdot \left[\frac{K_2}{s+b} + \cdots\right]$$

Applying this method to the previous example gives the residues quite easily:

$$Y(s) = \frac{-3s+1}{(s+2)(s+4)} = \frac{K_1}{s+2} + \frac{K_2}{s+4}$$

$$K_1 = \left.\frac{-3s+1}{s+4}\right|_{s=-2} = \frac{7}{2}$$

$$K_2 = \left.\frac{-3s+1}{s+2}\right|_{s=-4} = \frac{13}{-2}$$

For a repeated root, the above residue evaluation method works only for the highest-order repeated term in the expansion. For example, with

$$Y(s) = \frac{4s^2-1}{(s+2)^3} = \frac{K_1}{s+2} + \frac{K_2}{(s+2)^2} + \frac{K_3}{(s+2)^3}$$

evaluation gives

$$\frac{(4s^2-1)(s+2)^3}{(s+2)^3} = K_1(s+2)^2 + K_2(s+2) + K_3$$

$$K_3 = (s+2)^3 Y(s)|_{s=-2} = (4s^2-1)|_{s=-2} = 15$$

Multiplying both sides of the expansion by $(s+2)^2$ or $(s+2)$ will leave $(s+2)$ denominator factors, however, so K_1 and K_2 cannot be determined in this manner.

An alternate method to that of cross-multiplying and equating coefficients is to multiply both sides of the expansion by the repeated root term and differentiate with respect to s:

$$\frac{d}{ds}\left[\frac{(4s^2-1)(s+2)^3}{(s+2)^3}\right]=\frac{d}{ds}\left[K_1(s+2)^2+K_2(s+2)+K_3\right]$$

$$8s=2K_1(s+2)+K_2 \qquad (1\text{-}5)$$

Evaluating at $s=-2$ gives

$$8s|_{s=-2}=K_2=-16$$

Differentiating (1-5) a second time with respect to s and evaluating,

$$\left.\frac{d}{ds}(8s)\right|_{s=-2}=2K_1$$

$$K_1=4$$

In general, the coefficients of the repeated root terms of a partial fraction expansion, repeated p times,

$$Y(s)=\frac{K_1}{s+a}+\frac{K_2}{(s+a)^2}+\cdots+\frac{K_p}{(s+a)^p}+\cdots \text{ terms for other different roots}$$

are given by

$$K_i=\frac{1}{(p-i)!}\frac{d^{p-i}}{ds^{p-i}}\{(s+a)^p Y(s)\}|_{s=-a} \qquad i=1, 2, \ldots, p$$

For complex root terms, the residues are complex numbers. In fact, the residues of complex conjugate roots are complex conjugates of one another. For example, the expression

$$Y(s)=\frac{10}{s(s^2+4s+13)}=\frac{10}{s(s+2+j3)(s+2-j3)}$$

$$=\frac{K_1}{s}+\frac{K_2}{s+2+j3}+\frac{K_3}{s+2-j3}$$

and the residues are as follows:

$$K_1=\left.\frac{10}{s^2+4s+13}\right|_{s=0}=\frac{10}{13}=0.77$$

$$K_2=\left.\frac{10}{s(s+2-j3)}\right|_{s=-2-j3}=\frac{10}{(-2-j3)(-j6)}$$

$$=\frac{10}{-18+j12}=-\frac{15}{39}+j\left(-\frac{10}{39}\right)=0.46e^{-j146.3^\circ}$$

$$K_3 = -\frac{15}{39} + j\left(\frac{10}{39}\right) = 0.46e^{j146.3^\circ}$$

The last entry in the Laplace transform pair in Table 1-7 may be used to find the inverse Laplace transform of the pair of complex conjugate root terms:

$$Y(s) = \frac{0.77}{s} + \frac{0.46e^{-j146,3^\circ}}{s+2+j3} + \frac{0.46e^{j146,3^\circ}}{s+2-j3}$$

$$y(t) = [0.77 + 0.92e^{-2t}\cos(3t + 146.3^\circ)]u(t)$$

1.5.5 Time Response

The type of time function corresponding to each partial fraction expansion term for a Laplace-transformed signal depends upon the term's root location in the complex plane and upon whether or not the root is repeated. Table 1-8 shows representative time functions associated with various transform denominator root locations.

TABLE 1-8
Time functions corresponding to various Laplace transform denominator polynomial root locations.

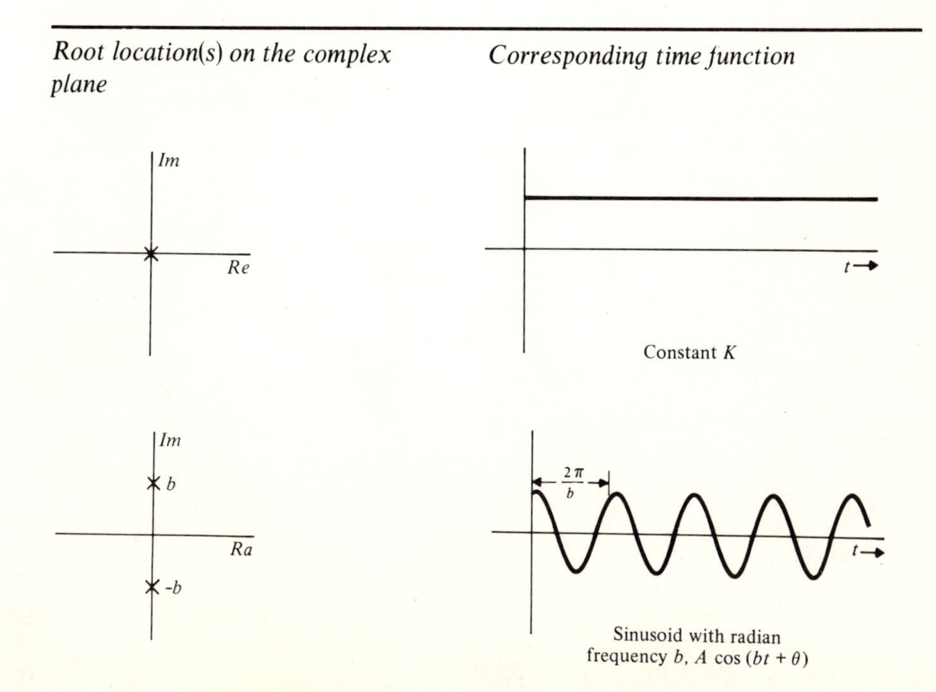

TABLE 1-8

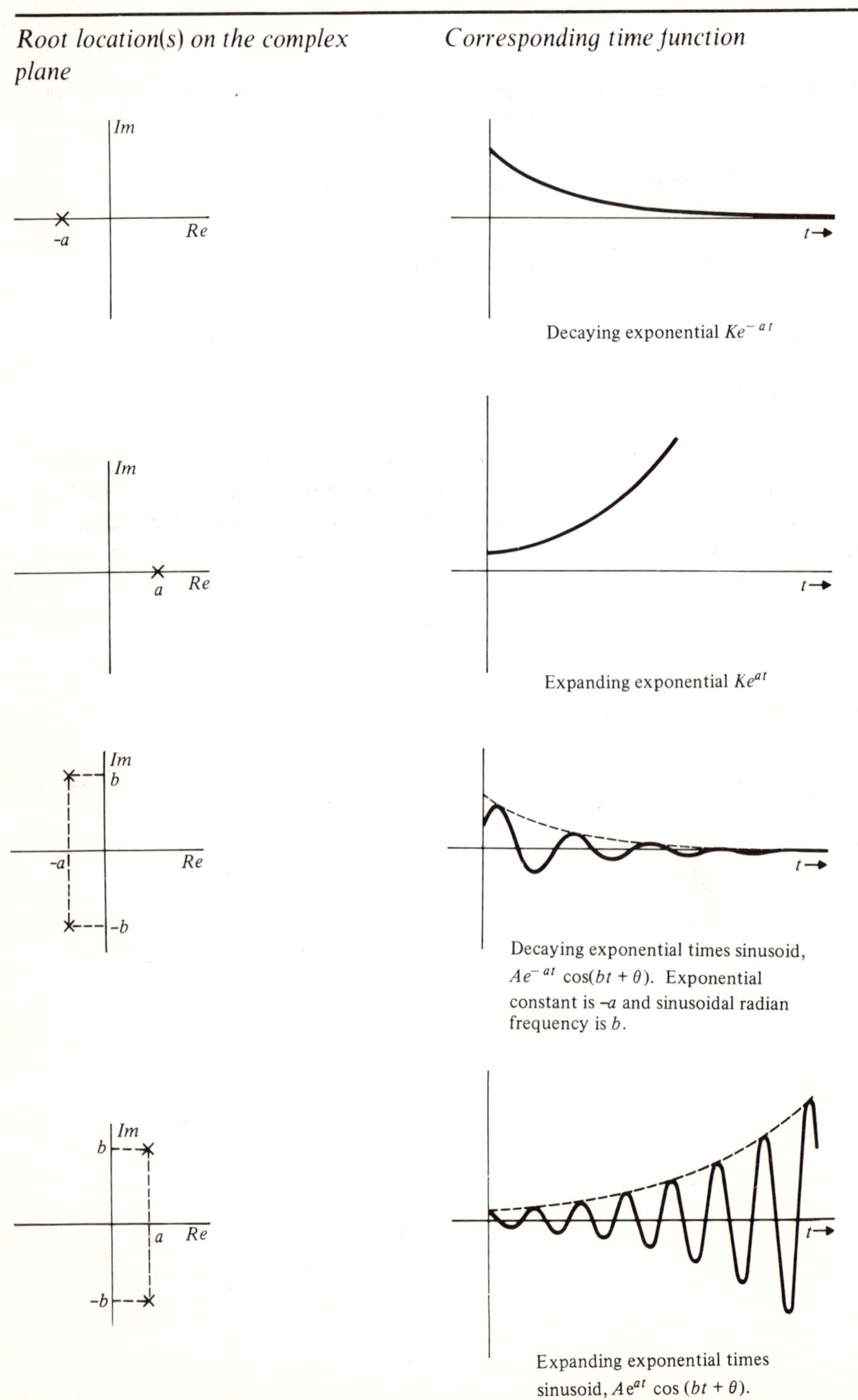

Decaying exponential Ke^{-at}

Expanding exponential Ke^{at}

Decaying exponential times sinusoid, $Ae^{-at}\cos(bt + \theta)$. Exponential constant is $-a$ and sinusoidal radian frequency is b.

Expanding exponential times sinusoid, $Ae^{at}\cos(bt + \theta)$.

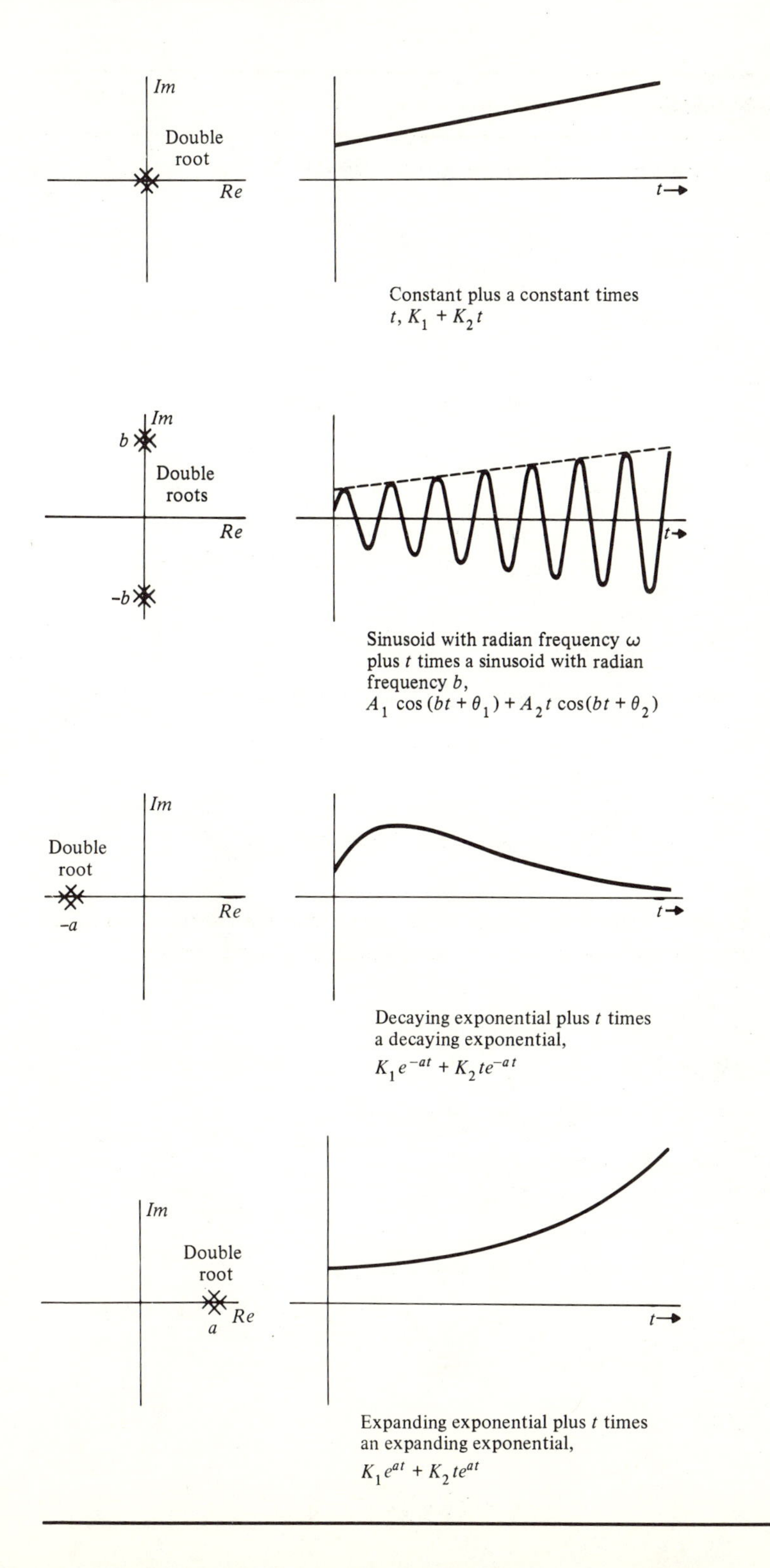

Im
Double
root
Re
t→
Constant plus a constant times
t, K1 + K2t
Im
b
Double
roots
Re
-b
t→
Sinusoid with radian frequency ω
plus t times a sinusoid with radian
frequency b,
A1 cos (bt + θ1) + A2t cos(bt + θ2)
Im
Double
root
-a
Re
t→
Decaying exponential plus t times
a decaying exponential,
K1e^-at + K2te^-at
Im
Double
root
a
Re
t→
Expanding exponential plus t times
an expanding exponential,
K1e^at + K2te^at

DRILL PROBLEMS

D1-10. For the following Laplace transformed signals, find $y(t)$ for $t \geqslant 0$:

(a) $Y(s)=\dfrac{s}{s+2}$

ans. $\delta(t)-2\,e^{-2t}$

(b) $Y(s)=\dfrac{3s-5}{s^2+4s+2}$

ans. $5.39\,e^{-3.414t}-2.39\,e^{-0.586t}$

(c) $Y(s)=\dfrac{3-6\,e^{-2s}}{(s+2)(s+3)}$

ans. $3\,e^{-2t}-3\,e^{-3t} - [6\,e^{-2(t-2)}-6\,e^{-3(t-2)}]u(t-2)$

(d) $Y(s)=\dfrac{10}{s^3+2s^2+5s}$

ans. $2+e^{-t}(-2\cos 2t-\sin 2t) = 2+2.24e^{-t}\cos(2t+153^\circ)$

(e) $Y(s)=\dfrac{4(s+1)}{(s+2)(s+3)^2}$

ans. $-4\,e^{-2t}+4\,e^{-3t}+8t\,e^{-3t}$

D1-11. Use Laplace transform methods to solve the following differential equations for $t \geqslant 0$ with the indicated boundary conditions:

(a) $\dfrac{dy}{dt}+4y=6\,e^{2t}$

$y(0^-)=3$

ans. $e^{2t}+2\,e^{-4t}$

(b) $\dfrac{dy}{dt}+y=3\cos 2t$

$y(0^-)=0$

ans. $-\frac{3}{5}e^{-t}+\frac{3}{5}\cos 2t+\frac{6}{5}\sin 2t = -\frac{3}{5}e^{-t}+1.34\cos(2t-63.4^\circ)$

(c) $\dfrac{d^2y}{dt^2}+7\dfrac{dy}{dt}+12y=10$

$y(0^-)=3$

$y'(0^-)=0$

ans. $\frac{10}{12}-\frac{13}{2}e^{-4t}+\frac{26}{3}e^{-3t}$

(d) $\dfrac{d^2y}{dt^2}+4\dfrac{dy}{dt}+20y=4$

$y(0^-)=-2$

$y'(0^-)=0$

ans. $\frac{1}{5}+e^{-2t}(-\frac{11}{5}\cos 4t-\frac{11}{10}\sin 4t)$
$=\frac{1}{5}+2.46e^{-2t}\cos(4t+153^\circ)$

(e) $\dfrac{d^3y}{dt^3}+5\dfrac{d^2y}{dt^2}+6\dfrac{dy}{dt}=0$

$y(0^-)=3$

$y'(0^-)=-2$

$y''(0^-)=7$

ans. $\frac{15}{6}+e^{-3t}-\frac{1}{2}e^{-2t}$

1.6 TRANSFER FUNCTIONS AND STABILITY

1.6.1 Transfer Functions

One of the most powerful tools of control system analysis and design is the transfer function representation, which is a generalization of the impedance concept in electrical and mechanical networks. For a single-input, single-output system with input $r(t)$ and output $y(t)$, the transfer function relating the output to the input is defined as

$$T(s)=\left.\frac{Y(s)}{R(s)}\right|_{\text{when all initial conditions are zero}}$$

To find the transfer function, Laplace-transform the system equations, with zero initial conditions, and form the ratio of output transform to input transform.

Consider the system described by

$$\frac{d^2y}{dt^2}+6\frac{dy}{dt}+8y=-\frac{dr}{dt}+5r \tag{1-6}$$

Laplace-transforming with zero initial conditions,

$$s^2Y(s)+6sY(s)+8Y(s)=-sR(s)+5R(s)$$

and

$$T(s)=\frac{Y(s)}{R(s)}=\frac{-s+5}{s^2+6s+8} \tag{1-7}$$

In systems described by linear, constant-coefficient integrodifferential equations, every Laplace-transformed signal is related to every other such signal by a transfer function. To avoid confusion, the term *transfer function* is reserved to describe the input-output relation and *transmittance* is used here to denote the similar relation between a pair of signals other than the input and output.

1.6.2 Impedances

In Laplace-transforming electrical network equations, the network source functions become their Laplace transforms, and the basic element voltage-current relations transform as follows:

$$v_R(t) = R i_R(t) \qquad V_R(s) = R I_R(s)$$

$$v_L(t) = L \frac{di_L}{dt} \qquad V_L(s) = sLI_L(s) - Li_L(0^-)$$

$$v_C(t) = \frac{1}{C} \int_{-\infty}^{t} i_C dt$$

$$= \frac{1}{C} \int_{-\infty}^{0^-} i_C dt + \frac{1}{C} \int_{0^-}^{t} i_C\, dt$$

$$= v_C(0^-) + \frac{1}{C} \int_{0^-}^{t} i_C\, dt \qquad V_C(s) = \frac{1}{sC} I_C(s) + \frac{v_C(0^-)}{s}$$

The transmittances which relate element voltages to their currents are termed the impedances of the elements:

$$Z(s) = \left. \frac{V(s)}{I(s)} \right|_{\text{when initial conditions are zero}}$$

The basic elements have impedances R, sL, and $(1/sC)$.

Similar relations may be applied to mechanical and other elements.

1.6.3 Zero-State and Zero-Input Response Components

The system output when the initial conditions are all zero is termed the *zero-state* response component. Its Laplace transform is given simply by the product of the transfer function and the input transform. If the system of (1-6) and (1-7) has zero initial conditions and if the input is

$$r(t) = 7e^{-3t}$$

the system output is given by

$$Y_{\text{zero state}}(s) = T(s)R(s) = \frac{7(-s+5)}{(s^2+6s+8)(s+3)}$$

If the system initial conditions are not zero, there is an additional output component present, the *zero-input* part of the response. The form of the zero-input response can be found from the transfer function denominator, provided that there have been no unwarranted cancellations of terms made between numerator and denominator of the transfer function. It is possible, although perhaps unlikely, that system parameters are just the right numbers so that a factor in the transfer function numerator cancels a denominator factor, causing the corresponding term in the system's zero-input response to be overlooked.

For a transfer function

$$T(s) = \frac{b_m s^m + b_{m-1} s^{m-1} + \cdots + b_1 s + b_0}{a_n s^n + a_{n-1} s^{n-1} + \cdots + a_1 s + a_0}$$

a differential equation relating the output $y(t)$ to the input $r(t)$ is

$$a_n \frac{d^n y}{dt^n} + a_{n-1} \frac{d^{n-1} y}{dt^{n-1}} + \cdots + a_1 \frac{dy}{dt} + a_0 y = b_m \frac{d^m r}{dt^m} + b_{m-1} \frac{d^{m-1} r}{dt^{m-1}} + \cdots + b_1 \frac{dr}{dt} + b_0 r$$

Recall that $T(s)$ was computed by assuming that all initial conditions are zero. The above differential equation can be Laplace-transformed again so that initial condition terms can be restored if desired. Including the initial condition terms in the Laplace transformed equations gives

$$\begin{aligned} &a_n[s^n Y(s) - s^{n-1} y(0^-) - s^{n-2} y'(0^-) - \cdots] \\ &\quad + a_{n-1}[s^{n-1} Y(s) - s^{n-2} y(0^-) - s^{n-3} y'(0^-) - \cdots] + \cdots + a_0 Y(s) \\ &= b_m[s^m R(s) - s^{m-1} r(0^-) - s^{m-2} r'(0^-) - \cdots] \\ &\quad + \cdots + b_0 R(s) \\ &(a_n s^n + a_{n-1} s^{n-1} + \cdots + a_1 s + a_0) Y(s) = (b_m s^m + \cdots b_0) R(s) \\ &\quad + \text{polynomial in } s \text{ with coefficients dependent upon initial conditions} \end{aligned}$$

$$Y(s) = \underbrace{T(s)R(s)}_{\substack{\text{zero-state} \\ \text{component}}} + \underbrace{\frac{\substack{\text{polynomial in } s \text{ with} \\ \text{coefficients dependent upon initial conditions}}}{a_n s^n + a_{n-1} s^{n-1} + \cdots + a_1 s + a_0}}_{\substack{\text{zero-input} \\ \text{component}}}$$

The zero-input component transform has the same denominator polynomial as $T(s)$.

The transfer function denominator polynomial is the system *characteristic polynomial*, and the roots of the characteristic polynomial, that is, the solutions of the characteristic equation

$$a_n s^n + a_{n-1} s^{n-1} + \cdots + a_1 s + a_0 = 0$$

are known as the system's *characteristic roots*.

For the example system of (1-6), the zero-input response is given by

$$Y_{\text{zero input}}(s) = \frac{c_1 s + c_2}{s^2 + 6s + 8}$$

where c_1 and c_2 are constants which depend upon the specific initial conditions. The complete response for general initial conditions is of the form

$$\begin{aligned} Y(s) &= Y_{\text{zero state}}(s) + Y_{\text{zero input}}(s) \\ &= \frac{7(-s+5)}{(s^2+6s+8)(s+3)} + \frac{c_1 s + c_2}{s^2+6s+8} \end{aligned} \tag{1-8}$$

1.6.4 Forced and Natural Response Components

An alternative to the zero-input/zero-state decomposition of system response is to separate the response into *natural* and *forced* parts.* The natural component consists of all the characteristic root terms in the partial fraction expansion for the response. The forced response component is the remainder of the response and is composed of the terms associated with the input transform.

For example, for the system (1-6), (1-7), the characteristic roots are $s = -2$ and $s = -4$. The response (1-8) expands in partial fractions as

$$\begin{aligned} Y(s) &= \underbrace{\frac{-7s+35}{(s+2)(s+4)(s+3)}}_{\substack{\text{zero-state}\\ \text{component}}} + \underbrace{\frac{c_1 s + c_2}{(s+2)(s+4)}}_{\substack{\text{zero-input}\\ \text{component}}} \\ &= \underbrace{\frac{-56}{s+3}}_{\substack{\text{forced}\\ \text{component}}} + \underbrace{\frac{K_1}{s+2} + \frac{K_2}{s+4}}_{\substack{\text{natural}\\ \text{component}}} \end{aligned}$$

*Years ago, the terms *transient* and *steady state* were used. There exist in practice, however, many interesting systems for which the natural response component is not transient and/or the forced component is not steady.

where the constants K_1 and K_2 depend upon the initial conditions. Both the zero-input and zero-state response components generally contribute to the natural response component.

1.6.5 Stability

A system described by linear, constant-coefficient integrodifferential equations is stable if all of its characteristic roots are to the left of the imaginary axis on the complex plane. For a stable system, the natural response component and the zero-input component decay with time.

Referring to Table 1-8, characteristic roots in the right half of the complex plane (to the right of the imaginary axis) correspond to terms which expand in time in the natural response, as do repeated roots along the imaginary axis. If *any* of the roots of the characteristic polynomial are in the right half-plane or are repeated and on the imaginary axis (including at the origin), one or more of the natural response terms expands with time, and the system is unstable. If *all* characteristic roots are in the left half-plane, the natural response terms all decay with time, and the system is stable.

For example, a system with transfer function

$$T(s)=\frac{-8s}{(s+3)(s+4)}$$

has characteristic roots $s=-3$ and $s=-4$, and thus is stable. The natural part of the system's response is of the form

$$y_{\text{natural}}(t)=K_1e^{-3t}+K_2e^{-4t}$$

which decays with time.

A system with transfer function

$$T(s)=\frac{-s^2+8}{(s^2+2s+5)(s-4)}$$

has characteristic roots $s=-1+j2$, $s=-1-j2$, and $s=4$. This system, because of the right half-plane root $s=4$, is unstable. The natural part of the response is of the form

$$y_{\text{natural}}(t)=Ae^{-t}[\cos(2t+\theta)]+Ke^{4t}$$

and the e^{4t} term expands with time.

Roots on the imaginary axis, if not repeated, give response terms which neither expand nor decay with time. A single root at the origin of the complex plane results in a constant response term, and a conjugate pair of imaginary roots give a steady sinusoidal term. A system is said to be *marginally stable* if it has no right half-plane or repeated imaginary axis roots, but there are nonrepeated imaginary axis characteristic roots.

It is common practice to characterize systems by the mutually exclusive terms *stable* (all characteristic roots in the left half-plane), *marginally stable* (one or more nonrepeated imaginary axis roots but no right half-plane roots), and *unstable* (repeated imaginary axis roots and/or right half-plane roots). A marginally stable system is neither stable nor unstable.

DRILL PROBLEMS

D1-12. For systems with input $r(t)$ and output $y(t)$ which are described by the following equations, find the system transfer functions:

(a) $$\frac{d^2y}{dt^2}+3\frac{dy}{dt}+7y=6r$$

ans. $6/(s^2+3s+7)$

(b) $$\frac{d^3y}{dt^3}+6\frac{d^2y}{dt^2}+2\frac{dy}{dt}+4y=-5\frac{d^2r}{dt^2}+8\frac{dr}{dt}$$

ans. $(-5s^2+8s)/(s^3+6s^2+2s+4)$

(c) $y(t)=8r(t-3)$

ans. $8e^{-3s}$

(d) $$\begin{cases} \dfrac{dx_1}{dt}=-3x_1+x_2+4r \\ \dfrac{dx_2}{dt}=-2x_1-r \\ y=x_1-2x_2 \end{cases}$$

ans. $(6s+21)/(s^2+3s+2)$

D1-13. Find the zero-state response for $t\geqslant 0$ of the systems with the following transfer functions and inputs:

(a) $$T(s)=\frac{4}{s+3}$$
$r(t)=u(t)$

ans. $\frac{4}{3}-\frac{4}{3}e^{-3t}$

(b) $T(s)=\dfrac{3s}{s+2}$

$r(t)=\delta(t)$

ans. $3\delta(t)-6e^{-2t}$

(c) $T(s)=\dfrac{-5s}{s^2+4s+3}$

$r(t)=6u(t)e^{-2t}$

ans. $15e^{-t}-60e^{-2t}+45e^{-3t}$

(d) $T(s)=\dfrac{4}{s+3}$

$r(t)=3u(t)\cos 2t$

ans. $-\frac{36}{13}e^{-3t}+\frac{36}{13}\cos 2t+\frac{24}{13}\sin 2t$
$=-\frac{36}{13}e^{-3t}+3.32\cos(2t-33.7^\circ)$

D1-14. Find the complete response for $t\geqslant 0$ of each of the following systems with the indicated input and initial conditions. Identify the zero-state and zero-input response components and the forced and natural response components.

(a) $T(s)=\dfrac{4}{s+3}$

$r(t)=u(t)$

$y(0^-)=-2$

ans. $\frac{4}{3}-\frac{10}{3}e^{-3t}$

(b) $T(s)=\dfrac{10}{s+4}$

$r(t)=\delta(t)$

$y(0^-)=0$

ans. $10e^{-4t}$

(c) $T(s)=\dfrac{s-5}{s^2+3s+2}$

$r(t)=u(t)$

$y(0^-)=-3$

$y'(0^-)=4$

ans. $-\frac{5}{2}+4e^{-t}-\frac{9}{2}e^{-2t}$

(d) $T(s)=\dfrac{s-5}{s^2+2s+3}$

$r(t)=6u(t)e^{2t}$

$y(0^-)=3$

$y'(0^-)=-4$

ans. $-\frac{18}{11}e^{2t}+e^{-t}[\frac{51}{11}\cos\sqrt{2}t+(109/11\sqrt{2})\sin\sqrt{2}t]$
$=(-\frac{18}{11})e^{2t}+8.4e^{-t}\cos(\sqrt{2}t-56.5^\circ)$

1.7 MULTIPLE-INPUT, MULTIPLE-OUTPUT SYSTEMS

1.7.1 Multiple Transfer Functions

If a system has several inputs $r_1(t)$, $r_2(t)$, ... and/or several outputs $y_1(t)$, $y_2(t)$, ..., there is a transfer function which relates each one of the outputs to each one of the inputs, when all other inputs are zero:

$$T_{ij}(s)=\left.\frac{Y_i(s)}{R_j(s)}\right|_{\text{when all initial conditions are zero and when all inputs except } R_j \text{ are zero}}$$

j — input number
i — output number

In general, when the system initial conditions are zero, the outputs are given by

$$\begin{aligned}
Y_1(s)&=T_{11}(s)R_1(s)+T_{12}(s)R_2(s)+T_{13}(s)R_3(s)+\cdots\\
Y_2(s)&=T_{21}(s)R_1(s)+T_{22}(s)R_2(s)+T_{23}(s)R_3(s)+\cdots\\
Y_3(s)&=T_{31}(s)R_1(s)+T_{32}(s)R_2(s)+T_{33}(s)R_3(s)+\cdots\\
&\vdots
\end{aligned}$$

A multiple-input and/or multiple-output system is said to be stable only if all characteristic roots of *all* its transfer functions are in the left half of the complex plane.

Suppose that a two-input, two-output system is described by the following differential equations, where r_1 and r_2 are the inputs and y_1 and y_2 are the outputs:

$$\begin{cases}
\dfrac{dy_1}{dt}+2y_1=r_1+5r_2\\
y_1+\dfrac{dy_2}{dt}+3y_2=4r_2+\dfrac{dr_2}{dt}
\end{cases}$$

Laplace-transforming with zero initial conditions,

$$\begin{cases} (s+2)Y_1(s) = R_1(s) + 5R_2(s) \\ Y_1(s) + (s+3)Y_2(s) = (s+4)R_2(s) \end{cases} \tag{1-9}$$

To find the transfer function which relates Y_1 to R_1, set R_2 to zero and solve for Y_1:

$$\begin{cases} (s+2)Y_1(s) = R_1(s) \\ Y_1(s) + (s+3)Y_2(s) = 0 \end{cases}$$

$$Y_1(s) = \frac{1}{s+2} R_1(s)$$

Then

$$T_{11} = \frac{Y_1(s)}{R_1(s)} = \frac{1}{s+2}$$

Solving, instead, for Y_2,

$$Y_2(s) = \frac{-1}{(s+2)(s+3)} R_1(s)$$

giving

$$T_{21}(s) = \frac{Y_2(s)}{R_1(s)} = \frac{-1}{(s+2)(s+3)}$$

Similarly, setting all inputs but R_2 to zero in (1-9),

$$\begin{cases} (s+2)Y_1(s) = 5R_2(s) \\ Y_1(s) + (s+3)Y_2(s) = (s+4)R_2(s) \end{cases}$$

$$Y_1(s) = \frac{5}{s+2} R_2(s)$$

$$T_{12}(s) = \frac{Y_1(s)}{R_2(s)} = \frac{5}{s+2}$$

And

$$Y_2(s) = \frac{s^2+6s+3}{(s+2)(s+3)} R_2(s)$$

$$T_{22}(s) = \frac{Y_2(s)}{R_2(s)} = \frac{s^2+6s+3}{(s+2)(s+3)}$$

For zero initial conditions and inputs,

$$r_1(t)=6 \sin 4t$$
$$r_2(t)=10$$

the outputs are given by

$$Y_1(s)=T_{11}(s)R_1(s)+T_{12}(s)R_2(s)$$
$$=\left(\frac{1}{s+2}\right)\left(\frac{24}{s^2+16}\right)+\left(\frac{5}{s+2}\right)\left(\frac{10}{s}\right)$$

and

$$Y_2(s)=T_{21}(s)R_1(s)+T_{22}(s)R_2(s)$$
$$=\frac{-1}{(s+2)(s+3)}\left(\frac{24}{s^2+16}\right)+\frac{s^2+6s+3}{(s+2)(s+3)}\left(\frac{10}{s}\right)$$

The natural part of the output y_1 will have an e^{-2t} contribution from T_{11} and another e^{-2t} contribution from T_{12} and so is of the form

$$y_{1\,\text{natural}}=K_1e^{-2t}$$

The natural response in y_2 has e^{-2t} and e^{-3t} terms from each of T_{21} and T_{22}, so is of the form

$$y_{2\,\text{natural}}=K_2e^{-2t}+K_3e^{-3t}$$

All characteristic roots in all transfer functions being in the left half-plane, this system is stable.

Multiple-input, multiple-output systems are commonly also called *multivariable systems.*

1.7.2 Accommodating Initial Conditions as Inputs

Control system designers seldom explicitly calculate natural components of system response because of their dependence upon the specific initial conditions, which are often unknown or of little concern. It is usually enough to know the natural response form and to know that the system is stable.

If desired, however, system initial conditions may be considered to be system inputs, and transfer functions may be found which relate the outputs to the initial condition inputs. For a system described by

$$\frac{d^2y}{dt^2}+3\frac{dy}{dt}+2y=4r(t)$$

Laplace-transforming with the initial conditions included gives

$$s^2Y(s) - sy(0^-) - y'(0^-) + 3[sY(s) - y(0^-)] + 2Y(s) = 4R(s)$$

$$(s^2 + 3s + 2)Y(s) = 4R(s) + (s+3)y(0^-) + y'(0^-)$$

$$Y(s) = \left[\frac{4}{s^2+3s+2}\right]R(s) + \left[\frac{s+3}{s^2+3s+2}\right]y(0^-) + \left[\frac{1}{s^2+3s+2}\right]y'(0^-)$$

$$= T_1(s)R(s) + T_2(s)y(0^-) + T_3(s)y'(0^-)$$

Although an unstable system cannot be made to be stable by any choice of initial conditions, the amplitudes of system response components may be strongly influenced by the initial conditions.

DRILL PROBLEMS

D1-15. A three-input, two-output system has the following transfer functions and inputs. Find the two outputs for $t \geqslant 0$ if all system initial conditions are zero:

$$T_{11}(s) = \frac{3}{s+2}$$

$$T_{12}(s) = \frac{s}{s+3}$$

$$T_{13}(s) = \frac{s+7}{s^2+5s+6}$$

$$T_{21}(s) = \frac{10}{(s+2)(s+3)}$$

$$T_{22}(s) = \frac{-6s+4}{s^2+5s+6}$$

$$T_{23}(s) = \frac{s+2}{s+3}$$

$$r_1(t) = u(t)$$

$$r_2(t) = \delta(t)$$

$$r_3(t) = 6u(t)e^{-2t}$$

ans. $\delta(t) + \frac{3}{2} - \frac{51}{2}e^{-2t} + 21e^{-3t} + 30te^{-2t}$,
$\frac{10}{6} + 11e^{-2t} - \frac{38}{3}e^{-3t}$

D1-16. For the following systems with outputs y and inputs r, find the transfer functions:

(a) $$\frac{d^2y}{dt^2}+3\frac{dy}{dt}+7y=6r_1+5r_2-4\frac{dr_2}{dt}$$

ans. $T_{11}(s)=6/(s^2+3s+7)$
$T_{12}(s)=(-4s+5)/(s^2+3s+7)$

(b) $$\begin{cases} \dfrac{d^2y_1}{dt^2}+6\dfrac{dy_1}{dt}+2y_1=\dfrac{dr}{dt}-3r \\ \dfrac{dy_2}{dt}+6y_2=4\dfrac{dr}{dt} \end{cases}$$

ans. $T_{11}(s)=(s-3)/(s^2+6s+2)$
$T_{21}(s)=4s/(s+6)$

(c) $$\begin{cases} \dfrac{d^3y_1}{dt^3}+7\dfrac{d^2y_1}{dt^2}+6\dfrac{dy_1}{dt}+y_1=\dfrac{d^2r_1}{dt^2}+3r_1+r_2 \\ \dfrac{d^3y_2}{dt^3}+7\dfrac{d^2y_2}{dt^2}+6\dfrac{dy_2}{dt}+y_2=4\dfrac{dr_2}{dt} \end{cases}$$

ans. $T_{11}(s)=(s^2+3)/(s^3+7s^2+6s+1)$
$T_{12}(s)=1/(s^3+7s^2+6s+1)$
$T_{21}(s)=0$
$T_{22}(s)=4s/(s^3+7s^2+6s+1)$

(d) $$\begin{cases} \dfrac{dx_1}{dt}=4x_1-3x_2+r_1 \\ \dfrac{dx_2}{dt}=-x_1+x_2+r_1-r_2 \\ y_1=x_1-x_2 \\ y_2=8x_2 \end{cases}$$

ans. $T_{11}(s)=1/(s^2-5s+1)$
$T_{12}(s)=(s-1)/(s^2-5s+1)$
$T_{21}(s)=(8s-40)/(s^2-5s+1)$
$T_{22}(s)=(-8s+32)/(s^2-5s+1)$

1.8 A POSITIONING SERVO

1.8.1 Basic System Arrangement

A simple but practical feedback control system is diagramed in Fig. 1-12. It is a positioning system or *position servo* for a large microwave antenna.

The antenna is modeled as a mass having a large moment of inertia, J.

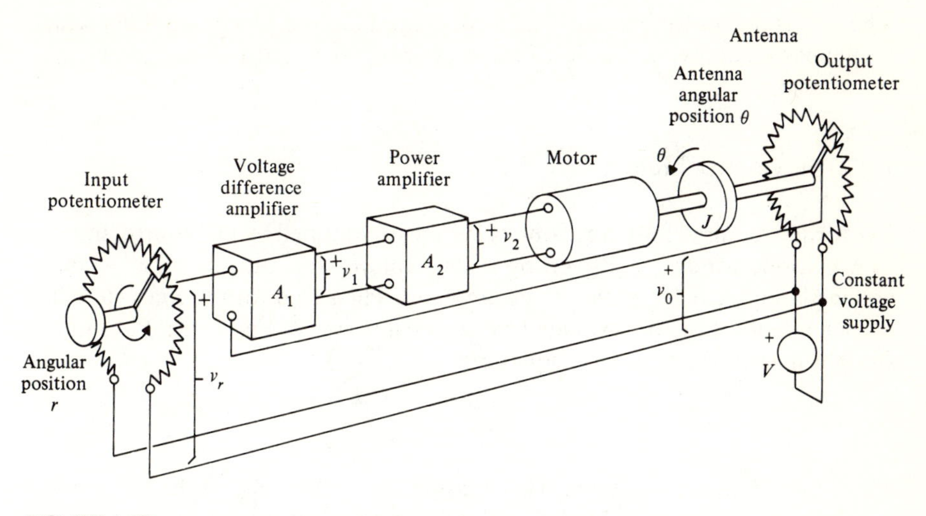

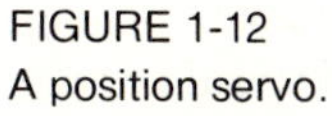
FIGURE 1-12
A position servo.

An output potentiometer measures the output shaft position, converting the position to a proportional voltage according to

$$v_o = K_p\theta$$

where θ is the output shaft angle in radians and v_0 is the output potentiometer voltage; K_p, the constant of proportionality between shaft position and potentiometer voltage, is the total voltage V divided by the maximum rotation of the potentiometer:

$$K_p = \frac{V}{\theta_{\max}} \qquad \text{V/rad}$$

The input potentiometer slider position r is converted to a voltage with a potentiometer identical to the output potentiometer:

$$v_r = K_p r$$

The difference between the two potentiometer signals is then amplified with gain A_1,

$$v_1 = A_1(v_r - v_o) = A_1 K_p(r - \theta)$$

where v_1 is the error voltage output of the difference amplifier. This voltage is then further amplified with gain A_2 and is applied to the motor terminals,

$$v_2 = A_2 v_1 = A_1 A_2 K_p(r - \theta) \qquad (1\text{-}10)$$

where v_2 is the motor voltage. The second amplifier is a power amplifier which is capable of supplying the electrical power necessary to drive the motor.

1.8.2 DC Control Motors

The torque required for many control systems is supplied by DC control motors in which one winding is driven by a constant voltage and the other winding voltage is supplied by an electronic amplifier. The speed-torque characteristics of such a motor are approximated by the ideal curves of Fig. 1-13a. Each of the straight lines in the figure has equation

$$\tau = k_1 v_a - k_2 \omega$$

where k_1 and k_2 are constants. This relation is a special case of the DC motor-generator equations of Table 1-5, with negligible armature inductance.

The constant k_2 is the negative slope of each of the linearized speed-torque curves:

$$k_2 = -\frac{\Delta\tau}{\Delta\omega}$$

As shown in Fig. 1-13b, the constant k_2 may be found from manufacturer's literature by forming the ratio

$$k_2 = \frac{\text{stall torque at rated voltage}}{\text{no-load speed at rated voltage}}$$

The constant k_1 relates the applied motor voltage to its torque when the rotor

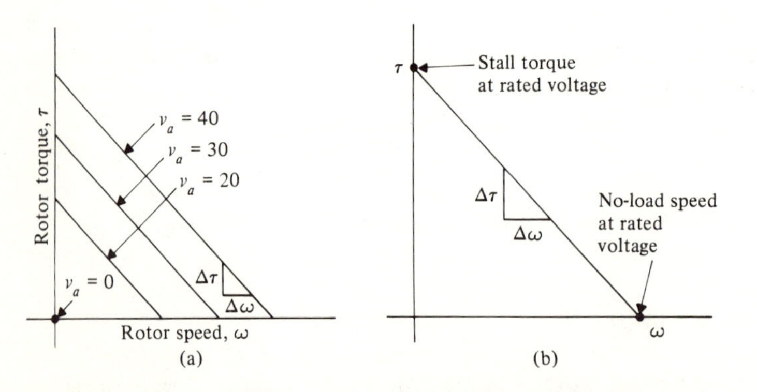

FIGURE 1-13
Idealized speed/torque curves for a certain control motor.
(a) Representative curves. (b) Stall torque and no-load speed.

speed ω is zero (or stalled):

$$k_1 = \frac{\text{stall torque at rated voltage}}{\text{rated voltage}}$$

1.8.3 AC Control Motors

If in-phase sinusoidal AC voltages are applied to the armature and field coils of a control motor,

$$v_a(t) = A \cos \omega_s t$$
$$v_f(t) = F \cos \omega_s t$$

the instantaneous torque developed has the following form:

$$\begin{aligned}\tau &= k_3 v_f(t) v_a(t) - k_4 \omega \\ &= k_3 AF \cos^2 \omega_s t - k_4 \omega\end{aligned}$$

The *average* torque developed is

$$\begin{aligned}\tau_{\text{ave}} &= k_3 AF(\tfrac{1}{2}) - k_4 \omega \\ &= \frac{k_3 F}{\sqrt{2}}\left(\frac{A}{\sqrt{2}}\right) - k_4 \omega \\ &= k_3' V_{\text{rms}} - k_4 \omega\end{aligned}$$

the average of the squared cosine term being $\frac{1}{2}$. The average torque, RMS armature voltage, and shaft speed are related in a similar way to those quantities for a DC control motor, Fig. 1-13.

The major advantage of using an AC control motor instead of a DC one is that the signals to be amplified in the control system are sinusoidal, with zero average value. The precise amplification of constant (or DC) signals is much more difficult since small inaccuracies can easily cause an unwanted constant error, or offset, in the amplifier output which is indistinguishable from an offset of the input. Also, transformers may be used in AC systems, for isolation, load splitting, and impedance matching.

In the arrangement of Fig. 1-12, replacing the constant potentiometer supply by a sinusoidal one, using AC amplifiers and an AC control motor, would convert that system to an AC type. Of course, the frequency of the AC signals must be high enough so that the rapid torque fluctuations developed by the motor are sufficiently smoothed by the shaft inertia. In some applications, an AC frequency of 60 Hz is used; in others such as in aircraft and some shipboard systems, 400 Hz is commonly used.

An AC control motor may also be used in arrangements where a controllable phase shift between the two sinusoidal coil voltages of fixed amplitude determines motor torque:

$$v_a(t) = A\cos(\omega_s t + \theta)$$
$$v_f(t) = F\cos\omega_s t$$
$$\tau = k_3 AF\cos(\omega_s t + \theta)\cos\omega_s t - k_4\omega$$
$$= k_3 AF[\tfrac{1}{2}\cos\theta + \tfrac{1}{2}\cos(2\omega_s t + \theta)] - k_4\omega$$
$$\tau_{\text{ave}} = k_3 AF(\tfrac{1}{2})\cos\theta - k_4\omega$$

Special three-phase AC control motors are termed *synchros* and have been in use in specialized applications for many years.

1.8.4 System Equations and Transfer Functions

For the DC system of Fig. 1-12, the output of the power amplifiers is given by (1-10):

$$v_2 = A_1 A_2 K_p(r - \theta)$$

where r is the desired angular position and θ is the actual shaft angular position. The control motor develops a torque of the form

$$\tau = k_1 v_2 - k_2\omega = k_1 v_2 - k_2\frac{d\theta}{dt}$$
$$= k_1 A_1 A_2 K_p(r - \theta) - k_2\frac{d\theta}{dt}$$

Equating the applied torque on the rotational antenna mass to the inertial force, there results

$$\tau = J\frac{d^2\theta}{dt^2} = k_1 A_1 A_2 K_p(r - \theta) - k_2\frac{d\theta}{dt} \tag{1-11}$$

Rearranging,

$$J\frac{d^2\theta}{dt^2} + k_2\frac{d\theta}{dt} + k_1 A_1 A_2 K_p\theta = k_1 A_1 A_2 K_p r(t)$$

Some of the equation coefficients, and thus some of the system properties, may be selected by the designer by appropriately choosing the control components. Other parameters, such as the moment of inertia of the load, J, cannot be changed.

The transfer function relating the input position $R(s)$ to the output position $\theta(s)$ is given by

$$Js^2\theta(s)+k_2s\theta(s)+k_1A_1A_2K_p\theta(s)=k_1A_1A_2K_pR(s)$$

$$T_1(s)=\frac{\theta(s)}{R(s)}=\frac{k_1A_1A_2K_p}{Js^2+k_2s+k_1A_1A_2K_p}$$

If the antenna experiences an external torque, say from wind, in addition to the motor's torque, Eq. (1-12) becomes, instead,

$$\tau=J\frac{d^2\theta}{dt^2}-\tau_{\text{wind}}=k_1A_1A_2K_p(r-\theta)-k_2\frac{d\theta}{dt}$$

$$J\frac{d^2\theta}{dt^2}+k_2\frac{d\theta}{dt}+k_1A_1A_2K_p\theta=k_1A_1A_2K_pr(t)+\tau_{\text{wind}}(t)$$

where τ_{wind} is the external torque in the sense of increasing θ. The transfer function relating $\theta(s)$ to $R(s)$ is unchanged, but now a disturbance input $\tau_{\text{wind}}(t)$ has been included in the model. The transfer function relating this input to θ is found by setting r to zero and Laplace-transforming with zero initial conditions:

$$Js^2\theta(s)+k_2s\theta(s)+k_1A_1A_2K_p\theta(s)=\mathcal{T}_{\text{wind}}(s)$$

$$T_2(s)=\frac{\theta(s)}{\mathcal{T}_{\text{wind}}(s)}=\frac{1}{Js^2+k_2s+k_1A_1A_2K_p}$$

In general, with zero initial conditions,

$$\theta(s)=T_1(s)R(s)+T_2(s)\mathcal{T}(s)$$

$$=\frac{k_1A_1A_2K_p/J}{s^2+(k_2/J)s+(K_1A_1A_2K_p/J)}R(s)+\frac{1/J}{s^2+(k_2/J)s+(k_1A_1A_2K_p/J)}\mathcal{T}(s)$$

1.9 SUMMARY

Control systems generally involve a plant or process with inputs and outputs. Controllers are designed to generate control inputs to the plant which will produce a desired plant behavior. Feedback, if used, can result in improved accuracy and speed of response and reduced dependence upon specific components and sensitivity to disturbances.

Determining system equations is the starting point in most control system analysis. Relations for electrical, electronic, translational mechanical, rotational mechanical, and electromechanical components are summarized in Tables 1-1 through 1-5.

Laplace transformation is advantageous when system models are linear, time-invariant integrodifferential equations. Transforms and properties are summarized in Tables 1-6 and 1-7, and partial fraction expansion is an important transform inversion technique. The time response corresponding to various partial fraction terms is summarized in Table 1-8.

The transfer function of a single-input, single-output linear, time-invariant system is the ratio of the Laplace transform of the output to the Laplace transform of the input, when all initial conditions are zero. Systems described by integrodifferential equations have transfer functions which may be expressed as the ratio of two polynomials in s.

System response is commonly considered to be composed of component parts in either of two ways:

1. zero-state and zero-input; or
2. forced and natural.

The zero-state response component is the system response when all initial conditions are zero,

$$Y_{\text{zero-state}}(s) = T(s)R(s)$$

where $T(s)$ is the system transfer function and $R(s)$ is the Laplace transform of the input. The zero-input response component is the system response due to nonzero initial conditions when the input is zero. Forced and natural components divide the response into those terms, due to the inputs, which do not correspond to system characteristic roots and the characteristic root terms.

A system is stable if, for any initial conditions, all of the terms in its natural (and thus its zero-input) response decay with time. Such will be the case if and only if all system characteristic roots are in the left half of the complex plane. Nonrepeated characteristic roots on the imaginary axis correspond to natural response terms which neither decay nor expand in time. A system with all characteristic roots in the left half-plane except for one or more nonrepeated imaginary axis roots is termed marginally stable. If any characteristic roots are in the right half of the complex plane or are repeated on the imaginary axis, the system is unstable.

For multiple-input, multiple-output systems, there is a transfer function for each combination of input and output:

$$T_{ij}(s) = \left.\frac{Y_i(s)}{R_j(s)}\right|_{\substack{\text{all initial conditions and all inputs} \\ \text{except } r_j \text{ set to zero}}}$$

where the y_i terms are outputs and the r_j terms are inputs. For zero initial conditions, the outputs are given by

$$\begin{aligned} Y_1(s) &= T_{11}(s)R_1(s) + T_{12}(s)R_2(s) + \cdots \\ Y_2(s) &= T_{21}(s)R_1(s) + T_{22}(s)R_2(s) + \cdots \\ &\vdots \end{aligned}$$

The form of the natural components of any output is governed by the collection of characteristic roots in all the transfer functions associated with that output.

Initial conditions may be related to system outputs with transfer functions in much the same way as are system inputs.

REFERENCES

The references given here and in the following chapters trace the history of topics presented in the text. While by no means comprehensive, they give a series of milestones in the development and understanding of these ideas. This, too, is our way of acknowledging those works from which we learned and to which we all owe a great deal.

Feedback

Blackman, R. B. "Effect of Feedback on Impedance." *Bell Syst. Tech. J.* 22 (October 1943).

Bode, H. W. "Feedback—The History of an Idea," in *Selected Papers on Mathematical Trends in Control Theory*. New York: Dover, 1964.

Maxwell, J. C. "On Governors." *Proc. Roy. Soc.* 16 (1868).

Mayr, O. "The Origins of Feedback Control." *Sci. Amer.* (October 1970).

Nyquist, H. "Regeneration Theory." *Bell Syst. Tech. J.* 11 (January 1932).

Wolf, A. *A History of Science, Technology and Philosophy in the Eighteenth Century*. New York: McGraw-Hill, 1939.

System Equations

Close, C. M. *The Analysis of Linear Circuits*. New York: Harcourt, Brace and World, 1966.

Cruz, J. B., Jr., and Van Valkenburg, M. E. *Signals in Linear Circuits*. Boston: Houghton-Mifflin, 1974.

Harmon, W. W., and Lytle, D. W. *Electrical and Mechanical Networks*. New York: McGraw-Hill, 1962.

Hostetter, G. H. *Fundamentals of Network Analysis*. New York: Harper & Row, 1980.

Van Valkenburg, M. E. *Network Analysis*. Englewood Cliffs, N.J.: Prentice-Hall, 1974.

Laplace Transformation, Transfer Functions, and Stability

Aseltine, J. A. *Transform Methods in Linear System Analysis*. New York: McGraw-Hill, 1958.

Bracewell, R. N. *The Fourier Transform and Its Applications*. New York: McGraw-Hill, 1978.

DiStefano, J. J. III; Stubberud, A. R.; and Williams, I. J. *Feedback and Control Systems* (*Schaum's Outline*). New York: McGraw-Hill, 1967.

Gardner, M. F., and Barnes, J. L. *Transients in Linear Systems*. New York: Wiley, 1942.

LePage, W. R. *Complex Variables and the Laplace Transform for Engineers*. New York: McGraw-Hill, 1961.

Ley, B. J.; Lutz, S. G.; and Rehberg, C. F. *Linear Circuit Analysis*. New York: McGraw-Hill, 1959.

Savant, C. J., Jr. *Fundamentals of the Laplace Transformation*. New York: McGraw-Hill, 1962.

PROBLEMS

1. List five examples of open-loop control systems and draw illustrative diagrams for each.
2. List five examples of closed-loop control systems and draw illustrative diagrams for each.
3. Identify the input(s), output(s), and major parts of the following control systems. Which are open-loop and which are closed-loop?
   (a) A heater with thermostat
   (b) A toaster
   (c) A human being reaching to touch an object
   (d) A human being steering an automobile
   (e) An electric power generating station
4. Using Laplace transform methods, find the solutions for $t \geqslant 0$ of the following differential equations, subject to zero initial conditions:

   (a) $\dfrac{dy}{dt}+3y=4$

   (b) $3\dfrac{dy}{dt}-y=\cos 2t$

   ans. $0.027e^{(1/3)t}+0.16\cos(2t-99.5^{\circ})$

   (c) $\dfrac{d^2y}{dt^2}+5\dfrac{dy}{dt}+4y=e^{-2t}$

   (d) $\dfrac{d^2y}{dt^2}+5\dfrac{dy}{dt}+4y=\sin 3t$

   ans. $\frac{1}{10}e^{-t}-\frac{1}{25}e^{-4t}+0.064\cos(3t-199^{\circ})$

   (e) $\dfrac{d^3y}{dt^3}+4\dfrac{d^2y}{dt^2}+4\dfrac{dy}{dt}=e^{-t}$

5. Find the Laplace transforms of the solutions, $Y(s)$:

   (a) $\dfrac{dy}{dt}-6y=1$

   $y(0^-)=5$ *ans.* $\dfrac{5s+1}{s^2-6s}$

   (b) $\dfrac{dy}{dt}+5y+6\displaystyle\int_0^t y\,dt=4t$

   $y(0^-)=-2$

   (c) $\dfrac{dy}{dt}-y+\displaystyle\int_{-\infty}^t y(t)\,dt=1$

   $y(0^-)=0$

   $\displaystyle\int_{-\infty}^{0^-} y\,dt=-2$ *ans.* $\dfrac{3}{s^2-s+1}$

   (d) $\dfrac{d^2y}{dt^2}+2\dfrac{dy}{dt}+5y=\delta(t)$

   $y(0^-)=0$

   $y'(0^-)=3$

(e) $2\frac{d^2y}{dt^2}+8\frac{dy}{dt}+8y=0$

$y(0^-)=-3$

$y'(0^-)=4$ *ans.* $\frac{-3s-8}{s^2+4s+4}$

6. Solve for the Laplace-transformed signals $Y_1(s)$ and $Y_2(s)$:

$$\begin{cases} \frac{d^2y_1}{dt^2}-3\frac{dy_1}{dt}+2y_1+2\frac{dy_2}{dt}+4y_2=0 \\ -2\frac{dy_1}{dt}-4y_1+\int_0^t y_2\,dt=8u(t) \end{cases}$$

$y_1(0^-)=5$

$y_1'(0^-)=-6$

$y_2(0^-)=7$

7. Solve for $y(t)$, $t \geqslant 0$, and sketch the solution:

(a) $\frac{dy}{dt}+4y=\delta(t)$

$y(0^-)=3$ *ans.* $4e^{-4t}$

(b) $\frac{dy}{dt}+3y=u(t)$

$y(0^-)=0$

(c) $2\frac{dy}{dt}-y=6e^{-4t}$

$y(0^-)=0$

(d) $\frac{d^2y}{dt^2}+5\frac{dy}{dt}+6y=0$

$y(0^-)=3$

$y'(0^-)=-4$ *ans.* $5e^{-2t}-2e^{-3t}$

(e) $\frac{d^2y}{dt^2}+5\frac{dy}{dt}+6y=\delta(t)$

$y(0^-)=0$

$y'(0^-)=0$

(f) $\frac{d^2y}{dt^2}+5\frac{dy}{dt}+6y=7e^{-4t}$

$y(0^-)=3$

$y'(0^-)=0$

(g) $\frac{d^2y}{dt^2}+2\frac{dy}{dt}+5y=0$

$y(0^-)=0$

$y'(0^-)=4$ *ans.* $2e^{-t}\cos(2t-90^\circ)$

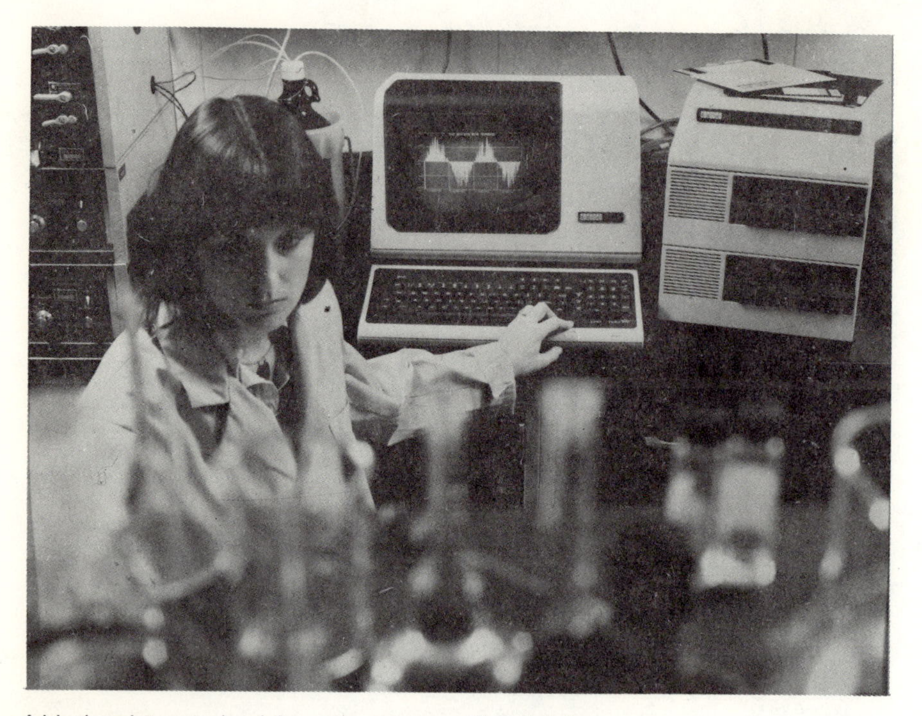

A biochemist controls a laboratory process with a digital control system. The video display in the background shows data in graphical form. (Photo courtesy of Digital Equipment Corporation.)

(h) $\dfrac{d^2y}{dt^2}+2\dfrac{dy}{dt}+5y=\delta(t)$

$y(0^-)=0$

$y'(0^-)=0$

8. Write simultaneous loop equations for the electrical networks of Fig. P1-1. Then Laplace-transform the equations, taking all initial conditions to be zero.

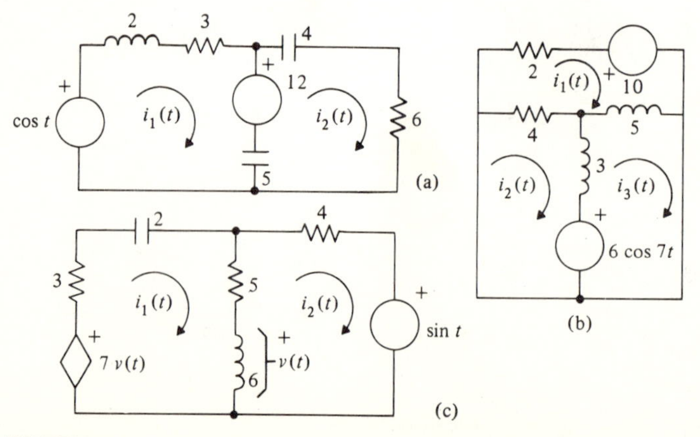

FIGURE P1-1

$$ans.\ (c)\ \begin{cases}(6s+8+1/2s)I_1(s)-(6s+5)I_2(s)-7V(s)=0\\ -(6s+5)I_1(s)+(6s+9)I_2(s)=-1/(s^2+1)\\ 6sI_1(s)-6sI_2(s)-V(s)=0\end{cases}$$

9. If all initial conditions of the networks of Fig. P1-2 are zero, find $v(t)$, $t \geqslant 0$.

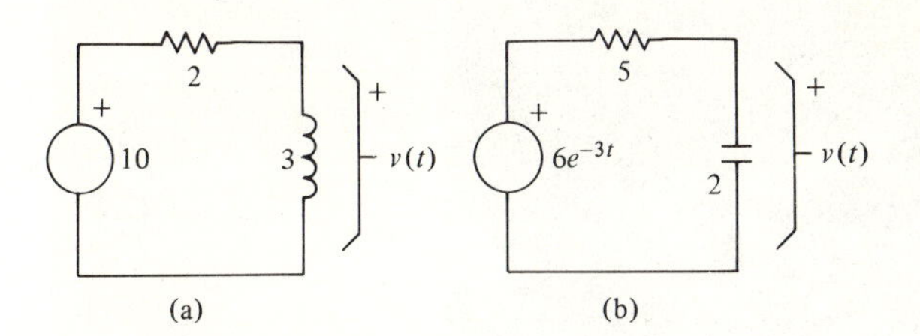

FIGURE P1-2

$$ans.\ (b)\ \tfrac{6}{29}e^{-(1/10)t}-\tfrac{6}{29}e^{-3t}$$

10. Write simultaneous nodal equations for the electrical networks of Fig. P1-3. Then Laplace-transform the equations, taking all initial conditions to be zero.

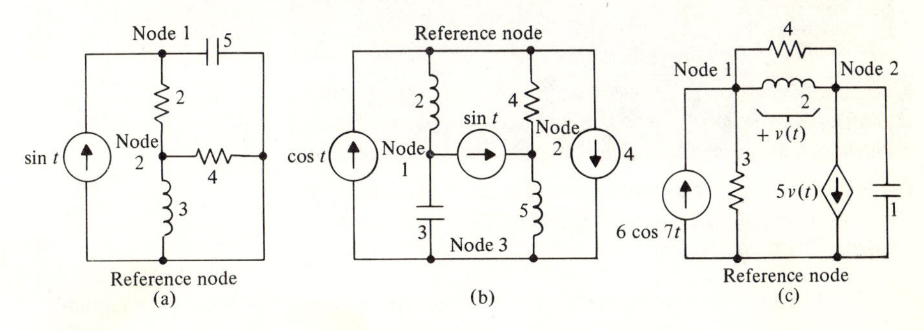

FIGURE P1-3

$$ans.\ (b)\ \begin{cases}(1/2s+3s)V_1(s)-3sV_3(s)=-1/(s^2+1)\\ (\tfrac{1}{4}+1/5s)V_2(s)-(1/5s)V_3(s)=1/(s^2+1)\\ -3sV_1(s)-(1/5s)V_2(s)+(3s+1/5s)V_3(s)=(4/s)\\ \qquad\qquad\qquad\qquad\qquad\qquad\qquad -s(s^2+1)\end{cases}$$

11. All initial conditions in the networks of Fig. P1-4 are zero. Find $I(s)$, then find $i(t)$ for $t \geqslant 0$.

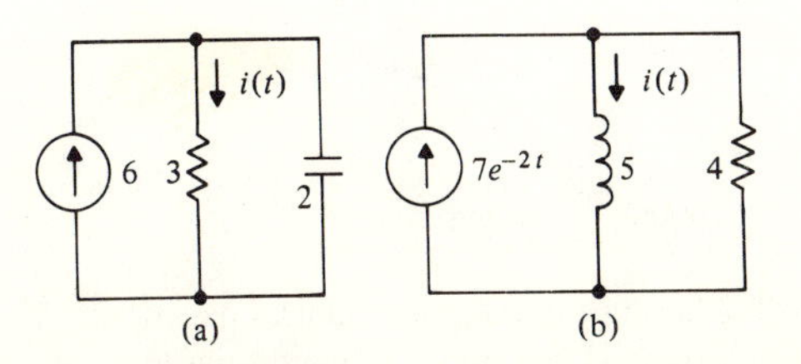

FIGURE P1-4

$$ans.\ (a)\ 1/s(s+\tfrac{1}{6}),\ 6-6e^{-(1/6)t}$$

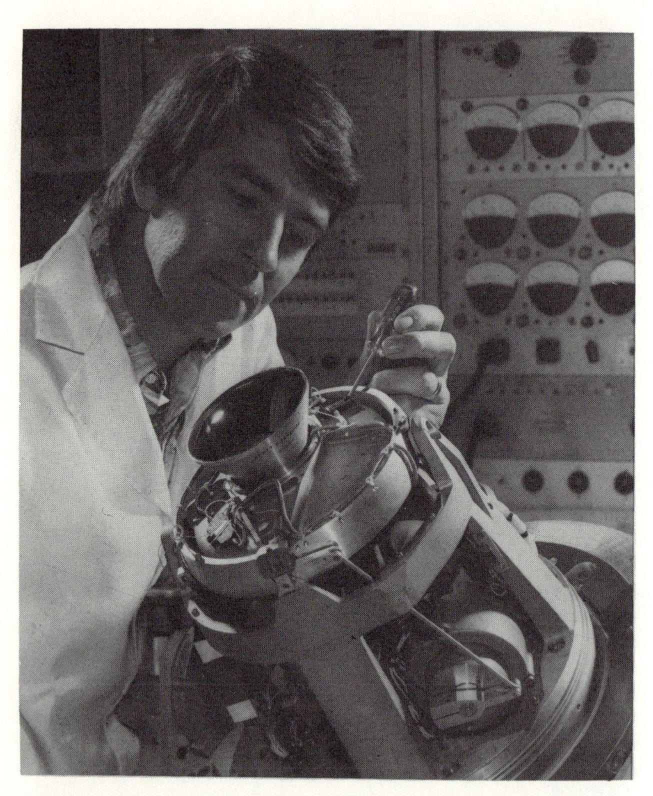

A technician adjusts the infrared sensor system on the head of a missile. Navigation system components are visible below the sensor. (Photo courtesy of Hughes Aircraft Company.)

12. Find $V(s)$ and find $v(t)$, $t \geqslant 0$ for the electronic network of Fig. P1-5. Assume zero initial conditions.

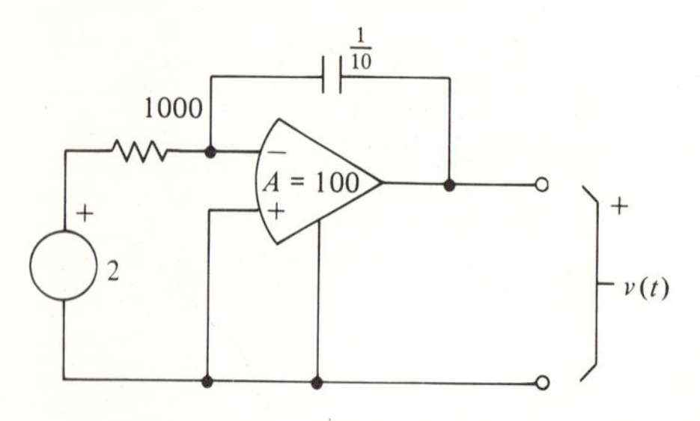

FIGURE P1-5

ans. $-0.0198/s(s+0.000099)$, $-200+200e^{-0.000099t}$

13. Write simultaneous differential equations for the translational mechanical networks of Fig. P1-6. Then Laplace-transform the equations, taking all initial conditions to be zero.

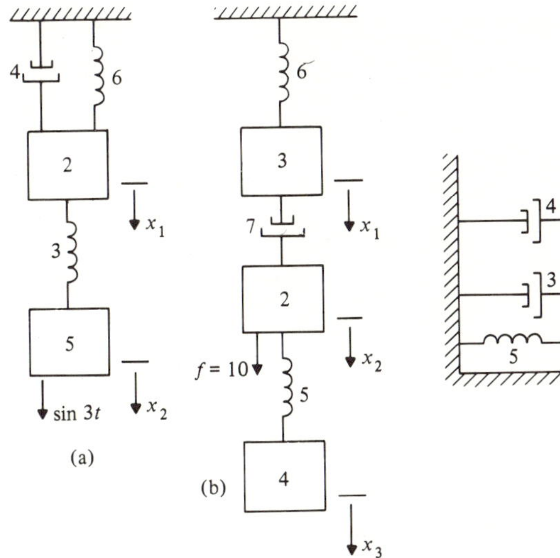

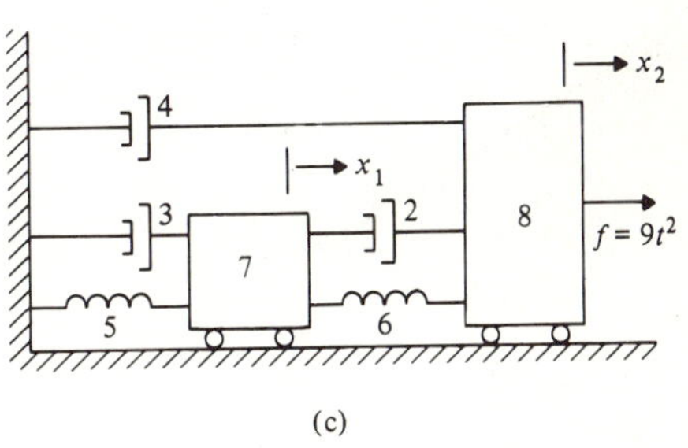

FIGURE P1-6

$$\textit{ans.}\ (b)\begin{cases}(3s^2+7s+6)X_1(s)-7sX_2(s)=0\\ -7sX_1(s)+(2s^2+7s+5)X_2(s)-5X_3(s)=10/s\\ -5X_2(s)+(4s^2+5)X_3(s)=0\end{cases}$$

An engineer works in the master control room of a television videotape center. Two reel-to-reel tape machines are in operation and a commercial automatic cassette player for announcements is at the left of the photograph. (Photo courtesy of Radio Corporation of America.)

14. Find $X(s)$ and find $x(t)$, $t \geqslant 0$ for the networks of Fig. P1-7, for which all initial conditions are zero.

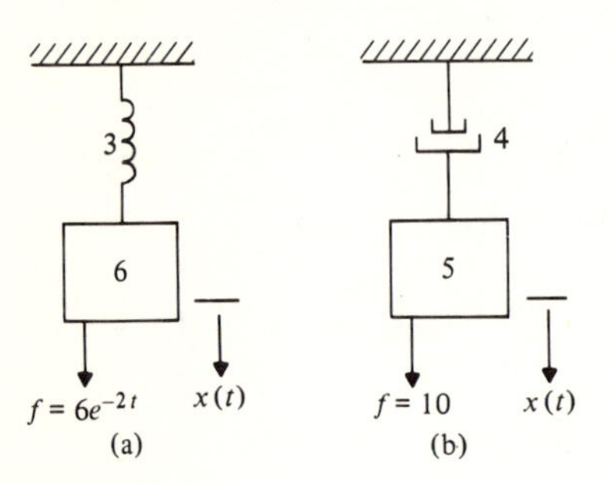

FIGURE P1-7

ans. (a) $1/(s+2)(s^2+\frac{1}{2})$, $0.22e^{-2t}+0.66\cos(0.707t-109.5^\circ)$

15. Write simultaneous differential equations for the rotational mechanical networks of Fig. P1-8. Then Laplace-transform the equations, taking all initial conditions to be zero.

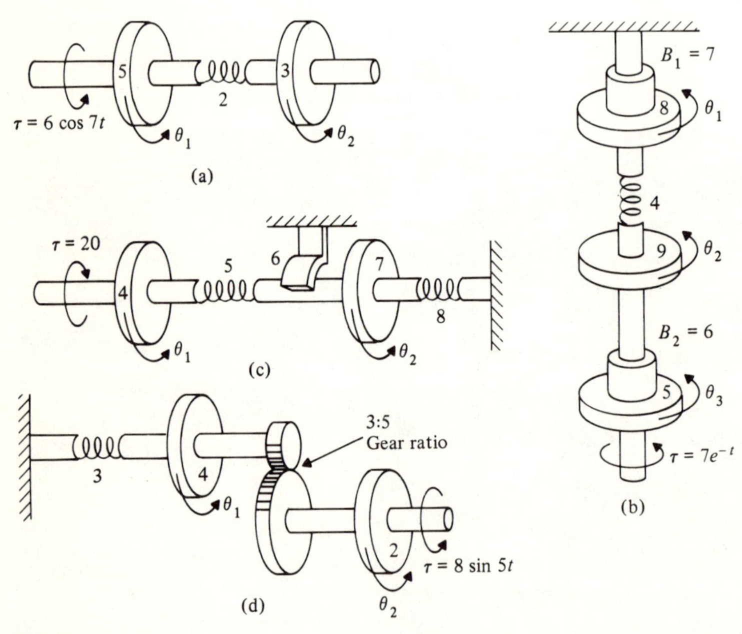

FIGURE P1-8

ans. (b)
$$\begin{cases} (8s^2+7s+4)\theta_1(s)-4\theta_2(s)=0 \\ -4\theta_1(s)+(9s^2+6s+4)\theta_2(s)-6s\theta_3(s)=0 \\ -6s\theta_2(s)+(5s^2+6s)\theta_3(s)=7/(s+1) \end{cases}$$

(d)
$$\begin{cases} \theta_2(s)=-\frac{3}{5}\theta_1(s) \\ \tau_2(s)=\frac{5}{3}\tau_1(s) \\ (4s^2+3)\theta_1(s)-\tau_1(s)=0 \\ 2s^2\theta_2(s)-\tau_2(s)=40/(s^2+25) \end{cases}$$

16. Find $\theta(s)$ and find $\theta(t)$, $t \geq 0$ for the networks of Fig. P1-9, for which all initial conditions are zero.

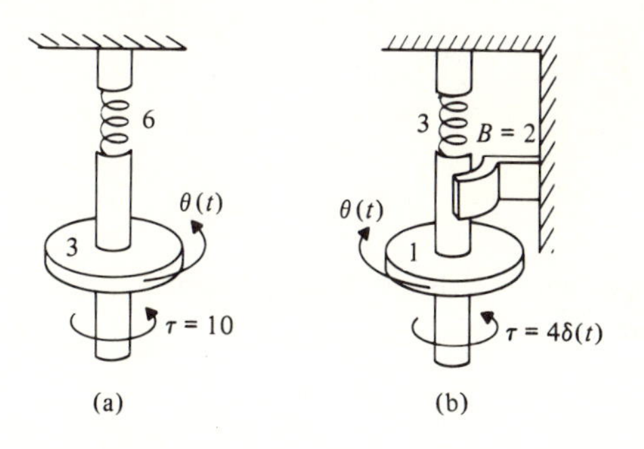

FIGURE P1-9

ans. (a) $10/3s(s^2+2)$, $\frac{5}{3} - \frac{5}{3}\cos(\sqrt{2}t)$

17. Write simultaneous integrodifferential equations for the electromechanical system of Fig. P1-10.

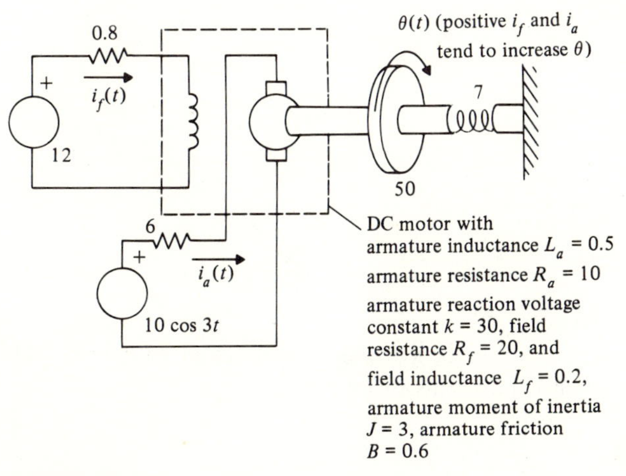

FIGURE P1-10

18. The following systems have input $r(t)$ and output $y(t)$ which are related by the given equations. Find the transfer function for each system.

(a) $$\frac{d^3y}{dt^3}+7\frac{d^2y}{dt^2}+5\frac{dy}{dt}+6y=-3\frac{dr}{dt}+2r$$

(b) $$\begin{cases} \dfrac{d^3x_1}{dt^3}+\dfrac{dx_2}{dt}+x_2=4r \\ \dfrac{dx_1}{dt}+2x_1-\dfrac{dx_2}{dt}=\dfrac{dr}{dt} \\ y=x_1+x_2 \end{cases}$$

ans. $$\frac{-s^4+s^2+9s+8}{s^4+s^2+3s+2}$$

(c) $\begin{cases} \dfrac{dx_1}{dt}=3x_1-2x_2+r(t) \\ \dfrac{dx_2}{dt}=2x_1+x_3-4r(t) \\ \dfrac{dx_3}{dt}=x_2 \\ y=x_1-x_3 \end{cases}$

19. Transfer functions and inputs $r(t)$ for systems are given below. Find each system output after $t=0$ if all initial conditions are zero.

(a) $T(s)=\dfrac{s-3}{s+4}$

$r(t)=u(t)$ *ans.* $-\frac{3}{4}+\frac{7}{4}e^{-4t}$

(b) $T(s)=\dfrac{s}{s+1}$

$r(t)=\delta(t)$

(c) $T(s)=\dfrac{1}{s^2+4}$

$r(t)=u(t)$ *ans.* $\frac{1}{4}-\frac{1}{4}\cos 2t$

(d) $T(s)=\dfrac{2s+3}{3s+2}$

$r(t)=3\cos 2t$

(e) $T(s)=\dfrac{10}{s^2+2s+10}$

$r(t)=\delta(t)$ *ans.* $\frac{10}{3}e^{-t}\sin 3t$

20. Find the transfer functions for each of the following multiple-input, multiple-output systems. The r terms are inputs and y terms are outputs.

(a) $\dfrac{d^2y}{dt^2}+3\dfrac{dy}{dt}+2y=2r_1+3\dfrac{dr_1}{dt}-6r_2$

(b) $\begin{cases} \dfrac{dy_1}{dt}-y_2=r_1-4\dfrac{dr_2}{dt} \\ y_1-2\dfrac{dy_2}{dt}+3y_2=r_2 \end{cases}$

ans. $\dfrac{-2s+3}{-2s^2+3s+1} \quad \dfrac{8s^2-12s+1}{-2s^2+3s+1}$

$\dfrac{-1}{-2s^2+3s+1} \quad \dfrac{5s}{-2s^2+3s+1}$

(c) $\begin{cases} \dfrac{dy_1}{dt}+3y_1-\displaystyle\int_0^t y_2\,dt=r_1 \\ \dfrac{dy_2}{dt}+2y_2+\dfrac{dy_1}{dt}-y_1=r_2-r_1 \end{cases}$

21. A two-input, two-output system has the following transfer functions:

$$\frac{Y_1(s)}{R_1(s)} = T_{11}(s) = \frac{-s}{s+1}$$

$$\frac{Y_1(s)}{R_2(s)} = T_{12}(s) = \frac{s+4}{s^2+3s+2}$$

$$\frac{Y_2(s)}{R_1(s)} = T_{21}(s) = \frac{3}{s+2}$$

$$\frac{Y_2(s)}{R_2(s)} = T_{22}(s) = -5$$

Find $y_1(t)$ and $y_2(t)$ for $t \geqslant 0$ if all initial conditions are zero and

$$r_1(t) = u(t)$$

$$r_2(t) = 2e^t$$

ans. $y_2(t) = (\frac{3}{2} - \frac{3}{2}e^{-2t} - 10e^t)u(t)$

22. Find the forms of the zero-input components of the response of systems with the transfer functions in Prob. 19.

ans. (a) Ke^{-4t}
(c) $A\cos(2t+\theta)$
(e) $Ae^{-t}\cos(3t+\theta)$

23. Find the transfer functions which relate the output y to the initial conditions in the system

$$3\frac{d^2y}{dt^2} + 2\frac{dy}{dt} + 6y = 0$$

24. Draw diagrams similar to Figs. 1-3 or 1-6 for the following systems:
    (a) Control of human skin temperature by sweating
    (b) An automatic traffic light system at an intersection which varies the ratio of green light times in the two cross-directions with the traffic ratio in the two directions
    (c) A simple economic model which includes as inputs government tax level, private business investment, and consumer spending and has per capita income after taxes as the output
    (d) Control of a nuclear reactor
    (e) The teaching and learning process with feedback

25. A useful model for a voltage amplifier is shown in Fig. P1-11a. The incoming voltage $v_i(t)$ is amplified with gain A and produces the open-circuit output voltage $Av_i(t)$. When feedback is added to an amplifier, desirable properties may result.

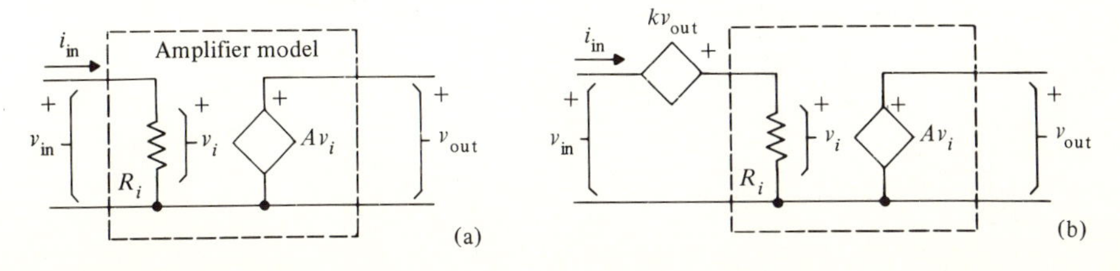

FIGURE P1-11

For the amplifier of Fig. P1-11a without feedback, the voltage gain is

$$G=\frac{v_{\text{out}}}{v_{\text{in}}}=A$$

Show that the voltage gain of the amplifier with feedback is

$$\frac{A}{1-kA}$$

For positive values of kA less than unity, the gain is increased by the feedback, a circumstance which was exploited (as the "regenerative" receiver) in the early days of radio when high-gain amplifiers were very difficult to obtain otherwise. Unfortunately, all of the minor shortcomings of the basic amplifier are emphasized by positive feedback, and $kA \geq 1$ results in instability. For negative values of kA, the voltage gain is reduced, but in return performance in other respects is improved.

The amplifier of Fig. P1-11a without feedback has input resistance

$$R_{\text{in}}=\frac{v_{\text{in}}}{i_{\text{in}}}=R_i$$

Show that the input resistance of the amplifier with feedback is $R_i(1-kA)$. For negative kA, the input resistance is increased.

26. The connection diagram for a bipolar transistor feedback AC amplifier is given in Fig. P1-12a. For sufficiently small signals composed of midrange frequencies, the network of Fig. P1-12b is an accurate model. In terms of the various parameters, find the transfer function which relates $V_{\text{out}}(s)$ to $V_{\text{in}}(s)$. As only sources and resistors are involved in this model, the transfer function will be a constant.

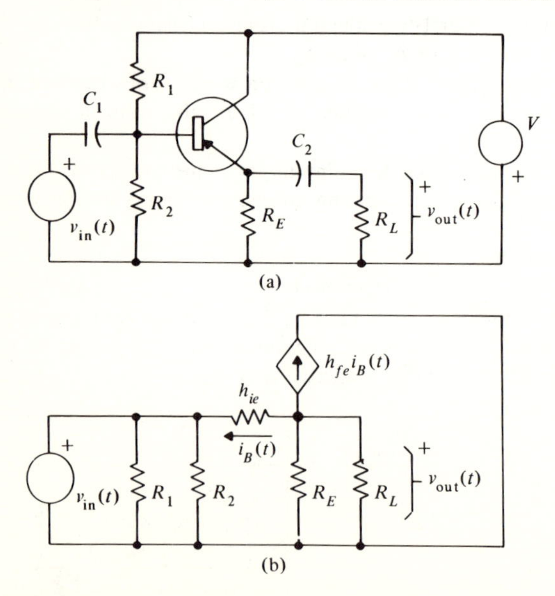

FIGURE P1-12

27. In vibration studies, the human body is often modeled by springs, masses, and dampers. For the model of a seated body with applied force f, Fig. P1-13, find the system equations.

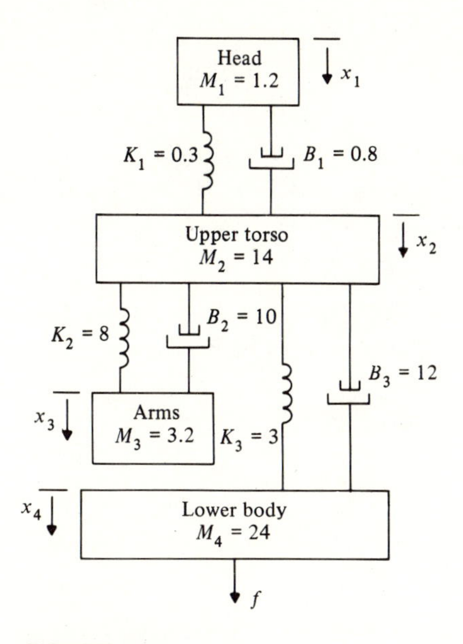

FIGURE P1-13

two

TRANSFER FUNCTIONS AND SYSTEM RESPONSE

2.1 PREVIEW

Block diagrams and signal flow graphs are graphical representations of the relationships between component parts of a system. Block diagrams are especially convenient for the manipulation of signals and system components, while signal flow graphs offer an economy of notation and advantages in analyzing and synthesizing the overall relations between system inputs and outputs.

After some experience with block diagram and signal flow graph system descriptions, the response of first-, second-, and higher-order systems is examined in some detail. While much of this material may be a review to the reader, its inclusion serves to establish notation and terminology and to bring our viewpoint and concerns into focus.

An example system for insulin delivery operates open-loop, and the required input necessary to produce a desired output is found. In another example system, an aircraft wing, resonance effects are examined.

2.2 BLOCK DIAGRAMS

2.2.1 Block Diagram Elements

Block diagrams are used to describe systems schematically. This graphical representation of Laplace-transformed equations may be manipulated with much the same ease as electric circuit diagrams, although the manipulation rules are different. Block diagrams may be used to describe the inner workings of a system (amplifiers, control motors, filters, etc.), and they offer a simple alternative to dealing with equations directly.

In a block diagram, a *block* is used to indicate a proportional relationship

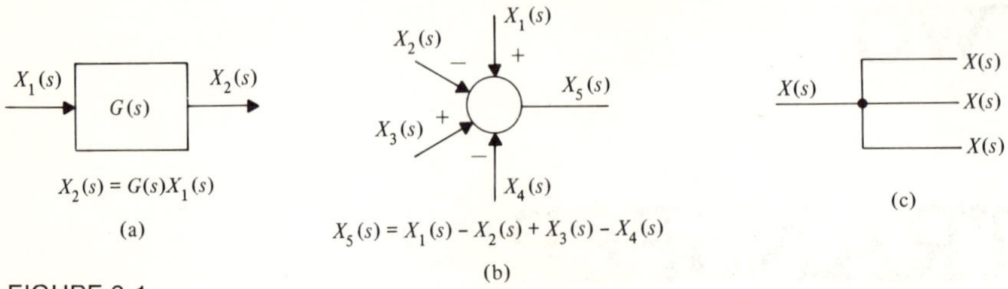

FIGURE 2-1
Basic block diagram elements. (a) Block. (b) Summer. (c) Junction or pickoff point.

between two Laplace-transformed signals. The proportionality function, or *transmittance*, relates incoming and outgoing signals. It is indicated within the block. A *summer* is used to indicate additions and subtractions of signals. A summer may have any number of incoming signals, but only one outgoing signal. The algebraic signs to be used in the summation are indicated next to the arrowhead for each incoming signal. A *junction* (sometimes termed a "pickoff point") indicates that the same signal is to go several places. Examples of each of these elements are shown in Fig. 2-1.

2.2.2 Reduction of Single-Input, Single-Output Diagrams

Rearranging system block diagrams to effect simplification or special structures is termed *block diagram algebra*. Since the block diagrams represent Laplace-transformed system equations, manipulating a diagram is equivalent to algebraic manipulation of the original equations, but diagram manipulation is generally far easier for human beings than is dealing with the equations directly.

For a single-input, single-output block diagram, *reduction* means simplifying the composite diagram to the point where it is a single block, displaying the transfer function relating the output to the input. In reducing a block diagram, it is helpful to proceed step by step, always maintaining the same overall relationship between input and output.

Some useful simplifications are the following. Two blocks in *cascade*, with no additional connections between them, are equivalent so far as the incoming and outgoing signals are concerned, to a single "product of transmittances" block, as indicated in Fig. 2-2a. Two blocks in *tandem*, Fig. 2-2b, are equivalent to a single "sum of transmittances" block. This result is modified if there are other signs besides pluses on the summer. Figure 2-2c shows one example.

For two blocks in a *feedback* configuration, Fig. 2-3, two possible algebraic signs on the summer are considered. $G(s)$ is termed the *forward transmittance* and $H(s)$ is the *feedback transmittance* in this arrangement. The relationships between the signals are

$$\begin{cases} Y(s) = G(s)E(s) \\ E(s) = R(s) \mp H(s)Y(s) \end{cases}$$

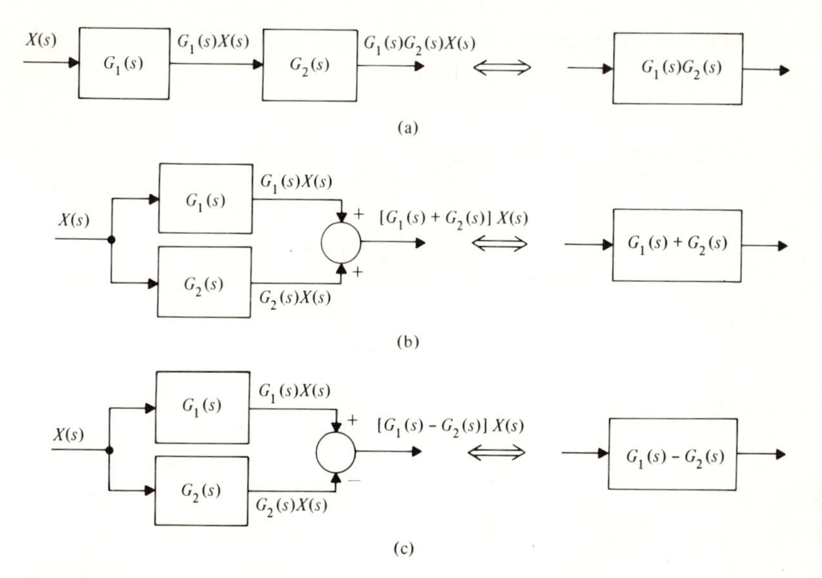

FIGURE 2-2
Equivalents of blocks (a) in cascade; (b) in tandem with positive summer signs; and (c) in tandem with negative summer sign.

Solving for $Y(s)$ in terms of $R(s)$ by eliminating $E(s)$,

$$
\begin{aligned}
Y(s) &= G(s)[R(s) \mp H(s)Y(s)] \\
&= G(s)R(s) \mp G(s)H(s)Y(s)
\end{aligned}
$$

$$
T(s) = \frac{Y(s)}{R(s)} = \frac{G(s)}{1 \pm G(s)H(s)} \tag{2-1}
$$

A negative sign on the feedback summation in Fig. 2-3 results in a positive algebraic sign in the denominator of $T(s)$; a plus sign on the summer gives a minus sign in (2-1).

Other useful equivalences are given in Fig. 2-4.

An example of using equivalences to reduce a single-input, single-output block diagram is shown in Fig. 2-5. At each step, a combination of elements with a single incoming signal and a single outgoing signal is simplified, until a single block showing the transfer function remains.

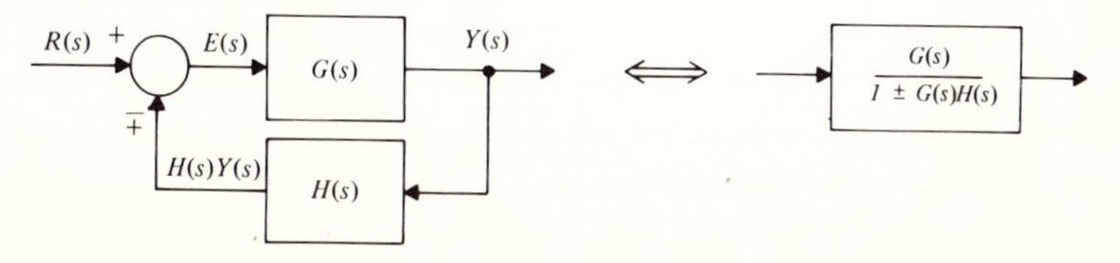

FIGURE 2-3
The feedback configuration.

This result can alternatively be obtained by eliminating the variables E, E_1, and E_2 from the following equations from Fig. 2-5:

$$E = R - \frac{5}{s}E_2 - 4E_1 \qquad E_2 = \frac{3s}{s+4}E_1$$

$$E_1 = \frac{1}{s+2}E \qquad Y = E_2 + \frac{s}{s+3}E_1$$

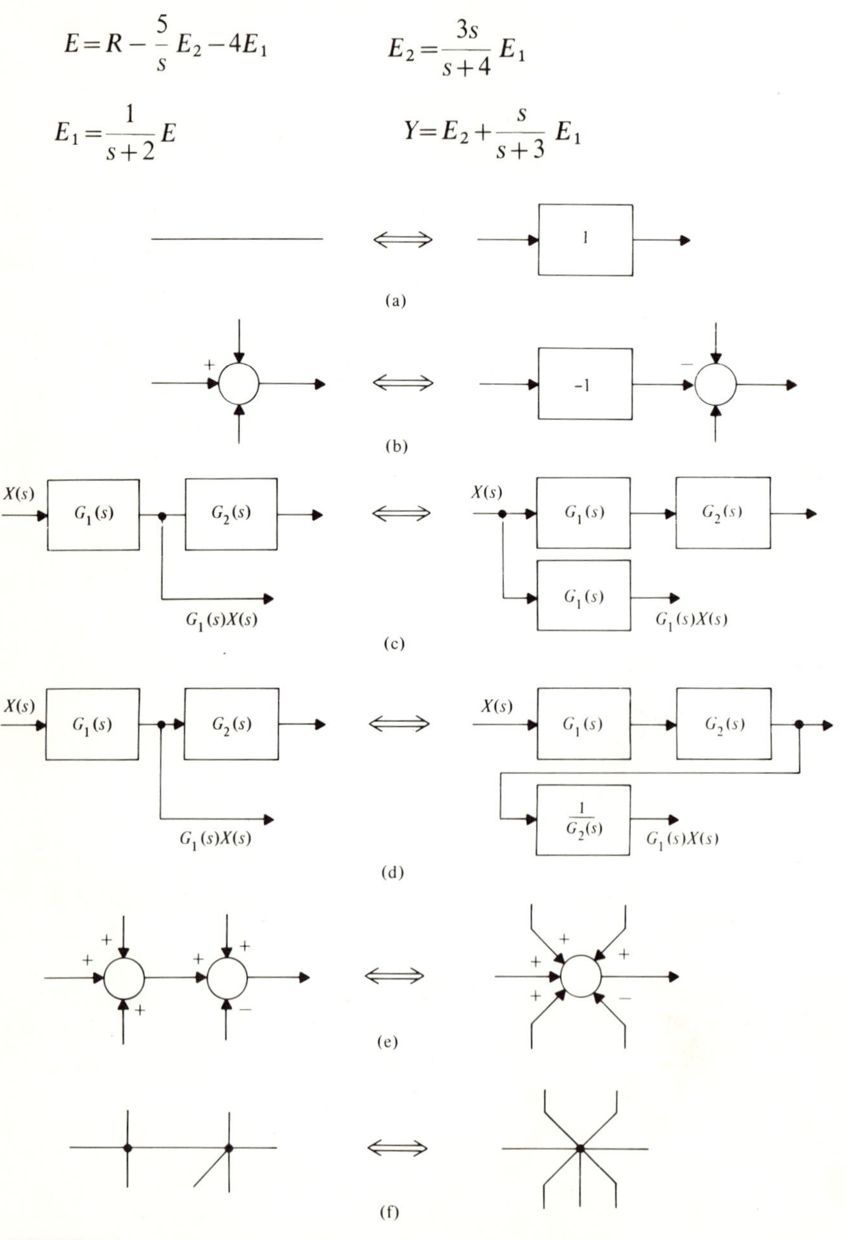

FIGURE 2-4
Other useful block diagram equivalences. (a) Insertion or removal of unity gain. (b) Changing a summer sign. (c) Moving a pickoff point back. (d) Moving a pickoff forward. (e) Combining or expanding summations. (f) Combining or expanding junctions.

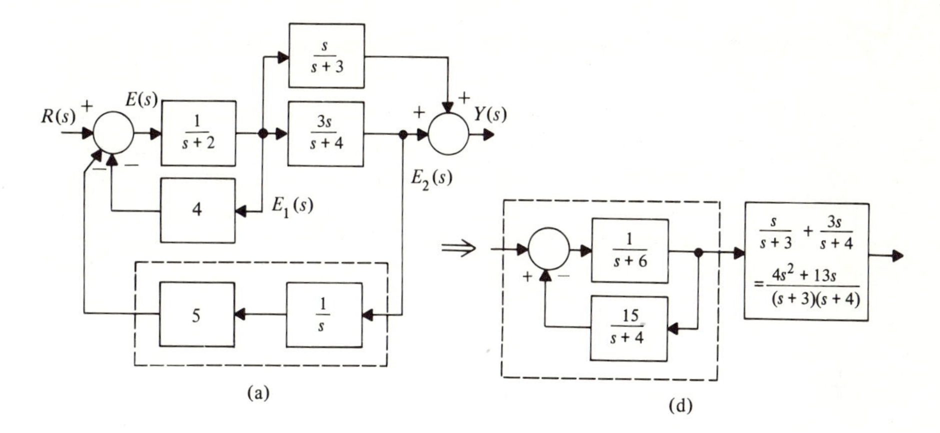

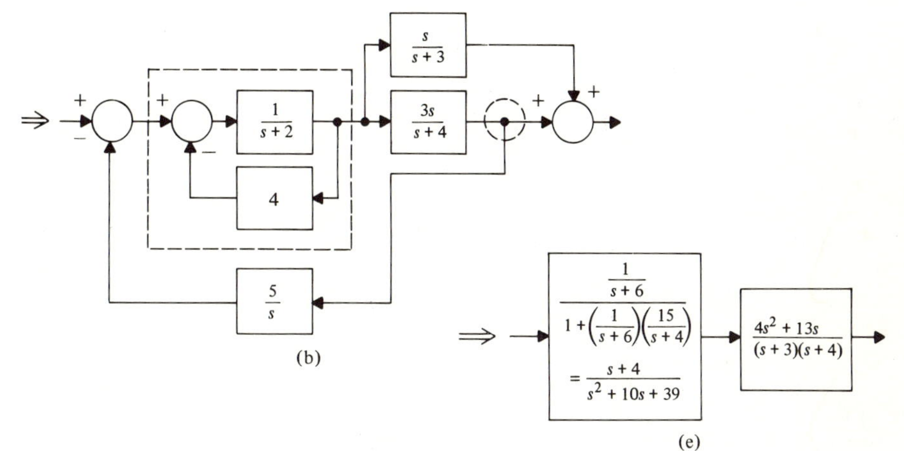

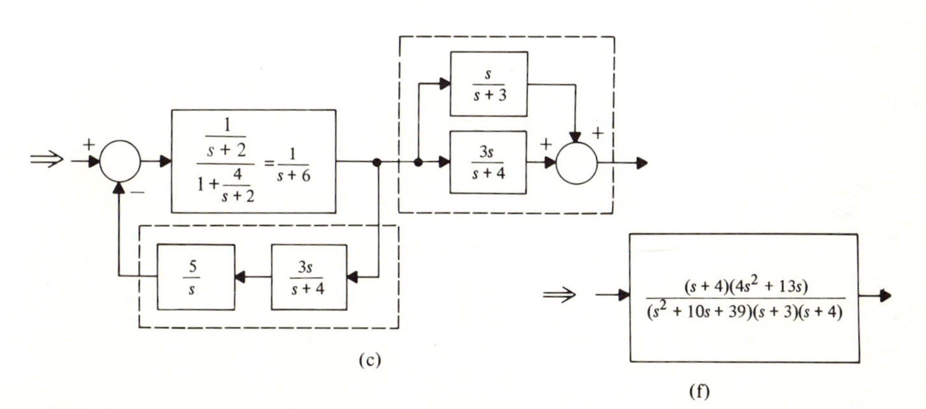

FIGURE 2-5
Example of a block diagram reduction. (a) Original block diagram. (b) Cascade blocks combined. (c) Feedback relationship applied and pickoff point moved. (d) Cascade blocks combined and tandem blocks combined. (e) Feedback relationship applied. (f) Cascade blocks combined, yielding the overall transfer function.

2.2.3 Reduction of Multiple-Input, Multiple-Output Diagrams

For a multiple-input, multiple-output system, block diagram reduction involves finding each of the system transfer functions. This is done by considering only one output at a time and reducing the relationship between each of the input signals to the one output. All but one input is set to zero to determine the transfer function relating an output to that input. For example, the four transfer functions for the two-input, two-output system of Fig. 2-6 are as follows:

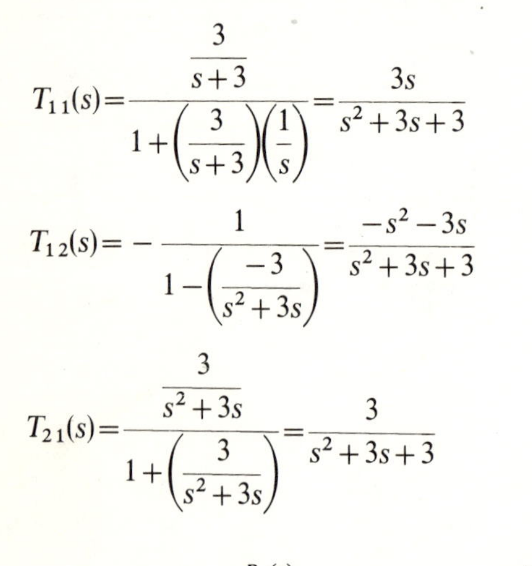

$$T_{11}(s)=\frac{\dfrac{3}{s+3}}{1+\left(\dfrac{3}{s+3}\right)\left(\dfrac{1}{s}\right)}=\frac{3s}{s^2+3s+3}$$

$$T_{12}(s)=-\frac{1}{1-\left(\dfrac{-3}{s^2+3s}\right)}=\frac{-s^2-3s}{s^2+3s+3}$$

$$T_{21}(s)=\frac{\dfrac{3}{s^2+3s}}{1+\left(\dfrac{3}{s^2+3s}\right)}=\frac{3}{s^2+3s+3}$$

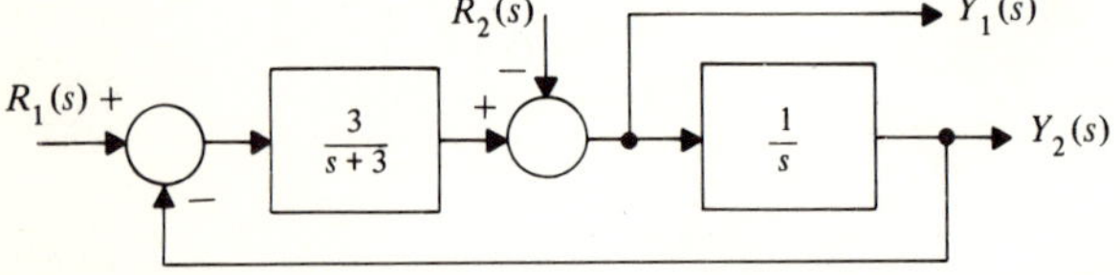

FIGURE 2-6
Block diagram for a two-input, two-output system.

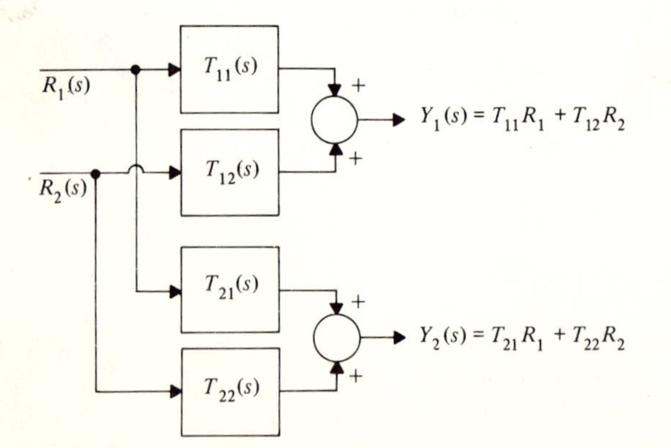

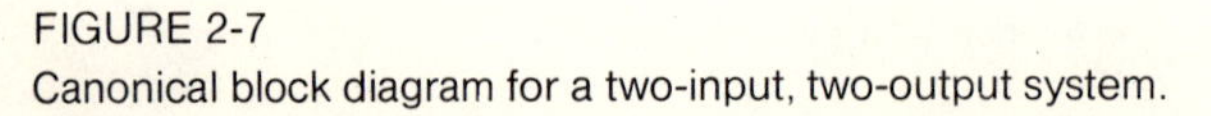

FIGURE 2-7
Canonical block diagram for a two-input, two-output system.

$$T_{22}(s) = -\frac{\dfrac{1}{s}}{1-\left(\dfrac{1}{s}\right)\left(\dfrac{-3}{s+3}\right)} = \frac{-s-3}{s^2+3s+3}$$

The input and output signals in a two-input, two-output system are related as shown in the canonical block diagrams of Fig. 2-7. Every block diagram is reducible to a similar equivalent form, where the transfer functions are placed in evidence.

DRILL PROBLEMS

D2-1. Reduce the following block diagrams, obtaining the system transfer function.

ans. (a) $-s/(s+20)$
(b) $3/(s^2+6s+11)$
(c) $20(s+1)/(s^4+s^3+14s^2+14s+20)$
(d) $\dfrac{-3s^2(s+2)+8s(s+2)(s+1)}{s^4+28s^3+71s^2+224s+180}$

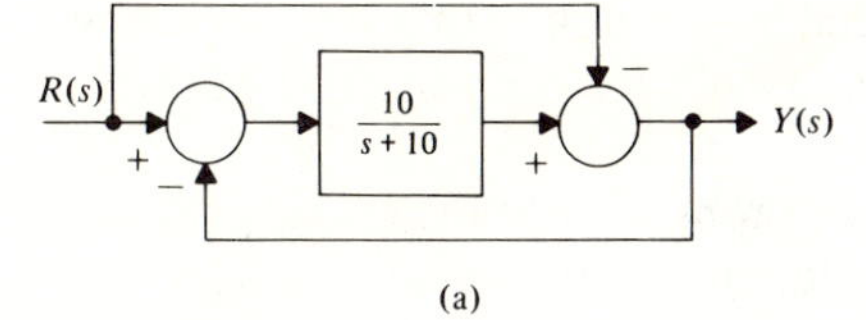

(a)

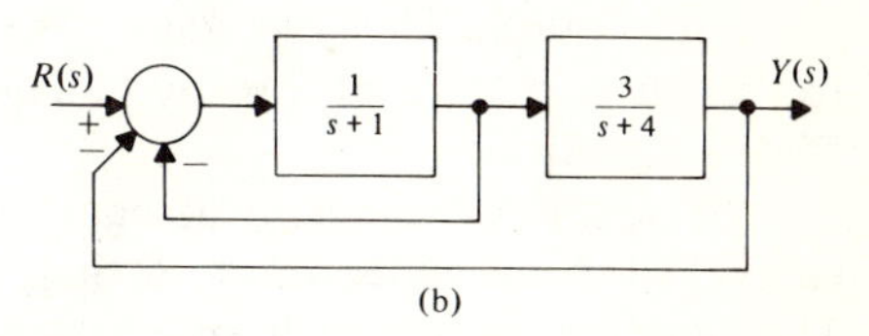

(b)

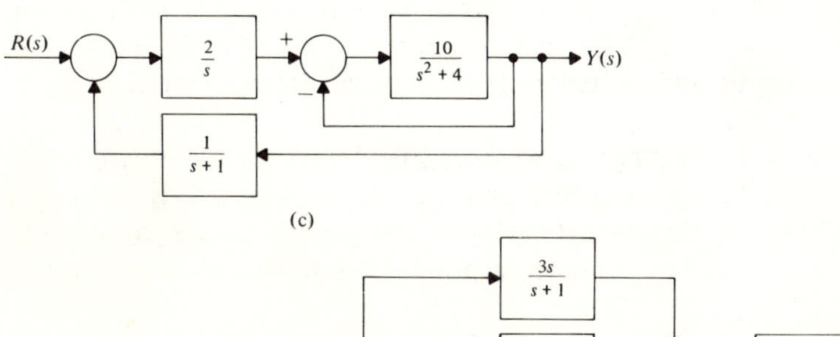

(c)

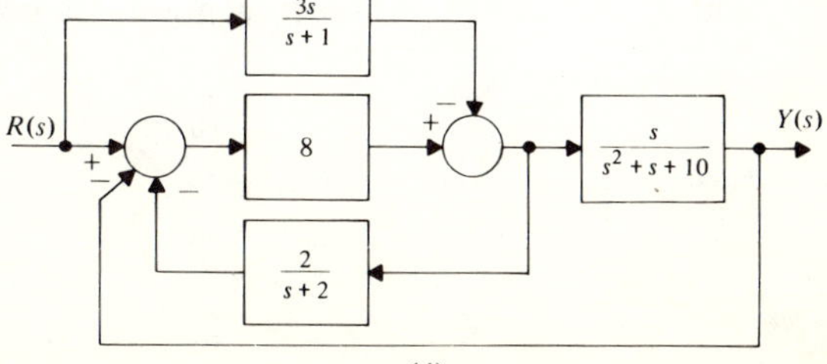

(d)

D2-2. Find the six system transfer functions of the following system.

ans. $10(s+3)/(s^2+15s+6)$,
$10(s+2)(s+3)/(s^2+15s+6)$,
$10s/(s^2+15s+6)$,
$10s/(s^2+15s+6)$,
$10s(s+2)/(s^2+15s+6)$,
$-s(s+2)/(s^2+15s+6)$

2.3 SIGNAL FLOW GRAPHS

2.3.1 Signal Flow Graph Elements

The same information conveyed by a block diagram is also portrayed by a signal flow graph. The advantage of a signal flow graph is that it is well suited to the determination of transfer functions in one step by a method known as Mason's gain rule.

There are two elements in signal flow graphs. One is the *branch*, which is equivalent to the block in block diagram language, as indicated in Fig. 2-8a. The signal exiting a branch equals the incoming branch signal times the branch transmittance $G(s)$. The other element is the *node*, Fig. 2-8b, which is equivalent to the summer followed by a junction. The signal at a node, $X(s)$ in the figure, is the sum (with all plus signs) of the signals coming into the node from branches. The node signal may be transmitted from the node via any number of outgoing branches.

Figure 2-9a is a block diagram, whereas Fig. 2-9b is the equivalent signal flow graph in which the signals at each node have been labeled. It is customary (although not absolutely necessary) to bring input signals R_1 and R_2 and output signals Y_1 and Y_2 into and out of the flow graph via branches with unity trans-

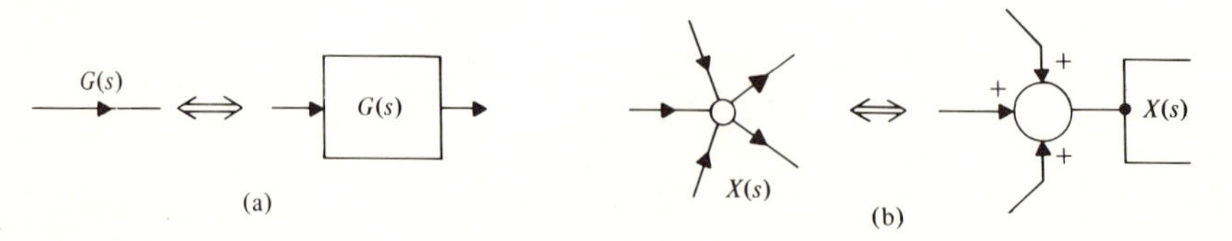

FIGURE 2-8
Signal flow graph elements. (a) Branch. (b) Node.

mittance. This practice sets off the inputs and outputs from the rest of the signal flow graph.

Signal flow graphs, like block diagrams, represent the Laplace-transformed equations of a system. To write the set of simultaneous equations represented by a signal flow graph, first label the signals at each node except for the redundant signals at input and output nodes where signals have been set off from the graph with unity transmittance branches. Then write one equation for each node, equating the signal at that node to the sum of the incoming signals, through branches, from other nodes.

For the example in Fig. 2-9, the systematic simultaneous equations represented are

$$\begin{cases} X_1(s)=R_1(s)+\dfrac{1}{s+2}X_2(s) \\ X_2(s)=-4X_1(s)+R_2(s)-7Y_1(s)-\dfrac{1}{s+4}X_2(s) \\ Y_1(s)=\dfrac{s}{s^2+3}X_2(s) \\ Y_2(s)=10X_1(s)-sY_1(s) \end{cases}$$

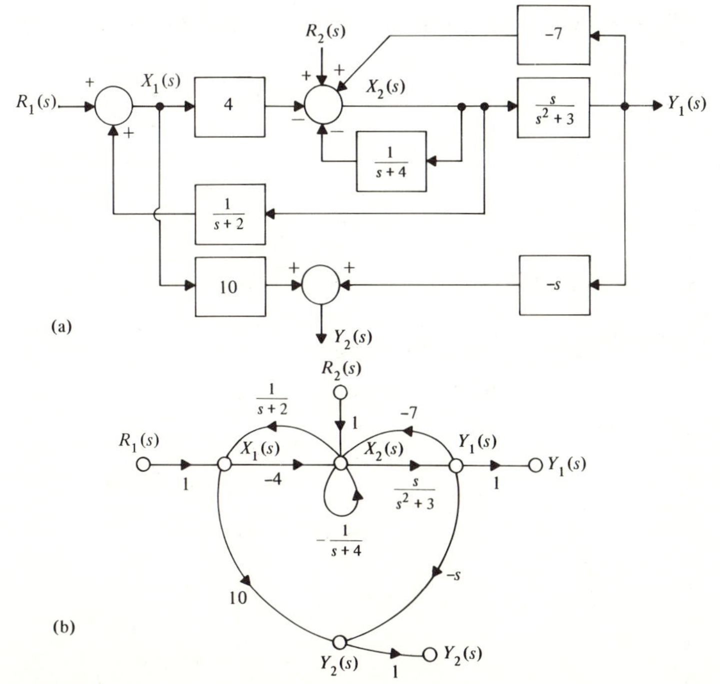

FIGURE 2-9

A block diagram and an equivalent signal flow diagram. (a) Block diagram. (b) Equivalent signal flow graph.

Note that the "self-loop," the branch with transmittance $-1/(s+4)$, which starts and ends at the same node, contributes a term involving the node signal X_2 itself on the right side of the equation for X_2. The sense of a self-loop, clockwise or counterclockwise, is of no consequence.

2.3.2 Mason's Gain Rule

The relations represented by signal flow graphs, like those of block diagrams, are linear algebraic equations with coefficients dependent upon the variable s. Finding a transfer function from a block diagram or signal flow graph is to solve those equations in terms of the input. In block diagram reduction, the equation solution is done in a series of transformations to equivalent, simpler sets of equations until a single equation is obtained. Signal flow graph expressions of sets of equations are not well suited to solution by reduction, but they are ideal for a systematic solution, which is equivalent to the use of determinants and Cramer's rule.

Mason's gain rule is a formula for finding the transfer function of a single-input, single-output system from its signal flow graph. It may be applied repeatedly to a multiple-input, multiple-output system to obtain each of the transfer functions. To explain the application of Mason's gain rule, some terms will first be defined and illustrated with the example signal flow graph of Fig. 2-10.

A *path* is any succession of branches, from input to output, in the direction

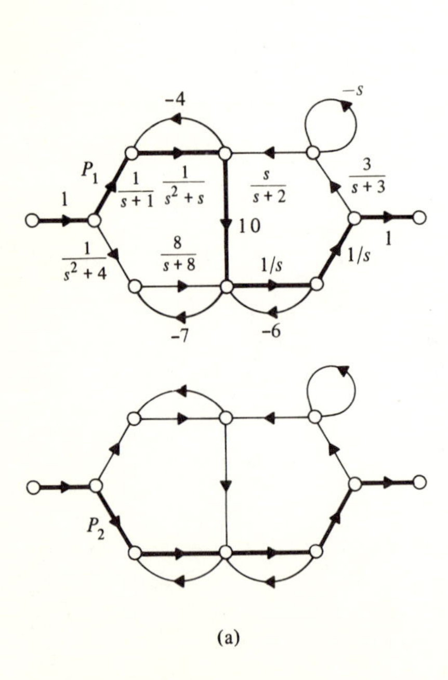

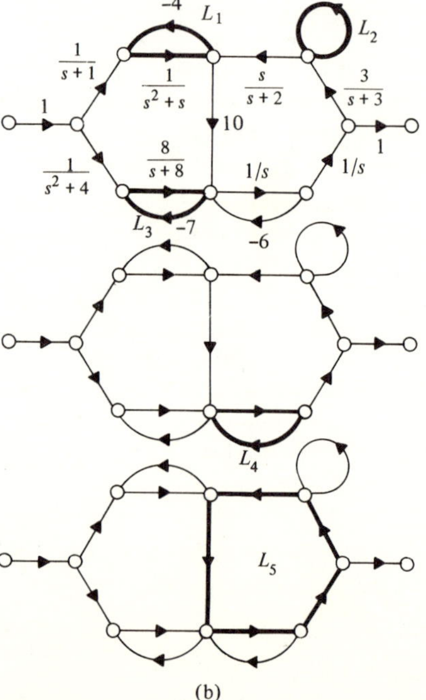

FIGURE 2-10
(a) Paths, and (b) loops in the Mason's gain rule example.

of the arrows, which does not pass any node more than once, as indicated for the example in Fig. 2-10a. The *path gain* is the product of the transmittances of the branches comprising the path. For the example,

$$P_1=\left(\frac{1}{s+1}\right)\left(\frac{1}{s^2+s}\right)(10)\left(\frac{1}{s}\right)\left(\frac{1}{s}\right)$$
$$P_2=\left(\frac{1}{s^2+4}\right)\left(\frac{8}{s+8}\right)\left(\frac{1}{s}\right)\left(\frac{1}{s}\right)$$

A *loop* is any closed succession of branches, in the direction of the arrows, which does not pass any node more than once, as in Fig. 2-10b. The *loop gain* is the product of the transmittances of the branches comprising the loop. For the example,

$$L_1=\frac{-4}{s^2+s}$$

$$L_2=-s$$

$$L_3=\frac{-56}{s+8}$$

$$L_4=\frac{-6}{s}$$

$$L_5=(10)\left(\frac{1}{s}\right)\left(\frac{1}{s}\right)\left(\frac{3}{s+3}\right)\left(\frac{s}{s+2}\right)$$

Two loops are said to be *touching* if they have any node in common. Otherwise, they are nontouching. Similarly, a loop and a path are touching if they have any node in common.

The *determinant* of a signal flow graph is

$\Delta=1-$(sum of all loop gains)$+$(sum of products of gains of all combinations of 2 nontouching loops)$-$(sum of products of gains of all combinations of 3 nontouching loops)$+\cdots$

For the example,

$$\Delta=1-(L_1+L_2+L_3+L_4+L_5)+(L_1L_2+L_1L_3+L_1L_4+L_2L_3+L_2L_4)$$
$$-(L_1L_2L_3+L_1L_2L_4)$$

The *cofactor* of a path is the determinant of the signal flow graph formed by deleting all loops touching the path. In the example,

$$\Delta_1=1-L_2=1+s$$

$$\Delta_2=1-(L_1+L_2)+(L_1L_2)=1+\frac{4}{s^2+s}+s+\frac{4s}{s^2+s}$$

Mason's gain rule is as follows: The transfer function of a single-input, single-output signal flow graph is

$$T(s)=\frac{P_1\Delta_1+P_2\Delta_2+P_3\Delta_3+\cdots}{\Delta}$$

For the example system, since there are two paths,

$$T(s)=\frac{P_1\Delta_1+P_2\Delta_2}{\Delta}$$

where P_1, P_2, Δ, Δ_1, and Δ_2 are given above.

2.3.3 Example Applications of Mason's Gain Rule

A less involved example is shown in Fig. 2-11, for which

$$\begin{aligned}T(s)&=\frac{P_1\Delta_1+P_2\Delta_2+P_3\Delta_3}{\Delta}\\&=\frac{6\left[1+\dfrac{3}{s+1}+\dfrac{5s}{s+2}+\dfrac{15s}{(s+1)(s+2)}\right]+\left(\dfrac{-4}{s+1}\right)\left(1+\dfrac{5s}{s+2}\right)+\left[\left(\dfrac{3}{s+1}\right)\left(\dfrac{s}{s+2}\right)\right]}{1+\dfrac{3}{s+1}+\dfrac{5s}{s+2}+\dfrac{15s}{(s+1)(s+2)}}\\&=\frac{36s^2+133s+44}{6s^2+26s+8}\end{aligned}$$

The signal flow graph of Fig. 2-12a is of a system with two inputs and two outputs. The corresponding single-input, single-output signal flow graphs for calculating the four associated transfer functions of the system are shown in Fig. 2-12b–e. Using Mason's gain rule, the system transfer functions are as follows:

$$\begin{aligned}T_{11}(s)=\frac{Y_1(s)}{R_1(s)}&=\frac{\left(\dfrac{s}{s+2}\right)(1)+\dfrac{10}{s(s+3)}(1)}{1-\left(\dfrac{-4s}{s+2}\right)-\left[\dfrac{-40}{s(s+3)}\right]}\\&=\frac{s^3+3s^2+10s+20}{5s^3+17s^2+46s+80}\end{aligned}$$

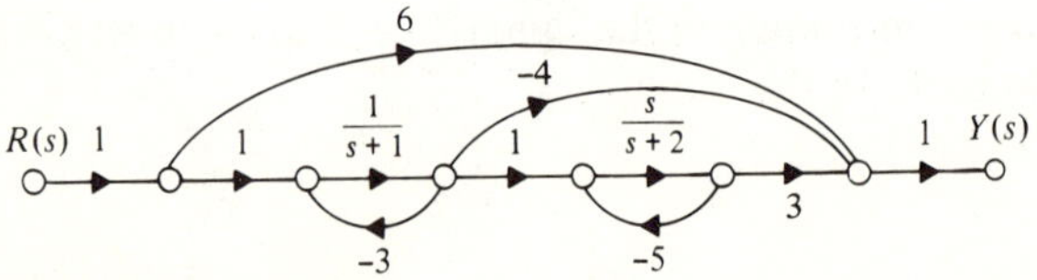

FIGURE 2-11
Example signal flow graph.

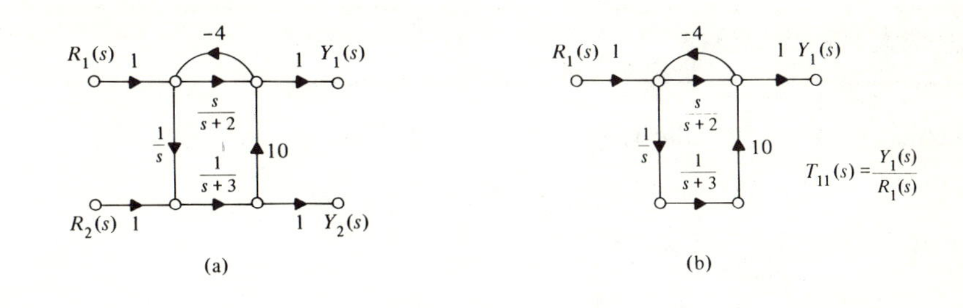

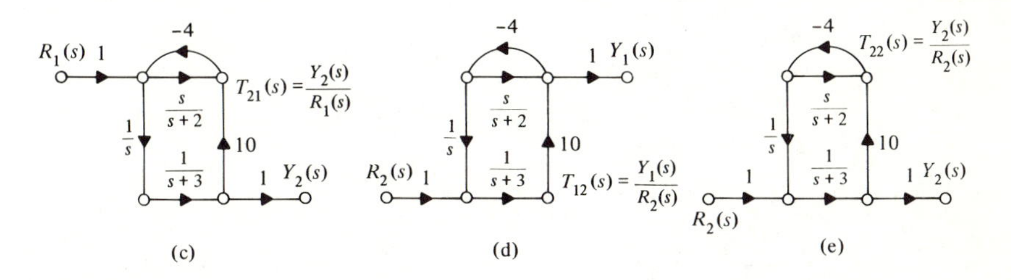

FIGURE 2-12
A multiple-input, multiple-output signal flow graph and its transfer functions.

$$T_{21}(s) = \frac{Y_2(s)}{R_1(s)} = \frac{\left[\frac{1}{s(s+3)}\right](1)}{1 - \left(\frac{-4s}{s+2}\right) - \left[\frac{-40}{s(s+3)}\right]}$$

$$= \frac{s+2}{5s^3 + 17s^2 + 46s + 80}$$

$$T_{12}(s) = \frac{Y_1(s)}{R_2(s)} = \frac{\left(\frac{10}{s+3}\right)(1)}{1 - \left(\frac{-4s}{s+2}\right) - \left[\frac{-40}{s(s+3)}\right]}$$

$$= \frac{10s^2 + 20s}{5s^3 + 17s^2 + 46s + 80}$$

$$T_{22}(s) = \frac{Y_2(s)}{R_2(s)} = \frac{\left(\frac{1}{s+3}\right)\left[1 - \left(\frac{-4s}{s+2}\right)\right]}{1 - \left(\frac{-4s}{s+2}\right) - \left[\frac{-40}{s(s+3)}\right]}$$

$$= \frac{5s^2 + 2s}{5s^3 + 17s^2 + 46s + 80}$$

DRILL PROBLEMS

D2-3. Convert the following block diagrams to equivalent signal flow graphs:

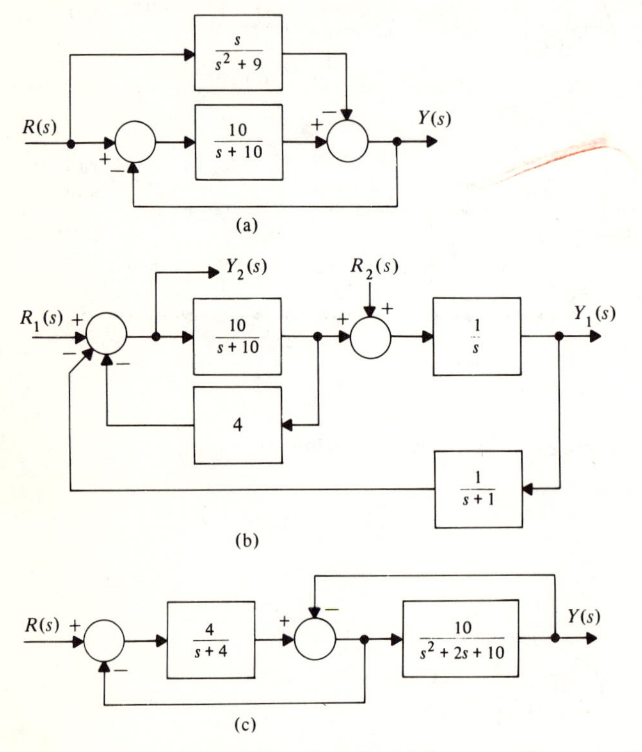

D2-4. Convert the following signal flow graphs to equivalent block diagrams:

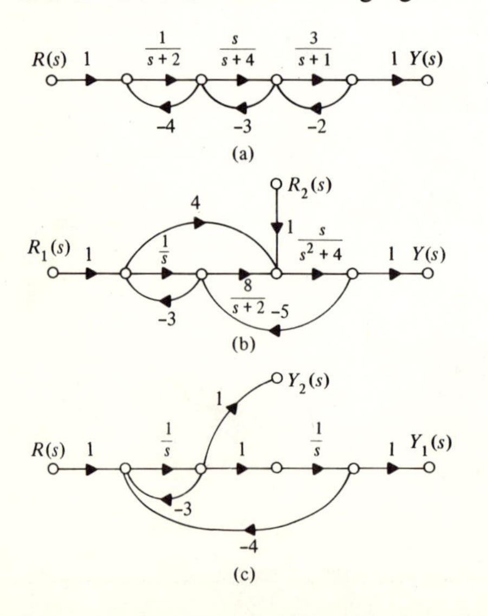

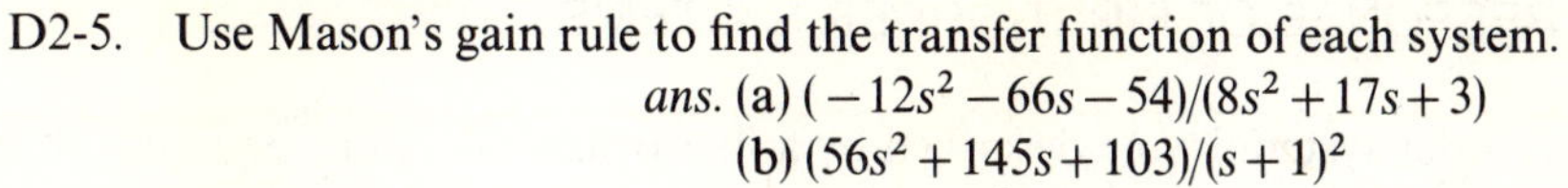

D2-5. Use Mason's gain rule to find the transfer function of each system.

ans. (a) $(-12s^2-66s-54)/(8s^2+17s+3)$
(b) $(56s^2+145s+103)/(s+1)^2$
(c) $(3s+2)/(s^2+4s+5)$
(d) $(3s+2)/(s^2+4s+5)$

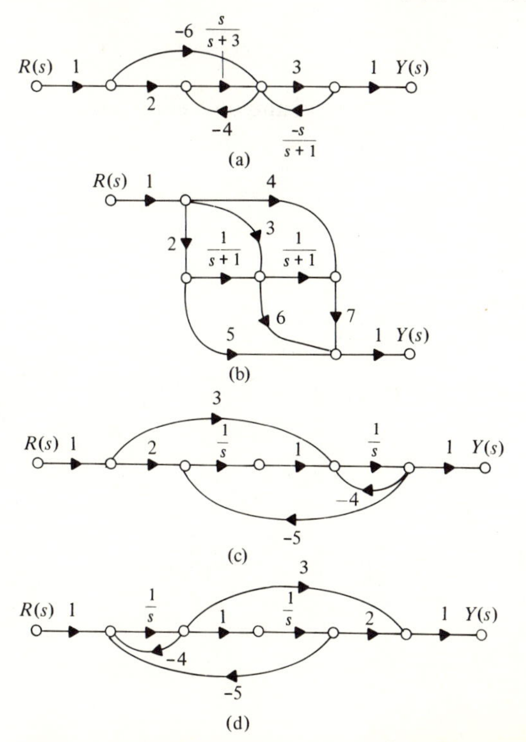

D2-6. Use Mason's gain rule to find the six transfer functions of the following system.

ans. $6/(s^2+29s+6)$,
$6s/(s^2+29s+6)$,
$s(s+2)/(s^2+29s+6)$,
$s(s+27)/(s^2+29s+6)$,
$-s(2s+6)/(s^2+29s+6)$,
$8s^2/(s^2+29s+6)$

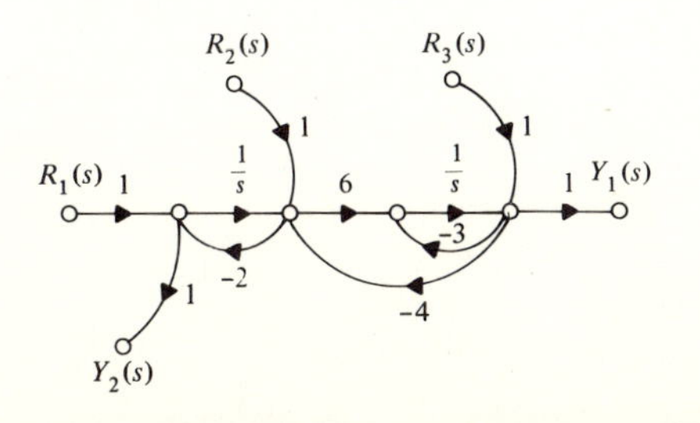

2.4 RESPONSE OF FIRST-ORDER SYSTEMS

A first-order system is one described by a first-order differential equation. The simplest such system has the equation

$$\frac{dy}{dt}+a_0 y(t)=b_0 r(t)$$

where a_0 and b_0 are constants. Laplace-transforming and collecting terms,

$$sY(s)-y(0^-)+a_0 Y(s)=b_0 R(s)$$

$$Y(s)=\underbrace{\frac{b_0}{s+a_0}R(s)}_{\substack{\text{zero-state}\\ \text{component}}} + \underbrace{\frac{y(0^-)}{s+a_0}}_{\substack{\text{zero-input}\\ \text{component}}}$$

The transfer function which relates $Y(s)$ to $R(s)$ is thus

$$T(s)=\frac{b_0}{s+a_0}$$

Consider a step input signal

$$r(t)=Au(t)$$

for which

$$R(s)=\frac{A}{s}$$

If the initial condition is zero, the system output is then given by

$$Y(s)=T(s)R(s)=\frac{b_0 A}{s(s+a_0)}$$

A partial fraction expansion gives

$$Y(s)=\frac{(b_0 A/a_0)}{s}+\frac{(-b_0 A/a_0)}{s+a_0}$$

or

$$y(t)=\left[\frac{b_0 A}{a_0}-\frac{b_0 A}{a_0}e^{-a_0 t}\right]u(t)$$

which is sketched in Fig. 2-13. If the characteristic root $s=-a_0$ is negative

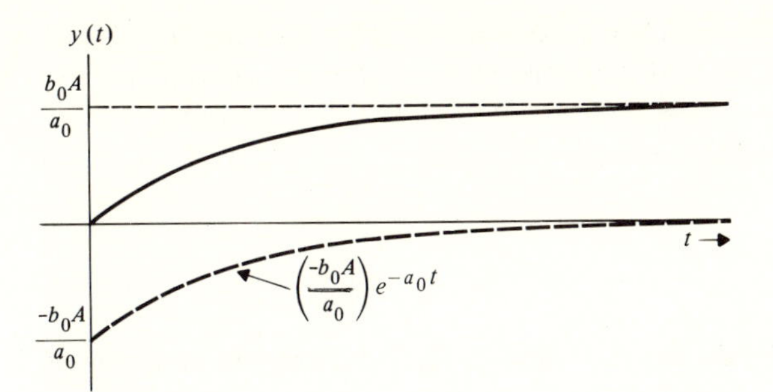

FIGURE 2-13
First-order system step response with zero initial condition.

(a_0 positive), the system is stable and the exponential natural term decays with time, leaving the constant forced (or "steady state") term. For a positive characteristic root (negative a_0), the exponential term expands with time, and the system is unstable.

If the system initial condition is not zero,

$$
\begin{aligned}
Y(s) &= T(s)R(s) + \frac{y(0^-)}{s+a_0} \\
&= \frac{b_0 A}{s(s+a_0)} + \frac{y(0^-)}{s+a_0} \\
&= \frac{(b_0 A/a_0)}{s} + \frac{y(0^-) - (b_0 A/a_0)}{s+a_0}
\end{aligned}
$$

giving

$$
y(t) = \left\{ \frac{b_0 A}{a_0} + \left[y(0^-) - \frac{b_0 A}{a_0} \right] e^{-a_0 t} \right\} u(t)
$$

The amplitude of the exponential term is changed, as illustrated in Fig. 2-14.

The response to other inputs is calculated similarly, using the Laplace transform of the input signal. The natural exponential term in the response

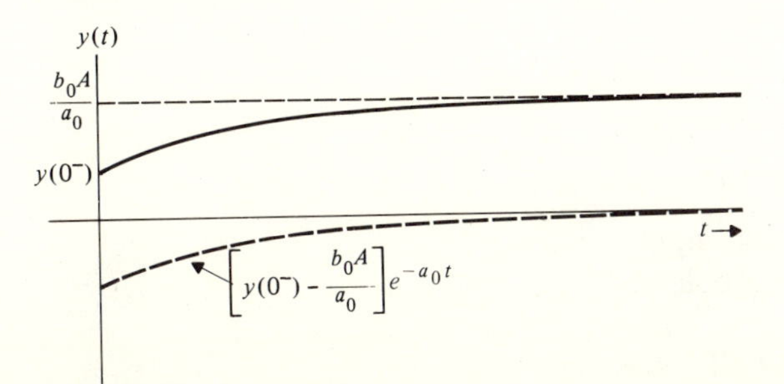

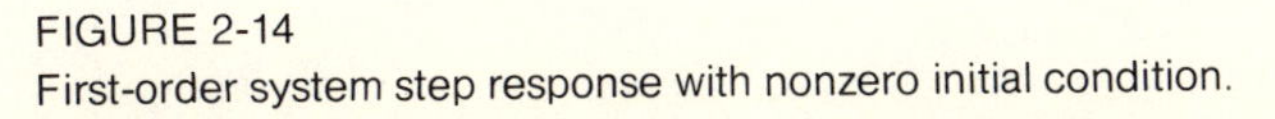

FIGURE 2-14
First-order system step response with nonzero initial condition.

consists of contributions from both the zero-state and the zero-input parts of the response. In general, the natural and zero-input responses die out in time when the system is stable.

The *time constant* of a stable first-order system is

$$\tau = \frac{1}{a_0}$$

It is the time interval over which the exponential $Ke^{-a_0 t}$ decays by a factor of $1/e = 0.37$. First-order system transfer functions (and similar terms in higher-order transfer functions) are sometimes placed in the form

$$T(s) = \frac{\text{numerator polynomial}}{s + a_0} = \frac{(\text{numerator polynomial})/a_0}{(1/a_0)s + 1}$$
$$= \frac{(\text{numerator polynomial})/a_0}{\tau s + 1}$$

to show the time constant explicitly.

The most general first-order system involves the input $r(t)$ and possibly several of its derivatives. These terms form the driving function $f(t)$:

$$\frac{dy}{dt} + a_0 y(t) = b_m \frac{d^m r}{dt^m} + \cdots + b_1 \frac{dr}{dt} + b_0 r(t) = f(t)$$

The corresponding transfer function is

$$T(s) = \frac{b_m s^m + \cdots + b_1 s + b_0}{s + a_0}$$

As a practical matter, a transfer function numerator seldom has an order greater than that of the denominator because such systems would emphasize higher frequencies without bound.

For the first-order system

$$\frac{dy}{dt} + a_0 y = b_1 \frac{dr}{dt} + b_0 r$$

$$Y(s) = \underbrace{\frac{b_1 s + b_0}{s + a_0} R(s)}_{\substack{\text{zero-state} \\ \text{component}}} + \underbrace{\frac{y(0^-) - b_1 r(0)}{s + a_0}}_{\substack{\text{zero-input} \\ \text{component}}}$$

The transfer function relating $Y(s)$ to $R(s)$ is

$$T(s) = \frac{b_1 s + b_0}{s + a_0}$$

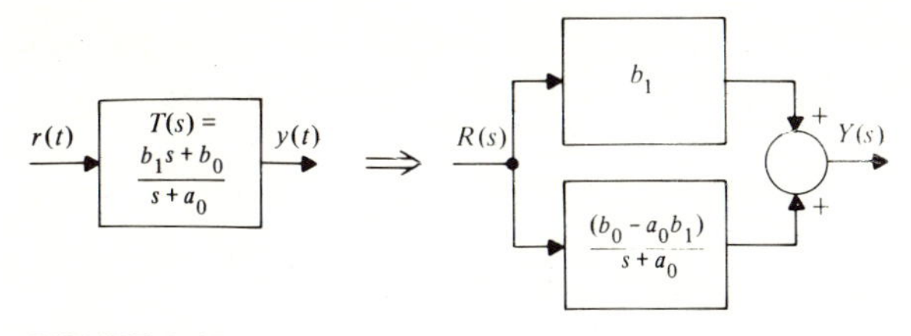

FIGURE 2-15
Tandem representation of a first-order system.

which may be expressed as

$$T(s)=b_1+\frac{(b_0-a_0b_1)}{s+a_0}$$

Such a system may be considered to be composed of a gain b_1 in tandem with a first-order system with constant transfer function numerator, as indicated in Fig. 2-15. This system output thus contains a component proportional to the input, in addition to a component of the previous type.

Growth of bacteria and many other biological activities are commonly modeled by a first-order differential equation and have time responses with a single exponential term. Electrical RL and RC circuits and radioactive decay also exhibit first-order behavior.

DRILL PROBLEMS

D2-7. Find and also sketch the response of the systems with the following transfer function, input, and initial condition:

(a) $T(s)=\dfrac{3}{s+3}$

$r(t)=6u(t)$

$y(0^-)=10$

ans. $6+4e^{-3t}$

(b) $T(s)=\dfrac{1}{s+10}$

$r(t)=3u(t)\cos 10t$

$y(0^-)=0$

ans. $0.212\cos(10t-45^\circ)-0.15e^{-10t}$

(c) $T(s)=\dfrac{s}{s+1000}$

$r(t)=7u(t)$

$y(0^-)=4$

ans. $11e^{-1000t}$

(d) $T(s)=\dfrac{20s}{s+300}$

$r(t)=8u(t)\sin 100t$

$y(0^-)=-10$

ans. $50.6\cos(100t-18.4^6)-58e^{-300t}$

(e) $T(s)=\dfrac{-4s+20}{s+300}$

$r(t)=10u(t)$

$y(0^-)=0$

ans. $0.67-40.67e^{-300t}$

2.5 RESPONSE OF SECOND-ORDER SYSTEMS

Second-order systems are described by second-order differential equations of the form

$$\frac{d^2y}{dt^2}+a_1\frac{dy}{dt}+a_0y(t)=b_m\frac{d^mr}{dt^m}+\cdots+b_1\frac{dr}{dt}+b_0r(t)$$

Laplace-transforming and collecting terms,

$$Y(s)=\underbrace{\frac{b_ms^m+\cdots+b_1s+b_0}{s^2+a_1s+a_0}R(s)}_{\text{zero-state component}}+\underbrace{\frac{\substack{\text{first-order numerator}\\ \text{polynomial involving initial conditions}}}{s^2+a_1s+a_0}}_{\text{zero-input component}}$$

The transfer function relating $Y(s)$ to $R(s)$ is

$$T(s)=\frac{b_ms^m+\cdots+b_1s+b_0}{s^2+a_1s+a_0}$$

and the system has characteristic polynomial

$$s^2+a_1s+a_0=(s-s_1)(s-s_2)$$

with roots s_1 and s_2 given by the quadratic formula

$$s_1, s_2=\frac{-a_1\pm\sqrt{a_1^2-4a_0}}{2}$$

If the characteristic roots s_1 and s_2 are different real numbers, the natural terms in the partial fraction expansion of the response are of the form

$$Y_{\text{natural}}(s)=\frac{\text{first-order numerator polynomial}}{s^2+a_1s+a_0}$$
$$=\frac{\text{numerator polynomial}}{(s-s_1)(s-s_2)}=\frac{K_1}{s-s_1}+\frac{K_2}{s-s_2}$$

which, as a function of time after $t=0$, consists of the sum of two real exponential terms of the first-order type:

$$y_{\text{natural}}(t)=K_1e^{s_1t}+K_2e^{s_2t}$$

Such a system is said to be *overdamped*.

For example, the overdamped second-order system with transfer function

$$T(s)=\frac{1}{s^2+3s+2}$$

with unit step input

$$R(s)=\frac{1}{s}$$

and zero initial conditions has response given by

$$Y(s)=\frac{1}{s(s^2+3s+2)}=\frac{\frac{1}{2}}{s}+\frac{-1}{s+1}+\frac{\frac{1}{2}}{s+2}$$
$$y(t)=[\tfrac{1}{2}-e^{-t}+\tfrac{1}{2}e^{-2t}]u(t)$$

This response is sketched in Fig. 2-16. Nonzero initial conditions would result in different amplitudes for the two exponential, natural terms.

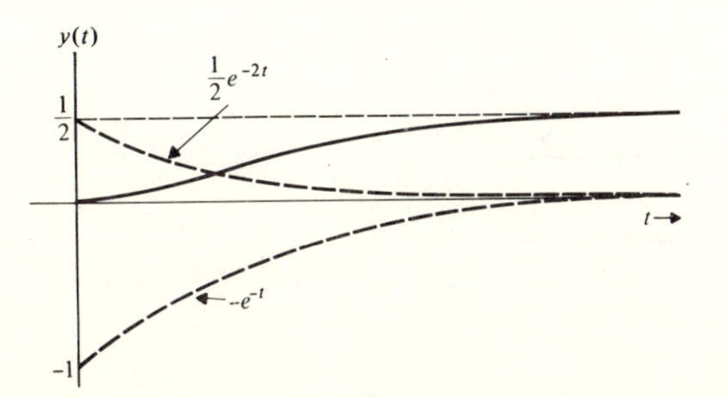

FIGURE 2-16
Step response of an overdamped second-order system.

If the characteristic roots s_1 and s_2 are equal, the natural terms are of the form

$$Y_{\text{natural}}(s)=\frac{\text{numerator polynomial}}{s^2+a_1s+a_0}=\frac{\text{numerator polynomial}}{(s-s_1)^2}$$

$$Y_{\text{natural}}(s)=\frac{K_1}{s-s_1}+\frac{K_2}{(s-s_1)^2}$$

for which the corresponding time function after $t=0$ is

$$y_{\text{natural}}(t)=K_1e^{s_1t}+K_2te^{s_1t}$$

Such a second-order system is said to be *critically damped.*

The critically damped system with transfer function

$$T(s)=\frac{10s+8}{s^2+4s+4}$$

has unit step response, with zero initial conditions, given by

$$Y(s)=\frac{10s+8}{s(s^2+4s+4)}=\frac{2}{s}+\frac{-2}{s+2}+\frac{6}{(s+2)^2}$$

for which

$$y(t)=[2-2e^{-2t}+6te^{-2t}]u(t)$$

This response is sketched in Fig. 2-17. Other initial conditions would result in different amplitudes for the e^{-2t} and te^{-2t} natural terms, but the same character of response.

If the roots of the characteristic polynomial are complex numbers, they are complex conjugates of one another,

$$s_1,\, s_2=a\pm j\omega$$

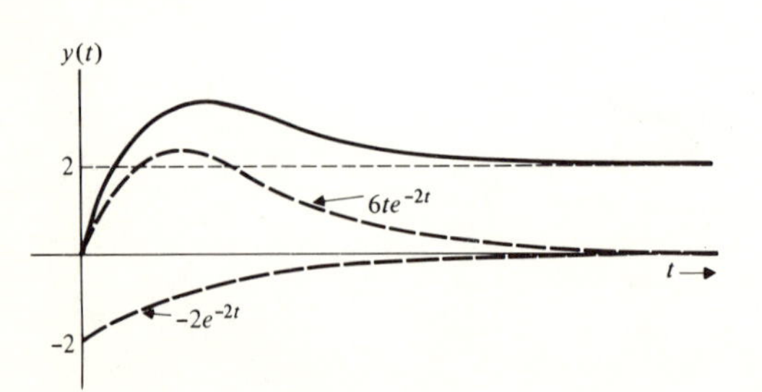

FIGURE 2-17
Step response of a critically damped second-order system.

and the natural response component is of the form

$$Y_{\text{natural}}(s) = \frac{\text{numerator polynomial}}{s^2 + a_1 s + a_0} = \frac{\text{numerator polynomial}}{(s - a - j\omega)(s - a + j\omega)}$$

$$= \frac{\text{numerator polynomial}}{(s+a)^2 + \omega^2} \tag{2-2}$$

corresponding to the time behavior

$$y_{\text{natural}}(t) = [Ae^{-at}\cos(\omega t + \theta)]u(t) \tag{2-3}$$

This type of second-order system is termed *underdamped*.

The underdamped system with transfer function

$$T(s) = \frac{-3s + 17}{s^2 + 2s + 17}$$

has unit step response

$$Y(s) = \frac{-3s + 17}{s(s^2 + 2s + 17)} = \frac{1}{s} + \frac{-(s+5)}{(s+1)^2 + (4)^2}$$

$$= \frac{1}{s} + \frac{Me^{j\theta}}{s + 1 - j4} + \frac{Me^{-j\theta}}{s + 1 + j4}$$

Using the evaluation method to find M and θ, there results

$$Me^{j\theta} = \left.\frac{-s-5}{s+1+j4}\right|_{s=-1+j4}$$

$$= \frac{1 - j4 - 5}{j8} = -\frac{1}{2} + j\frac{1}{2}$$

$$= \frac{\sqrt{2}}{2}e^{j135^\circ}$$

From Table 1-7,

$$y(t) = [1 + \sqrt{2}e^{-t}\cos(4t + 135^\circ)]u(t)$$

$$= [1 - \sqrt{2}e^{-t}\cos(4t - 45^\circ)]u(t)$$

This response is sketched in Fig. 2-18. Other initial conditions give different constants A and θ.

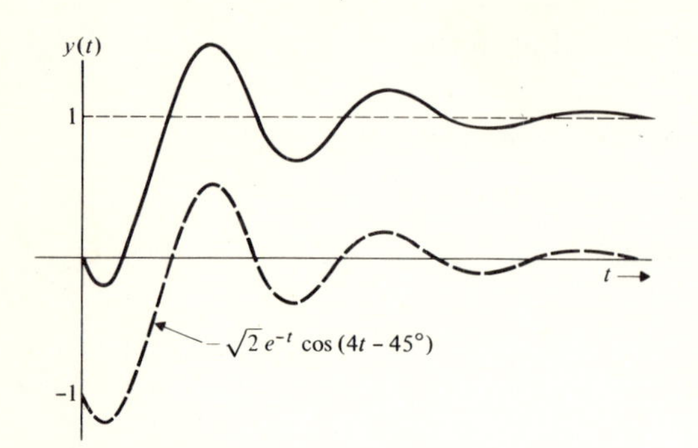

FIGURE 2-18
Step response of an underdamped second-order system.

DRILL PROBLEMS

D2-8. Find and also sketch the response of the systems with the following transfer function, input, and initial conditions:

(a) $T(s)=\dfrac{s}{s^2+7s+8}$

$r(t)=u(t)$
$y(0^-)=7$
$y'(0^-)=-4$

ans. $8.71e^{-1.44t}-1.71e^{-5.6t}$

(b) $T(s)=\dfrac{4}{s^2+4s+4}$

$r(t)=0$
$y(0^-)=-3$
$y'(0^-)=2$

ans. $-3e^{-2t}-4te^{-2t}$

(c) $T(s)=\dfrac{3}{2s^2+s+8}$

$r(t)=4u(t)$
$y(0^-)=0$
$y'(0^-)=0$

ans. $1.5+1.51e^{-0.25t}\cos(1.98t+173°)$

(d) $T(s)=\dfrac{4s-20}{s^2+4s+29}$

$r(t)=10\delta(t)$
$y(0^-)=0$
$y'(0^-)=6$

ans. $67.85e^{-2t}\cos(5t+54°)$

(e) $T(s)=\dfrac{s^2}{s^2+2s+17}$

$r(t)=0$
$y(0^-)=10$
$y'(0^-)=0$

ans. $10.3e^{-t}\cos(4t-14°)$

D2-9. Determine which of the following second-order systems are underdamped, which are critically damped, and which are overdamped:

(a) $T(s)=\dfrac{9s^2+3s+10}{s^2+5s+2}$

(b) $T(s)=\dfrac{s^2-2s}{s^2+6s+9}$

(c) $T(s)=\dfrac{64}{3s^2+4s+5}$

(d) $T(s)=\dfrac{19s-20}{s^2+s+100}$

(e) $T(s)=\dfrac{s^2+2s+100}{s^2+7s+49}$

ans. (a) overdamped; (b) critically damped;
(c) underdamped; (d) underdamped;
(e) underdamped

2.6 UNDAMPED NATURAL FREQUENCY AND DAMPING RATIO

The natural response of an underdamped second-order system, Eqs. (2-2) and (2-3), is described by the oscillation radian frequency ω and the exponential constant a, which may be found from the system's characteristic polynomial:

$$s^2+a_1s+a_0=(s+a)^2+\omega^2$$

An alternative description is in terms of the undamped natural frequency

ω_n and the damping ratio ζ. These quantities are related to the characteristic polynomial as follows:

$$s^2+a_1s+a_0=s^2+2\zeta\omega_n s+\omega_n^2$$

Equating the two descriptions,

$$(s+a)^2+\omega^2=s^2+2as+a^2+\omega^2=s^2+2\zeta\omega_n s+\omega_n^2$$

gives

$$a=\zeta\omega_n$$
$$\omega=\omega_n\sqrt{1-\zeta^2}$$

For ζ between 0 and 1, the characteristic roots lie on a circle of radius ω_n about the origin of the complex plane, as shown in Fig. 2-19. For $\zeta=0$, the roots are on the imaginary axis; and for $\zeta=1$, both roots are on the negative real axis, repeated.

The undamped natural frequency ω_n is the radian frequency at which the oscillations would occur if the damping ratio ζ were zero. If ζ were zero, the system natural response would have the form

$$Y_{\text{natural}}(s)=\frac{\text{numerator polynomial}}{s^2+\omega_n^2}$$
$$y_{\text{natural}}(t)=\{A\cos(\omega_n t+\theta)\}u(t)$$

Consider the second-order system with transfer function

$$T(s)=\frac{100}{s^2+3s+13}$$

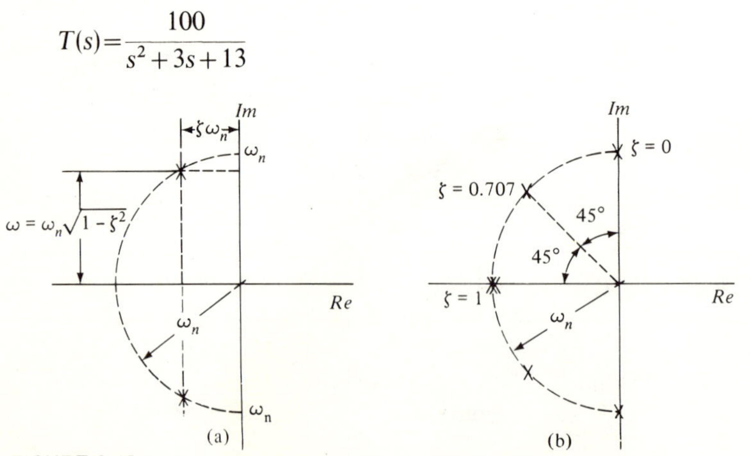

FIGURE 2-19
Characteristic roots of an underdamped second-order system, on the complex plan. (a) Relations between the various quantities. (b) Root locations for several values of damping ratio.

The undamped natural frequency of the system is

$$\omega_n = \sqrt{13} = 3.6$$

and the damping ratio is given by

$$2\zeta\omega_n = 3$$

$$\zeta = \frac{3}{2\omega_n} = \frac{3}{7.2} = 0.42$$

The true oscillation frequency is

$$\omega = \omega_n\sqrt{1-\zeta^2} = 3.6(0.9) = 3.27$$

A set of normalized step response curves for underdamped second-order systems of the form

$$T(s) = \frac{\omega_n^2}{s^2 + 2\zeta\omega_n s + \omega_n^2} \tag{2-4}$$

is shown in Fig. 2-20. For $\zeta = 0$, the oscillations continue forever. Larger values of ζ give more rapid decay of the oscillations but a slower rise of the response. For $\zeta = 1$, the system is critically damped.

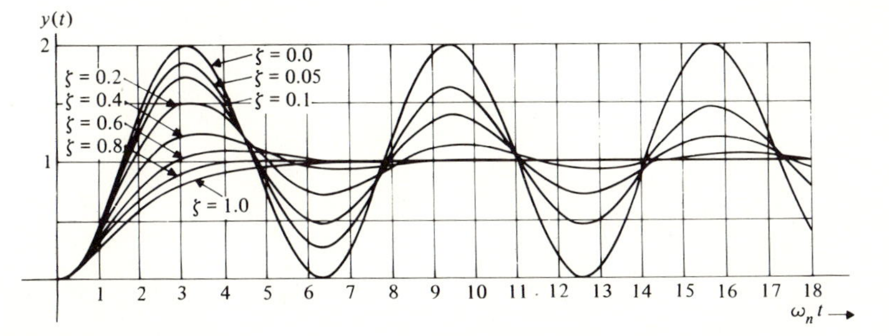

FIGURE 2-20
Normalized second-order system step response curves.

DRILL PROBLEMS

D2-10. For second-order systems with the following transfer functions, determine the undamped natural frequency, the damping ratio, and the oscillation frequency:

(a) $T(s) = \dfrac{100}{s^2 + s + 100}$

ans. 10, 0.05, 9.99

(b) $T(s)=\dfrac{3s-49}{s^2+3s+49}$

ans. 7, 0.214, 6.84

(c) $T(s)=\dfrac{s^2+9s}{s^2+4s+10}$

ans. 3.16, 0.632, 2.45

(d) $T(s)=\dfrac{s^2+20}{s^2+2s+20}$

ans. 4.47, 0.224, 4.36

(e) $T(s)=\dfrac{-3s+0.7}{s^2+0.3s+4}$

ans. 2, 0.075, 1.99

D2-11. Find the constant k for which the system with transfer function $T(s)$ has the given second-order response property.

(a) $T(s)=\dfrac{10}{s^2+40s+k}$

$\xi=0.7$

ans. 816

(b) $T(s)=\dfrac{ks+6}{s^2+ks+49}$

$\omega=4$

ans. 11.49

(c) $T(s)=\dfrac{20s}{3s^2+2s+k+4}$

$\xi=0.1$

ans. 29.33

(d) $T(s)=\dfrac{s^2-6}{ks^2+s+6}$

$\omega_n=2$

ans. 1.5

2.7 OVERSHOOT, RISE TIME, AND SETTLING TIME

The *percent overshoot*

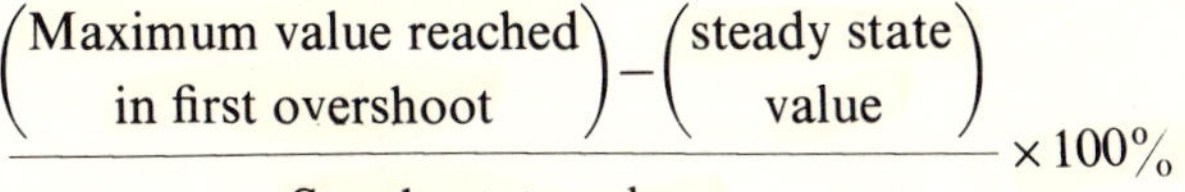

$$\frac{\begin{pmatrix}\text{Maximum value reached}\\ \text{in first overshoot}\end{pmatrix}-\begin{pmatrix}\text{steady state}\\ \text{value}\end{pmatrix}}{\text{Steady state value}}\times 100\%$$

in the step response for the system with transfer function (2-4) is plotted in Fig. 2-21. Over the years a damping ratio of about 0.7 has been found to be satisfactory in a wide range of applications such as positioning systems and aircraft autopilots. A convenient numerical value is $\sqrt{2}/2=0.707$ for which a and ω are equal. Response is more rapid than with critical damping, and overshoot is only about 5%.

Rise time is defined as the time required for the step response of a system to rise from 10 to 90% of the final value. Rise times for the second-order system of Eq. (2-4) may be read from the curves of Fig. 2-20. For example, for a damping ratio $\xi=0.2$, the rise time is approximately

$$T_r \cong \frac{1.4}{\omega_n}$$

as indicated in Fig. 2-22.

Normalized rise time is plotted versus damping ratio for the example second-order system in Fig. 2-23. Second-order systems which have other than a constant transfer function numerator polynomial will generally have a different relationship between damping ratio and rise time.

Settling time is the time required for a system step response to settle to

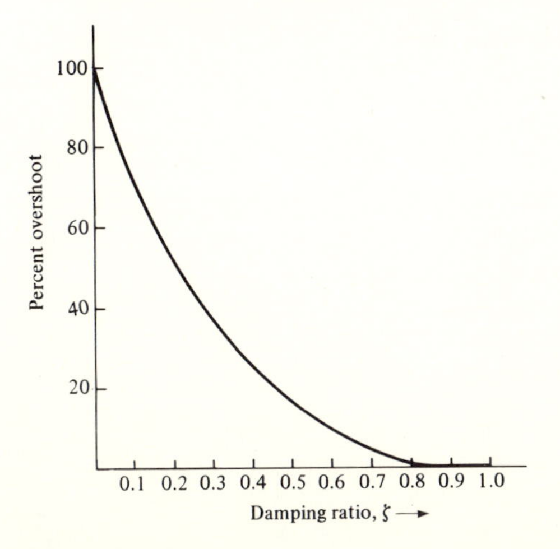

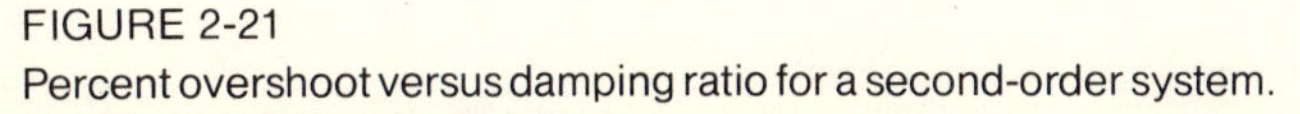

FIGURE 2-21
Percent overshoot versus damping ratio for a second-order system.

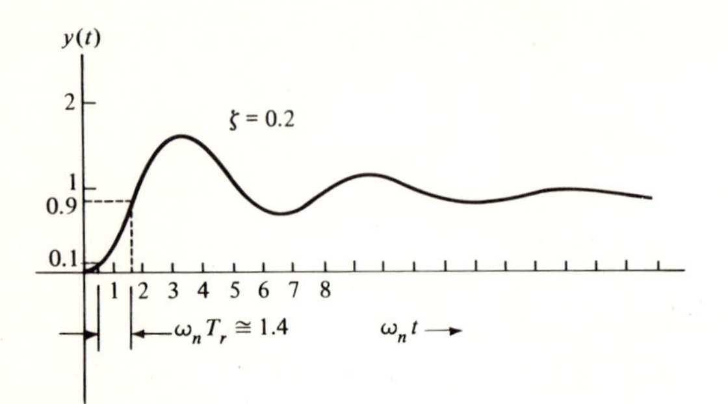

FIGURE 2-22
Finding the rise time for a certain second-order system.

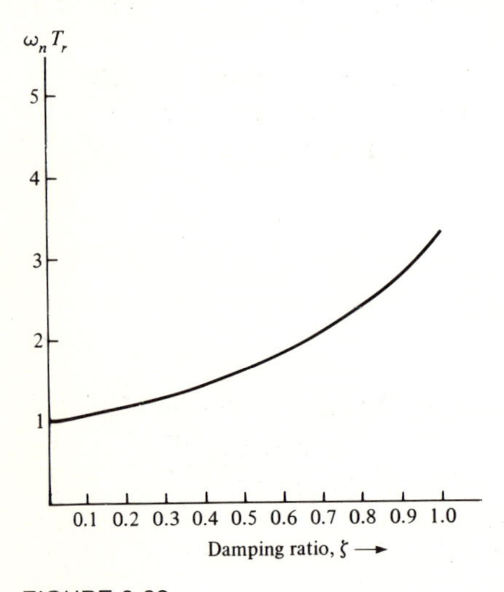

FIGURE 2-23
Rise time versus damping ratio for a second-order system.

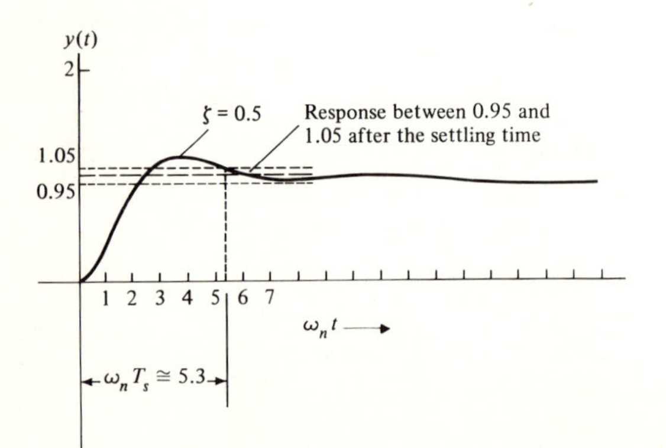

FIGURE 2-24
Finding the settling time for a certain second-order system.

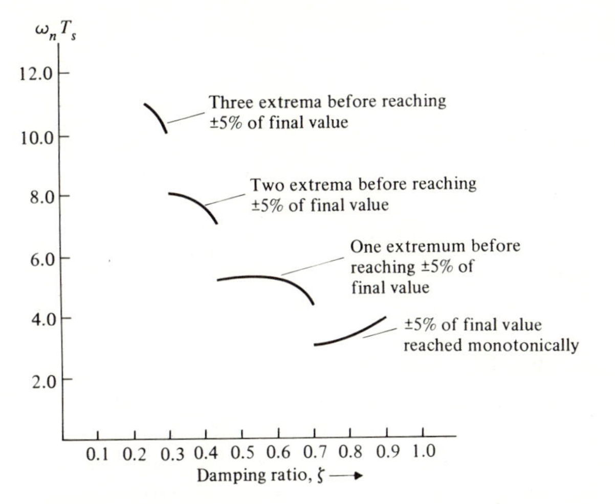

FIGURE 2-25
Settling time versus damping ratio for a second-order system.

within some specified band of values, usually $\pm 5\%$, about the final value. Using the curves of Fig. 2-20, the $\pm 5\%$ settling time for $\zeta = 0.5$ is approximately

$$T_s = \pm 5\% \text{ settling time} \cong \frac{5.3}{\omega_n}$$

as indicated in Fig. 2-24.

Normalized settling time is plotted versus damping ratio for the example second-order system in Fig. 2-25.

DRILL PROBLEMS

D2-12. Using the curves given in the text, find approximately the percent overshoot, rise time, and settling time of the following systems when driven by a step input:

(a) $T(s) = \dfrac{100}{s^2 + 4s + 100}$

ans. 54%, 0.12, 1.1

(b) $T(s) = \dfrac{49}{s^2 + 4s + 49}$

ans. 40%, 0.18, 1.5

(c) $T(s)=\dfrac{60}{2s^2+8s+30}$

ans. 18%, 0.44, 0.55

(d) $T(s)=\dfrac{75}{s^2+3s+20}$

ans. 32%, 0.34, 1.7

2.8 HIGHER-ORDER SYSTEM RESPONSE

The natural response of third- and higher-order systems consists of a sum of terms, one term for each characteristic root. For each distinct real characteristic root, there is a real exponential term in the system natural response. For each pair of complex conjugate roots there is a pair of complex exponential terms which are better expressed as an exponential times a sinusoid. Repeated roots give additional terms involving powers of time times the exponential.

For example, a system with transfer function

$$T(s)=\frac{-8s^2+5}{s^4+9s^3+37s^2+81s+52}=\frac{-8s^2+5}{(s+1)(s+4)(s^2+4s+13)}$$

has a natural response of the form

$$Y_{\text{natural}}(s)=\frac{K_1}{s+1}+\frac{K_2}{s+4}+\frac{K_3s+K_4}{(s+2)^2+(3)^2}$$

or

$$y_{\text{natural}}(t)=K_1e^{-t}+K_2e^{-4t}+Ae^{-2t}\cos(3t+\theta)$$

Rise time, overshoot, and settling time of the step response are also useful characterizations for higher-order systems. These quantities are not very easily calculated and tabulated, however.

DRILL PROBLEMS

D2-13. Find the form of the natural response of systems with the following transfer functions:

(a) $T(s)=\dfrac{100}{(s^2+4s+4)(s^2+4s+5)}$

ans. $K_1e^{-2t}+K_2te^{-2t}+K_3e^{-2t}\cos(t+\theta)$

(b) $T(s)=\dfrac{3s-12}{s^3+4s^2+13s}$

ans. $K_1+K_2e^{-2t}\cos(3t+\theta)$

(c) $T(s)=\dfrac{s^2}{3(s+3)^2(s^2+2s+10)}$

ans. $K_1e^{-3t}+K_2te^{-3t}+K_3e^{-t}\cos(3t+\theta)$

(d) $T(s)=\dfrac{5(s^2+2s+1)(s^2+2s+2)}{(s+1)^2(s^2+4)(s+8)}$

ans. $K_1e^{-t}+K_2te^{-t}+K_3\cos(2t+\theta)+K_4e^{-8t}$

(e) $T(s)=\dfrac{6s^3-4s^2+2s+400}{(s^2+s+10)^2(s^2+s+20)}$

ans. $K_1e^{-0.5t}\cos(3.12t+\theta_1)$
$+K_2e^{-0.5t}\cos(4.44t+\theta_2)$

2.9 AN INSULIN DELIVERY SYSTEM

Control system methods have been applied to the biomedical field to create an implanted insulin delivery system for diabetics. When food is eaten and digested, sugars, mainly glucose, are absorbed into the bloodstream. Normally, the pancreas secretes insulin into the bloodstream to metabolize the sugar. A diabetic's pancreas secretes insufficient insulin to metabolize blood sugar; blood sugar levels can thus become high enough to threaten damage to the body organs.

One solution to this problem is for the diabetic to take one injection of insulin each day. In Fig. 2-26a, typical blood sugar and insulin concentration histories for one day are shown for a normal person. In Fig. 2-26b, the blood sugar and insulin concentration histories for one day are shown for a diabetic who takes one insulin injection in the morning. Notice that blood sugar is often higher than normal, but the sugar concentration is far less than would be the case if no insulin had been injected. A higher dose of insulin could be taken in the morning to counteract the low insulin residual after dinner, but blood sugar concentration might be driven unacceptably low in the morning (hypoglycemia), causing weakness, trembling, and possibly fainting. Three injections a day, one before each meal, are generally infeasible due to damage to the veins and skin tissue.

One automatic control system of interest consists of a tiny insulin reservoir, control motor, and pump which is implanted in the body below the diaphragm. This electronic pancreas delivers insulin into the peritoneum using preprogrammed commands intended to establish insulin levels close to the levels of a normal individual. The pump runs at higher rates after meals than otherwise.

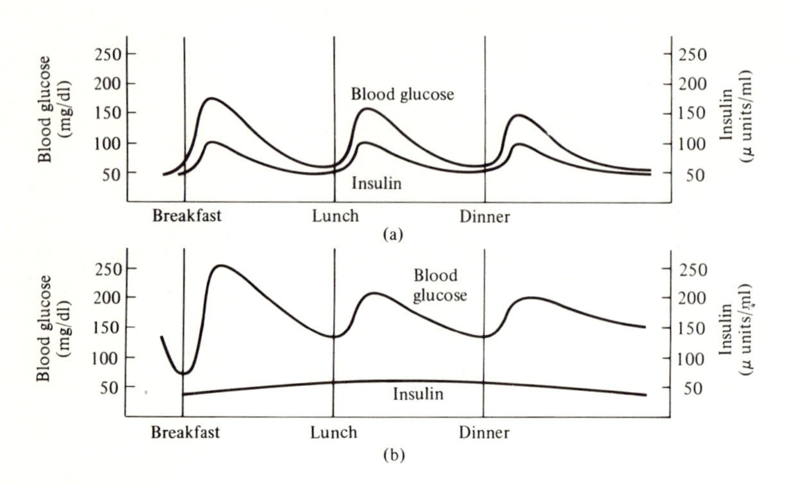

FIGURE 2-26
Typical blood sugar and insulin concentrations. (a) Normal person. (b) Diabetic with one daily insulin injection.

The diabetic must time meals to complement the behavior of the implanted system, but injections are required only every few weeks, to refill the insulin reservoir. Systems of this type have operated for many years without malfunction.

The methods of this chapter can be used to design such an insulin delivery system, which is the open-loop one shown in Fig. 2-27. A signal generator is programmed to drive the motor pump in such a way that the insulin delivery rate $i(t)$ approximates a desired delivery rate $i_D(t)$.

Figure 2-28a shows an approximate desired insulin rate $i_D(t)$ in cm^3/sec for one-third day, beginning with a meal. Figure 2-28b shows a similar function,

$$i(t) = Ate^{-at}u(t)$$

which has the particularly simple Laplace transform

$$I(s) = \mathscr{L}[i(t)] = \frac{A}{(s+a)^2}$$

A good approximation of $i_D(t)$ by $i(t)$ occurs when the constants A and a are

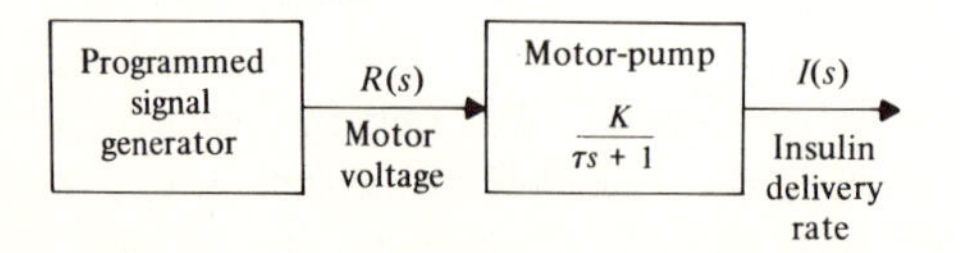

FIGURE 2-27
The implanted open-loop insulin delivery system.

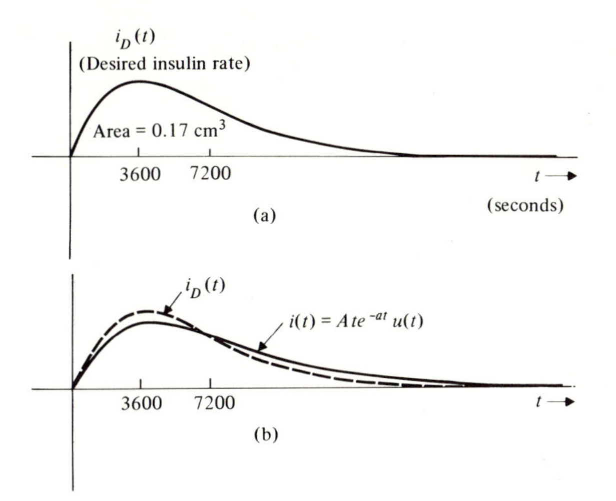

FIGURE 2-28
Approximating the required insulin rate with a time-weighted exponential. (a) Insulin rate required for a diabetic. (b) The function $i(t) = Ate^{-at}u(t)$.

selected so that $i(t)$ is maximum at $t = 3600$, as is $i_D(t)$, and so that the areas under the two curves are equal, with value 0.17 cm^3:

$$\frac{di}{dt} = -aAte^{-at} + Ae^{-at} = A(1 - at)e^{-at}$$

For the maximum of $i(t)$ to occur at $t = 3600$,

$$\left.\frac{di}{dt}\right|_{t=3600} = A(1 - 3600a)e^{-3600a} = 0$$

$$a = \tfrac{1}{3600} = 2.78 \times 10^{-4}$$

The area under the $i(t)$ curve after $t = 0$ is

$$\int_0^\infty Ate^{-at}\,dt = A\left[-\frac{1}{a}te^{-at} - \frac{1}{a^2}e^{-at}\right]_0^\infty$$
$$= \frac{A}{a^2}$$

Equating to the desired area of 0.17 cm^3 gives

$$\frac{A}{a^2} = (3600)^2 A = 0.17$$

$$A = 1.31 \times 10^{-8}$$

If $i(t)$ is to be produced by the system input $r(t)$, one has

$$I(s)=\frac{A}{(s+a)^2}=\frac{K}{\tau s+1}R(s)$$

giving the required programmed signal for each mealtime:

$$R(s)=\frac{A(\tau s+1)/K}{(s+a)^2}$$

For a motor pump with

$$\tau=5 \text{ sec}$$
$$K=2.3\times 10^{-6} \text{ cm}^3\text{/volt-sec}$$

and with the delivery rate for which

$$a=2.78\times 10^{-4} \text{ sec}^{-1}$$
$$A=1.31\times 10^{-8} \text{ cm}^3$$

then

$$R(s)=\frac{(1.31\times 10^{-8})(5s+1)/2.3\times 10^{-6}}{(s+2.78\times 10^{-4})^2}$$
$$=\frac{28.5\times 10^{-3}s+5.7\times 10^{-3}}{(s+2.78\times 10^{-4})^2}$$
$$=\frac{K_1}{(s+2.78\times 10^{-4})}+\frac{K_2}{(s+2.78\times 10^{-4})^2}$$
$$=\frac{K_1 s+(K_2+2.78\times 10^{-4}K_1)}{(s+2.78\times 10^{-4})^2}$$

$$K_1=28.5\times 10^{-3}$$

$$K_2=5.7\times 10^{-3}-2.78\times 10^{-4}K_1=5.69\times 10^{-3}$$

The programmed motor drive signal is thus to be

$$R(s)=\frac{28.5\times 10^{-3}}{s+2.78\times 10^{-4}}+\frac{5.69\times 10^{-3}}{(s+2.78\times 10^{-4})^2}$$

$$r(t)=[28.5e^{-(2.78\times 10^{-4})t}+5.69te^{-(2.78\times 10^{-4})t}]10^{-3}u(t) \qquad \text{volts}$$

which is sketched in Fig. 2-29a. Repetition, three times a day, of the motor drive signal will provide insulin delivery for many meals, as is shown in Fig. 2-29b.

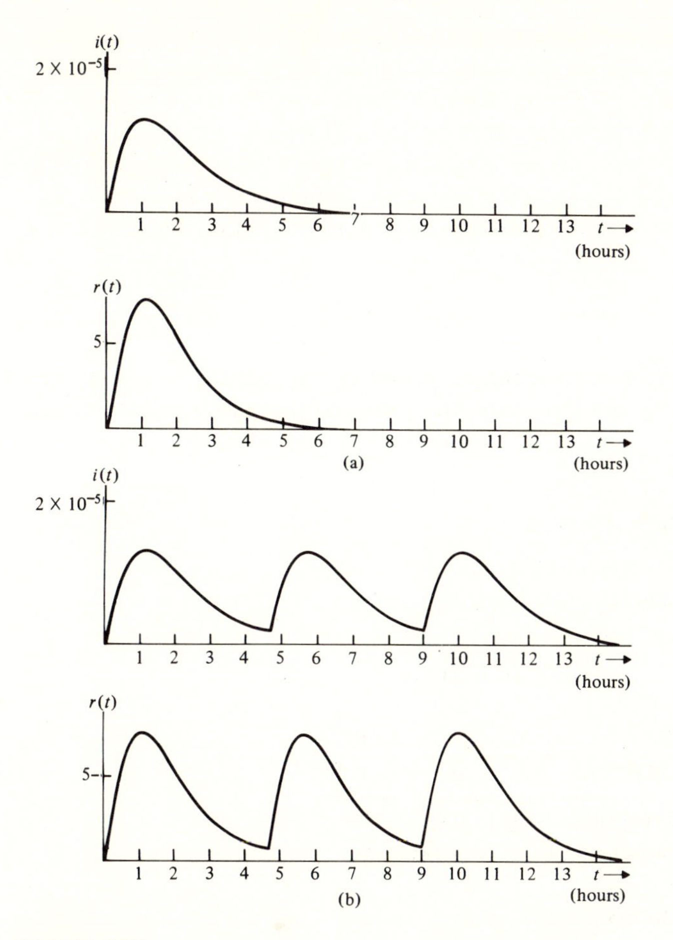

FIGURE 2-29
Response of the insulin delivery system. (a) Motor voltage and resulting insulin rate for one mealtime. (b) Repetitive motor voltage and insulin rate for three meals a day.

2.10 ANALYSIS OF AN AIRCRAFT WING

Students commonly perform an experiment in physics where a tuning fork is struck above an empty tube, resulting in a resonating tone. One can obtain the same effect by blowing into a bottle at just the right angle. What occurs is that the blowing or the tuning fork is at a natural frequency of the hollow glass chamber, causing a reinforcement or resonating effect. This effect can be catastrophic when dealing with certain mechanical systems. For example, in 1939 a bridge over Puget Sound at the Tacoma Narrows began to sway and twist due to a wind which whistled down the Narrows at just the right speed. In a mechanical sense the bridge was excited at a resonance, resulting in deflections which exceeded its structural limit and led to collapse.

A similar effect must be considered in aircraft wing design. Structural failure of wings on certain turbine-driven jet aircraft in the 1960s was traced to a mechanical resonance excited by the jet turbines. The following problem illustrates the problem of mechanical resonance for an aircraft wing.

Figure 2-30a depicts an aircraft wing with an applied force $f(t)$ and resulting deflection at the wingtip $x(t)$. Let us ignore aerodynamic and accelerative forces and consider only a sinusoidal excitation of the wing caused by the jet engine's turbine:

$$f(t) = D\cos\omega t$$

Figure 2-30b shows a simple mechanical model of the aircraft wing, a spring-mass system excited by the force $f(t)$. The differential equation which relates $f(t)$ to $x(t)$ is

$$M\frac{d^2x}{dt^2} + Kx = f(t) = D\cos\omega t$$

Laplace-transforming,

$$Ms^2X(s) - Msx(0^-) - Mx'(0^-) + KX(s) = \frac{Ds}{s^2+\omega^2}$$

$$X(s) = \underbrace{\frac{Ds}{(s^2+\omega^2)(Ms^2+K)}}_{\text{zero-state response component}} + \underbrace{\frac{Msx(0^-)+Mx'(0^-)}{Ms^2+K}}_{\text{zero-input response component}}$$

For zero initial conditions,

$$X(s) = \frac{(D/M)s}{(s^2+\omega^2)(s^2+K/M)} = \frac{K_1s+K_2}{s^2+\omega^2} + \frac{K_3s+K_4}{s^2+K/M}$$

$$= \frac{(K_1+K_3)s^3 + (K_2+K_4)s^2 + (K_1K/M + K_3/\omega^2)s + (K_2K/M + K_4/\omega^2)}{(s^2+\omega^2)(s^2+K/M)}$$

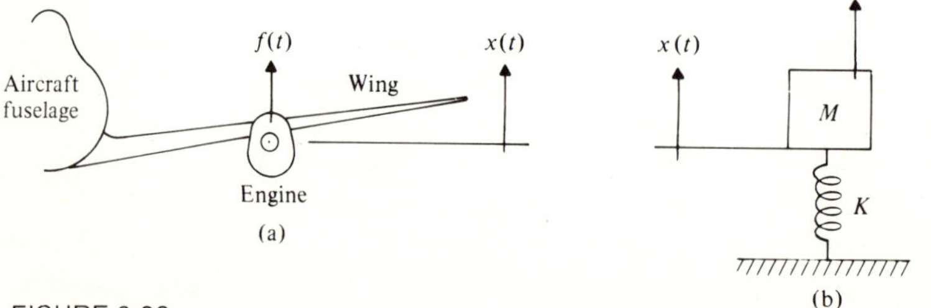

FIGURE 2-30
Deflection of an aircraft wing due to jet turbine vibrational force.
(a) The aircraft wing. (b) Model of aircraft wing.

giving

$$\begin{cases} K_1 + K_3 = 0 \\ K_2 + K_4 = 0 \\ \dfrac{K}{M}K_1 + \omega^2 K_3 = \dfrac{D}{M} \\ \dfrac{K}{M}K_2 + \omega^2 K_4 = 0 \end{cases}$$

$$K_2 = K_4 = 0 \qquad K_1 = -K_3 = \frac{D/M}{(K/M)-\omega^2}$$

$$X(s) = \frac{\dfrac{D/M}{(K/M)-\omega^2}\,s}{s^2+\omega^2} + \frac{\dfrac{-D/M}{(K/M)-\omega^2}}{s^2+(K/M)}$$

$$x(t) = \frac{D/M}{(K/M)-\omega^2}\cos\omega t - \frac{D/M}{(K/M)-\omega^2}\cos\sqrt{\frac{K}{M}}\,t \qquad t \geqslant 0$$

The following data are for an aircraft wing where the frequency ω of the turbine and the natural frequency $\sqrt{K/M}$ are close to one another:

$M = 1000$ lb of mass

$D = 3000$ lb of force

$K = 400{,}800.4$ lb of force/ft of deflection

$\omega = 20$ rad/sec

Deflections exceeding 5 ft are considered to be beyond the wing's structural limit. With these numbers,

$$x(t) = 3.748(\cos 20t - \cos 20.02t) \qquad t \geqslant 0$$

Maximum deflection will obviously occur at times, assuming they exist, when

$$\cos 20t = 1$$

and

$$\cos 20.02t = -1$$

These are times for which

$$20t = n2\pi \qquad n \text{ an integer}$$

and

$$20.02t = \pi + m2\pi \qquad m \text{ an integer}$$

one such solution being $n = m = 500$, $t = 50\pi$. The amount of maximum deflection will be

$$x_{max} = 3.748(2) = 7.496 \text{ ft}$$

To reduce the deflection amplitude, the wing may be stiffened to increase K to obtain a natural frequency $\sqrt{K/M}$ much larger than ω, giving a much smaller value for the amplitudes

$$K_2 = -K_3 = \frac{D/M}{(K/M) - \omega^2}$$

Stiffening the wing slightly to give

$K = 441{,}000$ lb of force/ft of deflection results instead in

$$K_2 = -K_3 = \frac{D/M}{(K/M) - \omega^2} = 0.073$$

$$x(t) = 0.073(\cos 20t - \cos 21t)$$

for which

$$x_{max} = 0.073(2) = 0.146 \text{ ft}$$

This is the solution which was used to prevent wing fatigue subsequent to the aircraft wing failures in the 1960s.

2.11 SUMMARY

Block diagram system representations consist of blocks, summers, and junctions. They are used to graphically describe system components and their interconnections.

To reduce a block diagram of a single-input, single-output system means to find, using block diagram equivalences, the single-block diagram containing the overall transfer function of the system. Multiple-input, multiple-output block diagrams are reduced by reducing a series of single-input, single-output diagrams, one for each different combination of input and output.

The signal flow graph representation of a system consists of nodes and branches. The signal at any node is the sum of the incoming signals through branches. A branch in a signal flow graph is equivalent to a block in a block diagram, and a node is equivalent to a summer, with all plus signs, followed by a junction.

Mason's gain rule states that the transfer function of a single-input, single-output system is

$$T(s)=\frac{\sum_i P_i\Delta_i}{\Delta}$$

where the P_i terms are the path gains, Δ_i is the cofactor of the ith path, and Δ is the determinant of the signal flow graph.

First-order systems are characterized by transfer functions with one real characteristic root. The natural component of the response of the system is a real exponential term which is described by its time constant.

Second-order systems have transfer functions with a second-order characteristic equation having two characteristic roots. The natural component of system response is overdamped, critically damped, or underdamped according to whether the system's characteristic roots are real and distinct, real and equal, or complex conjugate, respectively. Underdamped oscillatory signals are conveniently described in terms of their undamped natural frequency and damping ratio.

System step response is an important description of performance in many applications, and is commonly measured in terms of overshoot, rise time, and settling time.

In general, systems have natural response which consists of a real exponential term for each real characteristic root and an oscillatory term for each pair of complex conjugate characteristic roots.

REFERENCES

Signal Flow Graphs

Dertouzos, M. L.; Athans, M.; Spann, R. N.; and Mason, S. J. *Systems, Networks and Computation.* New York: McGraw-Hill, 1973.

Mason, S. J. "Feedback Theory: Some Properties of Signal Flow Graphs." *Proc. IRE* 41 (September 1953).

______. "Feedback Theory: Further Properties of Signal Flow Graphs." *Proc. IRE* 44 (July 1956).

Insulin Delivery and Biomedical Applications

Albisser, A. M. "Review of Artificial Pancreas Research." *Arch. Int. Med.* 137 (May 1977): 639–649.

Blackshear, P. J. "Implantable Drug-Delivery Systems." *Sci. Amer.*, December 1979, pp. 66–73.

Spencer, W. J. "For Diabetics: An Electronic Pancreas." *IEEE Spec.* 15, no. 6 (June 1978): 38–42.

Aircraft Dynamics

Etkin, B. B. *Dynamics of Atmospheric Flight*. New York: Wiley, 1972.
Kolk, R. W. *Modern Flight Dynamics*. Englewood Cliffs, N.J.: Prentice-Hall, 1961.

PROBLEMS

1. Using equivalences, reduce the block diagrams in Fig. P2-1 to single blocks, obtaining each system transfer function.

ans. (b) $-(1+s/2)s/(2s+1)$

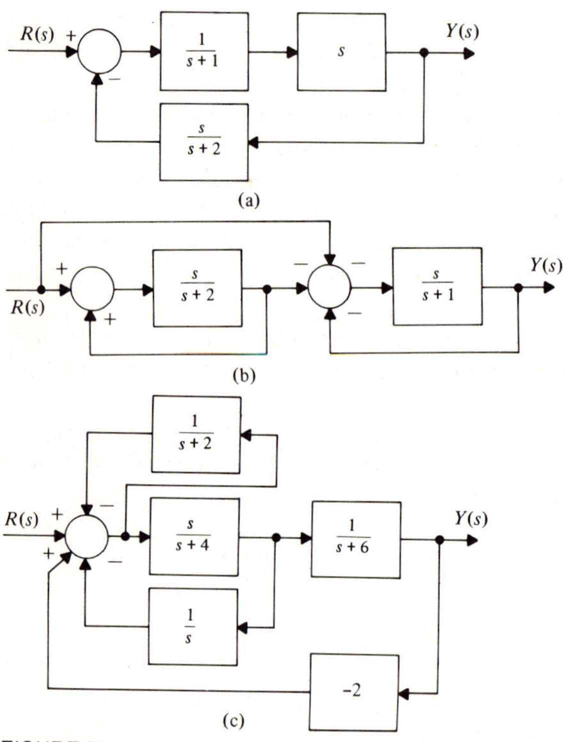

FIGURE P2-1

2. Using equivalences, find the four transfer functions for the system in Fig. P2-2.

$$ans.\ T_{12} = \frac{-s}{s^3+s^2+5s+4}$$

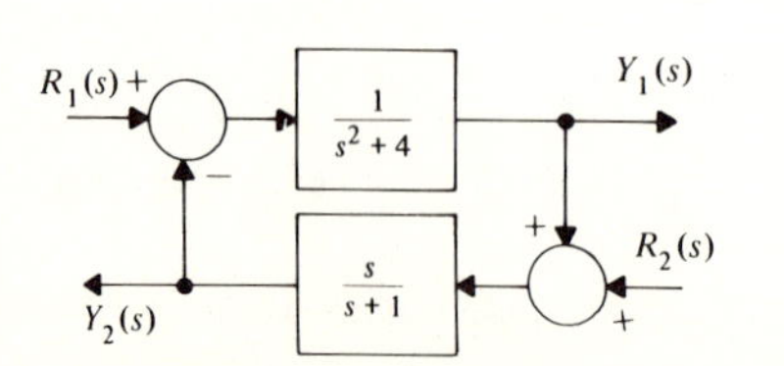

FIGURE P2-2

3. Write Laplace-transformed equations for the block diagram of Fig. P2-2, then solve the equations for the four system transfer functions.
4. Find the transfer function relating output to input for each of the signal flow graphs of Fig. P2-3 using Mason's gain rule.

$$ans. \text{(b)} \ \frac{1}{s^3+s^2+2s+3}$$

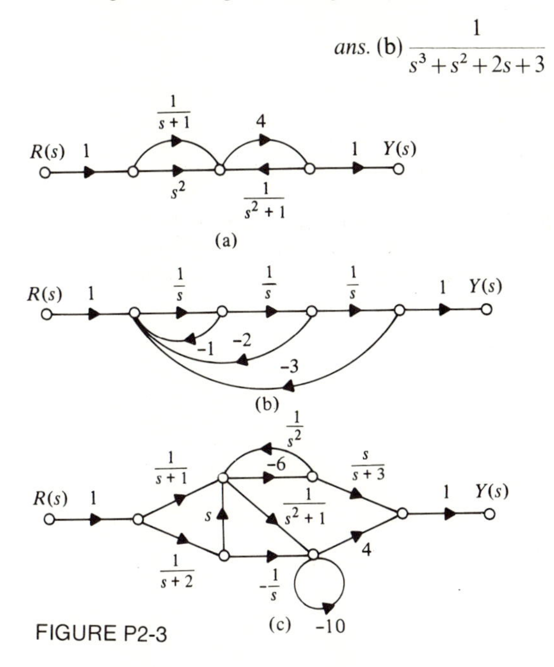

FIGURE P2-3

5. Using Mason's gain rule, find the four transfer functions of the system of Fig. P2-4.

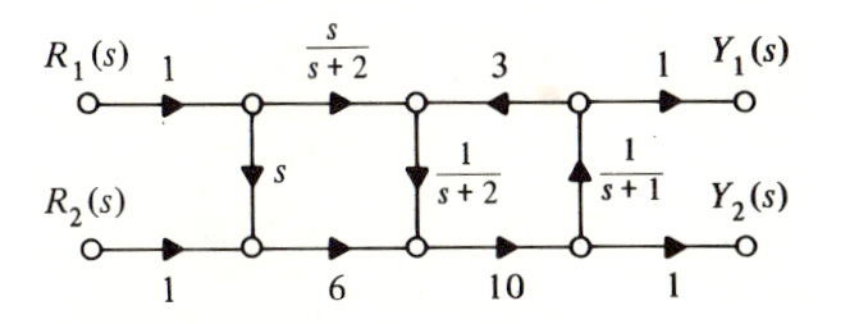

FIGURE P2-4

6. Convert the block diagrams of Fig. P2-1 to equivalent signal flow graphs.
7. Convert the signal flow graphs of Fig. P2-3 to equivalent block diagrams.
8. Write systematic simultaneous equations for the system of Fig. P2-4, then solve the equations for the four system transfer functions.
9. Draw block diagrams to represent the following Laplace-transformed equations. The R signals are inputs, and the Y signals are outputs. This is a *synthesis* problem and has, in each case, many possible solutions.

(a)
$$\begin{cases} X_1(s)=\dfrac{s}{s+1}R(s)+X_2(s) \\ X_2(s)=R(s)+10X_1(s) \\ Y(s)=R(s)+\dfrac{4}{s^2+9}X_2(s) \end{cases}$$

(b) $$\begin{cases} \left(\dfrac{4}{s+1}\right) X_1 - 6X_2 + 7R + \left(\dfrac{1}{s}\right) Y_1 = Y_2 \\ 3X_1 + \left(\dfrac{s}{s+1}\right) X_2 - \left(\dfrac{s}{s+2}\right) R = 0 \\ \left(\dfrac{2}{s+2}\right) X_1 - \left(\dfrac{s}{s+2}\right) X_2 + \left(\dfrac{10}{s^2+4}\right) Y_1 = Y_2 \\ \left(\dfrac{1}{s+2}\right) X_1 + 7X_2 + \left(\dfrac{1}{s}\right) R + 10Y_2 = 0 \end{cases}$$

10. For the system equations of Prob. 9, draw instead representative signal flow graphs.
11. Find and sketch the response $y(t)$, $t \geqslant 0$, of the first-order systems with the following transfer functions, inputs $r(t)$, and initial conditions. Indicate the natural part of the response and find its time constant.

(a) $T(s) = \dfrac{5}{s+5}$

$r(t) = \delta(t)$, an impulse

$y(0^-) = -4$

(b) $T(s) = \dfrac{-2s}{3s+1}$

$r(t) = u(t)$, a step

$y(0^-) = 0$

ans. $-\frac{2}{3}e^{-(1/3)t}$

(c) $T(s) = \dfrac{4s+1}{2s+4}$

$r(t) = u(t)$, a step

$y(0^-) = 3$

(d) $T(s) = \dfrac{3}{s+3}$

$r(t) = 2\cos 4t$

$y(0^-) = 0$

ans. $-\frac{18}{25}e^{-3t} + \frac{6}{5}\cos(4t - 53^\circ)$

(e) $T(s) = \dfrac{1}{2s+1}$

$r(t) = 6e^{-(1/2)t}$

$y(0^-) = -7$

12. Find and sketch the response $y(t)$, $t \geqslant 0$, of the second-order systems with the following transfer functions, inputs, $r(t)$, and zero initial conditions. Indicate the natural part of the response.

(a) $T(s)=\dfrac{3}{s^2+4s+3}$

$r(t)=\delta(t)$, an impulse

(b) $T(s)=\dfrac{s-4}{s^2+6s+9}$

$r(t)=u(t)$, a step

ans. $-\frac{4}{9}+\frac{4}{9}e^{-3t}+\frac{7}{3}te^{-3t}$

(c) $T(s)=\dfrac{1}{s^2+4s+1}$

$r(t)=u(t)$, a step

(d) $T(s)=\dfrac{10}{s^2+2s+10}$

$r(t)=2t$

ans. $-\frac{2}{5}+2t+0.66e^{-t}\cos(3t-54^\circ)$

(e) $T(s)=\dfrac{s}{s^2+4s+13}$

$r(t)=3\sin 2t$

13. Find a controller transmittance $G(s)$ such that the overall system of Fig. P2-5 is second-order and critically damped. A solution to this problem is not unique.

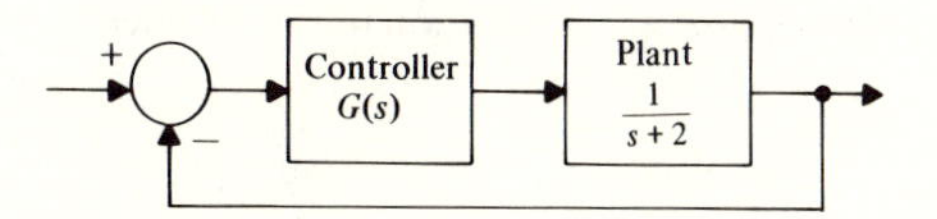

FIGURE P2-5

14. Identify the type of natural response (overdamped, critically damped, or underdamped) associated with each of the following characteristic polynomials:
   (a) $s^2+8s+12$
   (b) s^2+s+1

   ans. underdamped

   (c) $3s^2+9s+2$
   (d) $4s^2+s+40$

   ans. underdamped

   (e) s^2+4s+4

15. For the underdamped systems with the following transfer functions, find the undamped natural frequency ω_n, the damping ratio ζ, the exponential constant $a=\zeta\omega_n$, and the oscillation frequency ω:

   (a) $T(s)=\dfrac{-s^2}{s^2+6s+25}$

(b) $T(s)=\dfrac{s^2+2s+10}{4s^2+2s+100}$

ans. 5, 0.05, 0.25, 4.99

(c) $T(s)=\dfrac{20}{3s^2+s+20}$

(d) $T(s)=\dfrac{s-3}{s^2+8}$

ans. 2.83, 0, 0, 2.83

(e) $T(s)=\dfrac{1}{s^2+6s+30}$

16. For the system of Fig. P2-6, find the constant value of gain, K, for which the damping ratio of the overall system is 0.7. For this value of K, what is the system's undamped natural frequency?

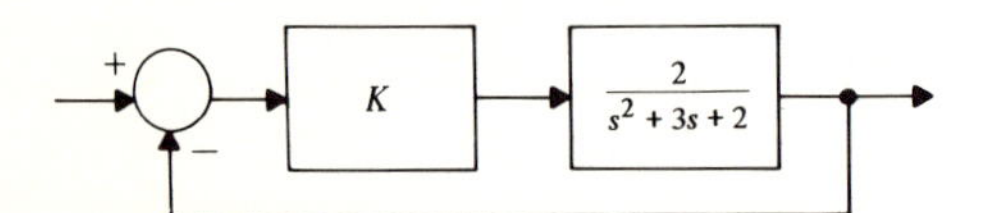

FIGURE P2-6

17. For the system of Fig. P2-7, find a (constant value of) gain, K, for which the natural component of the output decays at least as rapidly as e^{-10t}. A solution to this problem is not unique.

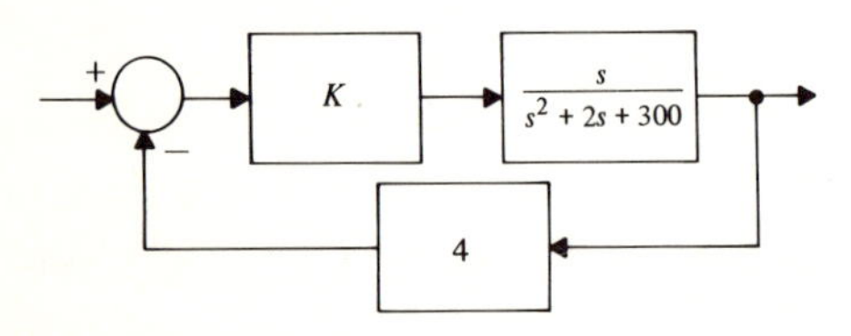

FIGURE P2-7

18. The required insulin delivery rate varies considerably from person to person because of differences in body chemistry. For the insulin delivery system, suppose that a peak delivery of insulin must occur at 2 hr (7200 sec) instead of at 1 hr. Let the total insulin volume to be delivered for each meal be 0.35 cm^3/sec instead of 0.17 cm^3/sec. Find a single cycle of the required motor drive signal.
19. For the aircraft wing problem (Sec. 2.10), instead of stiffening the wing with the mass remaining constant, suppose the stiffness remains constant but the mass M of the wing varies. K_1 represents half the maximum wing deflection. Plot K_1 versus M for M between 990 and 1000 lb of mass.
20. For the original M, D, and K of the aircraft wing problem (Sec. 2.10), plot K_1 versus ω for values of ω from 19.8 to 20 rad/sec.
21. The position control system for a spacecraft platform is governed by the following approximate equations:

$$\frac{d^2p}{dt^2}+\frac{dp}{dt}+2p=\theta$$

$$v_1=r-p$$

$$\frac{d\theta}{dt}=0.2v_2$$

$$v_2=4v_1$$

The variables involved are as follows:

$r(t)$ = desired platform position (input)
$p(t)$ = actual platform position (output)
$v_1(t)$ = amplifier input voltage
$v_2(t)$ = amplifier output voltage
$\theta(t)$ = motor shaft position

Draw a block diagram of the system, identifying the component parts and their transmittances. Then determine the system transfer function.

22. A savings account with an initial deposit of \$800 and 0.5% interest per month has a balance during the nth month of

$$b(n)=800(1.005)^n \qquad n=0, 1, 2, \ldots$$

This discrete function of the integer values n may be modeled by a continuous exponential function

$$f(t)=Ae^{at}$$

which has the same values as $b(n)$ for integer values of t, as indicated in Fig. P2-8. For convenience, t can be measured in months. For this model, find the constants A and a.

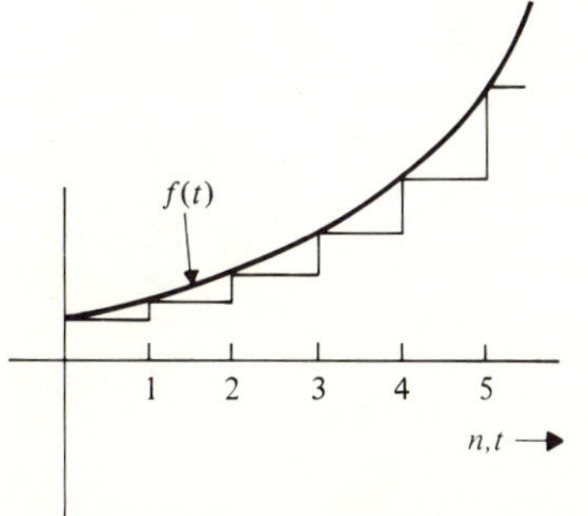

FIGURE P2-8

23. The mechanical device shown in Fig. P2-9 is an *accelerometer*. It is designed so that the position x_2 of the mass with respect to the case is approximately proportional to the

case acceleration, d^2x_1/dt^2. Find the transfer function that relates x_2 to the case acceleration,

$$r(t) = \frac{d^2x_1}{dt^2}$$

in terms of K, M, and B. Then, based upon your feeling for how such a device should behave in applications involving automobile accelerations, choose suitable values for K, M, and B.

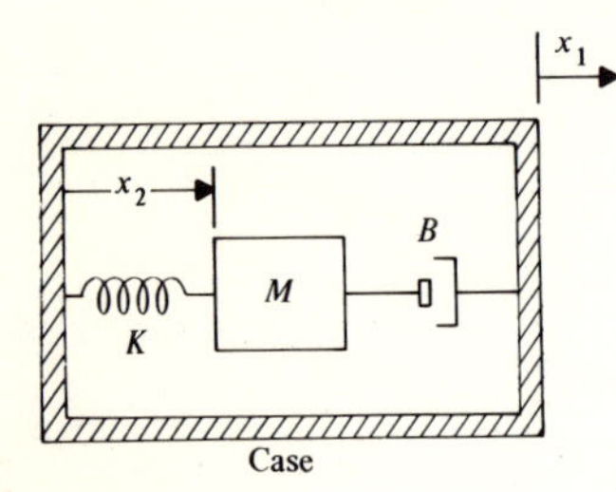

FIGURE P2-9

24. Power steering for an automobile is a feedback system which may be modeled as in Fig. P2-10. For a unit step input $A(s)$, find values of K_1 and K_2, if possible, for which the response $w(t)$ is critically damped and has a forced response of 0.4 unit. Repeat for a damping ratio of 0.707 and a forced response of 0.23 unit.

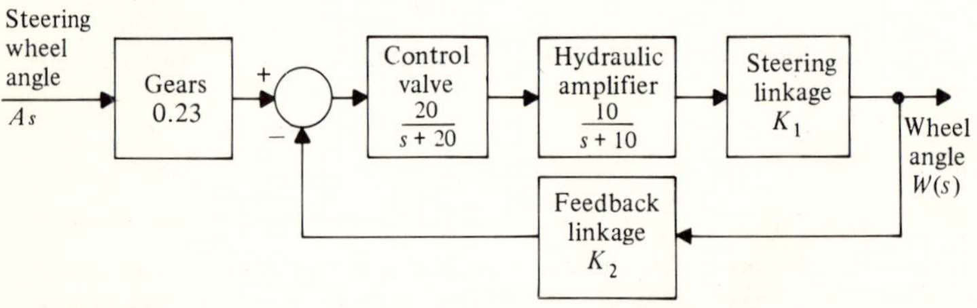

FIGURE P2-10

25. A feedback control system which is designed to maintain constant torque on a rotary shaft is modeled in Fig. P2-11. The torque sensor monitors strain on a section of the shaft, which is nearly proportional to the applied shaft torque.

For a step input of desired torque, choose the constants k_1 and k_2 in the controller, if possible, so that the system response is oscillatory with a damping ratio of 0.7 and an undamped natural frequency of 10 rad/sec.

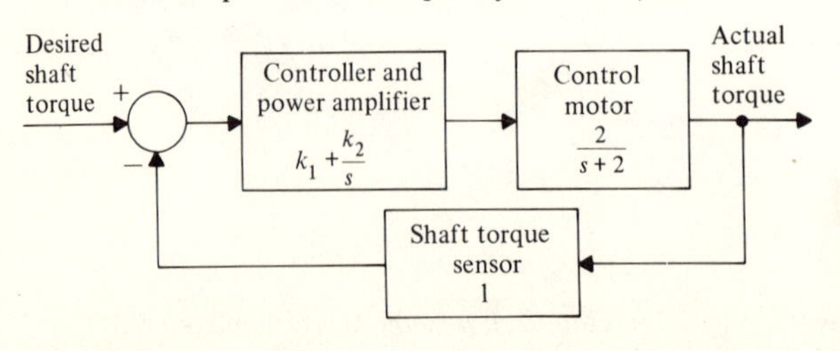

FIGURE P2-11

three

CHARACTERISTIC POLYNOMIAL STABILITY TESTING

3.1 PREVIEW

The stability of a linear, time-invariant system is determined by whether all of the roots of the system's characteristic polynomial are in the left half of the complex plane. In this chapter, it is shown that system instability is sometimes evident from just a consideration of the algebraic signs of the coefficients of its characteristic equation.

A definitive stability test is the Routh-Hurwitz test, a numerical procedure for determining the numbers of right half-plane and imaginary axis roots of a polynomial. The test does not give more specific root locations, as does factoring, but its application is far easier than factoring. Because of the importance of the test to a variety of applications, its treatment here is thorough. Methods are given by which it is always possible to determine the numbers of right half-plane (RHP), left half-plane (LHP), and imaginary axis roots of any polynomial.

Control system design generally involves an iterative procedure by which trial designs are analyzed again and again, as parameters are adjusted for desired performance. The tremendous saving in effort of Routh-Hurwitz testing over polynomial factoring, especially for high-order systems, is no less significant for computer-aided design than it is for design by hand calculation.

A system for tracking the position of a light source provides an example of the use of the coefficient tests and illustrates an iterative design in which improved system performance is obtained by increasing the complexity of the control system. An artificial limb control system demonstrates use of the Routh-Hurwitz test. Actions of the human brain and nervous system become part of the model, and mechanical system performance is aided by an additional "inner" feedback loop.

3.2 STABILITY CONCEPTS

Stability for systems in general may be defined in many different ways, depending upon the needs of a particular application. For linear, time-invariant systems, all common stability definitions are equivalent; each implies the others, and each holds if and only if all of the system's characteristic roots are in the left half of the complex plane. The most common general stability definitions are the following:

ASYMPTOTIC STABILITY. A system is asymptotically stable if for all possible initial conditions its zero-input response approaches zero with time. In a linear, time-invariant system with a rational transfer function $T(s)$, there is a zero-input response term for each denominator root of $T(s)$. A term expands or decays in time according to whether the root is in the RHP or the LHP. Hence such a system is asymptotically stable if and only if all of the characteristic denominator roots of $T(s)$ are in the LHP.

IMPULSE RESPONSE STABILITY. A system is said to be stable in the impulse response sense if its response to an impulse input approaches zero with time. For a unit impulse input to a linear, time-invariant system with transfer function $T(s)$, the Laplace transform of the output is

$$Y_{\text{impulse}}(s) = T(s) \cdot 1$$
$$y_{\text{impulse}}(t) = \mathscr{L}^{-1}[T(s)]$$

Expanding a rational transfer function $T(s)$ in partial fractions gives response terms which decay in time only if the roots of the characteristic denominator polynomial of $T(s)$ are all in the LHP.

For linear, time-varying systems, one must consider the impulse response as a function of the starting time of the impulse, because such systems behave differently at different times.

BOUNDED-INPUT, BOUNDED-OUTPUT (BIBO) STABILITY. A system is BIBO stable if, for every bounded input, its output is bounded. A necessary and sufficient condition for BIBO stability of a linear, time-invariant system with rational transfer function $T(s)$ is also that all of the characteristic denominator roots of $T(s)$ be in the LHP. This result may be established using the Laplace transform convolution relation. In convolving the system input with each exponential term arising from the partial fraction expansion of $T(s)$, it is straightforward to show that the corresponding response is always bounded only if the exponentials decay with time.

STABILITY IN THE SENSE OF LIAPUNOV. Physical systems, particularly those which are time-varying and nonlinear, may be considered to be stable provided energy is conserved in them. For their response to "blow up" would require a continual supply of energy. By considering a system's equations, whatever their origin, to represent a physical system, stability may be tested according to whether or not an equivalent physical system is conservative.

For systems which are not linear and/or not time-invariant, the above definitions give differing results. In practice, one chooses the stability definition most appropriate to the problem at hand. In linear, time-invariant systems, the subject of this text, all of these stability definitions are equivalent; such systems are stable according to any of these definitions if and only if all of the roots of the characteristic denominator polynomial of a rational transfer function $T(s)$ are in the left half of the complex plane.

3.3 COEFFICIENT TESTS FOR STABILITY

For first- and second-order systems, stability is determined by inspection of the characteristic polynomial. A first- or second-order polynomial has all roots in the left half of the complex plane if and only if all polynomial coefficients have the same algebraic sign. For example,

$$3s^2+s+10$$

is the characteristic polynomial of a stable system, while

$$3s^2+s-10$$

represents an unstable system.

For higher-order polynomials, representing higher-order systems, the algebraic signs of the polynomial coefficients may or may not yield information as to stability. A polynomial with all roots in the left half-plane (LHP) has factors of the form

$(s+a) \qquad a>0 \qquad$ real axis root in the LHP

and

$(s^2+bs+c) \qquad b>0$ and $c>0 \qquad$ two LHP roots, perhaps complex conjugate

When multiplied out, such a polynomial must have all coefficients of the same algebraic sign, all positive or all negative. No coefficient may be zero ("missing") in a system with LHP roots because there are no minus signs involved and thus no way for a coefficient to be canceled.

If imaginary axis roots exist in the polynomial, factors of the following forms may be present, in addition to the others:

(s) root at the origin

and

(s^2+a) $a>0$ complex conjugate roots on the imaginary axis

With such roots present, all polynomial coefficients must be of the same algebraic sign, but some coefficients may be zero.

Right half-plane (RHP) roots involve factors of the form

$(s-a)$ $a>0$ real axis root in the RHP

and

(s^2-as+b) $a>0$ and $b>0$ two roots in the RHP, perhaps complex conjugate

The presence of such factors may or may not cause differing algebraic signs of the coefficients and (by cancellation) zero coefficients.

Table 3-1 summarizes the information conveyed by these coefficient tests. For example, the polynomial

$$7s^6+5s^4-3s^3-2s^2+s+10$$

definitely has one or more RHP roots, indicated by the differing algebraic signs of the coefficients. Examination of the coefficient signs yields no information about root locations for the following polynomial:

$$8s^5+6s^4+3s^3+2s^2+7s+10$$

TABLE 3-1
Coefficient tests.

Properties of the Polynomial Coefficients	*Conclusion about Roots from the Coefficient Test*
All of same algebraic sign, none zero	No direct information about the roots; further testing (i.e., Routh-Hurwitz) necessary.
Differing algebraic signs, perhaps some zero coefficients also	At least one RHP root; possible imaginary axis roots also
All of same algebraic sign and one or more coefficients zero	Imaginary axis or RHP roots or both

The polynomial

$$s^6+3s^5+2s^4+8s^2+3s+17$$

has imaginary axis roots or RHP roots or both, indicated by the missing s^3 term. Imaginary axis roots in the above polynomial, if they are present, are complex conjugate since if there were an imaginary axis root at $s=0$, s would be a factor of the polynomial.

DRILL PROBLEMS

D3-1. What can be determined about the roots of the following polynomials from the coefficient tests?

(a) $-3s^4+2s^3+s+10$

ans. at least one RHP root

(b) $4s^4+3s^3+10s^2+8s+1$

ans. nothing

(c) s^5+4s^3+8

ans. imaginary axis (IA) or RHP roots or both

(d) $s^6+6s^4+3s^2+10$

ans. IA or RHP roots or both

3.4 ROUTH-HURWITZ TESTING

The Routh-Hurwitz test* is a numerical procedure to determine how many roots of a polynomial are in the RHP and how many are on the imaginary axis. It does not give specific root locations, as factoring schemes do, but performing the test is generally far easier than factoring.

Consider the polynomial

$$s^5+s^4+3s^3+9s^2+16s+10$$

as an example of Routh-Hurwitz array construction. First, form the initial part of the array from the polynomial. Write descending powers of s, starting with the highest power in the polynomial, through s^0, in a column to the left. Enter the

*Strictly speaking, Routh's array is described in this chapter. The Hurwitz criterion is an equivalent procedure involving determinants or continued-fraction expansion. The term *Routh–Hurwitz testing* is commonly used, however, in recognition of the independent contributions of both individuals.

coefficients of the polynomial

$$a_n s^n + a_{n-1} s^{n-1} + a_{n-2} s^{n-2} + a_{n-3} s^{n-3} + \cdots + a_1 s + a_0$$

in the first two rows, in the following order:

(first, a_n)	(third, a_{n-2})	(fifth)	⋮
(second, a_{n-1})	(fourth, a_{n-3})	(sixth)	⋮

s^5	1	3	16
s^4	1	9	10
s^3	◯		
s^2			
s^1			
s^0			

It is helpful to imagine the remaining elements in these first two rows to be zeros.

The Routh-Hurwitz test involves completion of the array of numbers in a certain way. The number which is to be placed in the first column, s^3 row (circled) location is related to numbers above it by

$$\frac{-\begin{vmatrix} a_n & a_{n-2} \\ a_{n-1} & a_{n-3} \end{vmatrix}}{a_{n-1}} = \frac{-\begin{vmatrix} 1 & 3 \\ 1 & 9 \end{vmatrix}}{1} = -6$$

It is the negative determinant of the indicated four numbers, divided by the first column, s^4 row number, a_{n-1}. The array is now as follows:

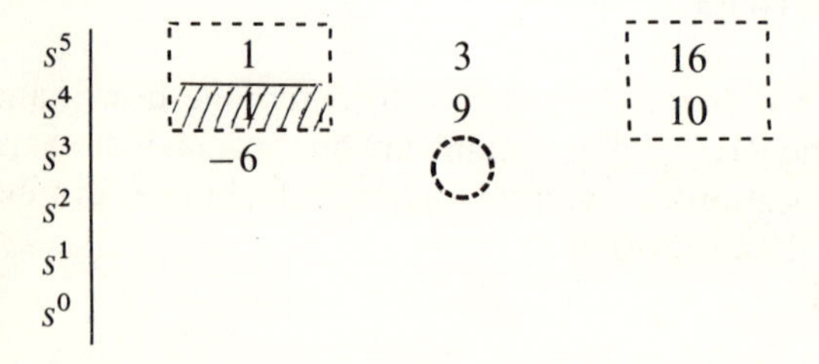

The next entry in the third row (circled above) is

$$\frac{-\begin{vmatrix} a_n & a_{n-4} \\ a_{n-1} & a_{n-5} \end{vmatrix}}{a_{n-1}} = \frac{-\begin{vmatrix} 1 & 16 \\ 1 & 10 \end{vmatrix}}{1} = 6$$

involving elements in the first and the third columns above, divided by the lower left of these elements:

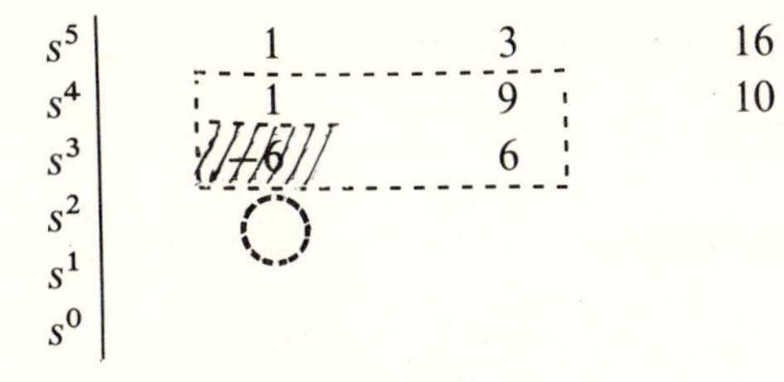

Further entries in the s^3 row would be zero.

Continuing this process for the s^2 row, the first entry (circled location above) will be

$$\frac{-\begin{vmatrix}1 & 9\\ -6 & 6\end{vmatrix}}{-6}=10$$

The next entry in the s^2 row will be

$$\frac{-\begin{vmatrix}1 & 10\\ -6 & 0\end{vmatrix}}{-6}=10$$

and further entries would be zeros:

s^5	1	3	16
s^4	1	9	10
s^3	−6	6	
s^2	10	10	
s^1			
s^0			

Continuing in this manner, the completed array is as follows:

s^5	1	3	16
s^4	1	9	10
s^3	−6	6	
s^2	10	10	
s^1	12		
s^0	10		

The number of RHP roots of the polynomial is equal to the number of algebraic sign changes in the left column of numbers, going from top to bottom. For the example polynomial, the number of left column sign changes is two, from 1 to -6 and from -6 to 10, so this polynomial has two RHP roots.

As another example, a system with transfer function

$$T(s)=\frac{5s^2-7s+2}{s^4+2s^3+3s^2+4s+1}$$

has characteristic polynomial

$$s^4+2s^3+3s^2+4s+1$$

which has the Routh–Hurwitz array below. There are no algebraic sign changes in the left column, so the polynomial has no RHP roots.

s^4	1	3	①
s^3	2	4	
s^2	1	①	
s^1	2		
s^0	①		

As one gains practice in completing Routh-Hurwitz arrays, it is easier to form the negative of the determinant by simply evaluating the difference of element products in reverse order.

Each array has properties which serve as a partial check on its correct completion. The number of nonzero row entries is normally reduced by one every two rows, with just one nonzero element in the s^1 row and in the s^0 row. And the last coefficient of the polynomial appears periodically as the last nonzero entry in every other row. The labor involved in hand calculation of the array may also be reduced with the aid of the following rule:

The elements of any row may be multiplied by any positive number without changing the test result.

DRILL PROBLEM

D3-2. How many roots of each of the following polynomials are in the right half of the complex plane?

(a) s^3+2s^2+3s+4

ans. 0

(b) $s^4-6s^3+7s^2+2s+4$

ans. 2

(c) $0.3s^4+1.1s^3+0.7s^2+s+2.1$

ans. 2

(d) $s^5 + s^4 + 2s^3 + 3s^2 + (1/2)$

ans. 2

(e) $2s^5 + s^4 + 2s^3 + 4s^2 + s + 6$

ans. 2

3.5 LEFT COLUMN ZEROS OF THE ARRAY

Consider the polynomial

$$s^4 + s^3 + 2s^2 + 2s + 3$$

A snag develops in the Routh-Hurwitz test:

s^4	1	2	3
s^3	1	2	
s^2	0	3	
s^1			
s^0			

The first s^1 row entry will be

$$\frac{-\begin{vmatrix} 1 & 2 \\ 0 & 3 \end{vmatrix}}{0}$$

which involves division by zero. This situation, where a zero occurs in the left column of the array, but where the entire row does not consist of zeros, is referred to as a *left column zero*.

The left column zero situation can be resolved through realizing that the zero is the result of the polynomial coefficients being *exactly* certain numbers. If the coefficients were only slightly different, the left column zero would instead be a small positive or negative number. Imagine that the polynomial coefficients have been altered very slightly, so that they are not the exact values which give the left column zero; instead they give some tiny nonzero number ε. ε may be considered to be either positive or negative, but it is usually easier to imagine it to be positive.

For the example, replacing the left column zero by ε and completing the remaining array entries in terms of ε gives the following:

s^4	1	2	3
s^3	1	2	
s^2	ε	3	
s^1	$\dfrac{2\varepsilon - 3}{\varepsilon}$		
s^0	3		

*In the limit as $\varepsilon \to 0$, using positive ε for convenience, the left column entries become as shown below, since

$$\lim_{\varepsilon \to 0} 2 - \frac{3}{\varepsilon} = -\infty$$

s^4	1
s^3	1
s^2	0^+
s^1	$-\infty$
s^0	3

* As ε goes to zero, any entries in the left column which approach zero are taken to be positive for the purpose of counting sign changes, if ε is taken to be positive. There are two algebraic sign changes in the left column as ε approaches zero in the example, so this polynomial has two RHP roots.

Should ε instead be taken to be negative, the limit will give the same number of left column sign changes (although not necessarily at the same locations in the array) provided that the polynomial has no imaginary axis roots. If imaginary axis roots are present, different perturbations of the polynomial coefficients may cause different numbers of RHP roots to be detected since a small polynomial coefficient change may move imaginary axis roots into either the LHP or the RHP, depending upon the nature of the perturbation.

When there are imaginary axis roots in the polynomial, a special circumstance termed *premature termination*, discussed in the next section, occurs. Except in this case, the ε method, with positive ε, may be safely used.

Another example of a Routh-Hurwitz array with a left column zero is shown below:

s^5	1	2	1
s^4	1	3	4
s^3	-1	-3	
s^2	ε	4	
s^1	$\dfrac{-3\varepsilon+4}{\varepsilon}$		
s^0	4		

$\rightarrow$

s^5	1
s^4	1
s^3	-1
s^2	0
s^1	$+\infty$
s^0	4

* An alternative method for circumventing left column zeros is to introduce additional, known roots to the polynomial, increasing its order but thereby changing the coefficients so that the left column zero does not occur. This method is especially useful in digital computer programming of the test. For example, the polynomial

$$s^4 + s^3 + 2s^2 + 2s + 3$$

was found previously to have a left column zero in the array. Incorporating an additional LHP root gives the polynomial

$$(s^4+s^3+2s^2+2s+3)(s+1)=s^5+2s^4+3s^3+4s^2+5s+3$$

This augmented polynomial's Routh-Hurwitz test does not involve a left column zero:

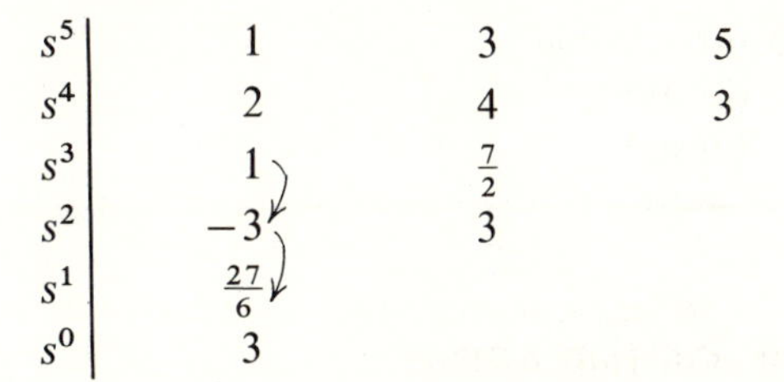

s^5	1	3	5
s^4	2	4	3
s^3	1	$\frac{7}{2}$	
s^2	-3	3	
s^1	$\frac{27}{6}$		
s^0	3		

It has two RHP roots, so the original polynomial must have two RHP roots also.

DRILL PROBLEMS

D3-3. How many roots of each of the following polynomials are in the right half of the complex plane?

(a) s^3+2s+3

ans. 2

(b) $3s^4+6s^3+2s^2+4s+5$

ans. 2

(c) $2s^4+2s^3+s^2+s-3$

ans. 1

(d) $s^5+s^4+3s^3+2s^2+4s+2$

ans. 2

D3-4. Are the following systems stable or unstable?

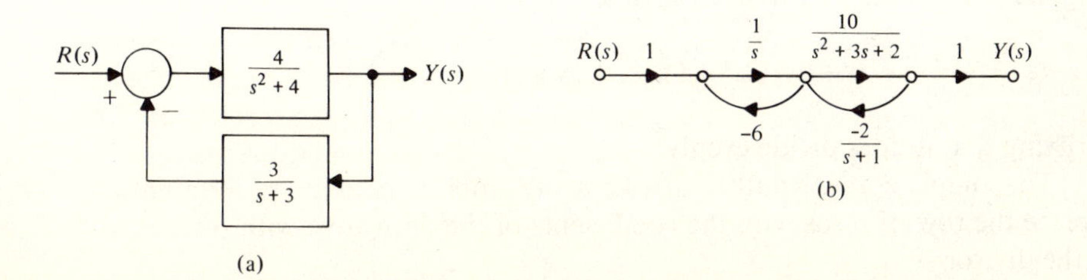

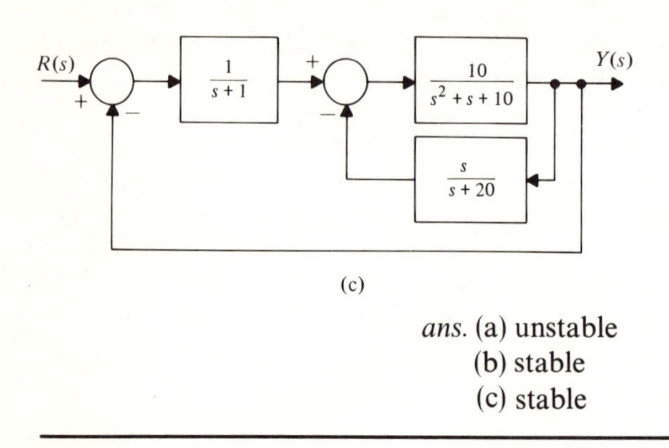

(c)

ans. (a) unstable
(b) stable
(c) stable

3.6 PREMATURE TERMINATION OF THE ARRAY

The polynomial

$$s^4+s^3+5s^2+3s+6$$

has a Routh-Hurwitz array which begins as follows:

s^4	1	5	6
s^3	1	3	
s^2	2	6	
s^1	0	0	
s^0	◯		

The s^1 row consists solely of zeros, so the circled entry would otherwise be indeterminant. When a row of zeros occurs, there exists an even or odd polynomial divisor of the original polynomial. The coefficients of this divisor polynomial are given by the previous nonzero row of the array. For the example polynomial, the coefficients in the s^2 row are 2 and 6, so the divisor is

$$2s^2+6$$

This divisor represents a conjugate pair of imaginary axis roots. Dividing the original polynomial by this even divisor gives

$$s^4+s^3+5s^2+3s+6=(2s^2+6)(\tfrac{1}{2}s^2+\tfrac{1}{2}s+1)$$

verifying that it does divide evenly.

To complete the Routh-Hurwitz array after a premature termination, replace the row of zeros with the coefficients of the derivative with respect to s of the divisor:

$$\frac{d}{ds}(2s^2+6)=4s$$

s^4	1	5	6
s^3	1	3	
s^2	2	6	
s^1	4		
s^0	6		

The example polynomial has no left column sign changes, and thus no RHP roots. It does have two imaginary axis roots, in the divisor.

DRILL PROBLEM

D3-5. For each of the following polynomials, complete the Routh-Hurwitz array and determine the number of roots in the right half of the complex plane:

(a) s^4+8s^2-7

ans. 1

(b) $s^4+2s^3+9s^2+4s+14$

ans. 0

(c) s^5+s^3+2s

ans. 2

(d) $s^5+3s^4+4s^3+7s^2+4s+2$

ans. 0

3.7 IMAGINARY AXIS ROOTS

An odd polynomial always has s as a factor. Whether or not s is a factor is evident from a glance at the polynomial, and the remainder is an even polynomial. For any even polynomial, replacing s by $(-s)$ leaves the polynomial unchanged. Thus the roots of an even polynomial, besides occurring in conjugate pairs, are also symmetrical about the imaginary axis.

There are three basic types of factors possible in an even polynomial. One type is

$$(s+ja)(s-ja)=(s^2+a^2)$$

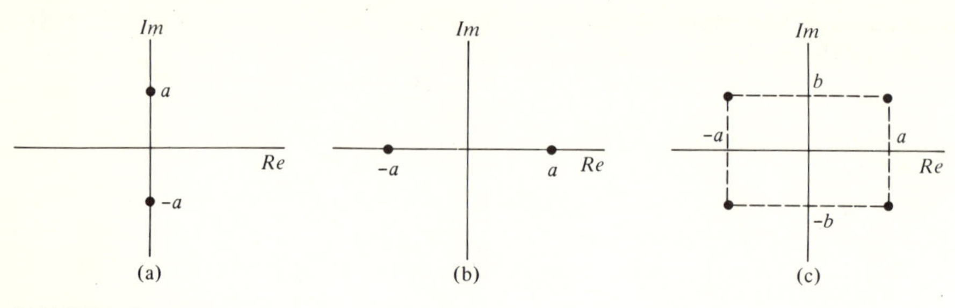

FIGURE 3-1
Even polynomial root locations.

which consists of complex conjugate roots on the imaginary axis, as indicated in Fig. 3-1a. Another is

$$(s+a)(s-a)=(s^2-a^2)$$

which are symmetrical roots on the real axis, one in the LHP and one in the RHP, as in Fig. 3-1b. The third type of factor, Fig. 3-1c, involves complex roots in quadrature, one pair in the LHP and one pair in the RHP:

$$(s+a+jb)(s+a-jb)(s-a+jb)(s-a-jb)=s^4+2(b^2-a^2)s^2+(a^2+b^2)^2$$

The additional symmetry of even polynomial roots about the imaginary axis allows determination of the number of imaginary axis roots. Each RHP root of an even polynomial must be matched by just one corresponding LHP root. Thus if an even polynomial is of sixth order and is known to have just one RHP root, it has just one LHP root, and the remaining four roots must be on the imaginary axis. If an even eighth-order polynomial has three RHP roots, it must have three LHP roots and two imaginary axis roots.

For the polynomial

$$s^6+s^5+5s^4+s^3+2s^2-2s-8$$

the coefficient tests indicate the presence of at least one RHP root. The Routh-Hurwitz test begins as follows:

s^6	1	5	2	−8
s^5	1	1	−2	
s^4	4	4	−8	
s^3	0	0		

So

$$4s^4+4s^2-8$$

is a divisor of the original polynomial. Replacing the row of zeros by the coefficients of the derivative of the divisor polynomial,

$$\frac{d}{ds}(4s^4+4s^2-8)=16s^3+8s$$

the completed array is as follows:

s^6	1	5	2	−8
s^5	1	1	−2	
s^4	4	4	−8	
s^3	16	8		
s^2	2	−8		
s^1	72			
s^0	−8			

The test of the even polynomial divisor is that portion of the array below the dashed line. There is one left column sign change in that portion, so the divisor has one RHP root, and thus one LHP root. The other two divisor roots must then be on the imaginary axis. When a polynomial has imaginary axis roots, the Routh-Hurwitz array will *definitely* contain a row of zeros, with all such roots contained within the divisor polynomial.

Since the entire array has one left column sign change, the entire polynomial has one RHP root (in the even divisor). There are exactly two imaginary roots because if there are any, they must be in the divisor. The remaining roots must then be in the LHP. It is concluded that the example polynomial has the following numbers of the various types of roots:

RHP = 1

LHP = 3

Imaginary axis = 2

This information can be found from the Routh-Hurwitz test for any polynomial.

Another array with a premature termination is the following:

s^4	1	9	20
s^3	6	24	
s^2	5	20	
s^1	10		
s^0	20		

For this polynomial, there are no RHP roots. The second-order even polynomial divisor has no RHP roots and thus no LHP roots. Hence there must be two roots on the imaginary axis. The remaining two roots are not on the imaginary axis; all such roots must be in the divisor. They can only be in the LHP.

The polynomial

$$s^5+s^4+6s^3+6s^2+25s+25$$

prematurely terminates at the s^3 row:

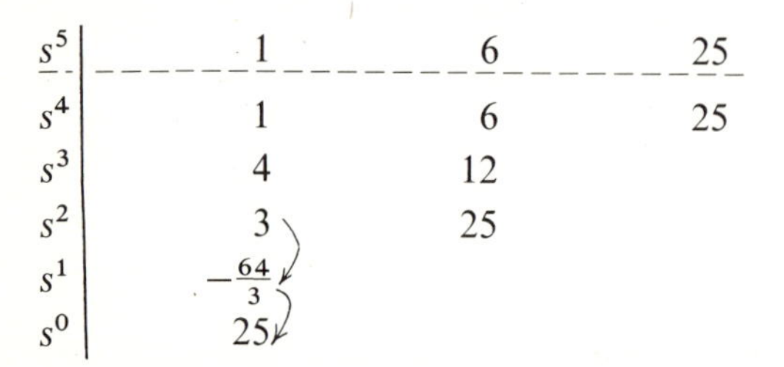

s^5	1	6	25
s^4	1	6	25
s^3	4	12	
s^2	3	25	
s^1	$-\frac{64}{3}$		
s^0	25		

The fourth-order even polynomial divisor has two RHP roots, and so its remaining two roots must be in the LHP. There can thus be no imaginary axis roots. The entire polynomial has two RHP roots, and the remaining three roots must be in the LHP.

In very special circumstances, a Routh-Hurwitz array will prematurely terminate more than once in the course of completing the array. Replacement of additional rows of zeros by the derivative of the immediately preceding divisor polynomial will allow completion of the array. Multiple premature terminations indicate the presence of repeated roots in the polynomial divisor.

If an array should involve both a left column zero and a premature termination, care should be taken. The premature termination indicates the possible presence of imaginary axis polynomial roots, for which the ε method of handling left column zeros may not give correct results. Perturbing the coefficients of a polynomial with imaginary axis roots may send those roots into either the RHP or the LHP, depending upon the nature of the perturbation. The simplest solution in such a situation is to introduce an additional known root to the polynomial, increasing its order but eliminating the left column zero.

DRILL PROBLEM

D3-6. For each of the following polynomials, how many roots are in the LHP, how many are in the RHP, and how many are on the imaginary axis?

(a) s^4+3s^2+4

ans. 2RHP, 2LHP

(b) $s^4+2s^3+5s^2-4s-14$

ans. 1RHP, 3LHP

(c) $s^5+2s^4+3s^3+6s^2+2s+4$

ans. 1LHP, 4IA

(d) $3s^5+2s^3+s$

ans. 2RHP, 2LHP, 1IA

(e) $2s^5+4s^4+s^3+2s^2+3s+6$

ans. 3LHP, 2RHP

3.8 ADJUSTABLE SYSTEMS

It is often desired to know for what range or ranges of an adjustable parameter K a system is stable. For example, suppose the transfer function of a system is, in terms of K,

$$T(s)=\frac{4}{s^2+2s+K}$$

This system is stable for all positive values of K because the roots of a quadratic are in the LHP if and only if all coefficients have the same algebraic sign. For $K=0$,

$$T(s)=\frac{4}{s(s+2)}$$

and the system is marginally stable because of the imaginary axis characteristic root at $s=0$. For negative values of K, the transfer function has characteristic roots in the RHP and the system is unstable.

The system with transfer function

$$T(s)=\frac{2s+K}{s^2+(2+K)s+4}$$

is, similarly, stable for

$$K>-2$$

For systems with characteristic polynomials of higher order, the Routh-Hurwitz test is a useful tool for determining stability in terms of a constant but adjustable parameter. Suppose

$$s^4+2s^3+4s^2+2s+K$$

is the denominator polynomial of a system transfer function, in terms of an adjustable constant K. The Routh-Hurwitz test in terms of K is as follows:

$$\begin{array}{c|ccc} s^4 & 1 & 4 & K \\ s^3 & 2 & 2 & \\ s^2 & 3 & K & \\ s^1 & \dfrac{6-2K}{3} & & \\ s^0 & K & & \end{array}$$

All of the left column array entries must be of the same algebraic sign if the polynomial is to have no RHP roots. Thus

$$\frac{6-2K}{3}>0$$

and

$$K>0$$

or

$$0<K<3$$

for system stability.

For $K=3$, the array contains a row of zeros, $3s^2+3$ is a factor of the polynomial, and the system is marginally stable. For $K=0$, s is a factor of the polynomial and, again, the system is marginally stable.

For the adjustable polynomial

$$s^4+2s^3+4s^2+Ks+6$$

the Routh-Hurwitz array is the following:

$$\begin{array}{c|ccc} s^4 & 1 & 4 & 6 \\ s^3 & 2 & K & \\ s^2 & 4-\dfrac{K}{2} & 6 & \\ s^1 & K-\dfrac{12}{4-K/2} & & \\ s^0 & 6 & & \end{array}$$

If the polynomial is to have all LHP roots,

$$4-\frac{K}{2}>0 \qquad K<8$$

and

$$K - \frac{12}{4 - K/2} > 0$$

Making use of the requirement that $4 - K/2$ be positive, from the first inequality, the second inequality may be multiplied by the positive number $4 - K/2$. If the inequality were multiplied by a negative number, its sense would be reversed. Then

$$\left(4 - \frac{K}{2}\right)\left(K - \frac{12}{4 - K/2}\right) = -\frac{1}{2}K^2 + 4K - 12 > 0$$

The quadratic function

$$-\tfrac{1}{2}K^2 + 4K - 12$$

is negative for large negative K and is negative for large positive K. To determine if there are intermediate values of K for which the function is positive, its roots are found using the quadratic formula:

$$K = \frac{-4 \pm \sqrt{16 - 24}}{2(-1)}$$

As the function's roots are complex, it is concluded that the inequalities cannot be satisfied for any (real) K. The original polynomial thus has RHP roots for all K.

DRILL PROBLEMS

D3-7. For what range(s), if any, of the adjustable constant K are all roots of the following polynomials in the left half of the complex plane?

(a) $s^3 + (2 + K)s^2 + (8 + K)s + 6$

ans. $-1.12 < K$

(b) $2s^3 + (6 - 2K)s^2 + (4 + 3K)s + 10$

ans. $-\frac{1}{3} < K < 2$

(c) $s^4 + (10 + K)s^3 + 3s^2 + 9s + 11$

ans. no value of K

(d) $s^4 + s^3 + 3s^2 + 2s + 4 + K$

ans. $-4 < K < -2$

D3-8. Find the range(s) of positive, constant K, if any, for which the following systems are stable.

ans. (a) $0 < K < \frac{32}{9}$
(b) $\frac{4}{3} < K$

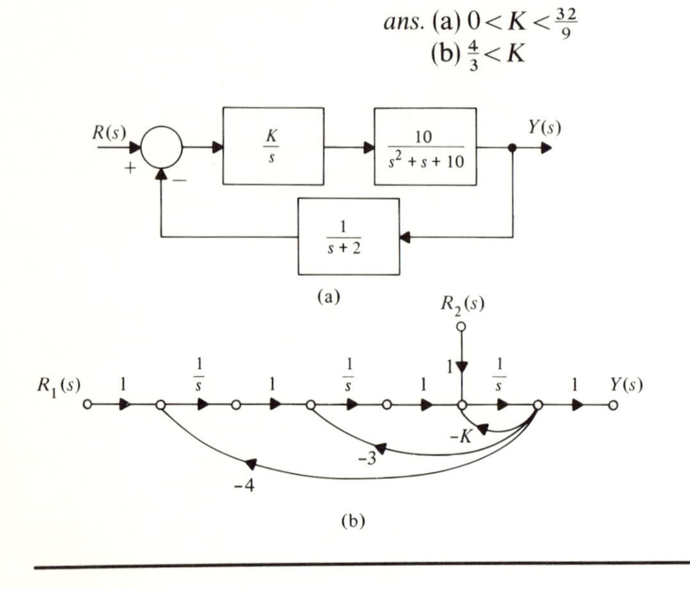

3.9 TESTING RELATIVE STABILITY

A system with all characteristic roots in the LHP but with one or more roots only slightly to the left of the imaginary axis has a natural response which decays very slowly. The longer the distance from the imaginary axis to the nearest characteristic root of a stable system, the faster the slowest decaying term in the system's natural response dies out. The distance on the complex plane between the nearest characteristic root and the imaginary axis is termed the *relative stability* of the system. Normally, the relative stability concept is used only in connection with stable systems.

For the characteristic polynomial

$$s^4 + 14s^3 + 73s^2 + 168s + 144$$

a Routh-Hurwitz test shows all roots to be in the LHP:

s^4	1	73	144
s^3	14	168	
s^2	61	144	
s^1	$\frac{8232}{61}$		
s^0	144		

Shifting the imaginary axis 2 units to the left, as indicated in Fig. 3-2, is

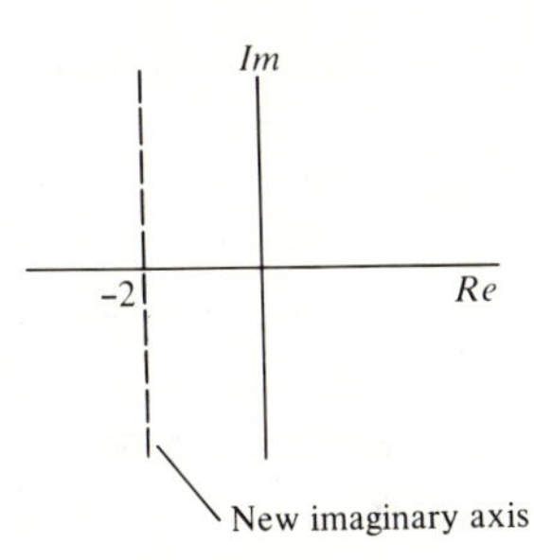

FIGURE 3-2
Axis shift on the complex plane.

accomplished by substituting $\sigma - 2$ for each s in the original polynomial:

$$\begin{aligned}
&(\sigma-2)^4 + 14(\sigma-2)^3 + 73(\sigma-2)^2 + 168(\sigma-2) + 144 \\
&\quad = (\sigma^4 - 8\sigma^3 + 24\sigma^2 - 32\sigma + 16) + 14(\sigma^3 - 6\sigma^2 + 12\sigma - 8) \\
&\qquad + 73(\sigma^2 - 4\sigma + 4) + 168\sigma - 336 + 144 \\
&\quad = \sigma^4 + 6\sigma^3 + 79\sigma^2 + 12\sigma + 4
\end{aligned}$$

A Routh-Hurwitz test on the polynomial with shifted axis is given below. There are no RHP roots of the shifted polynomial in σ, so the original polynomial has all roots to the left of $s = -2$. The system thus has a relative stability of at least 2 units:

$$\begin{array}{c|ccc}
\sigma^4 & 1 & 79 & 4 \\
\sigma^3 & 6 & 12 & \\
\sigma^2 & 77 & 4 & \\
\sigma^1 & \frac{900}{77} & & \\
\sigma^0 & 4 & &
\end{array}$$

The characteristic polynomial

$$s^4 + 2s^3 + 3s^2 + s + 1$$

has all roots in the LHP, as may be verified by a Routh-Hurwitz test:

$$\begin{array}{c|ccc}
s^4 & 1 & 3 & 1 \\
s^3 & 2 & 1 & \\
s^2 & \frac{5}{2} & 1 & \\
s^1 & \frac{1}{5} & & \\
s^0 & 1 & &
\end{array}$$

Suppose that it is desired to determine if all of the roots of this polynomial are to the left of $s=-1$ on the complex plane. Substituting $\sigma-1$ for each s in the polynomial gives

$$(\sigma-1)^4+2(\sigma-1)^3+3(\sigma-1)^2+(\sigma-1)+1=\sigma^4-2\sigma^3+3\sigma^2-3\sigma+2$$

The σ-polynomial is the s-polynomial with the imaginary axis shifted 1 unit to the left. The coefficient test shows the σ-polynomial to have RHP roots; therefore, there are roots of the s-polynomial to the right of $s=-1$.

A Routh-Hurwitz test of the σ-polynomial is as follows:

$$\begin{array}{c|ccc}
\sigma^4 & 1 & 3 & 2 \\
\sigma^3 & -2 & -3 & \\
\sigma^2 & \frac{3}{2} & 2 & \\
\sigma^1 & \frac{7}{3} & & \\
\sigma^0 & 2 & &
\end{array}$$

The original polynomial has no RHP roots and two roots, likely complex conjugate, to the right of $s=-1$.

Shifting the imaginary axis, instead, $\frac{1}{2}$ unit to the left on the complex plane gives

$$\begin{aligned}
&(\sigma-\tfrac{1}{2})^4+2(\sigma-\tfrac{1}{2})^3+3(\sigma-\tfrac{1}{2})^2+(\sigma-\tfrac{1}{2})+1 \\
&\quad=(\sigma^4-2\sigma^3+1.5\sigma^2-0.5\sigma+0.0625)+2(\sigma^3-1.5\sigma^2+0.75\sigma-0.125) \\
&\qquad+3(\sigma^2-\sigma+0.25)+\sigma-0.5+1 \\
&\quad=\sigma^4+3\sigma^2-\sigma+1.0625
\end{aligned}$$

which has the following Routh-Hurwitz test:

$$\begin{array}{c|ccc|c}
\sigma^4 & 1 & 3 & 1.0625 & 1 \\
\sigma^3 & \varepsilon & -1 & & 0^+ \\
\sigma^2 & \dfrac{3\varepsilon+1}{\varepsilon} & 1.0625 & \rightarrow & +\infty \\
\sigma^1 & -1-\dfrac{1.0625\varepsilon^2}{3\varepsilon+1} & & & -1 \\
\sigma^0 & 1.0625 & & & 1.0625
\end{array}$$

The original polynomial thus must have two roots to the left of $s=-1$, no roots between $s=-1$ and $s=-\frac{1}{2}$, and two roots between $s=-\frac{1}{2}$ and $s=0$.

Repeated shifts of the imaginary axis together with Routh-Hurwitz testing offer a powerful method for locating the real parts of the roots of a polynomial to any desired degree of accuracy. Some digital computer factoring programs use this method because of its speed and convergence properties.

DRILL PROBLEMS

D3-9. Determine the number of units of relative stability of the systems with the following transfer functions:

(a) $$T(s)=\frac{20}{(s+2)^2(s^2+5s+12)}$$

ans. 2

(b) $$T(s)=\frac{4(s+2)(s^2+9)^2}{(s+3)(s^2+s+8)^2}$$

ans. $\frac{1}{2}$

(c) $$T(s)=\frac{s^4}{(s+1)(s^2+2s+4)(s^2+2s+7)}$$

ans. 1

(d) $$T(s)=\frac{-3s^3+3s^2+s+7}{s(s^2+s+10)(s^2+3s+8)(s+4)(s+5)}$$

ans. 0

D3-10. Find the number of units of relative stability of each of the following systems:

ans. (a) 11.4
(b) 1.21

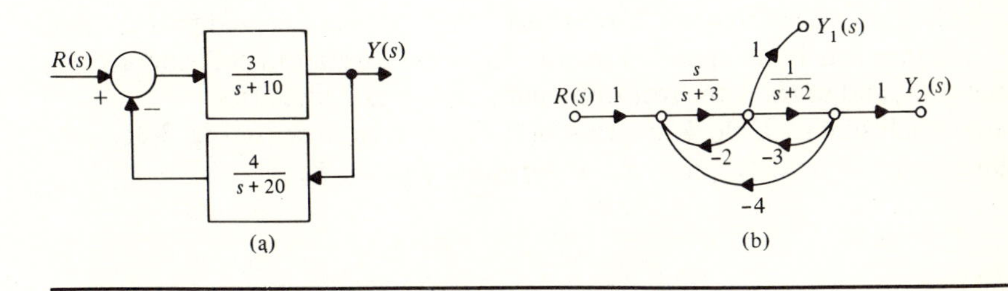

3.10 A LIGHT SOURCE TRACKING SYSTEM

This example system is designed to follow, in one dimension, a moving light source. As pictured in Fig. 3-3a, when equal light intensities are detected by the two photodiodes, the electrical bridge is balanced, and zero voltage is applied to the drive motor. When one photodiode receives more light than the other, the bridge is unbalanced, and a nonzero voltage is amplified and applied to the drive motor which then moves the photodiodes toward the equal-light-intensity position.

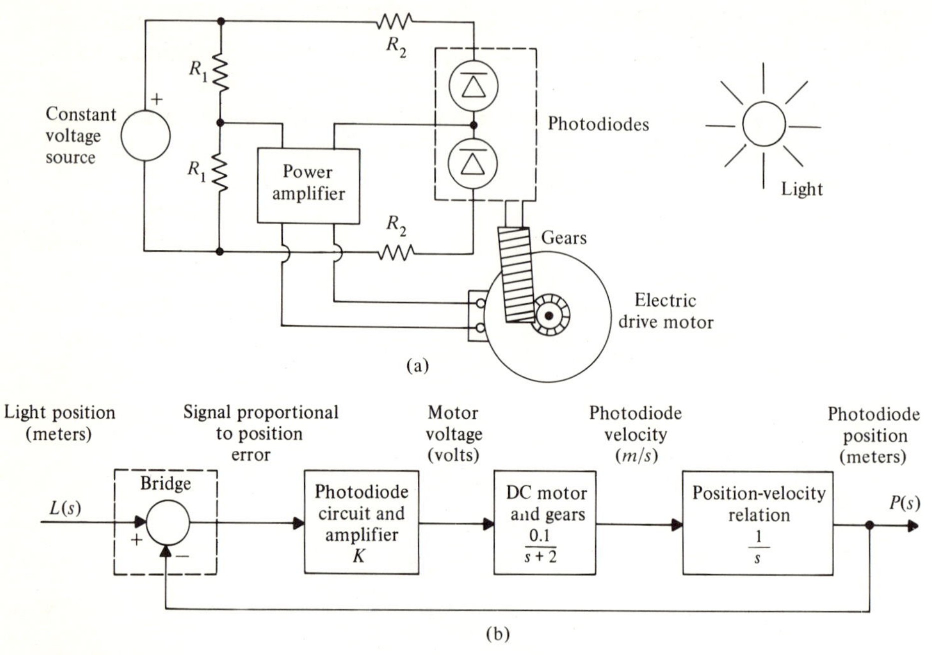

FIGURE 3-3
A light source tracking system. (a) Physical arrangement. (b) Block diagram model.

Similar systems are used for precision machine tool alignment, where the light is reflected from a calibrated scale or the light is transmitted through a tiny hole in the tool or the work. Variations on this system are used to track the sun or another star in navigation systems, to follow aircraft in collision avoidance systems, and to track the recording path on optical videodiscs.

For small signals, a block diagram of the system is shown in Fig. 3-3b. The system transfer function is, in terms of the gain constant K,

$$T(s)=\frac{\dfrac{0.1K}{s(s+2)}}{1+\dfrac{0.1K}{s(s+2)}}=\frac{0.1K}{s^2+2s+0.1K}$$

which is stable for all $K>0$.

A relative stability of 2 units for a system means that the natural component of system response decays with time as e^{-2t}, that is, with a $\frac{1}{2}$-sec time constant. This degree of stability cannot be achieved with the example system. Shifting the imaginary axis to the left 2 units by replacing s by $\sigma-2$ in the characteristic polynomial gives

$$\begin{aligned}s^2+2s+0.1K&=(\sigma-2)^2+2(\sigma-2)+0.1K\\&=\sigma^2-2\sigma+0.1K\end{aligned}$$

The shifted polynomial, having differing algebraic signs for its coefficients, indicates that at least one root of the original characteristic equation is to the right of $s=-2$ on the complex plane, regardless of the value of K.

System response to a unit step change in light position, for various representative values of K, are shown in Fig. 3-4. For each, the settling time is relatively long, in consequence of the small degree of relative stability.

The performance of this system may be improved substantially by the addition of velocity feedback as well as the position feedback. A tachometer coupled to the drive motor shaft will produce a voltage nearly proportional to the motor speed, which in turn is proportional to the photodiode velocity. Adding a fraction of this voltage to the bridge voltage (which is amplified to drive the motor) results in the block diagram of Fig. 3-5a. Using Mason's gain rule on the system's signal flow graph in Fig. 3-5b,

$$T'(s)=\frac{\left(\frac{0.1K}{s+2}\right)\left(\frac{1}{s}\right)}{1-(-4K')\left(\frac{0.1K}{s+2}\right)-(-1)\left(\frac{0.1K}{s+2}\right)\left(\frac{1}{s}\right)}$$

$$=\frac{0.1K}{s^2+(2+0.4KK')s+0.1K}$$

Here it is seen that both coefficients of the characteristic polynomial may be chosen at will by the designer by selecting appropriate values of K and K'.

For example, a characteristic polynomial

$$(s+5)^2=s^2+10s+25=s^2+(2+0.4KK')s+0.1K$$

may be achieved with

$$K=250$$
$$K'=0.08$$

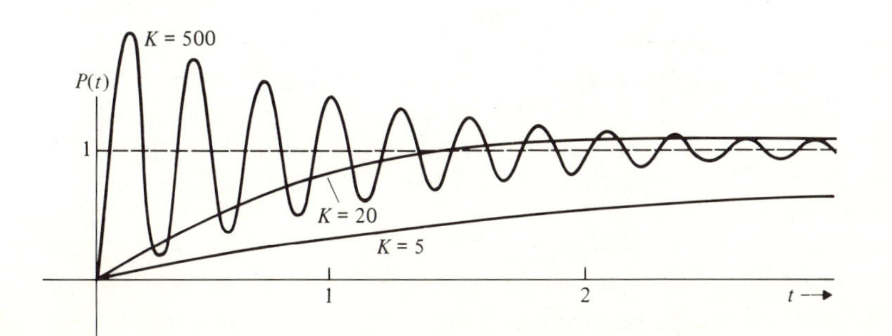

FIGURE 3-4
Unit-step response of the light source tracking system for various values of gain K.

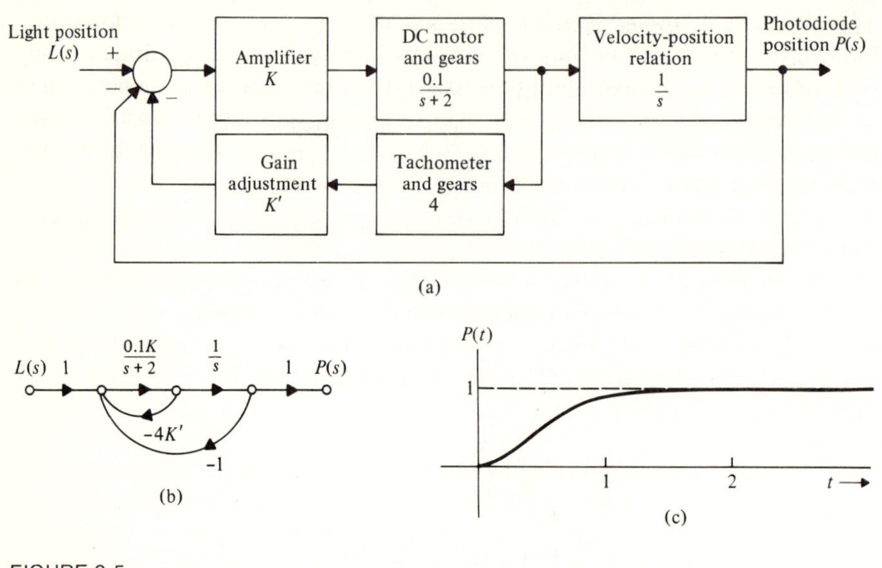

FIGURE 3-5
Light source tracking system with rate compensation. (a) Block diagram. (b) Signal flow graph. (c) Unit step response for $K = 250$, $K' = 0.08$.

With these values of K and K', the system's step response is critically damped and has a relative stability of 5 units. Its unit step response is shown in Fig. 3-5c.

3.11 AN ARTIFICIAL LIMB

Development of prosthetics (artificial limbs) has paralleled wars and natural calamities. Until well into the 1960s all but a very few prosthetics were non-active devices connected to what remained of the limb. Lower arm devices required the user to have an elbow joint, and lower leg devices required a usable knee joint.

In the early 1960s, a large number of birth defects were associated with use of the drug thalidomide by pregnant women. Among the birth defects were jointless limbs—for example, an upper arm but no elbow or lower arm. These highly publicized birth defects renewed interest in an active prosthetic such as an automatic lower arm which would react to signals from the muscles of the upper arm. The "Boston Arm" was an early such artificial limb which involved a torque motor and velocity feedback.

The human body has a natural time delay between decisions of the brain and reception of signals by the muscles, so rational limits must be placed upon the capabilities of an automatic artificial limb. If the device is too strong and sensitive, it could operate too rapidly for the brain and body to control, making the human-machine team a bionic pretzel. On the other hand, too little sensitivity could result in a bionic statue. Between the two extremes lie useful designs.

Figure 3-6a shows a block diagram for a bionic arm in a closed-loop system with the body. For simplicity, motion is considered in one direction only. The brain monitors the desired position and the sensed position, generating an error signal to the nervous system. Special sensors pick up the electric muscle impulses (myoelectric signals), and an amplifier produces a voltage which drives a DC control motor. The motor circuit involves tachometer feedback, as shown. The output of the motor circuit is the velocity of the limb in one direction which, when integrated, is the limb position.

A simplified model for the system is given in Fig. 3-6b. The action of the brain is approximated by the transmittance

$$G_B(s) = 1 + \frac{0.1}{s}$$

which involves consideration of both the position error and its integral. The nervous system is modeled by the first-order system with transmittance

$$G_N(s) = \frac{(1/T)}{s + (1/T)} = \frac{4}{s+4}$$

where $T = \frac{1}{4}$, the time constant, is approximately the neuromuscular delay time.

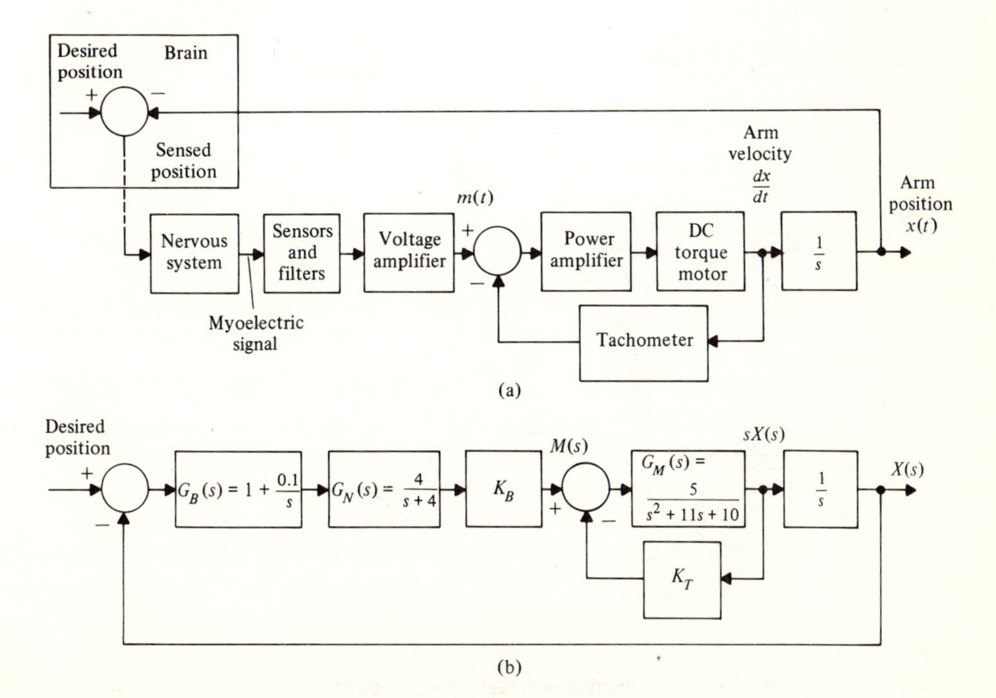

FIGURE 3-6
Prosthetic control system model. (a) Prosthetic and human control system. (b) Simplified model for the system.

The myoelectric signal is sensed and amplified with a gain K_B to form the amplified voltage $m(t)$.

The power amplifier, control motor, and mechanical load have a second-order transmittance, relating motor control voltage $m(t)$ to arm velocity, modeled as

$$G_M(s) = \frac{5}{s^2 + 11s + 10} = \frac{5}{(s+1)(s+10)}$$

This block exhibits time constants of 1 sec, associated with mechanical inertia, and $\frac{1}{10}$ sec due primarily to the motor itself. The step response of $G_M(s)$ is sketched in Fig. 3-7a.

The tachometer, with transmittance K_T, provides feedback for the motor and arm, giving the following overall transmittance of those components:

$$G_T(s) = \frac{G_M(s)}{1 + K_T G_M(s)} = \frac{\dfrac{5}{s^2 + 11s + 10}}{1 + \dfrac{5K_T}{s^2 + 11s + 10}}$$

$$= \frac{5}{s^2 + 11s + (10 + 5K_T)}$$

A characteristic equation

$$s^2 + 11s + 30.25$$

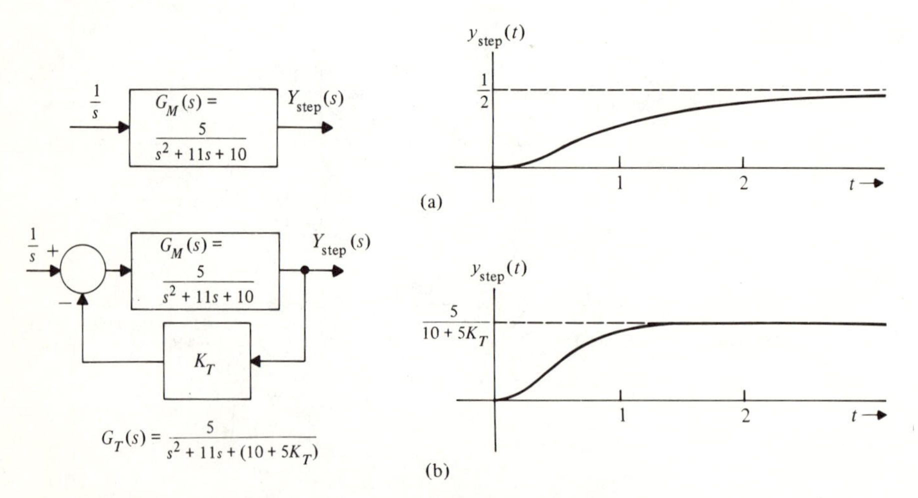

FIGURE 3-7
Improving the motor-arm performance with tachometer feedback and critical damping. (a) Step response of the motor and arm without tachometer feedback. (b) Step response with tachometer feedback.

is obtained for

$$10+5K_T=30.25 \qquad K_T=4.05$$

and gives critically damped step response as shown in Fig. 3-7b.

The entire system is redrawn in Fig. 3-8 for a tachometer gain $K_T=4.05$ which gives critical damping of the motor-arm block. The transfer function of the entire system is

$$T(s)=\frac{G_B(s)G_N(s)K_BG_T(s)(1/s)}{1+G_B(s)G_N(s)K_BG_T(s)(1/s)}$$

$$=\frac{\dfrac{20(s+0.1)K_B}{s^2(s+4)(s^2+11s+30.25)}}{1+\dfrac{20(s+0.1)K_B}{s^2(s+4)(s^2+11s+30.25)}}$$

$$=\frac{20(s+0.1)K_B}{s^5+15s^4+74s^3+121s^2+20K_Bs+2K_B}$$

A Routh-Hurwitz test for the stability of $T(s)$ is as follows:

$$\begin{array}{c|ccc} s^5 & 1 & 74 & 20K_B \\ s^4 & 15 & 121 & 2K_B \\ s^3 & 65.9 & 19.86K_B & \\ s^2 & 121-4.52K_B & 2K_B & \\ s^1 & \dfrac{2271K_B-89.76K_B^2}{121-4.52K_B} & & \\ s^0 & 2K_B & & \end{array}$$

For stability,

$$\begin{cases} 121-4.52K_B>0 \\ K_B(2271-89.76K_B)>0 \\ 2K_B>0 \end{cases}$$

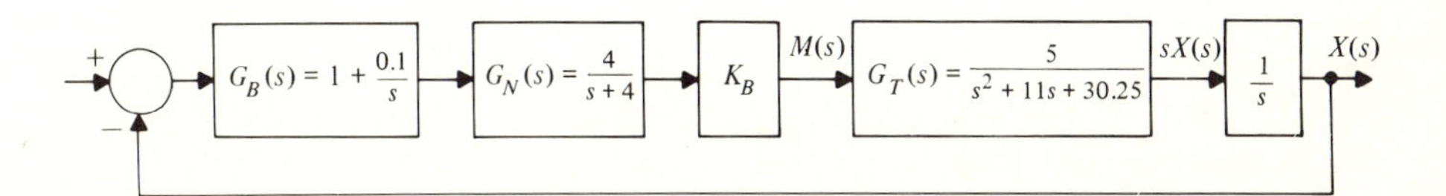

FIGURE 3-8
Block diagram for the artificial limb system where the tachometer feedback has been incorporated into the transmittance G (s).

The first of these requirements gives

$$K_B < 26.8$$

The second requirement gives, for positive K_B,

$$K_B < 25.3$$

and the third requires that

$$K_B > 0$$

Thus the system will be stable for

$$0 < K_B < 25.3$$

3.12 SUMMARY

A system is stable if and only if all of its characteristic roots are in the left half of the complex plane (LHP). First- and second-order polynomials have all roots in the LHP if and only if all polynomial coefficients have the same algebraic sign. Occasionally, the presence of RHP and imaginary roots of a higher-order polynomial may be detected by an examination of the algebraic signs of the polynomial coefficients, as is summarized in Table 3-1.

The Routh-Hurwitz test provides a convenient, definitive method of ascertaining system stability in general. It is much easier than factoring a characteristic polynomial. The number of algebraic sign changes in the left column of the Routh-Hurwitz array equals the number of RHP roots of the polynomial tested.

A left column zero situation is when there is a zero entry at the left of a row but there is at least one nonzero entry in the rest of the row. To complete an array with a left column zero, replace the zero with ε and take the limit of the left column entries as ε approaches zero. Alternatively, additional known root terms may be multiplied into the original polynomial to give a polynomial of higher order but without the left column zero.

A premature termination of a Routh-Hurwitz array is when any row, through the s^0 row, consists solely of zeros. A premature termination indicates that the tested polynomial has an even or odd polynomial divisor. The coefficients of the divisor polynomial are given by the entries in the row above the row of zeros.

To complete a prematurely terminated array, replace the row of zeros with the coefficients of the derivative with respect to s of the divisor polynomial and complete the array as usual. The completed array includes the test of the divisor polynomial which must have equal numbers of RHP and LHP roots. Any remaining roots of an even or odd divisor polynomial must be on the imaginary axis.

An adjustable polynomial has coefficients dependent upon a parameter K. The Routh-Hurwitz test may be performed in terms of K to determine the range(s) of K for which all polynomial roots are in the LHP.

To shift the imaginary axis on the complex plane a units to the left, replace the complex variable by $\sigma - a$. The Routh-Hurwitz test may be used to determine the number of roots of a polynomial which are to the right of $s = -a$ by replacing each s in the polynomial by $\sigma - a$ and testing the shifted polynomial. The relative stability of a stable system is the distance from the imaginary axis to the nearest (LHP) characteristic root.

REFERENCES

Routh-Hurwitz Testing

Clark, R. N. *Introduction to Automatic Control Systems.* New York: Wiley, 1962.

Fuller, A. T., ed. *Stability of Motion.* London: Taylor and Francis, and New York: Halsted Press, 1975.

Guillemin, E. A. *The Mathematics of Circuit Analysis.* New York: Wiley, 1949.

Hostetter, G. H. "Using the Routh-Hurwitz Test to Determine the Numbers and Multiplicities of Real Roots of a Polynomial." *IEEE Trans. Circ. Syst.*, August 22, 1975.

Hurwitz, A. "On the Conditions Under Which an Equation Has Only Roots with Negative Real Parts." *Math. Analen* 46 (1895).

Routh, E. J. *Stability of a Given State of Motion.* London: Macmillan, 1877.

———. *Dynamics of a System of Rigid Bodies.* New York: Macmillan, 1892.

Savant, C. J. Jr. *Basic Feedback Control System Design.* New York: McGraw-Hill, 1958.

Truxal, J. C. *Control System Synthesis.* New York: McGraw-Hill, 1955.

Prosthetics

Allan, R. "Electronics Aids the Disabled." *IEEE Spec.* 13, no. 11 (November 1976): 36–40.

"Brain Controls Use of Artificial Arm." *Prod. Eng.*, October 7, 1968, p. 23.

"Novel Artificial Arm Gives Wearer Six Degrees of Freedom." *Prod. Eng.*, October 20, 1969, p. 15.

PROBLEMS

1. What can be determined about each of the following polynomial's roots from the coefficient tests?
   (a) $s^3 - 2s^2 + s + 3$
   (b) $s^4 + 2s^2 + 3s + 6$

   ans. at least one imaginary axis or RHP root

   (c) $s^5 + 3s^4 + 2s^3 + s^2 + 4s + 8$
   (d) $-3s^5 + 2s^4 + 5s^3 + 6s + 12$

   ans. at least one RHP root

2. For the following polynomials, how many roots are in the RHP?
   (a) $s^3 + 4s^2 + 1$
   (b) $2s^4 + 3s^3 + 14s^2 + 2s + 6$

   ans. 0

(c) $s^4+2s^3+3s^2+4s+5$
(d) $s^5+3s^4+2s^3+s^2+2s+1$

ans. 2

(e) $s^4-s^3-2s^2+2s+3$
(f) $s^5+2s^4+s^3+2s^2+s+4$

ans. 2

(g) $s^5+2s^4+3s^3+s^2+6s+4$
(h) $2s^4+s^3+2s^2+s-4$

ans. 1

(i) $s^6+2s^5+3s^4+4s^3+5s^2+8s+1$
(j) $2s^5+8s^3+4s^2+s+2$

ans. 2

3. For the following polynomials, determine how many roots are in the LHP, how many are in the RHP, and how many are on the imaginary axis:
   (a) s^4+2s^2+10
   (b) $2s^4+2s^2+3s+4$

   ans. 2, 2, 0

   (c) $s^6+2s^4+3s^2+1$
   (d) $s^5+s^4+4s^3+3s^2+3s$

   ans. 2, 0, 3

4. Use the Routh-Hurwitz test to show that all roots of a cubic,

$$s^3+a_2s^2+a_1s+a_0$$

are in the LHP if and only if a_2, a_1, and a_0 are positive and

$$a_1a_2>a_0$$

5. Determine which of the systems of Fig. P3-1 are stable.

   ans. (b) stable

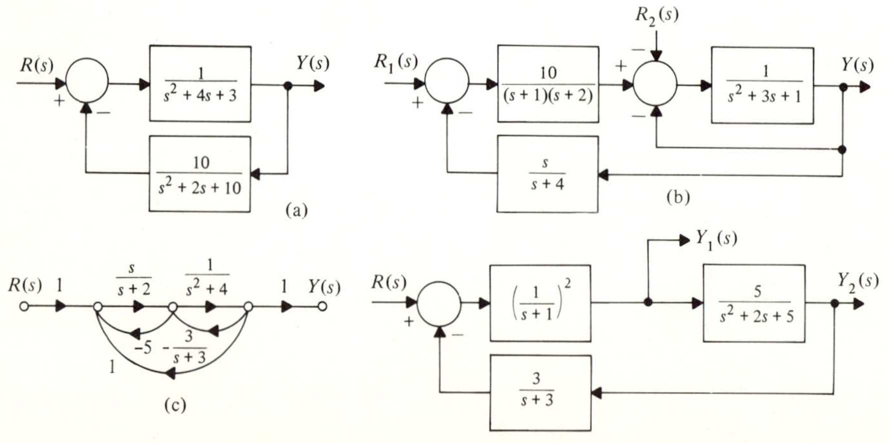

FIGURE P3-1

Test equipment is used to monitor the operation of a telephone switching center near Denver, Colorado. System errors are automatically detected then traced by this system. (Photo courtesy of American Telephone and Telegraph Company.)

6. For what range(s) of the adjustable parameter K do the following polynomials have all roots in the LHP?
   (a) Ks^3+2s^2+3s+1
   (b) s^3+Ks^2+3s+1

   ans. $K>\frac{1}{3}$

   (c) s^3+2s^2+Ks+K
7. Find the range(s) of the adjustable parameter $K>0$ for which the systems of Fig. P3-2 are stable.

   ans. (b) $K>0.49$

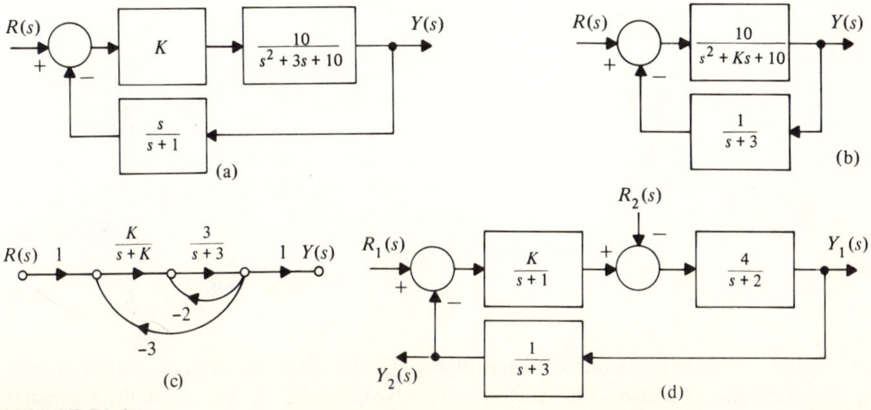

FIGURE P3-2

8. Are all roots of the polynomial

$$s^3+10s^2+30s+29$$

to the left of $s=-2$ in the complex plane?

9. For what range(s) of the adjustable constant K are all roots of the polynomial

$$s^3+Ks^2+11s+14$$

to the left of $s=-1$ in the complex plane?

10. What is the relative stability of each of the following systems?

(a) $T(s)=\dfrac{-s+6}{4s^2+2s+6}$

(b) $T(s)=\dfrac{10(s^2+4)(s^2-4)}{3(s+1)^2(s^2+s+12)}$

ans. $\frac{1}{2}$

11. Determine if the system of Fig. P3-3 has a relative stability in excess of 3 units.

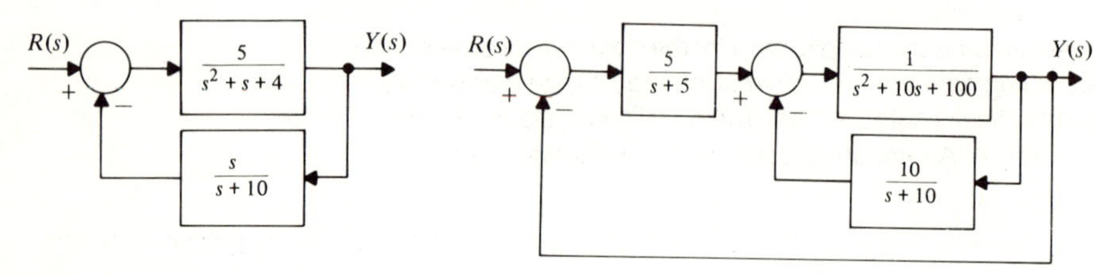

FIGURE P3-3

FIGURE P3-4

12. Determine the relative stability of the system of Fig. P3-4 to the nearest unit.

13. For what range(s) of the adjustable constant $K>0$, if any, is the relative stability of the system of Fig. P3-5 greater than 2 units?

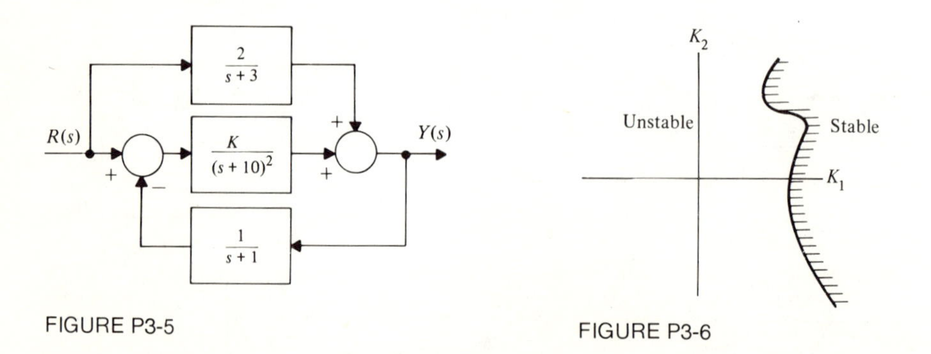

FIGURE P3-5

FIGURE P3-6

14. A plot such as the example sketch of Fig. P3-6 shows the range of values of two parameters K_1 and K_2 for which a system is stable. It is called a *stability boundary*

Automated systems in the control tower of a modern airport greatly improve efficiency and safety of the operation. (Photo courtesy of Digital Equipment Corporation.)

diagram. Draw such a diagram for a system with characteristic equation

$$s^2+(2+0.1K_1)s+(K_1+K_2)=0$$

15. For the light source tracking system, suppose an electrical compensator network is added instead of the tachometer, as in Fig. P3-7. Find gain constants K_1 and K_2, if possible, such that the system has a relative stability of at least 5 units.

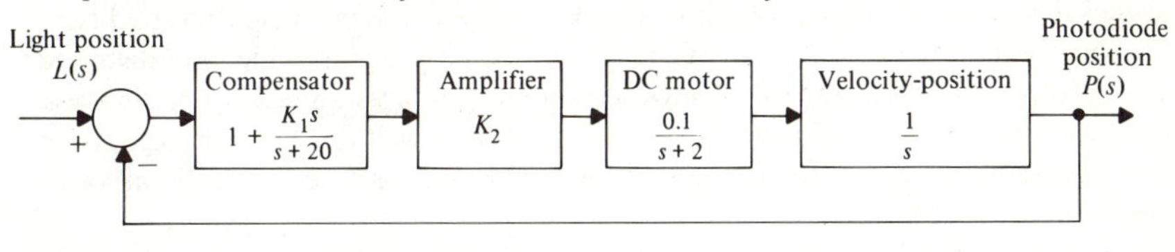

FIGURE P3-7

16. For the artificial limb system (Sec. 3.11), suppose the neuromuscular lag is modeled by $T=\frac{1}{2}$ sec instead of $T=\frac{1}{4}$ sec. Find the maximum value of K_B for which the overall system is stable.
17. For the artificial limb system (Sec. 3.11) with $T=\frac{1}{4}$ sec, suppose more tachometer feedback is used. Let $K_T=10$ instead of 4.05. Find the maximum value of K_B for which the overall system is stable.
18. A simple model for the roll stabilizer on a large ship is given in Fig. P3-8. Find the two system transfer functions in terms of the fin actuator gain K, and determine the range of $K>0$ for which the system is stable. Then, for $K=1$, use the Routh-Hurwitz test to determine the relative stability of the system to within 0.5 unit.

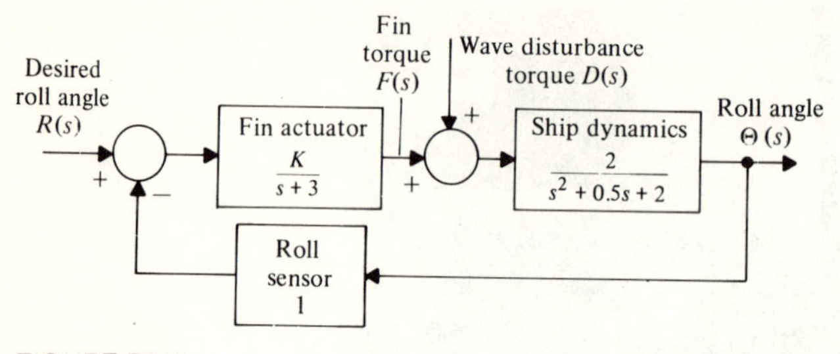

FIGURE P3-8

19. When paper is rolled in a paper mill, it is extremely important to maintain a specific tension as the paper is wound. A model for the control of this process is given in Fig. P3-9 in terms of various constants. On the basis of your imagination of the behavior of the components involved, choose nonzero values for K_1, K_2, a_1, a_2, a_3, and a_4, and determine if your system model is stable.

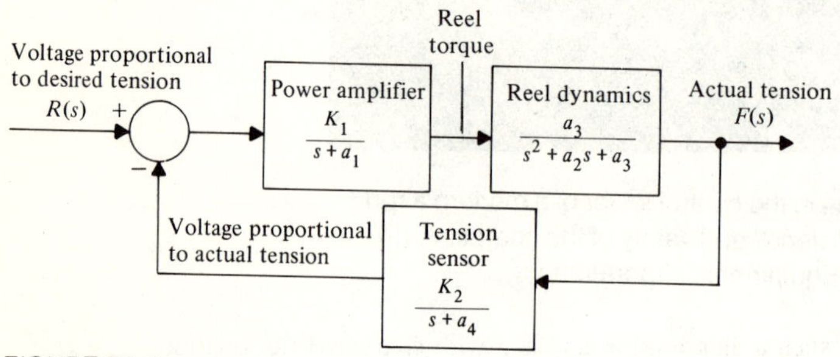

FIGURE P3-9

20. A linearized model of a frequency-locked loop is given in Fig. P3-10. The frequency difference between an incoming sinusoidal signal and an oscillator is sensed and integrated to produce a voltage which drives the voltage-controlled oscillator. The oscillator's frequency is proportional to the control voltage. When there is zero frequency difference, the integrator input is zero, its output is constant, and the oscillator frequency equals the incoming frequency. For incoming signals which are distorted and corrupted by noise, the frequency-locked loop produces a nearly pure sinusoid, locked in frequency to the incoming signal. The effect is as if the incoming signal had been processed by a very sophisticated filter which had removed virtually all of its noise and distortion.

Use the Routh-Hurwitz test to find a suitable value of $K > 0$.

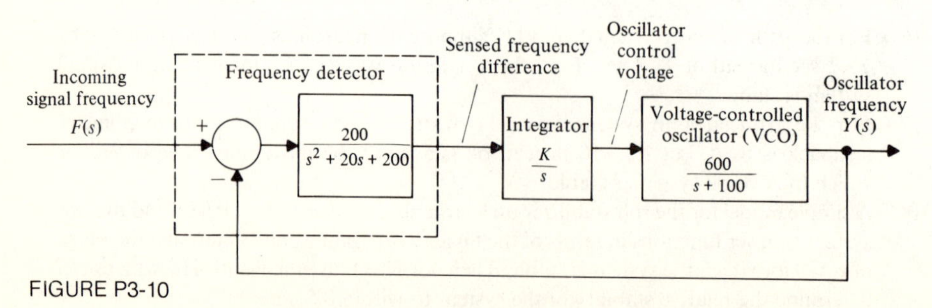

FIGURE P3-10

21. A motor shaft velocity control system model is given in Fig. P3-11. For what range of $K>0$ is this system's relative stability greater than 50 units?

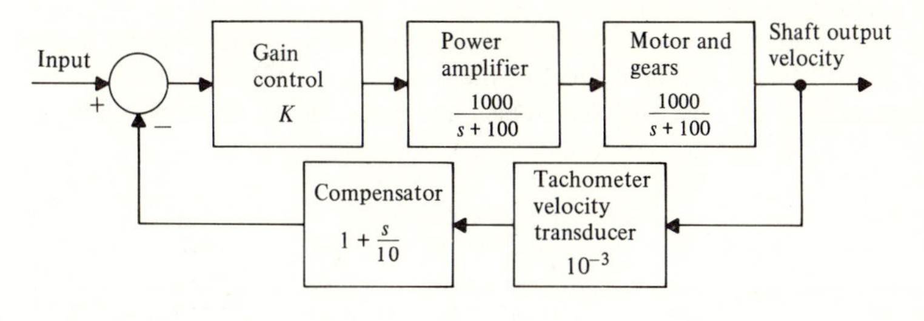

FIGURE P3-11

22. A motor shaft position control system is modeled in Fig. P3-12. Find the relative stability of the system as a function of the positive gain K.

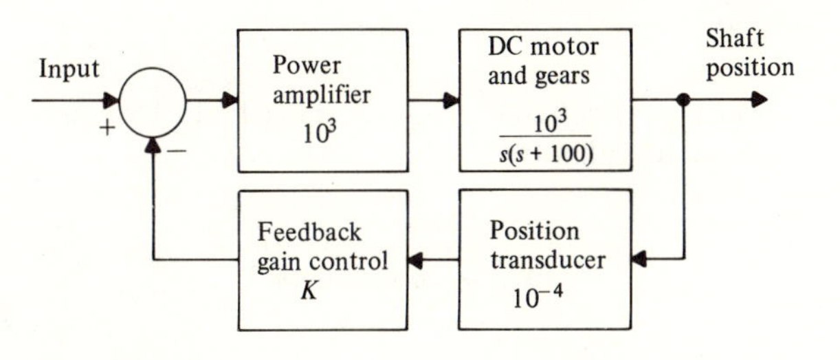

FIGURE P3-12

23. A system without an input which may be used to generate a sinusoidal output signal is shown in Fig. P3-13. When the differential equation describing $y(t)$ has a characteristic equation with roots on the imaginary axis of the complex plane, the system's zero-input response is sinusoidal. Find the value of the adjustable constant K, in terms of a, for which the system has a sinusoidal output. Also find the hertz frequency of the oscillations in terms of the constant a.

Achieving characteristic roots precisely on the imaginary axis is impossible for inexact K. If the characteristic roots are slightly in the RHP, the oscillation amplitude will increase exponentially, and if they are in the LHP, the oscillations will decay exponentially in time. In practice, another control system may be used to slowly adjust K to maintain nearly constant sinusoidal amplitude.

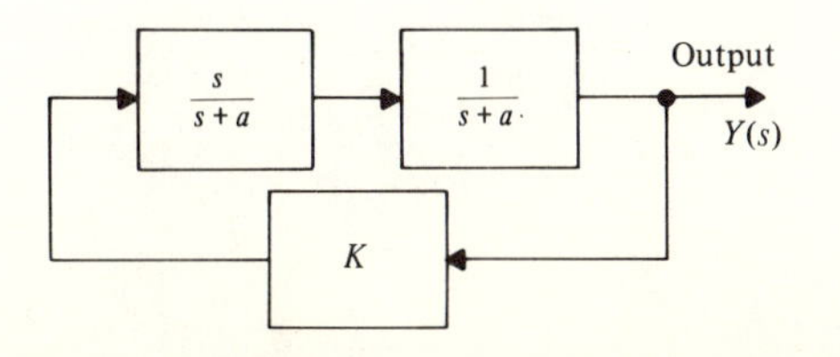

FIGURE P3-13

24. A position servosystem model is given in Fig. P3-14. The compensator is an imperfect attempt to integrate the sensed velocity to obtain position feedback.
   (a) For what positive values of the adjustable parameter K is the system which relates shaft *velocity* to the input stable?
   (b) For a stable input-to-velocity-output system, the input-to-position-output system is marginally stable. Carefully analyze the consequences of the latter system not being stable. Then give several suitable and several unsuitable applications for this system as a position controller.

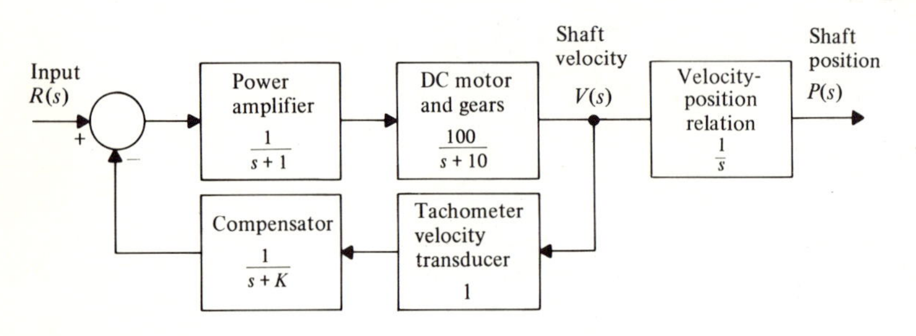

FIGURE P3-14

25. A simplified block diagram of an aircraft roll control is given in Fig. P3-15. Determine the relative stability of this system to within 10%.

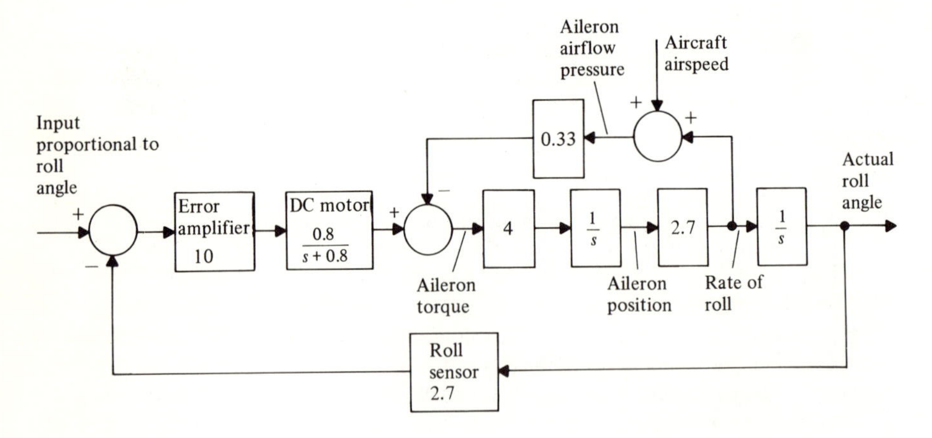

FIGURE P3-15

four

PERFORMANCE SPECIFICATIONS

4.1 PREVIEW

In an important class of control systems, it is desired that a plant output "track" or "follow" a controller input. The steady state response of these systems when driven by constant, ramp, parabolic, and higher power of time inputs is an important measure of their performance.

This chapter begins with a discussion of the initial- and final-value theorems and definitions of steady state errors of a system to power-of-time inputs. The use of forward path integrations in a feedback system to improve steady state errors arises naturally in the course of examining system type, equivalent unity feedback systems, and steady state error coefficients.

Two other important measures of system performance are also discussed. They are the sensitivity of the system to changes in system parameters and the degree to which the system is affected by disturbance inputs. For each, the advantages of feedback are pointed out and illustrated with examples.

The use of performance indices for the selection of certain system parameters is then introduced. Simple examples are given to show the idea of this approach, including the selection of a second-order system damping ratio.

An example transportation system is designed for accurate control of vehicle velocity and position. In another example system, a phase-locked loop for a citizen band radio receiver, the station acquisition characteristics are designed using steady state error considerations.

4.2 INITIAL- AND FINAL-VALUE THEOREMS

The *initial value* of a function of time $y(t)$ is denoted by $y(0)$. It is related to the signal's Laplace transform by

$$y(0) = \lim_{s \to \infty} [sY(s)]$$

For example, the initial value of the function with Laplace transform

$$Y(s)=\frac{-6s^3+3s^2-8}{7s^4+9s^3-5s^2+4s+1}$$

is

$$y(0)=\lim_{s\to\infty}[sY(s)]=-\tfrac{6}{7}$$

In general the values $y(0^-)$, $y(0)$, and $y(0^+)$ for a Laplace-transformable function may differ. If the function $y(t)$ is discontinuous at $t=0$, then

$$y(0^-)=\lim_{t\to 0^-} y(t)\neq y(0^+)=\lim_{t\to 0^+} y(t)$$

which of course is the definition of a discontinuity. If there is an impulse at $t=0$, $y(0)$ will be infinite. For functions with rational transforms, the initial-value theorem is especially easy to visualize. The partial fraction expansion of such a function can have the representative types of terms in Table 4-1. For each, and for other terms as well, multiplying by s and taking the limit as s goes to infinity gives the correct contribution to $y(0)$.

TABLE 4-1
Application of the initial- and final-value theorems to representative Laplace transform terms.

Transform $Y(s)$	*Time function* $y(t)$, $t\geqslant 0$	*Lim* $sY(s)$, $s\to\infty$	*Lim* $sY(s)$, $s\to 0$
K	$K\delta(t)$	∞	0
$\frac{K}{s}$	K	K	K
$\frac{K}{s^2}$	Kt	0	∞
$\frac{K}{s+a}$	Ke^{-at}	K	0
$\frac{Kb}{s^2+b^2}$	$K\sin bt$	0	0
$\frac{Kb}{(s+a)^2+b^2}$	$Ke^{-at}\sin bt$	0	0
$\frac{Ks}{s^2+b^2}$	$K\cos bt$	K	0
$\frac{K(s+a)}{(s+a)^2+b^2}$	$Ke^{-at}\cos bt$	K	0

The *final value* or *steady state value* of a function is, if it exists, the limit

$$\lim_{t \to \infty} y(t)$$

The final-value theorem states that

$$\lim_{t \to \infty} y(t) = \lim_{s \to 0} [sY(s)]$$

when the final value exists and is finite.

For signals with rational Laplace transforms, any denominator root in the RHP or repeated imaginary axis roots give an ever-expanding time function, which does not have a finite final value. A single complex conjugate pair of imaginary axis roots corresponds to a sinusoidal function, for which a final value will not exist. Hence, for a final value of $y(t)$ to exist, all denominator roots of $Y(s)$ must be in the LHP except possibly for one root at $s=0$. It is the root at $s=0$, corresponding to a constant term in $y(t)$ after $t=0$, which then contributes a nonzero final value.

Application of the final-value theorem is simply calculation of the residue of a K/s term in the function's partial fraction expansion:

$$K = [sY(s)]|_{s=0}$$

Unfortunately, this limit gives answers even when a final value does not exist, as is demonstrated by the representative entries in Table 4-1. The transform

$$Y(s) = \frac{-3s^2+4}{s^4+5s^3+8s^2+4s} = \frac{-3s^2+4}{s(s+1)(s+2)^2}$$

$$= \frac{K_1}{s} + \frac{K_2}{s+1} + \frac{K_3}{s+2} + \frac{K_4}{(s+2)^2}$$

represents a time function of the form

$$y(t) = [K_1 + K_2 e^{-t} + K_3 e^{-2t} + K_4 t e^{-2t}]u(t)$$

which has the final value K_1. The final value is correctly given by

$$\lim_{t \to \infty} y(t) = \lim_{s \to 0} [sY(s)] = \tfrac{4}{4} = 1 = K_1$$

The Laplace transform

$$Y(s) = \frac{s-6}{s^4+s^3+3s^2-5s} = \frac{s-6}{s(s-1)[(s+1)^2+4]}$$

$$= \frac{K_1}{s} + \frac{K_2}{s-1} + \frac{K_3 s + K_4}{(s+1)^2+(2)^2}$$

represents a time function of the form

$$y(t)=[K_1+K_2e^t+Ae^{-t}\cos(2t+\theta)]u(t)$$

which does not have a finite final value because of the e^t term. Application of the final-value theorem limit gives

$$\lim_{s\to 0}[sY(s)]=\tfrac{6}{5}$$

which is incorrect.

DRILL PROBLEMS

D4-1. For each of the following Laplace-transformed signals, find the initial value $y(0)$. For functions with discontinuities or impulses at $t=0$, these initial values may differ from the initial conditions $y(0^-)$.

(a) $Y(s)=\dfrac{4s-1}{s^2+3s}$

ans. 4

(b) $Y(s)=\dfrac{5}{s^3+2s^2+11}$

ans. 0

(c) $Y(s)=\dfrac{6e^{-2s}}{s+4}$

ans. 0

(d) $Y(s)=\dfrac{30s}{s^3+2s^2+11s+3}$

ans. 0

D4-2. For each of the following Laplace-transformed signals, find the final value if it exists. If a final value of the signal does not exist, so state.

(a) $Y(s)=\dfrac{6s^3-3s^2+2s-4}{s^4+3s^3+2s^2+s+10}$

ans. does not exist

(b) $Y(s)=\dfrac{s^2-2s+10}{(s^2+2s+10)(s^2+3s)}$

ans. $\frac{1}{3}$

(c) $Y(s)=7e^{-3s}$

ans. 0

(d) $Y(s)=\dfrac{6s^2+5}{s^3+2s^2+11s}$

ans. $\frac{5}{11}$

(e) $Y(s)=\dfrac{30}{s^3+2s^2+11s+3}$

ans. 0

D4-3. For each of the following systems, the input $R(s)$ is a unit step. Find the steady state output signal $Y(s)$ if it exists.

ans. (a) 1
(b) 0
(c) 1
(d) does not exist

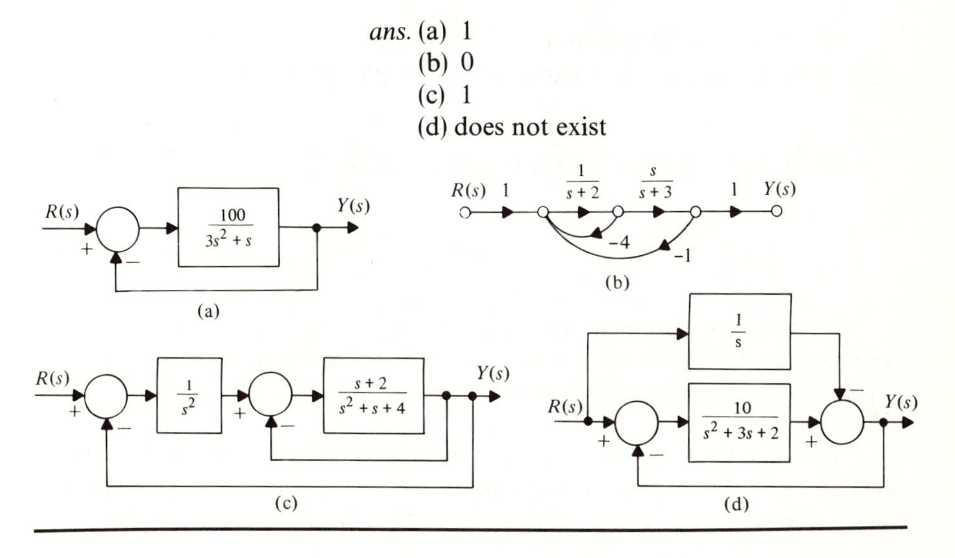

4.3 NORMALIZED STEADY STATE ERROR

A *tracking system* (or *servosystem*) is a control system in which the output, as nearly as possible, equals (or "tracks" or "follows") a reference input. One measure of how well such a system performs is its steady state response to a step input. If the system transfer function is $T(s)$, a step input

$$R(s)=\frac{A}{s}$$

with zero initial conditions produces the output

$$Y(s)=R(s)T(s)=\frac{A}{s}T(s)$$

The error between input and output is

$$E(s)=R(s)-Y(s)=\frac{A}{s}[1-T(s)]$$

Applying the final-value theorem, assuming that the error reaches a final value,

$$\lim_{t\to\infty} e(t)=\lim_{s\to 0} sE(s)=\lim_{s\to 0} A[1-T(s)]$$

The normalized result, independent of step amplitude, is

$$\text{Normalized steady state error to a step}=\frac{\lim_{t\to\infty} e(t)}{A}=\lim_{s\to 0}[1-T(s)]$$

Similarly, *when it is known that the final value exists*, normalized errors may be computed for ramp, parabolic, and higher powers of time t:

$$\text{Normalized steady state error to an input } \frac{At^n u(t)}{n!}=\frac{\lim_{t\to\infty} e(t)}{A}$$

$$=\lim_{s\to 0} s\cdot\frac{1}{s^{n+1}}[1-T(s)]=\lim_{s\to 0}\frac{1}{s^n}[1-T(s)]$$

The standard power-of-time inputs are $u(t)$, $tu(t)$, $\frac{1}{2}t^2u(t)$, $\frac{1}{6}t^3u(t)$, and so on. The first three of these are occasionally referred to as position, velocity, and acceleration inputs, descriptive of their function if the system output is a mechanical position.

Consider the unity feedback system of Fig. 4-1. The difference between the input and the output is formed, amplified, and applied to the plant input in such a way as to reduce this difference. The system transfer function is

$$T(s)=\frac{Y(s)}{R(s)}=\frac{K/(s+2)}{1+K/(s+2)}=\frac{K}{s+2+K}$$

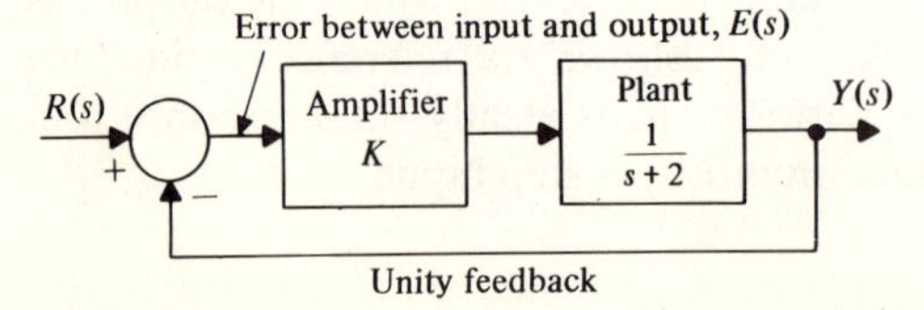

FIGURE 4-1
A unity feedback system.

For a step input

$$R(s)=\frac{A}{s}$$

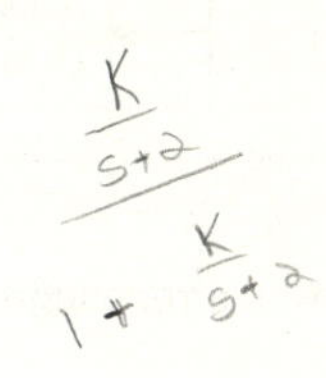

we have

$$Y(s)=T(s)\frac{A}{s}$$

and the error between input and output is

$$E(s)=R(s)-Y(s)=\frac{A}{s}[1-T(s)]=\frac{A}{s}\left[\frac{s+2}{s+2+K}\right]$$

The error reaches a steady state for any positive K (in fact, for any K larger than -2) and is

$$\lim_{t\to\infty} e(t)=\lim_{s\to 0} sE(s)=\lim_{s\to 0} A\left[\frac{s+2}{s+2+K}\right]$$

$$=\frac{2A}{2+K}$$

The normalized error, the ratio of error to the step amplitude A, is

$$\frac{2}{2+K}$$

which may be made arbitrarily small in this case by choosing a sufficiently large amplifier gain K.

For a ramp, parabolic, or higher power of t input, the error for the example system becomes infinite. For a ramp input

$$R(s)=\frac{A}{s^2}$$

we have

$$E(s)=\left(\frac{A}{s^2}\right)\left(\frac{s+2}{s+2+K}\right)=\frac{K_1}{s}+\frac{K_2}{s^2}+\frac{K_3}{s+2+K}$$

The term K_2/s^2 in the partial fraction expansion of $E(s)$ corresponds to the time function K_2t, which grows without bound.

A simple method of obtaining zero steady state error for a step input for the example system is to integrate the error to drive the plant. The addition of

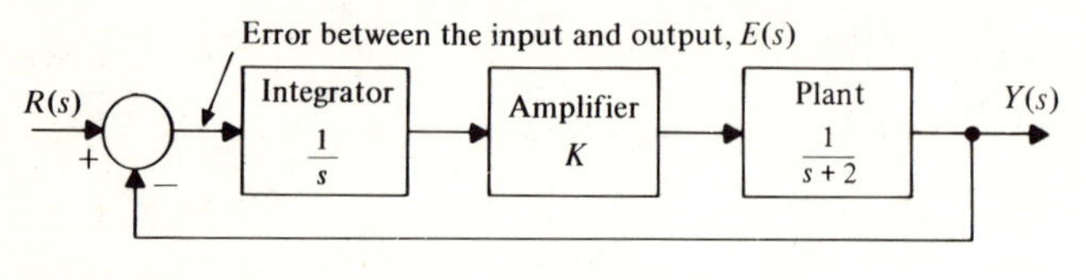

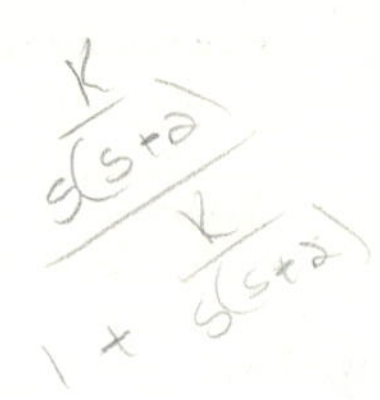

FIGURE 4-2
A tracking system with error integration.

an integrator to the previous example is shown in Fig. 4-2. For this system,

$$T(s)=\frac{K/s(s+2)}{1+K/s(s+2)}=\frac{K}{s^2+2s+K}$$

For a step input

$$R(s)=\frac{A}{s}$$

the new error signal is given by

$$E(s)=R(s)-Y(s)=\frac{A}{s}[1-T(s)]$$

$$=\frac{A}{s}\left[\frac{s^2+2s}{s^2+2s+K}\right]$$

The steady state error for any positive K is

$$\lim_{t\to\infty} e(t)=\lim_{s\to 0} sE(s)=0$$

This result occurred because of the factor of s in the numerator of $E(s)$. Without the integrator, the numerator of $E(s)$ contained no such factor and a finite steady state error resulted for a step input.

For a ramp input

$$R(s)=\frac{A}{s^2}$$

the normalized steady state error is

$$\lim_{t\to\infty} e(t)=\lim_{s\to 0} \frac{s+2}{s^2+2s+K}=\frac{2}{K}$$

DRILL PROBLEMS

D4-4. For each of the following systems, find the normalized steady state error, if it exists, between output and input when the input is a step.

ans. (a) 0
(b) 1
(c) $\frac{5}{3}$
(d) $\frac{2}{3}$

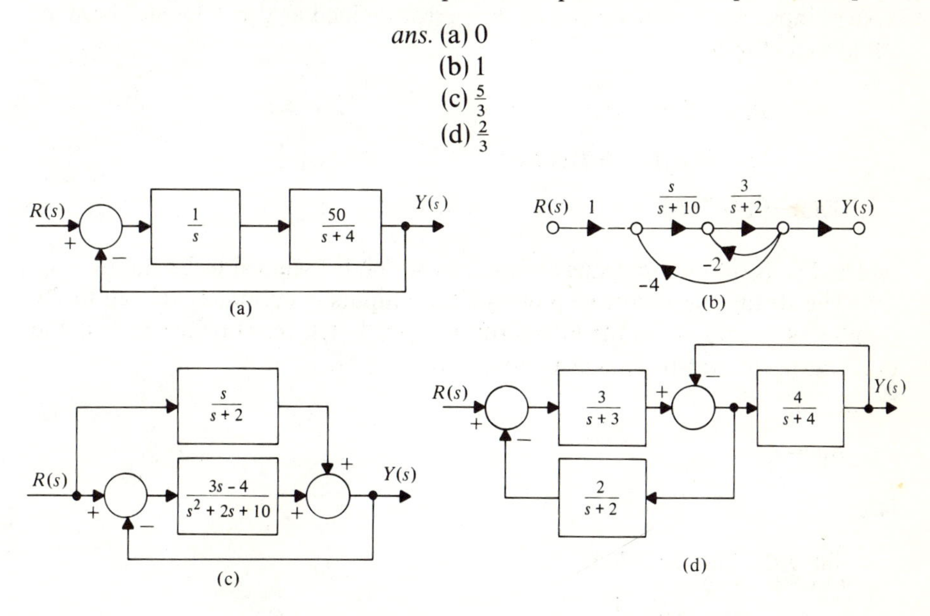

D4-5. For systems with the following transfer functions, find normalized steady state errors, if they exist, between output and input for step, ramp, and parabolic inputs:

(a) $T(s)=\dfrac{-3s^2+5}{s^3+3s^2+2s+5}$

ans. $0, \frac{2}{5}, \infty$

(b) $T(s)=\dfrac{2s^2+2s+5}{s^3+2s^2+2s+5}$

ans. infinite errors

(c) $T(s)=\dfrac{3s^2+2s+10}{(s+2)^2(s^2+2s+10)}$

ans. $\frac{3}{4}, \infty, \infty$

(d) $T(s)=\dfrac{2s-1}{2s^4+4s^3+4s^2+2s+1}$

ans. $2, \infty, \infty$

4.4 SYSTEM TYPE AND STEADY STATE ERROR

4.4.1 System Type

The error transmittance of a system, $T_E(s)$, is the transmittance relating the system input $R(s)$ to the error $E(s)$. For error defined as the difference between input and output,

$$E(s) = R(s) - Y(s) = R(s) - T(s)R(s)$$
$$= [1 - T(s)]R(s) = T_E(s)R(s)$$
$$T_E(s) = [1 - T(s)]$$

where $T(s)$ is the system transfer function, which is assumed to be stable.

The steady state error to power-of-time inputs is intimately related to the number of factors of s in the numerator of $T_E(s)$. If $T_E(s)$ has no factor of s in the numerator, the steady state error to a step

$$R(s) = \frac{A}{s}$$

is

$$\lim_{t\to\infty\ \text{step}} e(t) = \lim_{s\to 0} sT_E(s)R(s)$$
$$= \lim_{s\to 0} sT_E(s)\frac{A}{s}$$
$$= AT_E(0)$$

which is finite. Figure 4-3 illustrates how a system may respond to a step input to give a constant steady state error.

Higher-power-of-t inputs give infinite steady state error. For a ramp input

$$R(s) = \frac{A}{s^2}$$

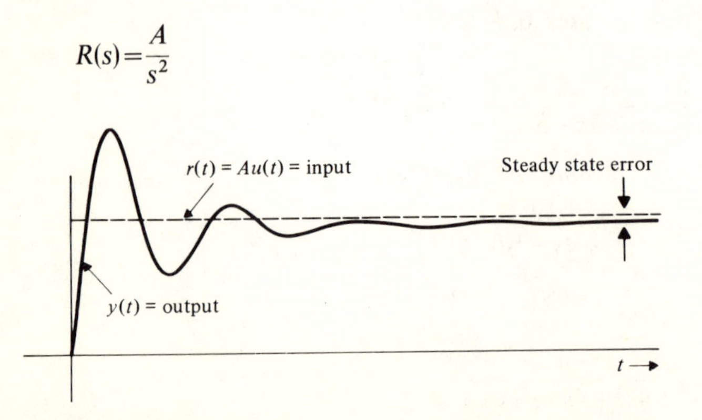

FIGURE 4-3
Type 0 system step response.

the error

$$E(s) = T_E(s)\frac{A}{s^2}$$

would have a repeated denominator root at $s=0$, indicative of a ramp term in $e(t)$. The final-value theorem does not apply in this case because $e(t)$ does not approach a final value. Taking the limit

$$\lim_{s \to 0} sT_E(s)\frac{A}{s^2} = \infty$$

does give the correct answer, though.

If the system is stable and $T_E(s)$ has one factor of s in the numerator, the steady state error to a step is zero:

$$\lim_{s \to \infty \text{ step}} e(t) = \lim_{s \to 0} sT_E(s)\frac{A}{s} = AT_E(0) = 0$$

To a ramp, the error is constant:

$$\lim_{s \to \infty \text{ ramp}} e(t) = \lim_{s \to 0} sT_E(s)\frac{A}{s^2} = A \lim_{s \to 0} \frac{T_E(s)}{s}$$

The constant steady state error results by cancellation with the s factor in the numerator of $T_E(s)$. Figure 4-4 illustrates how a system may respond to a ramp input to give a finite steady state error.

For higher-power-of-t inputs, the error is infinite, since

$$E(s) = T_E(s)\frac{A}{s^n}$$

has a repeated $s=0$ denominator root.

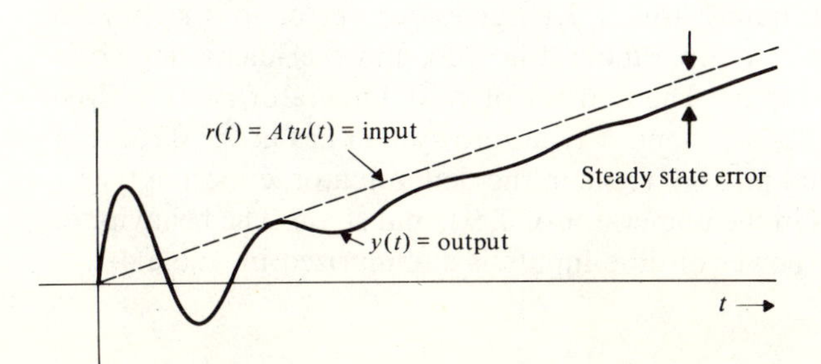

FIGURE 4-4
Type 1 system ramp response.

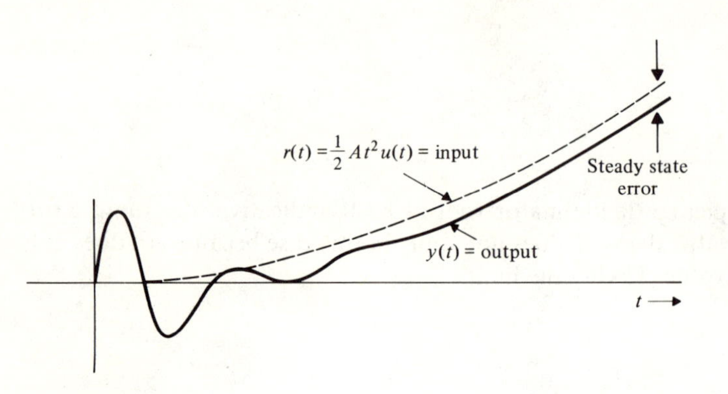

FIGURE 4-5
Type 2 system parabolic response.

TABLE 4-2
Steady state error.

System Type (Number of $s=0$ numerator roots of the error transmittance T_E)	*Steady State Error to Step Input* $r(t)=Au(t)$ $R(s)=A/s$	*Steady State Error to Ramp Input* $r(t)=Atu(t)$ $R(s)=A/s^2$	*Steady State Error to Parabolic Input* $r(t)=\frac{1}{2}At^2u(t)$ $R(s)=A/s^3$	*Steady State Error to Input* $r(t)=\frac{1}{6}At^3u(t)$ $R(s)=A/s^4$
0	$AT_E(0)$	∞	∞	∞
1	0	$A \lim_{s\to 0}\left[\frac{T_E(s)}{s}\right]$	∞	∞
2	0	0	$A \lim_{s\to 0}\left[\frac{T_E(s)}{s^2}\right]$	∞
3	0	0	0	$A \lim_{s\to 0}\left[\frac{T_E(s)}{s^3}\right]$
⋮				

A factor of s^2 in the numerator of $T_E(s)$ gives zero error to a step, zero error to a ramp, finite error to a parabola (Fig. 4-5), and ever-increasing error for higher-power-of-time inputs. The number of $s=0$ numerator roots in $T_E(s)$ is termed the *type number* of a system. A type zero system has no $s=0$ roots in the numerator of $T_E(s)$ (and no $s=0$ roots in the denominator, either); a type 1 system has one factor of s in the numerator of $T_E(s)$; and so on. The behavior of various system types for power-of-time inputs is summarized in Table 4-2.

4.4.2 Unity Feedback Systems

When a control system has unity feedback, the input r and output y are compared directly, as in Fig. 4-6. The error signal drives the forward transmittance

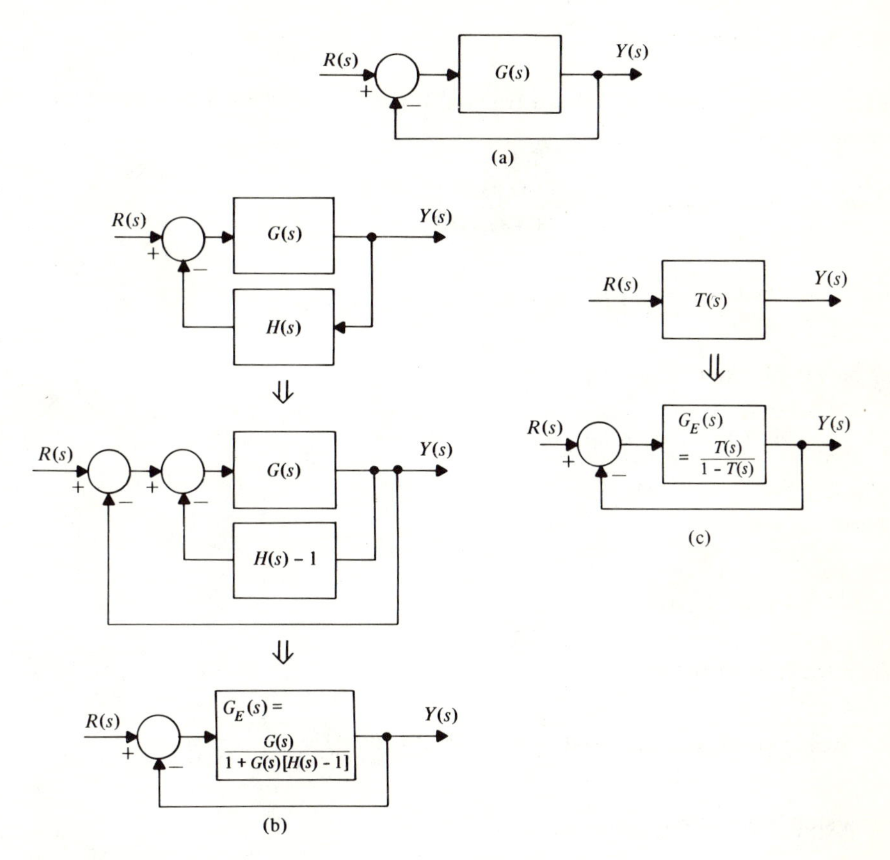

FIGURE 4-6
(a) A unity feedback system. (b) Converting a feedback system to an equivalent unity fedback system. (c) General conversion of a system to an equivalent unity feedback system.

$G(s)$, as in Fig. 4-6a. A system that does not have unity feedback can be converted into an equivalent unity feedback structure, as in the example of Fig. 4-6b. In general, a system with transfer function $T(s)$, Fig. 4-6c, is equivalent to a unity feedback system with forward transmittance

$$G_E(s) = \frac{T(s)}{1 - T(s)}$$

System type, being determined by the number of $s = 0$ numerator roots of

$$T_E(s) = 1 - T(s)$$

for a unity feedback system, is alternatively determined by the number of $s = 0$ denominator roots of the equivalent forward transmittance $G_E(s)$. Hence it is seen that adding factors of $(1/s)$ to the forward transmittance raises the system type number, so long as the resulting system is stable.

4.4.3 Error Coefficients

In terms of the equivalent forward transmittance $G_E(s)$, the error between input and output is

$$E(s) = R(s) - Y(s) = R(s)\left[1 - \frac{G_E(s)}{1 + G_E(s)}\right]$$

$$= R(s)\left[\frac{1}{1 + G_E(s)}\right]$$

For power-of-t inputs

$$R(s) = \frac{A}{s^n}$$

we get

$$E(s) = \frac{A}{s^n[1 + G_E(s)]}$$

When the limits exist,

$$\text{Steady state error to input } \frac{A}{s^n} = \lim_{s \to 0} sE(s) = \lim_{s \to 0} \frac{A}{s^{n-1}[1 + G_E(s)]}$$

For a step input, $n = 1$,

$$\text{Steady state error to input } \frac{A}{s} = \lim_{s \to 0} \frac{A}{1 + G_E(s)} = \frac{A}{1 + K_0}$$

where

$$K_0 = \lim_{s \to 0} G_E(s)$$

For a ramp input, $n = 2$,

$$\text{Steady state error to input } \frac{A}{s^2} = \lim_{s \to 0} \frac{A}{s[1 + G_E(s)]} = \lim_{s \to 0} \frac{A}{sG_E(s)} = \frac{A}{K_1}$$

where

$$K_1 = \lim_{s \to 0} sG_E(s)$$

For higher-power-of-t inputs,

$$\text{Steady state error to input } \frac{A}{s^n} = \lim_{s \to 0} \frac{A}{s^{n-1} G_E(s)} = \frac{A}{K_{n-1}} \qquad n = 2, 3, 4, \ldots$$

where

$$K_i = \lim_{s \to 0} s^i G_E(s)$$

* The constants K_i are known as the system *steady state error coefficients.* Table 4-2, rewritten in terms of the error coefficients, is Table 4-3. For the unity feedback type zero system of Fig. 4-1,

$$G_E(s) = G(s) = \frac{K}{s+2}$$

$$K_0 = \lim_{s \to 0} G(0) = \frac{K}{2}$$

giving

$$\text{Steady state error to a step input } \frac{A}{s} = \frac{A}{1+K_0} = \frac{A}{1+K/2} = \frac{2A}{2+K}$$

as was found earlier by using the system transfer function. This system has infinite error to ramp and higher-power-of-t inputs.

The related unity feedback type 1 system of Fig. 4-2 has

$$G_E(s) = G(s) = \frac{K}{s(s+2)}$$

TABLE 4-3
Steady state error using error coefficients.

System Type	*Steady State Error to Step Input* $R(s)=A/s$	*Steady State Error to Ramp Input* $R(s)=A/s^2$	*Steady State Error to Parabolic Input* $R(s)=A/s^3$	*Steady State Error to Input* $R(s)=A/s^4$
0	$\frac{A}{1+K_0}$	∞	∞	∞
1	0	$\frac{A}{K_1}$	∞	∞
2	0	0	$\frac{A}{K_2}$	∞
3	0	0	0	$\frac{A}{K_3}$
⋮				

An integrator was placed in the system to make it type 1. For this system,

$$K_0 = \lim_{s\to 0} G(s) = \infty$$

and the step error is

$$\text{Error to a step input } \frac{A}{s} = \frac{A}{1+K_0} = 0$$

For a ramp input,

$$K_1 = \lim_{s\to 0} sG(s) = \frac{K}{2}$$

$$\text{Error to a ramp input } \frac{A}{s^2} = \frac{A}{K_1} = \frac{2A}{K}$$

The steady state errors to higher-power-of-t inputs are infinite.

4.4.4 The Importance of Power-of-Time Inputs

If the input signal to a system is expanded in a power series about time $t = a$, there results

$$\begin{aligned} r(t) &= r(a) + \left.\frac{dr}{dt}\right|_{t=a} t + \frac{(d^2r/dt^2)|_{t=a}}{2!} t^2 + \frac{(d^3r/dt^3)|_{t=a}}{3!} t^3 + \cdots \\ &= A_0 + A_1 t + A_2 t^2 + A_3 t^3 + \cdots \end{aligned}$$

Thus sufficiently smooth input signals may be considered to be composed of step, ramp, parabolic, and higher-order components over intervals of time. For small intervals of time, such signals are usually approximated well by just the lower-order terms. For example, consider the input signal $r(t)$ in Fig. 4-7.

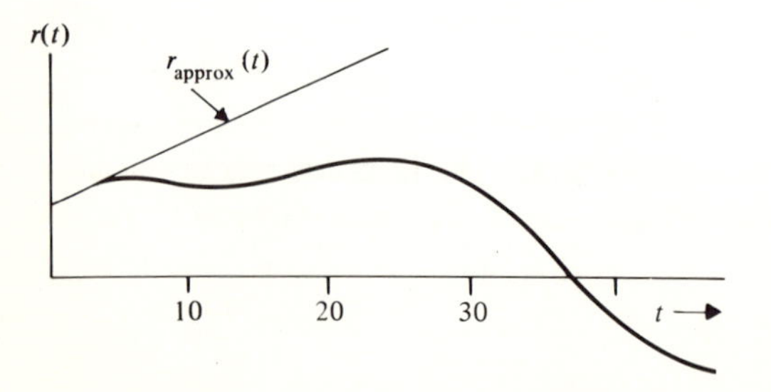

FIGURE 4-7
Approximating a function with powers of time.

It is approximated well by a signal of the form

$$r_{\text{approx}}(t) = A_0 + A_1 t$$

for several seconds, as is demonstrated in the figure. To the extent that a system with input $r(t)$ will track a step and a ramp well, it will track $r(t)$ well initially.

With experience, control system designers learn to accurately estimate the performance of tracking systems in general from their performance with power-of-t inputs.

DRILL PROBLEMS

D4-6. Find the output-input error transmittance of each of the following systems. Then determine the type number of each system and, if the response reaches steady state, normalized steady state errors to step and to ramp inputs.

ans. (a) $0, \frac{7}{15}, \infty$
(b) $0, -\frac{4}{3}, \infty$
(c) $1, 0, \frac{6}{25}$
(d) $0, \frac{23}{28}, \infty$

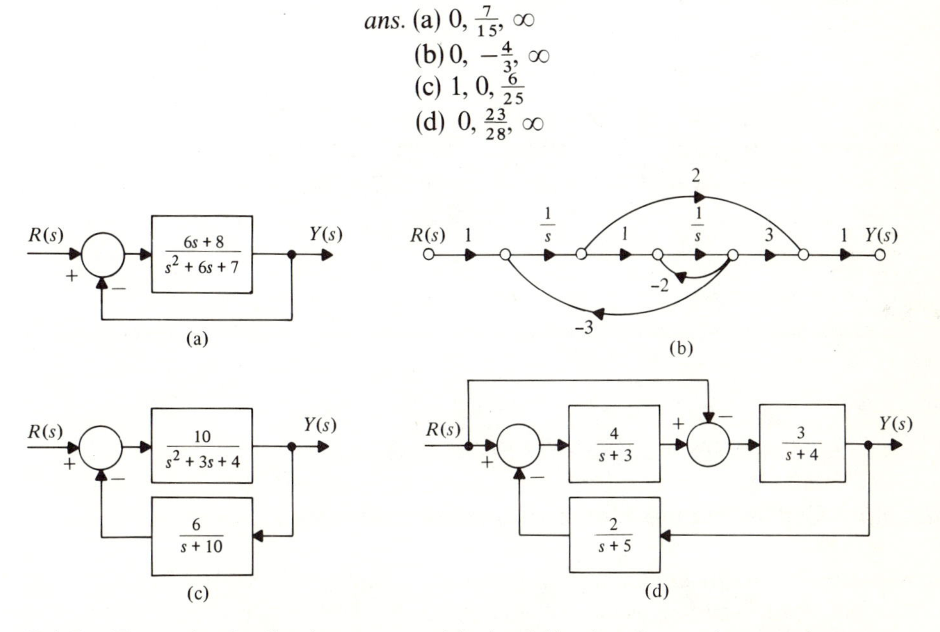

D4-7. For unity feedback systems with the following forward transmittances, determine each system type number and, if the response reaches steady state, normalized steady state output-input errors to step and to ramp inputs.

(a) $G(s) = \dfrac{10}{s^3 + 3s^2 + 2s}$

ans. $1, 0, \frac{1}{5}$

(b) $G(s)=\dfrac{1}{s^3+2s^2+s+3}$

ans. 0, infinite errors

(c) $G(s)=\dfrac{4(s+1)}{(s+2)(s+3)(s^2+s+10)}$

ans. $0, \frac{15}{16}, \infty$

(d) $G(s)=\dfrac{2s+1}{2s^4+4s^3+4s}$

ans. 1, infinite errors

D4-8. Find the error coefficients for each of the following systems. Then, if the responses reach steady state, find the normalized steady state output-input errors to step and to ramp inputs.

ans. (a) $\frac{8}{9}, \infty$
(b) 0, 5.9

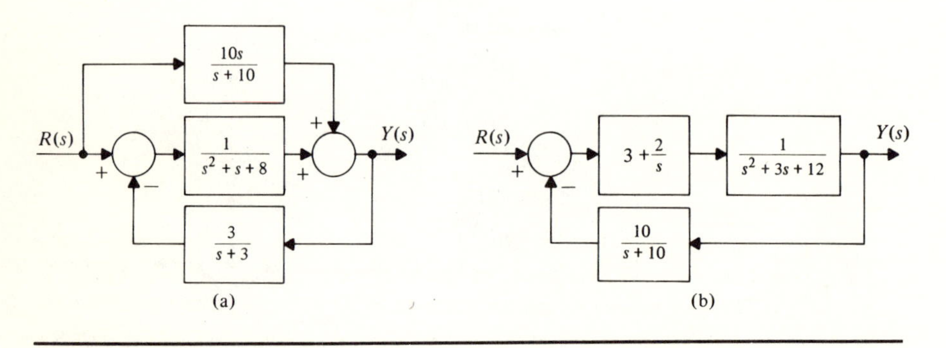

4.5 SENSITIVITY TO SYSTEM PARAMETERS

4.5.1 Calculating the Effects of Changes

One of the major advantages of feedback is that it may be used to make the response of a system relatively independent of certain types of changes or inaccuracies in the plant model. For example, in the system of Fig. 4-8a, the nominal system transfer function is

$$T(s)=\frac{400/(s+2)}{1+400/(s+2)}=\frac{400}{s+402}$$

Now suppose one of the plant parameters changes or is wrongly modeled, as in Fig. 4-8b. For $k_1=1$, the plant is the nominal one; other values of k_1 correspond

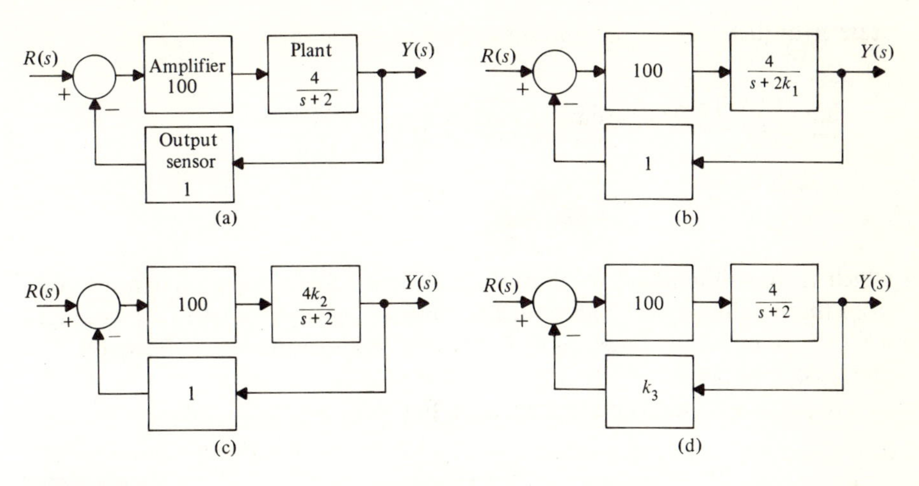

FIGURE 4-8
Determining the effects of changes in the parameters of a feedback system. (a) The nominal system. (b) Perturbation of one plant parameter. (c) Perturbation of another plant parameter. (d) Perturbation of the sensor gain.

to perturbations from the nominal plant. In terms of k_1,

$$T(s)=\frac{400/(s+2k_1)}{1+400/(s+2k_1)}=\frac{400}{s+400+2k_1}$$

and it is seen that even 50% changes in the parameter, corresponding to $k_1=\frac{1}{2}$ and $k_1=\frac{3}{2}$, result in a relatively minor change in $T(s)$. Even negative values of k_1 (say $k_1=-1$), for which the plant is unstable, give much the same, stable, overall transfer function $T(s)$.

*The system's steady state error to a unit step input is, in terms of k_1,

$$\lim_{s\to 0} s\left(\frac{1}{s}\right)[1-T(s)]=\lim_{s\to 0}\frac{s+2k_1}{s+400+2k_1}$$
$$=\frac{2k_1}{400+2k_1}$$

which is dominated by the factor of 400 and is nearly proportional to k_1.

For the parameter perturbed by k_2 in Fig. 4-8c,

$$T(s)=\frac{400k_2/(s+2)}{1+400k_2/(s+2)}=\frac{400k_2}{s+400k_2+2}$$

For this parameter, the overall transfer function coefficients are proportional or nearly proportional to the amount of perturbation k_2. The system's steady state error (between output and input) to a unit step input is

$$\lim_{s \to 0} s\left(\frac{1}{s}\right)[1-T(s)] = \lim_{s \to 0} \frac{s+2}{s+400k_2+2}$$

$$= \frac{2}{400k_2+2}$$

which is also dominated by the factor of $400k_2$ for moderate changes in k_2 from the nominal $k_2=1$ and is nearly inversely proportional to k_2. Changes from the nominal amplifier gain of 400 will produce the same effects on $T(s)$ and its step response.

If the sensor gain is perturbed, as in Fig. 4-8d,

$$T(s) = \frac{400/(s+2)}{1+400k_3/(s+2)} = \frac{400}{s+400k_3+2}$$

The steady state error to a unit step input is

$$\lim_{s \to 0} s\left(\frac{1}{s}\right)[1-T(s)] = \frac{s+400(k_3-1)+2}{s+400k_3+2}$$

$$= \frac{400(k_3-1)+2}{400k_3+2}$$

which may become quite large in comparison to the previous expressions for comparable percent parameter changes. In this case, the result is expected, since an error by the sensor in the perceived plant output is indistinguishable by the rest of the system from an actual output error.

4.5.2 Sensitivity Functions

In general, the *sensitivity* of a single-input, single-output system transfer function to changes in a specific system parameter a is defined as

$$S_a = \lim_{\Delta a \to 0} \frac{\Delta T/T}{\Delta a/a}$$

It is the limiting ratio of the fractional change in the transfer function to the fractional change in the parameter. Taking the limit,

$$S_a = \lim_{\Delta a \to 0} \frac{a}{T}\frac{\Delta T}{\Delta a} = \frac{a}{T}\frac{\partial T}{\partial a}$$

For example, the feedback system with constant "block" transmittances in

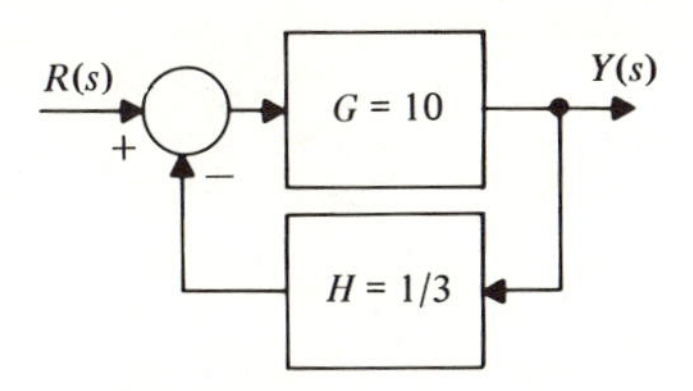

FIGURE 4-9
Finding sensitivity of a feedback system with constant transmittances.

Fig. 4-9 might represent a feedback amplifier over a range of frequencies. The transfer function of this system is the constant

$$T=\frac{G}{1+GH}=\frac{10}{1+\frac{10}{3}}=\frac{30}{13}$$

The sensitivity of T to changes in G is

$$S_G=\frac{G}{T}\frac{\partial T}{\partial G}=\frac{1}{1+GH}$$

which for $G=10$, $H=\frac{1}{3}$ is

$$S_G=\frac{1}{1+\frac{10}{3}}=\frac{3}{13}$$

This is to say that, with the feedback, the transfer function changes only $\frac{3}{13}$ as much with small changes in G as it would without feedback.

The sensitivity of T to changes in H is

$$S_H=\frac{H}{T}\frac{\partial T}{\partial H}=\frac{-GH}{1+GH}$$

which for $G=10$, $H=\frac{1}{3}$ is

$$S_H=\frac{-\frac{10}{3}}{1+\frac{10}{3}}=-\frac{10}{13}$$

The transfer function changes with changes in H much more than with changes in G. The minus sign in S_H indicates that T decreases with an increase in H.

For the feedback system of Fig. 4-10, suppose that the plant parameter a is nominally $a=2$ but is subject to small changes about the nominal value.

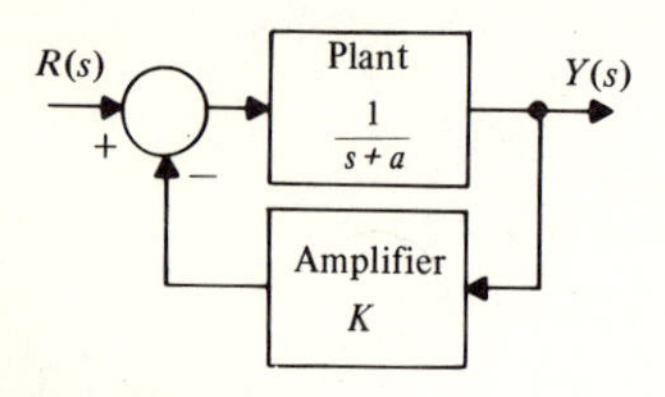

FIGURE 4-10
Finding the sensitivity of a feedback system to changes in a plant parameter.

The transfer function of the system is

$$T(s)=\frac{1/(s+a)}{1+K/(s+a)}=\frac{1}{s+a+K}$$

The sensitivity of $T(s)$ to changes in a is given by

$$S_a=\frac{a}{T}\frac{\partial T}{\partial a}=a(s+a+K)\frac{-1}{(s+a+K)^2}$$

$$=\frac{-a}{s+a+K}$$

Sensitivities are generally functions of the complex variable s. For $a=2$, the sensitivity is

$$S_a=\frac{-2}{s+2+K}$$

which may, for any s, be reduced by making K sufficiently large.

As a more involved example, consider the system of Fig. 4-11 for which the transfer function, as a function of the parameter a, is

$$T(s,a)=\frac{\left(\frac{s+3}{s}\right)\left(\frac{10}{s^2+as+10}\right)}{1+\left(\frac{s+3}{s}\right)\left(\frac{10}{s^2+as+10}\right)}$$

$$=\frac{10s+30}{s^3+as^2+20s+30}$$

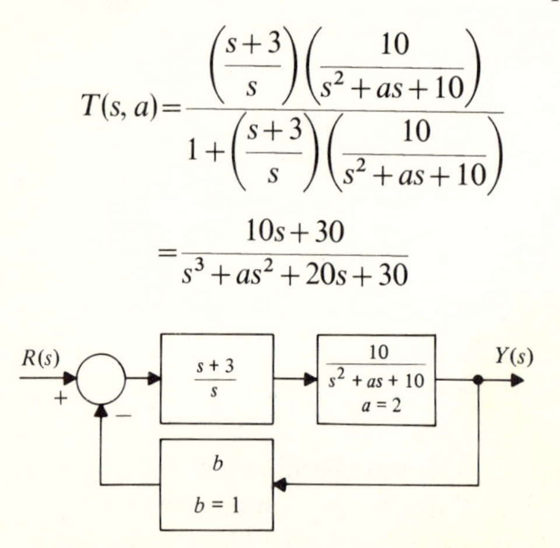

FIGURE 4-11
Calculating sensitivities to parameter changes for another system.

$$\frac{\partial T}{\partial a}=\frac{-s^2(10s+30)}{(s^3+as^2+20s+30)^2}$$

$$S_a=\frac{a}{T}\frac{\partial T}{\partial a}=\left[\frac{a}{\dfrac{10s+30}{s^3+as^2+20s+30}}\right]\left[\frac{-s^2(10s+30)}{(s^3+as^2+20s+30)^2}\right]$$

$$=\frac{-as^2}{s^3+as^2+20s+30}$$

For the nominal value of $a=2$,

$$S_a=\frac{-2s^2}{s^3+2s^2+20s+60}$$

For small changes in a about the nominal value of $a=2$,

$$S_a\cong\frac{a}{T}\frac{\Delta T}{\Delta a}$$

$$\frac{\Delta T(s)}{T(s)}\cong\frac{\Delta a}{a}S_a=\frac{\Delta a}{a}\frac{-2s}{s^3+2s^2+20s+30} \tag{4–1}$$

For a specific value of s, (4-1) relates fractional changes in the transfer function to fractional small changes in the parameter. In calculating the response to a step input to the system, for example, the transfer function is evaluated at $s=0$ to obtain the residue corresponding to the forced response. Equation (4-1) with $s=0$ gives

$$\frac{\Delta T(0)}{T(0)}\cong 0 \qquad \Delta T(0)\cong 0$$

for small changes Δa.

For the parameter b in the feedback path of the system of Fig. 4-9,

$$T(s,b)=\frac{\left(\dfrac{s+3}{s}\right)\left(\dfrac{10}{s^2+2s+10}\right)}{1+b\left(\dfrac{s+3}{s}\right)\left(\dfrac{10}{s^2+2s+10}\right)}$$

$$=\frac{10s+30}{s^3+2s^2+(10+10b)s+30b}$$

$$\frac{\partial T}{\partial b}=\frac{-(10s+30)(10s+30)}{(s^3+2s^2+10s+10bs+30b)^2}$$

For the nominal $b=1$,

$$\frac{\partial T}{\partial b}=\frac{-(10s+30)(10s+30)}{(s^3+2s^2+20s+30)^2}$$

$$S_b=\frac{b}{T}\frac{\partial T}{\partial b}=\left(\frac{1}{\dfrac{10s+30}{s^3+2s^2+20s+30}}\right)\left[\frac{-(10s+30)(10s+30)}{(s^3+2s^2+20s+30)^2}\right]$$

$$=\frac{-10s-30}{s^3+2s^2+20s+30}$$

For small changes Δb,

$$\frac{\Delta T(s)}{T(s)}\cong\frac{\Delta b}{b}S_b=\left(\frac{\Delta b}{b}\right)\frac{-10s-30}{s^3+2s^2+20s+30}$$

At $s=0$, for calculating the forced response to a step input,

$$\frac{\Delta T(0)}{T(0)}\cong\left(\frac{\Delta b}{b}\right)(-1)$$

Changes in T are proportional (in the opposite sense) to changes in b.

DRILL PROBLEM

D4-9. Find the sensitivities of the transfer functions of the following systems to small changes in k_1, k_2, and k_3 about the given nominal values:

ans. (a) $(s+2)/(s+5)$, $-2/(s+5)$, $-3/(s+5)$
(b) $(s^3+9s^2+26s+24)/(s^3+9s^2+4s+72)$,
$-3(s^2+6s+16)/(s^3+9s^2+4s+72)$,
$(3s^3+27s^2+128s+240)/$
$(s+4)(s^3+9s^2+4s+72)$

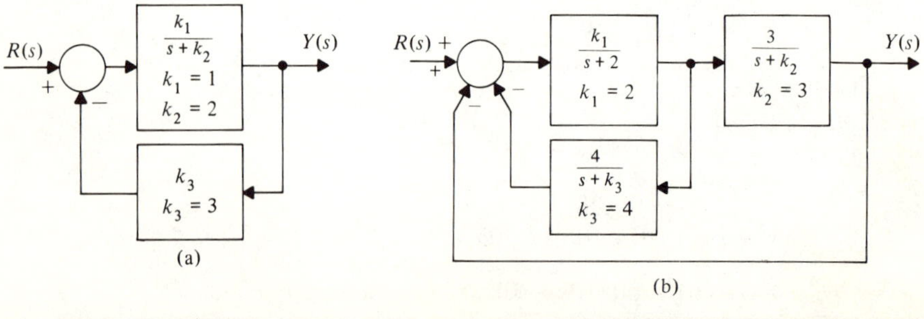

(a)

(b)

4.6 SENSITIVITY TO DISTURBANCE SIGNALS

Another major advantage of feedback is that it may be used to reduce the effects of disturbance inputs upon system response. For example, for the system of Fig. 4-12a, a disturbance signal $D(s)$ affects the plant but is not accessible to the designer. The transfer function relating $Y(s)$ to $D(s)$ is

$$T_D(s)=\frac{1}{s+2}$$

For a unit step disturbance input, the final value of the output due to the disturbance is given by

$$Y(s)=\frac{1}{s}\frac{1}{s+2}$$

$$\lim_{t\to\infty} y(t)=\lim_{s\to 0} sY(s)=\tfrac{1}{2}$$

If the plant is driven in the feedback arrangement of Fig. 4-12b, the transfer function relating $Y(s)$ to $D(s)$ becomes, instead,

$$T_D(s)=\frac{1/(s+2)}{1-[-K/(s+2)]}=\frac{1}{s+2+K}$$

For a unit step disturbance input to the feedback system, the resulting steady state output is given by

$$Y(s)=\frac{1}{s}\left(\frac{1}{s+2+K}\right)$$

$$\lim_{t\to\infty} y(t)=\lim_{s\to 0} sY(s)=\frac{1}{2+K}$$

which may be arbitrarily small by choosing K sufficiently large.

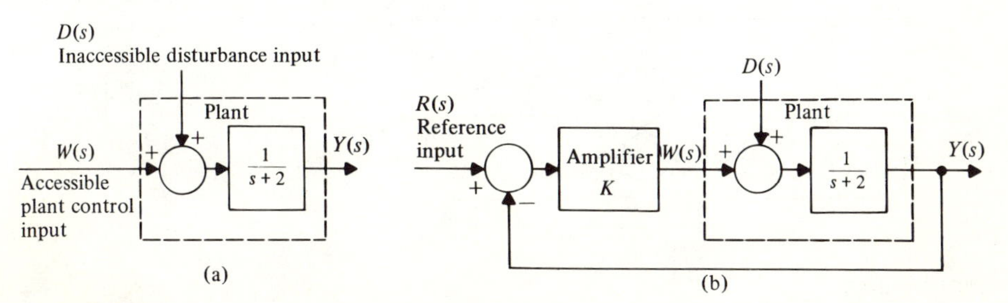

FIGURE 4-12
Disturbance input to a system with and without feedback. (a) Plant with control and disturbance inputs. (b) Plant with feedback.

Another example system is shown in Fig. 4-13a. The disturbance signal $D(s)$ represents an unwanted, largely unknown external effect upon the plant. The two system transfer functions are

$$T_R(s)=\frac{Y(s)}{R(s)}=\frac{\dfrac{10(s+3)}{s(s^2+2s+10)}}{1+\dfrac{10(s+3)}{s(s^2+2s+10)}}$$

$$=\frac{10(s+3)}{s^3+2s^2+20s+30}$$

$$T_D(s)=\frac{Y(s)}{D(s)}=\frac{\dfrac{10}{s^2+2s+10}}{1+\dfrac{10(s+3)}{s(s^2+2s+10)}}$$

$$=\frac{10s}{s^3+2s^2+20s+30}$$

The system is stable, as is easily determined from a Routh-Hurwitz test:

$$\begin{array}{c|cc} s^3 & 1 & 20 \\ s^2 & 2 & 30 \\ s^1 & 5 & \\ s^0 & 30 & \end{array}$$

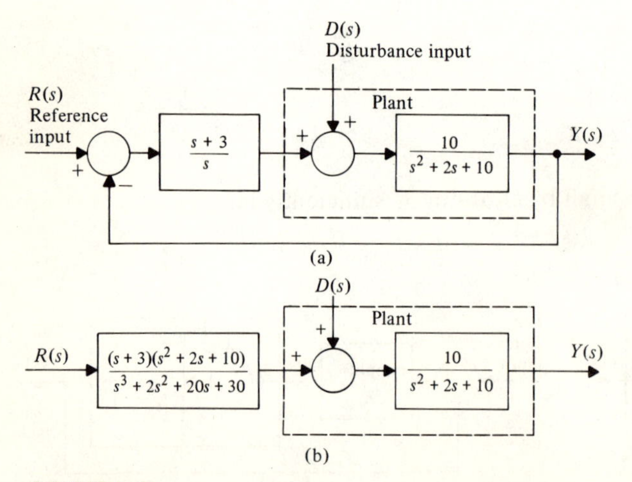

FIGURE 4-13
Reducing the effects of disturbance signals with feedback.
(a) A feedback control system. (b) An open-loop system with the same relation between $Y(s)$ and $R(s)$.

A unit step disturbance to this system will produce zero contribution to the steady state output since

$$\lim_{s\to 0} s\left(\frac{1}{s}\right) T_D(s) = \lim_{s\to 0} \frac{10s}{s^3+2s^2+20s+30} = 0$$

Now consider the open-loop (nonfeedback) system of Fig. 4-13b. It has the same transfer function relating $Y(s)$ and $R(s)$ as does the feedback system. For this system, however, the relationship between the output and the disturbance is not modified by feedback:

$$T_D(s) = \frac{Y(s)}{R(s)} = \frac{10}{s^2+2s+10}$$

A unit step disturbance of the open-loop system will produce a unit contribution to the steady state output:

$$\lim_{s\to 0} s\left(\frac{1}{s}\right) T_D(s) = \lim_{s\to 0} \frac{10}{s^2+2s+10} = 1$$

DRILL PROBLEM

D4-10. For each of the following systems, find the normalized steady state error to a step input $R(s)$ and the normalized steady state error to a disturbance input $D(s)$:

ans. (a) $\frac{1}{4}, \frac{5}{4}$
(b) 0, 1

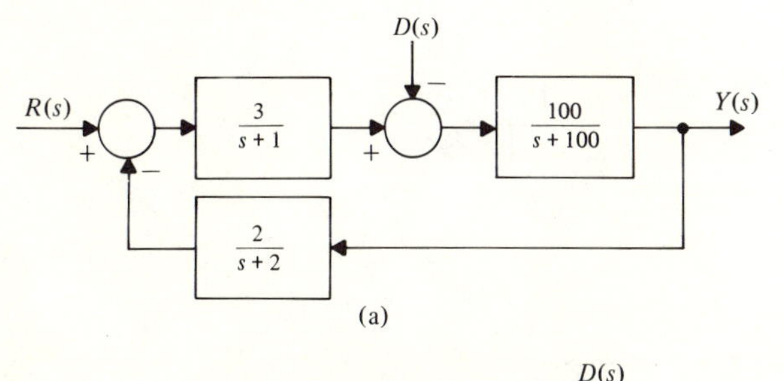

(a)

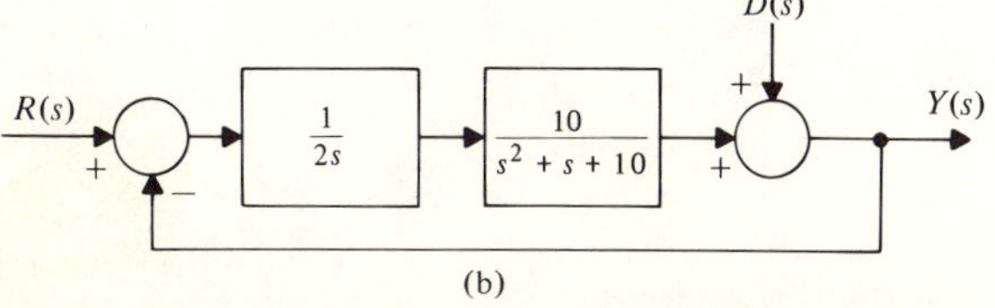

(b)

4.7 PERFORMANCE INDICES AND OPTIMAL SYSTEMS

It generally happens that a system design problem reaches the point where one or more parameters are to be selected to give the best performance. If a measure or index of performance can be expressed mathematically, the problem may be solved for the best choice of the adjustable parameters. The resulting system is termed *optimal* with respect to the selection criteria.

A commonly used performance index is the integral of the square of the error to a step input

$$I_S = \int_0^\infty e^2_{\text{step}}(t)\, dt$$

If the step error is expressed as a function of the adjustable parameters, the index may be minimized with respect to the parameters, yielding the optimal parameter values.

Consider the adjustable system of Fig. 4-14a, for which it is desired to choose the parameter k to give minimum integral square error to a step input. The system transfer function, using Mason's gain rule on the system signal flow graph in Fig. 4-14b, is

$$T(s) = \frac{2/s(s+3) - k/(s+3)}{1 + 2/s(s+3)} = \frac{2-ks}{s^2+3s+2}$$

The error transmittance is

$$T_E(s) = 1 - T(s) = \frac{s^2 + (3+k)s}{s^2+3s+2}$$

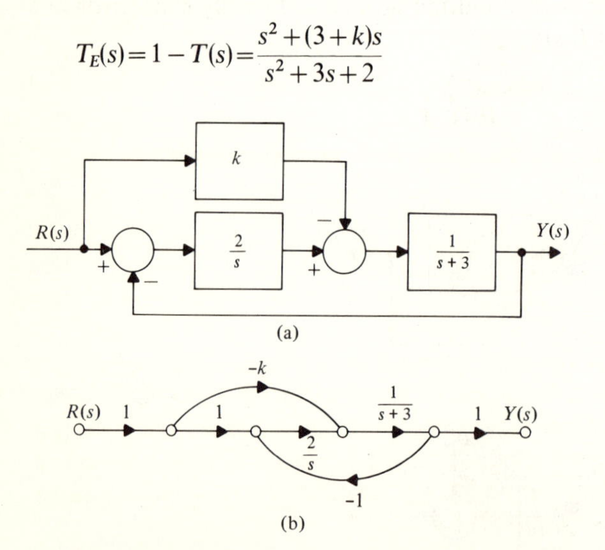

FIGURE 4-14
An adjustable feedback system. (a) System block diagram. (b) System signal flow graph.

and the error to a step input is given by

$$E(s) = \frac{A}{s} T_E(s) = A \frac{s+3+k}{s^2+3s+2}$$

$$= A\left(\frac{k+2}{s+1} + \frac{-k-1}{s+2}\right)$$

As a function of time, the error signal is

$$e(t) = \mathscr{L}^{-1}[E(s)] = A[(k+2)e^{-t} - (k+1)e^{-2t}]u(t)$$

The square of the error is

$$e^2(t) = A^2[(k+2)^2 e^{-2t} - 2(k+1)(k+2)e^{-3t} + (k+1)^2 e^{-4t}]u(t)$$

and the integral square error, in terms of k, is

$$I_S(k) = \int_0^\infty e^2(t)\,dt$$

$$= A^2\left[(k^2+4k+4)\left(\frac{e^{-2t}}{-2}\right) - 2(k^2+3k+2)\left(\frac{e^{-3t}}{-3}\right) + (k^2+2k+1)\left(\frac{e^{-4t}}{-4}\right)\right]_0^\infty$$

$$= A^2[(k^2+4k+4)(\tfrac{1}{2}) - 2(k^2+3k+2)(\tfrac{1}{3}) + (k^2+2k+1)(\tfrac{1}{4})]$$

$$= \frac{A^2}{12}[k^2+6k+11]$$

As a function of k, $I_S(k)$ is a parabola, with minimum given by

$$\frac{dI_S}{dk} = \frac{A^2}{12}[2k+6] = 0$$

$$k = -3$$

Thus the optimal system, in the sense of minimum integral square error to a step with all parameters but k fixed, is the one with this value of k.

Determination of the "best" damping ratio for a second-order system offers another example of the use of performance indices. Consider the stable second-order system with complex conjugate characteristic roots, with transfer function of the form

$$T(s) = \frac{\omega_n^2}{s^2 + 2\xi\omega_n s + \omega_n^2} \tag{4-2}$$

For a unit step input, the error between output and input is given by

$$E(s)=\frac{1}{s}[1-T(s)]=\frac{1}{s}\left[\frac{s^2+2\xi\omega_n s}{s^2+2\xi\omega_n s+\omega_n^2}\right]$$

$$=\frac{s+2\xi\omega_n}{s^2+2\xi\omega_n s+\omega_n^2}=\frac{s+2\xi\omega_n}{(s+\omega_n)^2+(\omega_n\sqrt{1-\xi^2})^2}$$

$$=\frac{s+\omega_n}{(s+\omega_n)^2+(\omega_n\sqrt{1-\xi^2})^2}+\left(\frac{\xi}{\sqrt{1-\xi^2}}\right)\frac{\omega_n\sqrt{1-\xi^2}}{(s+\omega_n)^2+(\omega_n\sqrt{1-\xi^2})^2}$$

$$e(t)=\mathscr{L}^{-1}[E(s)]$$

$$=e^{-\xi\omega_n t}\left[\cos(\omega_n\sqrt{1-\xi^2}t)+\frac{\xi}{\sqrt{1-\xi^2}}\sin(\omega_n\sqrt{1-\xi^2}t)\right]u(t)$$

The integral of the square of the error to a step input is as follows:

$$I_S=\int_0^\infty e^2(t)\,dt$$

$$=\int_0^\infty e^{-2\xi\omega_n t}\left[\cos(\omega_n\sqrt{1-\xi^2}t)+\frac{\xi}{\sqrt{1-\xi^2}}\sin(\omega_n\sqrt{1-\xi^2}t)\right]^2 dt$$

Letting

$$t'=\omega_n t \qquad dt'=\omega_n\,dt$$

$$I_S=\frac{1}{\omega_n}\int_0^\infty e^{-2\xi t'}\left[\cos(\sqrt{1-\xi^2}t')+\frac{\xi}{\sqrt{1-\xi^2}}\sin(\sqrt{1-\xi^2}t')\right]^2 dt'$$

A computer-generated plot of $\omega_n I_S(\xi)$ is given as Fig. 4-15. Minimal mean square error to a step input for the system (4-2) occurs for $\xi=0.5$.

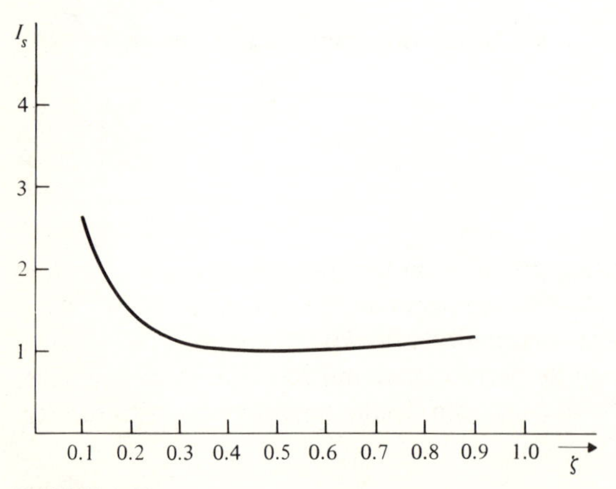

FIGURE 4-15
Integral square error performance measure for a certain second-order system with adjustable damping ratio.

Other useful performance indices may involve errors for other test signals, such as a ramp input, and include the integral of the magnitude of the error,

$$I_M = \int_0^\infty |e(t)|\, dt$$

and integrals where the square and magnitude are weighted with powers of time t to emphasize the behavior at large t:

$$I_{TS} = \int_0^\infty te^2(t)\, dt$$

$$I_{TM} = \int_0^\infty t|e(t)|\, dt$$

Figure 4-16 shows the performance measures I_M, I_{TS}, and I_{TM} for the second-order system (4-2). Clearly, the optimum value of ξ depends upon the definition of "goodness," the performance measure used. Minimum integral magnitude error for the example system occurs for $\xi = 0.67$. Minimum I_{TS} occurs for $\xi = 0.6$, and minimum I_{TM} for $\xi = 0.7$.

In some situations, such as in the control of a spacecraft for minimal fuel use, the performance index is clearly given by the design objectives. In others, an index, if used, must be chosen somewhat arbitrarily. In the latter case, choosing desired response characteristics such as rise time, overshoot, and steady state error directly may be much more sensible than choosing a performance index.

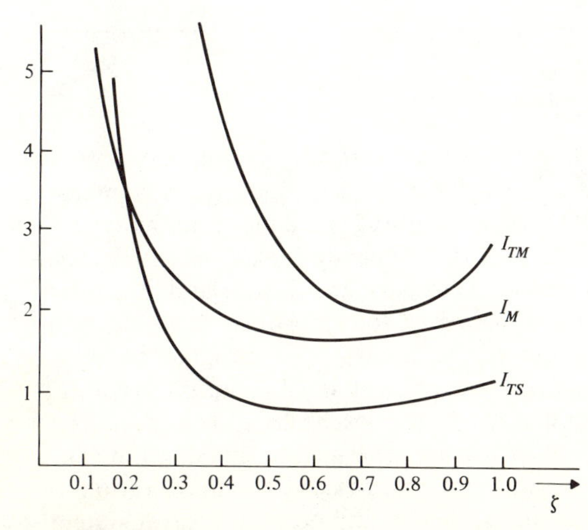

FIGURE 4-16
Other performance measures for a certain second-order system with adjustable damping ratio.

DRILL PROBLEM

D4-11. Use Figs. 4-15 and 4-16 to determine the optimum choice of the parameter k for the given transfer function with the indicated step response performance measure.

(a) $T(s)=\dfrac{100}{s^2+ks+100}$

Integral square error I_s

ans. 10

(b) $T(s)=\dfrac{k}{2s^2+s+k}$

Integral magnitude error I_M

ans. 0.278

(c) $T(s)=\dfrac{k}{s^2+ks+k}$

Integral time-weighted square error I_{TS}

ans. 1.44

(d) $T(s)=\dfrac{10k}{s^2+8s+10k}$

Integral time-weighted magnitude error I_{TM}

ans. 3.26

4.8 A TRANSPORTATION SYSTEM

The control of transportation systems is an interesting design area. Since many European trains are electric, velocity control through traction motor control is common in that area of the world. Desired speed inputs to the system are made by the operator, with an override occurring in case of emergency. Such systems are used in Britain, France, Germany, and Switzerland. A similar system is used for the 100-mi/hr Japanese Kyoto-to-Tokyo train. In San Francisco, the BART (Bay Area Rapid Transit) system is designed to automatically vary the speed as conditions warrant, without human intervention.

Manned aircraft and space flight are further examples of transportation systems where control inputs to the vehicle are generated by an autopilot. The same methodology can be applied to bus, car, and passenger train operation to improve performance. In this example, a velocity and position control system for a rail vehicle is examined. It is similar to systems employed for passenger trains in Germany and Switzerland.

Figure 4-17a shows the relation between motor drive and velocity for

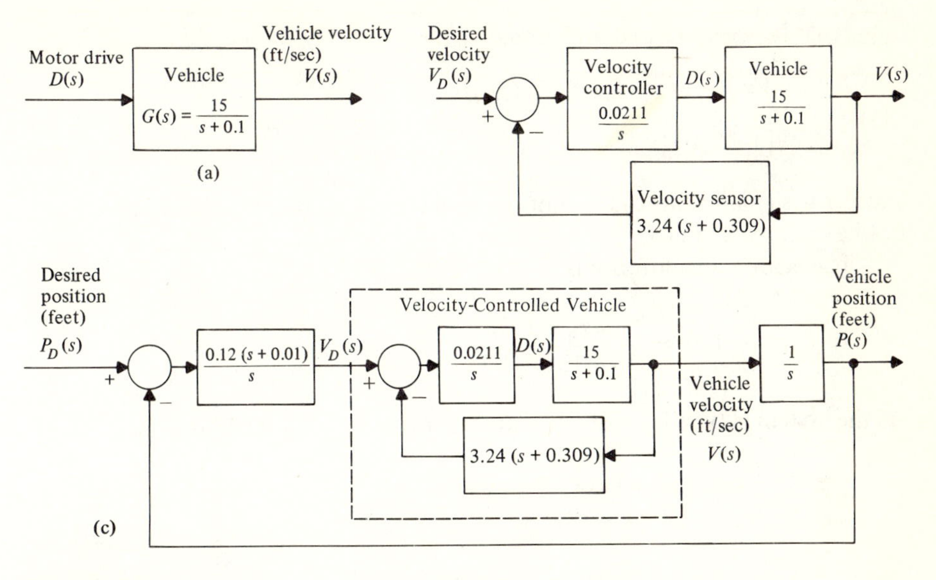

FIGURE 4-17
Controlling a transportation vehicle. (a) Vehicle model. (b) Vehicle with velocity control feedback loop. (c) Vehicle with velocity and position control.

an electric rail car. A unit step input $D(s)$ will produce a steady state car velocity given by

$$\lim_{s \to 0} s\left(\frac{1}{s}\right)\left(\frac{15}{s+0.1}\right) = 150 \text{ ft/sec}$$

with a time constant of 10 sec. In perhaps more familiar terms, in 10 sec the car will accelerate to

$$(150)(0.63) = 94.5 \text{ ft/sec}$$
$$(94.5)(3600/5280) = 64 \text{ mi/hr}$$
$$(94.5)(3600/3280) = 103.7 \text{ km/hr}$$

Automobile racing enthusiasts would characterize this vehicle by saying that it can accelerate from 0 to 60 mi/hr in 8.8 sec or from 0 to 100 km/hr in 9.3 sec.

In Fig. 4-17b, the vehicle is shown as part of a feedback system for velocity control, where V_D is the desired velocity input. This system has transfer function

$$T_V(s) = \frac{\left(\frac{0.0211}{s}\right)\left(\frac{15}{s+0.1}\right)}{1 + \left(\frac{0.0211}{s}\right)\left(\frac{15}{s+0.1}\right)(3.24)(s+0.309)}$$

$$= \frac{0.317}{s^2 + 1.125s + 0.317}$$

The error between desired and actual velocity is

$$E_V(s) = V_D(s) - V(s) = V_D(s) - T_V(s)V_D(s)$$
$$= [1 - T_V(s)]V_D(s)$$

The error signal in this case is not the summer signal because this system does not have unity feedback.

The error transmittance is

$$T_{VE}(s) = 1 - T_V(s) = \frac{s(s+1.125)}{s^2+1.125s+0.317}$$

As the system is type 1, its steady state error to a step input will be zero:

$$\lim_{s\to 0} s\left(\frac{1}{s}\right)\left[\frac{s(s+1.125)}{s^2+1.125s+0.317}\right] = 0$$

Its normalized steady state error to a ramp input is

$$\lim_{s\to 0} s\left(\frac{1}{s^2}\right)\left[\frac{s(s+1.125)}{s^2+1.125s+0.317}\right] = \frac{1.125}{0.317} = 3.55\text{ft/sec}$$

The velocity control feedback system has the characteristic equation

$$s^2+1.125s+0.317=0$$

and repeated characteristic roots $s_1, s_2 = -0.56$. Its relative stability is 0.56 unit, and its natural response dies out as fast as

$$e^{-0.56t}$$

compared to the natural response

$$e^{-0.1t}$$

of the vehicle alone, Fig. 4-17a.

To achieve control of the vehicle position, a second feedback loop has been added to the system in Fig. 4-17c. The transfer function of the complete system is

$$T(s) = \frac{\left[\dfrac{0.12(s+0.01)}{s}\right]\left(\dfrac{0.317}{s^2+1.125s+0.317}\right)\left(\dfrac{1}{s}\right)}{1+\left[\dfrac{0.12(s+0.01)}{s}\right]\left(\dfrac{0.317}{s^2+1.125s+0.317}\right)\left(\dfrac{1}{s}\right)}$$

$$= \frac{0.038s+0.00038}{s^4+1.125s^3+0.317s^2+0.038s+0.00038}$$

The system is stable, as a Routh-Hurwitz test easily shows:

$$\begin{array}{c|ccc}
s^4 & 1 & 0.317 & 0.00038 \\
s^3 & 1.125 & 0.038 & \\
s^2 & 0.283 & 0.00038 & \\
s^1 & 0.036 & & \\
s^0 & 0.00038 & &
\end{array}$$

The error between desired and actual position is

$$E(s) = P_D(s) - P(s) = [1 - T(s)]P_D(s)$$

giving the following error transmittance:

$$T_E(s) = 1 - T(s) = \frac{s^2(s^2 + 1.125s + 0.317)}{s^4 + 1.125s^3 + 0.317s^2 + 0.038s + 0.00038}$$

The system is type 2; it exhibits zero steady state error to a step and to a ramp input $P_D(s)$. A ramp desired position constitutes a step in desired velocity. In the steady state, this system will thus approach zero error in both position and velocity.

4.9 PHASE-LOCKED LOOP FOR A CB RECEIVER

The phase-locked loop (PLL) is an important component of many telecommunications systems. It is used to demodulate the stereo channel in FM broadcast receivers, to detect and maintain the color subcarrier in color television receivers, to generate precise frequencies in citizen band (CB) receivers, and in many other applications.

Figure 4-18a shows a pictorial diagram of the operation of a superhetrodyne receiver for the citizen band. The signal from the receiver's antenna and a sinusoidal mixer signal generated in the receiver are mixed in such a way as to produce sums and differences of the antenna signal frequencies with the sinusoidal mixing frequency signal. Those difference frequencies which are centered at 10.7 MHz are passed by the bandpass filter and detected. By changing the mixing signal frequency, different incoming channels are translated to the 10.7-MHz passband to be detected.

From a highly stable crystal-controlled oscillator, logic circuits produce a digital waveform at the proper mixing frequency for the desired channel to be detected. The mixer requires a smooth, sinusoidal mixing signal, and it is the purpose of the phase-locked loop to "lock" the frequency of a voltage-controlled oscillator to the digital rate. In some receivers, a PLL is also used to aid in the frequency synthesis, but that situation will not be considered here.

A phase-locked loop model is given in Fig. 4-18b. The phase of a sinusoidal signal is proportional to the integral of its frequency. The difference in phase

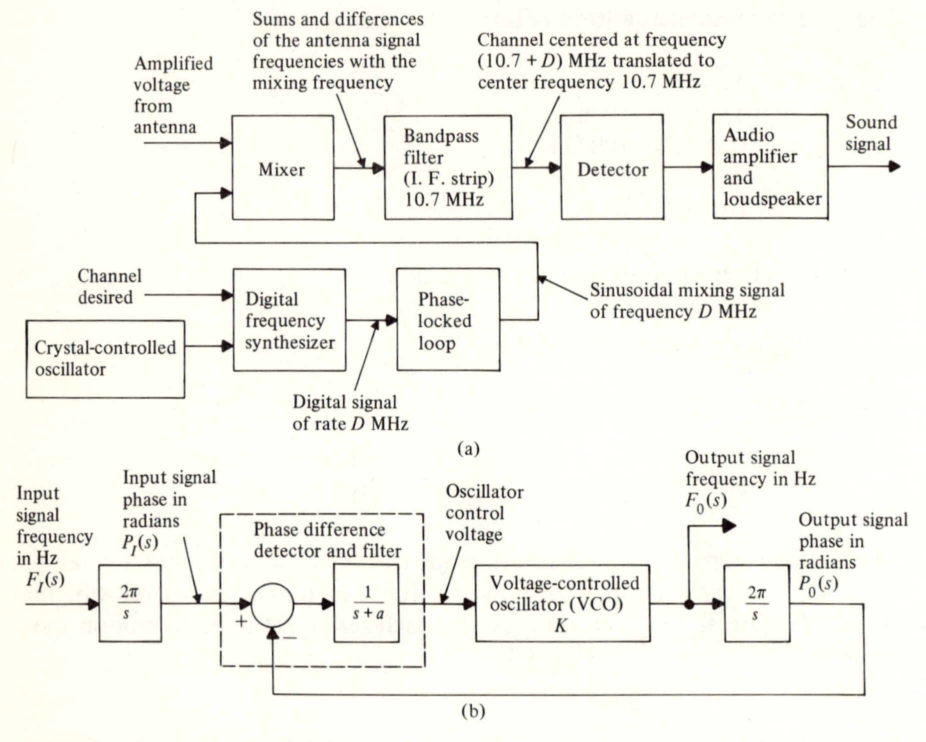

FIGURE 4-18
Phase-locked loop for a Citizen's Band receiver. (a) Description of a CB receiver. (b) Phase-locked loop model.

between the input and output signals is sensed and used to control the frequency of an oscillator in such a way that difference in phase (and thus in frequency) is reduced. Commercial phase-locked loops may differ from this arrangement in that the phase difference detector is nonlinear and capable of sensing phase differences only as large as π or 2π rad. Nonetheless, this simple model adequately predicts PLL performance for most applications.

The system block diagram of Fig. 4-18b is unusual, compared to previous examples, because it relates frequencies and phases of signals, not the signals themselves.

The transfer function of the system which relates output frequency to input frequency is

$$T(s)=\frac{F_O(s)}{F_I(s)}=\frac{\dfrac{2\pi}{s}\left(\dfrac{1}{s+a}\right)(K)}{1+\left(\dfrac{1}{s+a}\right)(K)\left(\dfrac{2\pi}{s}\right)}$$

$$=\frac{2\pi K}{s^2+as+2\pi K}$$

The error between input and output frequency is given by

$$E(s) = F_I(s) - F_O(s) = [1 - T(s)]F_I(s)$$

and has transmittance

$$T_E(s) = 1 - T(s) = \frac{s^2 + as}{s^2 + as + 2\pi K}$$

The system is type 1 and has zero steady state error to a step input frequency change.

The choice of filter time constant $(1/a)$ and voltage-controlled oscillator gain K is based upon the desired characteristics of the system when responding to a change in desired frequency. A reasonable requirement would be for the minimum mean square error damping ratio of 0.5 and a settling time of 0.1 sec.

The system transfer function

$$T(s) = \frac{2\pi K}{s^2 + as + 2\pi K} = \frac{\omega_n^2}{s^2 + 2\xi\omega_n s + \omega_n^2}$$

is of the form analyzed in detail in Chap. 2, with

$$\omega_n = \sqrt{2\pi K} \tag{4-3}$$

and

$$\xi = \frac{a}{2\omega_n} = \frac{a}{2\sqrt{2\pi K}} \tag{4-4}$$

Using Fig. 2-25, which shows normalized settling time as a function of damping ratio, for a 0.5 damping ratio,

$$\omega_n T_S = 5.3 \tag{4-5}$$

For a settling time $T_S = 0.1$, Eq. (4-5) gives

$$\omega_n = \frac{5.3}{0.1} = 53$$

Using (4-3),

$$K = \frac{\omega_n^2}{2\pi} = \frac{(53)^2}{6.28} = 447$$

and with $\xi = 0.5$, (4-4) gives

$$a = 2\omega_n \xi = 2(53)(0.5) = 53$$

This preliminary design is now examined for "worst case" behavior in changing from channel to channel. The lowest CB carrier frequency is 26.97 MHz, and the highest is 27.26 MHz, corresponding respectively to PLL input frequencies of

$$26.97 - 10.7 = 16.27 \text{ MHz}$$

and

$$27.26 - 10.7 = 16.56 \text{ MHz}$$

The largest change in PLL input frequency between stations will be from 16.27 to 16.56 MHz, a step change of 290,000 Hz. By design, the frequency settles to within 5%,

$$(0.05)(290{,}000) = 14{,}500 \text{ Hz}$$

in 0.1 sec, but it will take about 3 times that long for the frequency to settle to below about 50 Hz, which is necessary for intelligible reception. A 0.3-sec maximum time to change stations is probably quite acceptable.

When the device is first turned on, however, it is possible that it will have to respond to a step change of up to 16.56 MHz. Although the behavior of a PLL for such a change in input frequency will likely be nonlinear at first, the linear model will be used to predict an approximate initial length of time until the PLL output frequency has settled to within about 50 Hz. In the first 0.1 sec, the response will settle to

$$(0.05)(16{,}560{,}000) = 828{,}000 \text{ Hz}$$

As the envelope of the second-order oscillatory natural behavior is exponential, the response will settle to within 5% of this value in the next 0.1 sec:

$$(0.05)(828{,}000) = 41{,}400 \text{ Hz}$$

Continuing,

$$(0.05)(41{,}400) = 2070 \text{ Hz}$$
$$(0.05)(2070) = 103.5 \text{ Hz}$$

so an acceptable "worst case" initial tuning is predicted to be within about 0.5 sec.

4.10 SUMMARY

For any function $y(t)$,

$$y(0) = \lim_{s \to \infty} sY(s)$$

If a function $y(t)$ has a finite value, it is given by

$$\lim_{t \to \infty} y(t) = \lim_{s \to 0} [sY(s)]$$

If the function $y(t)$ does not have a finite final value, the above limit may give a misleading result.

Tracking systems are designed so that the system output, as nearly as possible, equals the input. One measure of their performance is the steady state errors between input and output for step, ramp, parabolic, and other power-of-t input signals. When the error has a final value, the normalized steady state error to an input $(A/n!)t^n u(t)$ is given by the final-value theorem:

$$\lim_{s \to 0} \frac{1}{s^n} [1 - T(s)]$$

A system which has finite steady state error for the nth-power-of-t input will have zero steady state error for lower powers of t and infinite steady state error for higher powers of t. With experience, the general performance of tracking systems can be accurately estimated from their performance with power-of-t inputs.

System type number is the number of $s = 0$ numerator roots in its output-input error transmittance

$$T_E(s) = [1 - T(s)]$$

Type number determines the steady state error properties of a system, which are summarized in Table 4-2.

Any single-input, single-output system may be represented with a unity feedback structure, if desired. The equivalent forward transmittance is related to the transfer function by

$$G_E(s) = \frac{T(s)}{1 - T(s)}$$

System type is thus also the number of $s = 0$ denominator roots of $G_E(s)$. To increase the type number of such a system, it is necessary to add $1/s$ factors to the forward transmittance, provided that the resulting system remains stable.

System error coefficients are defined as

$$K_i = \lim_{s \to 0} s^i G_E(s)$$

and these are related to steady state errors in Table 4-3.

Feedback may be used to make system response relatively independent of inaccuracies in some of the system's parameters. A feedback system may thus be designed to perform well even when the controlled plant parameters are not known accurately or when they drift with time. In general, the sensitivity of a transfer function T to a change in a parameter a is

$$S_a = \lim_{\Delta a \to 0} \frac{\Delta T/T}{\Delta a/a} = \frac{a}{T}\frac{\partial T}{\partial a}$$

It is the fractional rate of change of T with fractional change in a. Sensitivities may depend upon the complex variable s.

Feedback may also be used to reduce the effects of disturbance signals upon a system's response. A tracking feedback arrangement which compares the output with a desired reference signal and works to drive the error between the two to zero tends to lessen disturbance effects.

If the quality of performance of a system can be expressed with a performance measure I in terms of the adjustable system parameters, the selection of parameters may be reduced to the mathematical process of finding the set of parameters which maximize or minimize I. A common performance index for a tracking system is the integral square error to a step input:

$$I_S = \int_0^\infty e_{\text{step}}^2(t)\, dt$$

The mathematics of finding extrema of functions I of many variables is formidable in most cases, with results that can be quite dependent upon the specific performance measure used.

REFERENCES

Tracking Systems

Ahrendt, W. R., and Savant, C. J. Jr. *Servomechanism Practice.* New York: McGraw-Hill, 1960.

Bode, H. W. *Network Analysis and Feedback Amplifier Design.* Princeton, N. J.: Van Nostrand, 1945.

Cannon, R. H. Jr. *Dynamics of Physical Systems.* New York: McGraw-Hill, 1967.

Chestnut, H., and Mayer, R. W. *Servomechanisms and Regulating System Design,* vol. 1. New York: Wiley, 1959.

Newton, G. C.; Gould, L. A.; and Kaiser, J. F. *Analytical Design of Linear Feedback Controls.* New York: Wiley, 1957.

Savant, C. J. Jr. *Control System Design.* New York: McGraw-Hill, 1964.

Parameter Sensitivity

Cruz, J. B. Jr. *Feedback Systems.* New York: McGraw-Hill, 1972.

———, ed. *System Sensitivity Analysis.* Stroudsburg, Pa.: Dowden, 1973.

Horowitz, I. M. *Synthesis of Feedback Systems.* New York: Academic Press, 1963.

Kreindler, E. "On the Definition and Application of the Sensitivity Function." *J. Franklin Inst.* 285 (January 1968).

Tomovic, R. *Sensitivity Analysis of Dynamic Systems.* New York: McGraw-Hill, 1963.

Performance Measures

Athans, M. "The Status of Optimal Control Theory and Applications for Deterministic Systems." *IEEE Trans. Auto. Contr.*, July 1966.

Dorf, R. C. *Time-Domain Analysis and Design of Control Systems.* Reading, Mass.: Addison-Wesley, 1965.

McCausland, I. *Introduction to Optimal Control.* New York: Wiley, 1969.

Sage, A. P., and White, C. C. III. *Optimum Systems Control.* Englewood Cliffs, N.J.: Prentice-Hall, 1977.

Schultz, D. G., and Melsa, J. L. *State Functions and Linear Control Systems.* New York: McGraw-Hill, 1967.

Transportation Systems

Friedlander, G. D. "Electronics and Swiss Railways." *IEEE Spec.* 11, no. 9 (September 1974): 68–75.

Stefani, R. T. "Design and Simulation of an Automobile Guidance and Control System." *Computers Ed. (COED) Trans. Amer. Soc. Eng. Ed. (ASEE)*, January 1978.

Citizen Band Receivers and Phase-Locked Loops

International Telephone & Telegraph. *Reference Data for Radio Engineers.* Indianapolis: Sams, 1979.

Nash, G. "Phase-Locked Loop Design Fundamentals." Motorola Semiconductor Products Application Note AN-535, 1970.

PROBLEMS

1. Find the initial values $y(0)$ of the signals with the following Laplace transforms. For functions with discontinuities or impulses at $t=0$, these initial values may differ from the initial conditions $y(0^-)$.

   (a) $Y(s)=\dfrac{-4s^3+3s^2+10}{s^3+9s^2+8s+27}$

   (b) $Y(s)=\dfrac{6s+9se^{-4s}}{s^2+8s+10}$

   ans. 6

   (c) $Y(s)=\dfrac{4s^2-3s+12}{11s^4+6s^3+2s^2+17s}$

   (d) $Y(s)=\dfrac{18s^2+10}{s^5+9s^4-8s^3}$

   ans. 0

2. Determine if signals with the following Laplace transforms have a final value. If a finite final value exists, use the final-value theorem to find it.

   (a) $Y(s)=\dfrac{-3s^2+2s+10}{s^3+7s^2+8s+12}$

(b) $Y(s)=\dfrac{6s^2+8s+9}{s^4+6s^3+4s^2+9s+20}$

ans. no final value

(c) $Y(s)=\dfrac{12s^{-2s}}{3s^4+6s^3+2s^2+10s+12}$

(d) $Y(s)=\dfrac{3}{s^5+3s^4+7s^3+10s^2+12s}$

ans. $\frac{1}{4}$

3. For the systems of Fig. P4-1, find the normalized steady state error for a step input $r(t)=u(t)$. The error signal is $r(t)-y(t)$.

ans. (b) final value does not exist.

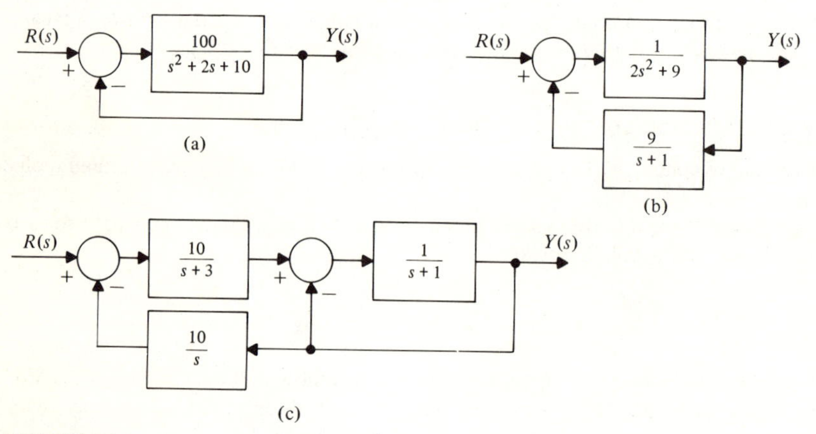

FIGURE P4-1

4. Find the type number and normalized steady state errors for step, ramp, and parabolic inputs for tracking systems with the following transfer functions:

(a) $T(s)=\dfrac{100}{s^2+4s+100}$

(b) $T(s)=\dfrac{4s+100}{s^2+4s+100}$

ans. 2, 0, 0, $\frac{1}{100}$

(c) $T(s)=\dfrac{3s+100}{s^2+4s+100}$

(d) $T(s)=\dfrac{s+5}{s^3+4s^2+s+5}$

ans. infinite errors

(e) $T(s)=\dfrac{2s^2+s+2}{s^4+3s^3+12s^2+s+2}$

5. Convert the systems of Fig. P4-2 to equivalent unity feedback systems.

ans. (b) $2s/(s^4+5s^3+8s^2+6s+2)$

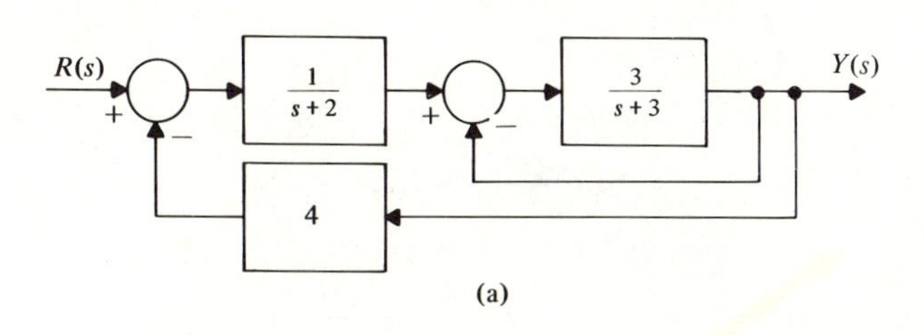

(a)

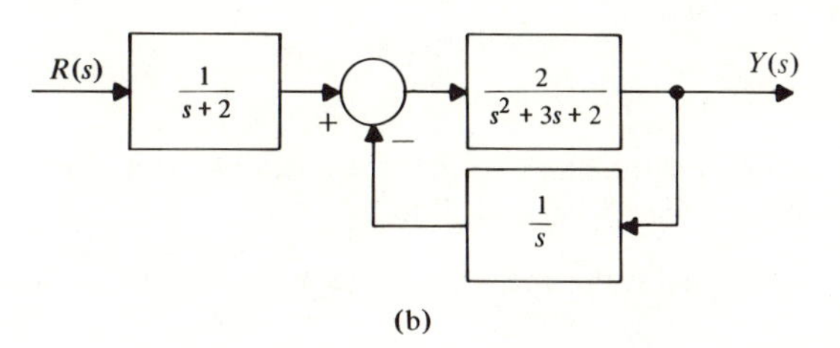

(b)

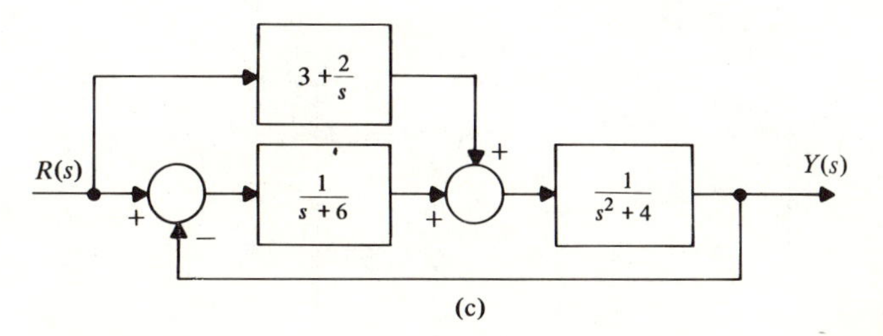

(c)

FIGURE P4-2

6. The forward transmittances of unity feedback tracking systems are given below. For each, find the type number of the system, the steady state error coefficients K_0, K_1, and K_2, and the normalized steady state errors to step, ramp, and parabolic inputs.

(a) $$G_E(s)=\frac{-3s+4}{s^4+3s^3+10s^2+10s}$$

(b) $$G_E(s)=\frac{4}{s^3+2s^2+3s}$$

ans. 1; ∞, $\frac{4}{3}$, 0; 0, $\frac{3}{4}$, infinite error

(c) $$G_E(s)=\frac{6s^2-3s+2}{s^4}$$

(d) $$G_E(s)=\frac{3s+4}{s^4+3s^3}$$

ans. infinite errors

(e) $$G_E(s)=\frac{4s^2+5s+2}{s^4+3s^3}$$

7. What is the type number of the system of Fig. P4-3? What are the normalized steady state errors for step, ramp, and parabolic inputs for this system? The error signal is $E(s)$.

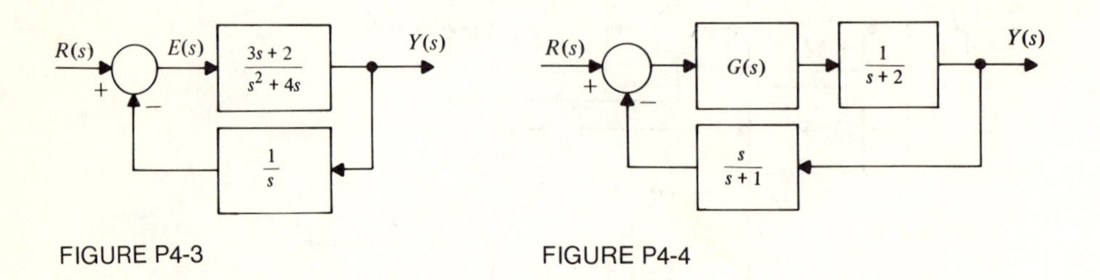

FIGURE P4-3

FIGURE P4-4

8. Choose a transmittance $G(s)$ for the block so that the overall system of Fig. P4-4 is type 2. Many different choices for $G(s)$ are possible.

9. For the system of Fig. P4-5, what is the normalized steady state error to a step input? The error signal is $E(s)$.

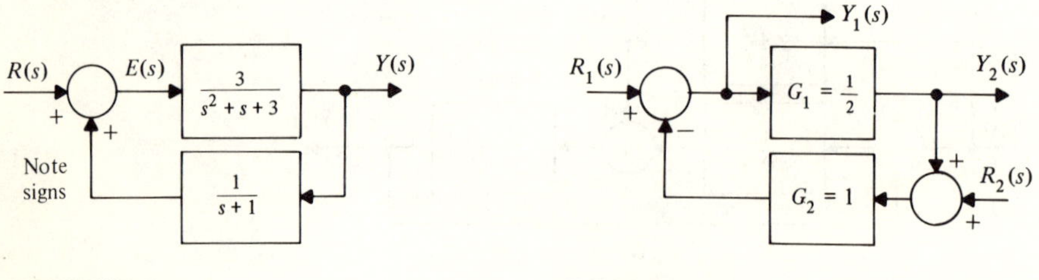

FIGURE P4-5

FIGURE P4-6

10. For the feedback system of Fig. P4-6 with constant "block" transmittances, find the sensitivity of each of the four transfer functions to small changes in G_1 and to small changes in G_2.

ans. sensitivity of T_{11} to $G_1 = -\frac{1}{3}$

11. Find the sensitivities of the following transfer functions to small changes in the parameter a about the given nominal values:

(a) $T(s) = \dfrac{2a}{s+a} \qquad a = 3$

(b) $T(s) = \dfrac{10}{s^2 + as + 10} \qquad a = 2$

ans. $-2s/(s^2 + 2s + 10)$

(c) $T(s) = \dfrac{as + 13}{s^2 + as + 13} \qquad a = 4$

12. It is sometimes possible to eliminate the effects of an inaccessible disturbance upon

the output altogether. When this is done, the disturbance is said to be *decoupled* from the output. For the system of Fig. P4-7 find a "block" transmittance $G(s)$, if possible, for which $D(s)$ is decoupled from $Y(s)$.

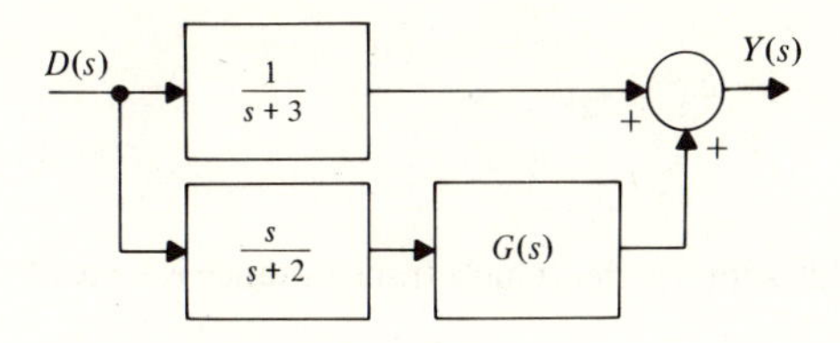

FIGURE P4-7

13. Tracking systems such as the one in Fig. P4-8, in which the reference input is zero (since it is missing), are called *regulators*. It is desired to keep the output as near to zero as is possible in the system, even when there are disturbances present. For what range of the adjustable constant K, if any, is the steady state value of $y(t)$ less than 0.1 when $d(t)$ is a unit step signal?

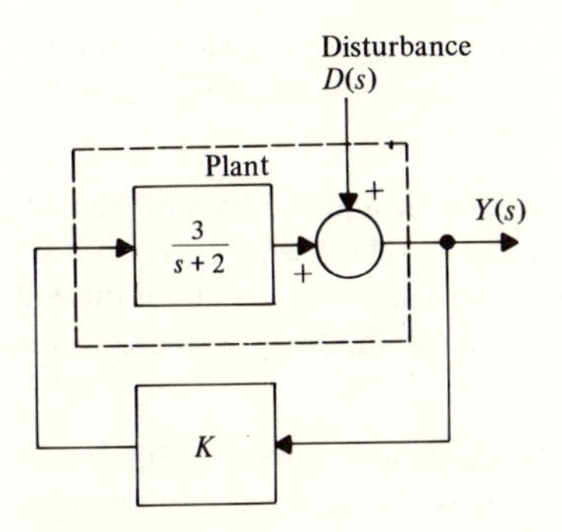

FIGURE P4-8

14. For signals with the following Laplace transforms, find values of the adjustable constants k, if possible, for which

$$I_S = \int_0^\infty e^2(t)\,dt$$

is smallest:

(a) $E(s) = \dfrac{ks+4}{s^2+3s+2}$

(b) $E(s) = \dfrac{-3s+k}{s^2+2s+1}$

ans. 0

15. For the system of Fig. P4-9, find the value of k, if it exists, such that for a step input $r(t)$ the integral square error between $y(t)$ and $r(t)$ is minimum.

ans. 1

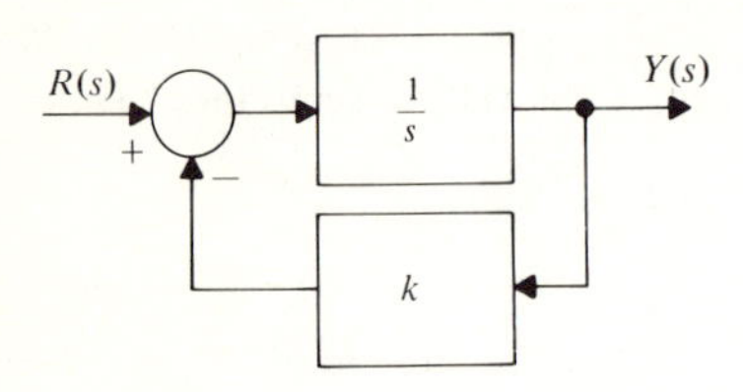

FIGURE P4-9

16. For the transportation system of Sec. 4.8, suppose the vehicle transmittance is instead

$$G(s)=\frac{15}{s+0.2}$$

Find the steady state position error for a 10,000-ft step change in desired position.

17. For the transportation system of Sec. 4.8 with the parameter values given in the text, suppose that an electrical malfunction causes a unit step to be added to $V_D(s)$ before the signal is applied to the velocity-controlled vehicle subsystem. Find the resulting steady state error in vehicle position.

18. For the phase-locked loop of Sec. 4.9, find a and K such that the system's natural response is critically damped and has suitable speed of response for the CB receiver application.

19. For the citizen band receiver of Sec. 4.9 with the values of K and a chosen in the text, suppose a listener is initially tuned to the station at 26.97 MHz. The listener suddenly tunes to the station at 27.26 MHz and remains tuned for 1 sec. Curious about the end of a message the listener had been tuned to, the receiver is suddenly returned to the station at 26.97 MHz. Sketch $f_0(t)$.

20. A power plant frequency control system has the block diagram given in Fig. P4-10. Find, in terms of K, relative stability, steady state errors due to power of t reference inputs, and steady state error due to a unit step change in load torque. Choose, on the basis of your feeling of how a large power generator should be controlled, a value for K. For this value of K, find the form of the system's natural response.

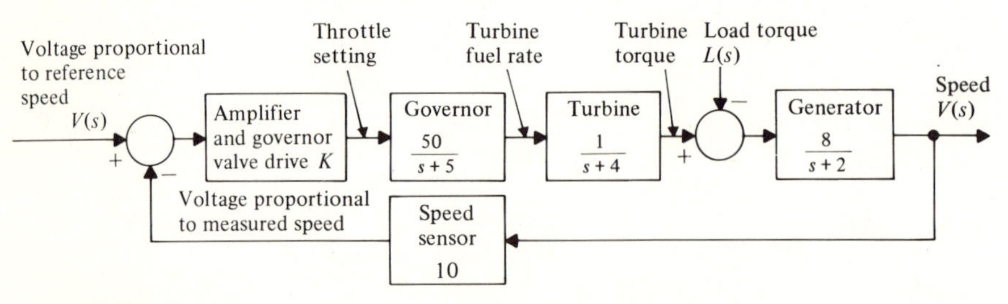

FIGURE P4-10

21. A simplified block diagram for a chemical process control system is shown in Fig. P4-11. The controller is called a PID (or *three-term*) type, because it develops a signal which is a linear combination of terms which are porportional to, the derivative of, and the integral of the incoming signal $f(t)$. If possible, choose values for k_1, k_2, and k_3 such that the resulting system has a relative stability of at least 5 units and zero steady state error to a step input.

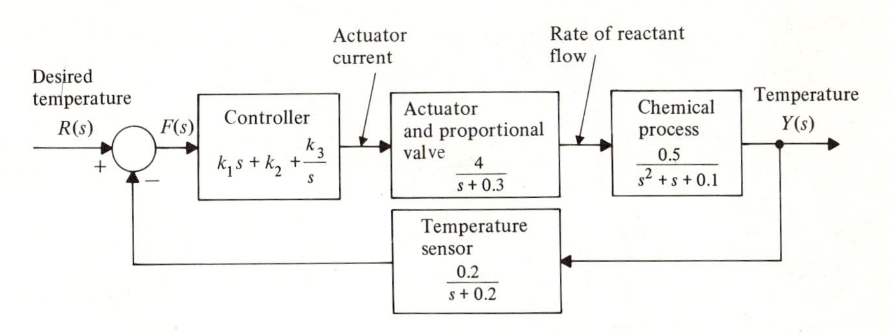

FIGURE P4-11

22. Figure P4-12 shows a simplified model of a submarine depth control when the submarine is submerged. The ship-settling dynamics transmittance is

$$G(s) = \frac{10^4}{s^2 + 2 \times 10^3 s + 10^6}$$

(a) Carefully explain the meaning of the "block" with transmittance ($1000/s$). Is this a component of the system in the same sense as an amplifier or a motor might be?
(b) Determine if the system is stable.
(c) Find the steady state change in actual depth due to a unit step change in desired depth.
(d) A sonar beacon is released by the ship, causing a unit step change in the weight of the vessel. Find the steady state change in actual depth.

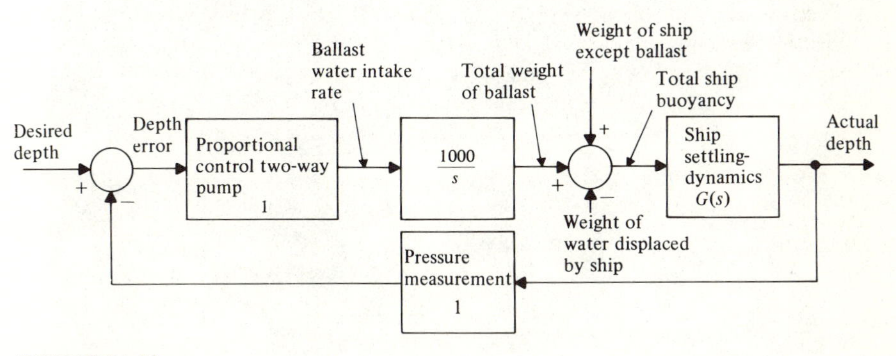

FIGURE P4-12

23. A tape loop positioning system for a digital computer tape drive is diagramed in Fig. P4-13. Find values of the constants k_1, k_2, and k_3, if possible, so that this system has a relative stability in excess of 100 units and a steady state error no greater than 5%. The motor time constant k_3 cannot be less than 0.2.

Discuss the importance of determining the maximum loop position excursions when the entire tape-handling system is operating and, in detail, how these could be determined.

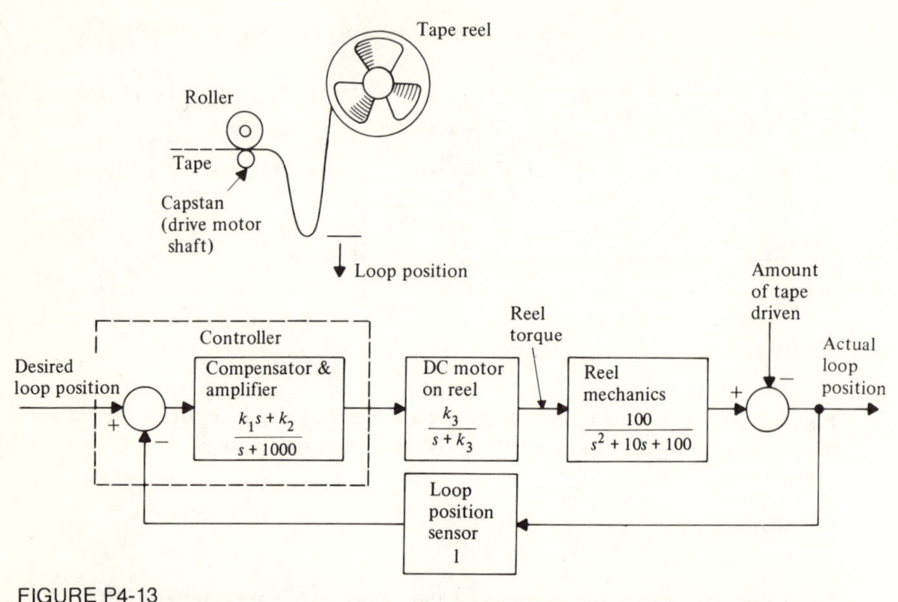

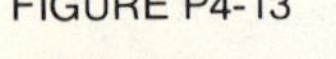
FIGURE P4-13

A petrochemical process is supervised by an engineer using a digital process control system. Advances in these systems have recently made large savings in energy possible. (Photo courtesy of Honeywell Corporation.)

24. Table 4-4 gives approximate measurements of the output $y(t)$ of a system at various times t when the system input is a step

$$r(t)=8.3u(t)$$

(a) Using the data, find an approximate first-order linear, time-invariant system model, specifying its transfer function.
(b) Repeat, but find an approximate second-order linear, time-invariant system model. Specify the approximate system transfer function.
(c) Carefully explain how a higher-than-second-order approximate model could be obtained from the data. Could the data in this table be used to develop a 30th-order model? How might one determine a "best" model of a given order?
(d) Suppose a second-order model for a different system is

$$T(s)=\frac{4}{s^2+3s+2}$$

Find an approximate *first*-order model for the same system. Compare the step responses of the two models.

TABLE 4-4
An input-output record for a system.

Time in seconds, t	*Output value y(t)*
0.0	0.0
0.15	1.1
0.3	2.2
0.45	2.8
0.6	3.3
0.75	3.5
0.9	3.5
1.05	3.4
1.2	3.3
1.35	3.3

25. An automatic gain control (AGC) system is used in applications where it is desired to keep the peak excursions of a signal within fixed limits. In amplitude modulation and frequency modulation (AM and FM) broadcasting and communication, for example, modulating signals with excessively large peak values will cause severe interference with adjacent transmission channels. AGC systems are commonly used to automatically vary the modulating signal volume so that it is as high as can be accommodated without the signal peaks exceeding a certain maximum value.

AGC systems are also termed automatic volume controls (AVC) and leveling amplifiers. They are also used to maintain a constant output amplitude in oscillators, to maintain picture contrast in television systems, and to vary the gain in radio receivers so that strong and weak stations (which vary with antenna position and orientation) are detected with nearly the same volume.

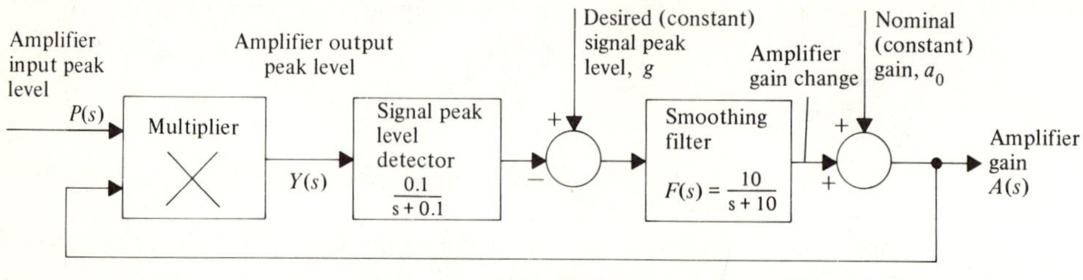

FIGURE P4-14

Figure P4-14 shows a block diagram of an AGC system. The peak level of the amplifier output is detected and compared with the desired peak level q. The peak level error is passed through a filter $F(s)$ which slows and smoothes the amplifier gain changes. The nominal amplifier gain a_0 is changed by the amount of the controlled gain change, which in turn changes the amplifier output peak level.

The model in Fig. P4-14 involves a multiplication operation, so the system is nonlinear. However, for intervals of time when the peak level is constant, the multiplier acts as a constant transmittance to the feedback signal, and linear analysis methods apply. For the multiplier,

$$y(t) = p(t)a(t)$$

(a) For what range of *constant* amplifier input peak levels p is this system stable? The values of the inputs q and a_0 do not affect system stability. However, a constant value of p does, since the multiplicative operation makes p the equivalent of a constant block transmittance.

(b) For *constant* $p = 2$ and $a_0 = 10$, find and sketch the amplifier gain $a(t)$ for

$$q(t) = 1 + u(t)$$

where $u(t)$ is the unit step function.

(c) Repeat (b) with $p = 0.5$, instead.

(d) Find and sketch $a(t)$ for $q = 1$, $a_o = 10$ and $p(t) = u(t)$.

five

ROOT LOCUS ANALYSIS AND DESIGN

5.1 PREVIEW

Root locus methods offer a sophisticated graphical technique for system analysis and design which is well suited to approximation and visualization of alternatives. This chapter begins with a discussion of pole-zero plots for rational functions and their graphical evaluation.

Root locus construction is summarized as a set of six rules, each of which is developed and illustrated with a variety of examples. Computer-aided root locus plotting is addressed in Sec. 5.5.6, where the alternatives of hand construction and calculator and computer plotting are seen to complement one another, each being important to a part of system design. Application is then made of root locus methods to system parameters other than gain and to systems with negative parameter ranges.

The later sections of this chapter introduce compensator design for feedback tracking systems using root locus. Cascade integral, integral plus proportional, lag, and lead compensation are each discussed in general, applied to an example system, and compared. Then potential advantages of including compensation in the feedback path are considered and illustrated with rate feedback compensation of the example system. The final section concerns proportional-integral-derivative compensation.

Although Sec. 5.8.4 discusses and illustrates multiparameter root locus design, for the most part compensator pole and zero locations are chosen here by visualizing the various alternatives on a root locus plot. Then a final root locus is used to select compensator gain for adequate relative stability. The drill problems, too, emphasize obtaining usable designs quickly, later refining them where desired.

The prospect of super performance with very-high-order compensation is discussed in Sec. 5.10.3, along with other practical compensator design considerations.

5.2 POLE-ZERO PLOTS

5.2.1 Poles and Zeros

✻ The *zeros* of a function are the values of the variable for which the function is zero. The values of the variable for which the function is infinite (or its inverse zero) are its *poles*. For a rational function, the zeros are the roots of the numerator polynomial and the poles are the roots of the denominator polynomial. For example, the function

$$F(s)=\frac{2s^2+2s-12}{s^2+7s+10}=\frac{2(s-2)(s+3)}{(s+2)(s+5)}$$

has zeros at $s=2$ and $s=-3$, and poles at $s=-2$ and $s=-5$.

A rational function

$$F(s)=\frac{b_m s^m+b_{m-1}s^{m-1}+\cdots+b_1 s+b_0}{a_n s^n+a_{n-1}s^{n-1}+\cdots+a_1 s+a_0}$$

may always be factored into the form

$$F(s)=\frac{k(s-z_1)(s-z_2)\cdots(s-z_m)}{(s-p_1)(s-p_2)\cdots(s-p_n)}$$

where its poles and zeros are evident. The zeros are the numbers $z_1, z_2, \ldots, z_m$, and the poles are the numbers $p_1, p_2, \ldots, p_n$. The constant k is known as the ✻ *multiplying constant* of the function and is the ratio of the leading coefficients of the numerator and denominator polynomials:

$$k=\frac{b_m}{a_n}$$

Examples of factoring rational functions are as follows:

$$F_1(s)=\frac{3s+4}{2s^2+6s+4}=\frac{\frac{3}{2}[s-(-\frac{4}{3})]}{[s-(-1)][s-(-2)]}$$

$$=\frac{\frac{3}{2}(s+\frac{4}{3})}{(s+1)(s+2)}$$

$$F_2(s)=\frac{-2s^3+32s}{5s^2+20s+65}=\frac{-\frac{2}{5}s[s-(-4)](s-4)}{[s-(-2+3j)][s-(-2-3j)]}$$

$$=\frac{-\frac{2}{5}s(s+4)(s-4)}{(s+2-3j)(s+2+3j)}$$

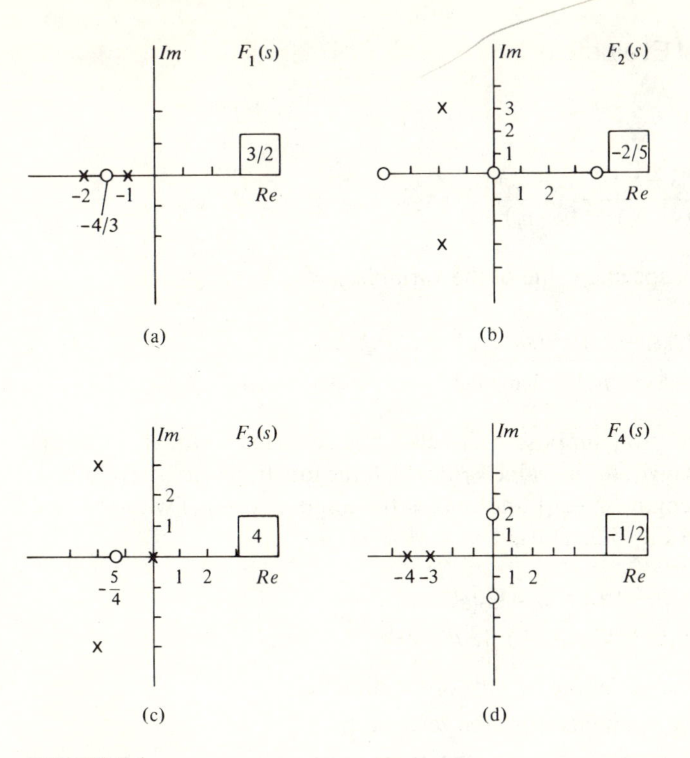

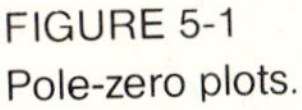
FIGURE 5-1
Pole-zero plots.

When the poles and zeros of a function are plotted on the complex plane, the result is a *pole-zero plot*, from which important properties of the function may be visualized. The zero values are indicated by ○ on the plot and pole values are indicated by **X**. Figure 5-1 shows pole-zero plots for the functions $F_1(s)$ and $F_2(s)$ above and for the following functions:

$$F_3(s)=\frac{4s+5}{s^3+4s^2+13s}=\frac{4(s+\frac{5}{4})}{s(s+2+3j)(s+2-3j)}$$

$$F_4(s)=\frac{-s^2-4}{2s^2+14s+24}=\frac{-\frac{1}{2}(s+2j)(s-2j)}{(s+3)(s+4)}$$

We use the notation of Clark (1962) and others whereby the multiplying constant of a rational function is placed in a box at the right of the pole-zero plot. The rational function is then entirely determined by the plot. This arrangement is by no means common in practice, but the alternatives of separately accounting for the multiplying constant or dealing only with functions with unity multiplying constant are apt to result in unnecessary complication, confusion, and error.

5.2.2 Graphical Evaluation

A rational function

$$F(s)=\frac{k(s-z_1)(s-z_2)\cdots(s-z_m)}{(s-p_1)(s-p_2)\cdots(s-p_n)}$$

when evaluated at a specific value of the variable, $s=s_0$, is

$$F(s_0)=\frac{k(s_0-z_1)(s_0-z_2)\cdots(s_0-z_m)}{(s_0-p_1)(s_0-p_2)\cdots(s_0-p_n)}$$

On a pole-zero plot, suppose a directed line segment is drawn from the position of a pole, say p_1, to the value s_0 at which the function is to be evaluated. The segment has length $|s_0-p_1|$ and makes the angle $\angle(s_0-p_1)$ with the real axis, as indicated in Fig. 5-2. Thus,

$$F(s_0)=\frac{k|s_0-z_1|e^{j\angle(s_0-z_1)}|s_0-z_2|e^{j\angle(s_0-z_2)}\cdots}{|s_0-p_1|e^{j\angle(s_0-p_1)}|s_0-p_2|e^{j\angle(s_0-p_2)}\cdots}$$

$$|F(s_0)|=\frac{|k|\begin{pmatrix}\text{product of the lengths of the directed}\\ \text{line segments from the zeros to } s_0\end{pmatrix}}{\begin{pmatrix}\text{product of the lengths of the directed}\\ \text{line segments from the poles to } s_0\end{pmatrix}}$$

$$\angle F(s_0)=\begin{bmatrix}\text{(sum of the angles of the directed line segments from the zeros}\\ \text{to } s_0)-\text{(sum of the pole angles)}+180^\circ \text{ if } k \text{ is negative}\end{bmatrix}$$

Should k be positive, the 180° is not added to the angle, and $|k|=k$. For example, for $F(s)$ with the pole-zero plot of Fig. 5-3a, a graphical evaluation, Fig. 5-3b, at $s=-2+j$ gives

$$|F(s_0)|=|F(-2+j)|\cong\frac{3(1)(1.4)}{(1.4)(3.2)}=0.94$$

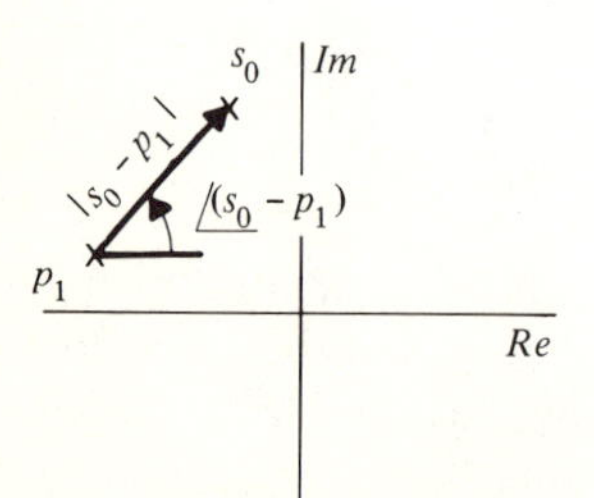

FIGURE 5-2
Evaluating a rational function at $s=s_0$.

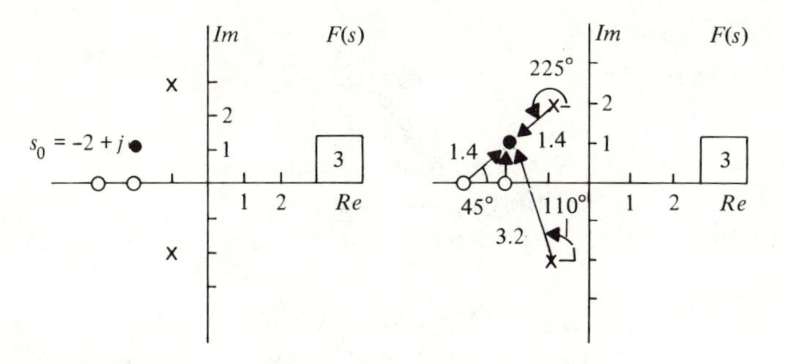

FIGURE 5-3
Example of graphical evaluation.

and

$$\begin{aligned}\angle F(s_0) &= \angle F(-2+j) = (90^\circ + 45^\circ) - (225^\circ + 110^\circ) \\ &= -200^\circ \quad \text{or} \quad +160^\circ\end{aligned}$$

If the multiplying constant had been negative, an additional angle of 180° would be added to obtain $\angle F(s_0)$.

The evaluation of residues in the partial fraction expansion of a rational function involves simply evaluating the function with a pole removed, at the previous location of that pole.

DRILL PROBLEMS

D5-1. Draw pole-zero plots for the following functions. Include the multiplying constant with the plot.

(a) $F(s) = \dfrac{3s-1}{s^2+2s}$

ans. zero at $s = \frac{1}{3}$; poles at $s = 0$ and $s = -2$; multiplying constant 3

(b) $F(s) = \dfrac{9s^2+1}{(s^2+8s+17)^2}$

ans. zeros at $s = \pm j\frac{1}{3}$; repeated poles at $s = -4 \pm j$; multiplying constant 9

(c) $F(s) = \dfrac{-2s^2+6s+3}{(s^2+3s+8)(s^2+6s+15)}$

ans. zeros at $s = 3.44$ and $s = -0.44$; poles at $-1.5 \pm j2.4$ and $-3 \pm j2.45$; multiplying constant -2

(d) $F(s)=\dfrac{(3s+1)(2s+1)}{(4s+1)(7s+1)^2}$

ans. zeros at $s=-\frac{1}{3}$ and $s=-\frac{1}{2}$; pole at $s=-\frac{1}{4}$ and repeated pole at $s=-\frac{1}{7}$; multiplying constant 3/98

D5-2. Find the rational functions represented by the following pole-zero plots.

ans. (a) $6(s+3)(s^2+2s+10)/s$
(b) $-3(s-2000)/(s+2000) \times (s^2+2000s+5\times 10^6)^2$
(c) $\frac{2}{3}(s^2+225)/(s+30)^2(s^2+20s+500)$

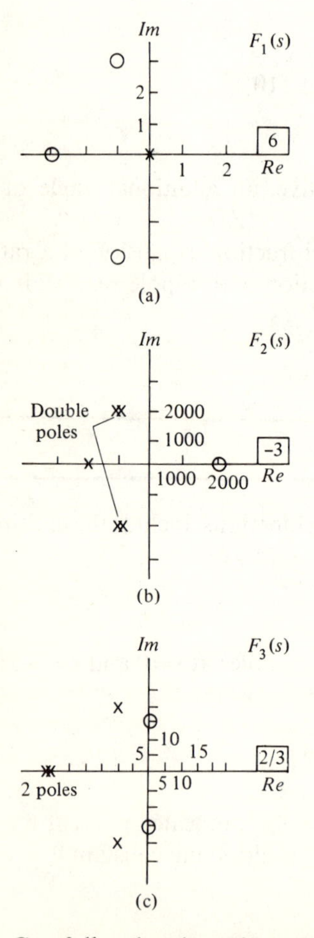

D5-3. Carefully sketch pole-zero plots for the following functions, then graphically evaluate the functions at the indicated value of the variable s. Do not *calculate* distances and angles; rather, estimate them approximately from the plot.

(a) $$F(s)=\frac{10(s-2)}{(s+1)(s+2)}$$

$s=j3$

ans. $3.2e^{-j5^\circ}$

(b) $$F(s)=\frac{4s^2+32}{(s^2+8s+20)(s+2)}$$

$s=2+j$

ans. $0.3e^{-j11^\circ}$

(c) $$F(s)=\frac{4(s^2-4s+5)^2}{(s+3)^2(s^2+6s+10)}$$

$s=j3$

ans. $2.0e^{-j34^\circ}$

5.3 ROOT LOCUS FOR FEEDBACK SYSTEMS

A root locus plot consists of the loci of the poles of a transfer function, or other rational function, as some parameter is varied. A basic system configuration is that of the simple feedback system of Fig. 5-4, with $G(s)$ and $H(s)$ rational functions, where the constant gain K is the parameter of interest. The transfer function of this basic system is

$$T(s)=\frac{KG(s)}{1+KG(s)H(s)}$$

and the poles of the transfer function are the roots of

$$1+KG(s)H(s)=0$$

which depend upon the parameter K. The product of the forward transmittance $KG(s)$ and the feedback transmittance $H(s)$ is termed the open-loop transmittance (or gain) of the system. The poles and zeros of $G(s)H(s)$ are called the

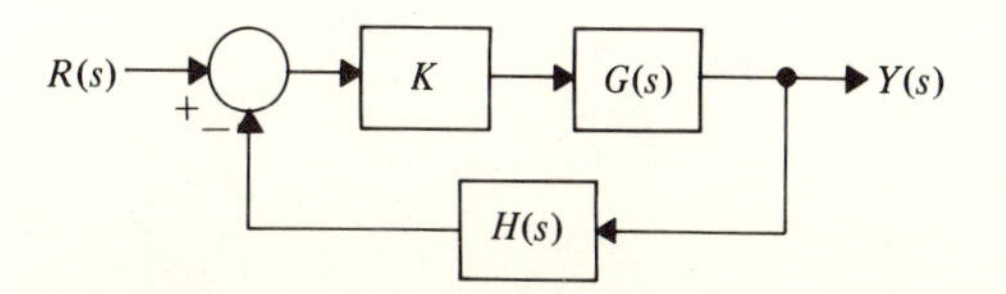

FIGURE 5-4
A simple feedback configuration.

open-loop poles and zeros, while the poles and zeros of $T(s)$ are the *closed-loop* poles and zeros.

If

$$1+KG(s)H(s)=0$$

$$G(s)H(s)=-\frac{1}{K}$$

and for positive K, this means that a point s which is a pole of $T(s)$ makes

$$|G(s)H(s)|=\frac{1}{K}$$

and

$$\angle G(s)H(s)=180^\circ$$

Suppose that there is a point s for which the second of these conditions is satisfied. Then whatever the magnitude of $G(s)H(s)$ for this value of s, there is a corresponding value of K. Thus any point s for which $\angle G(s)H(s)=180^\circ$ is a point on the root locus, for some value of K.

As $K\to 0$, $|G(s)H(s)|\to\infty$, and as $K\to\infty$, $|G(s)H(s)|\to 0$. This is to say that the poles of $T(s)$ are near the poles of $G(s)H(s)$ for small K and are near the zeros of $G(s)H(s)$ for large K. The loci begin on the poles of GH and end on the zeros of GH.

Figure 5-5 shows an example root locus plot. It consists of a pole-zero plot for $G(s)H(s)$, which is generally easy to construct because $G(s)$ and $H(s)$, being components of the system, are usually known in factored or partially factored form. Superimposed upon the pole-zero plot for $G(s)H(s)$ are curves which are the loci of the poles of $T(s)$ as K varies from zero to infinity. The locus segments are symmetrical about the real axis, and the sense of increasing K is indicated on each segment.

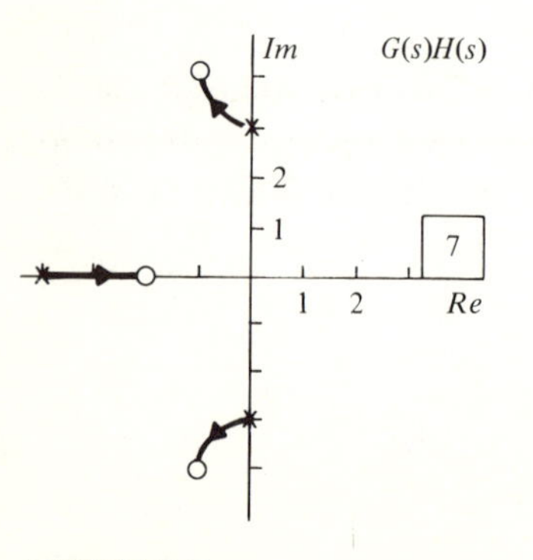

FIGURE 5-5
A root locus plot.

To determine if a given point is a point on the root locus for some value of K between zero and $+\infty$, it is only necessary to determine whether or not the angle of $G(s)H(s)$ is 180°. This determination is easily made graphically, using directed line segments:

$$\begin{aligned}\angle G(s_0)H(s_0) = &\text{ sum of zero angles to } s_0\\ &- \text{sum of pole angles to } s_0\\ &+180^\circ \text{ if the multiplying constant is negative}\end{aligned}$$

At first, only the most common case where the multiplying constant for GH is positive is considered here. In this case, contributions to the angle of GH are just from the directed line segments.

For the GH product with pole-zero plot given in Fig. 5-6a, the angle of $G(s)H(s)$ for the indicated value of s is (approximately)

$$60^\circ - (50^\circ + 40^\circ + 35^\circ) = -65^\circ$$

Thus the indicated point is not on the root locus. For the point tested in Fig. 5-6b,

$$180^\circ - (0^\circ + 110^\circ - 110^\circ) = 180^\circ$$

Hence that point is on the root locus. There is a pole of $T(s)$ there for some positive value of K.

Because, in practice, most transmittances have more poles than zeros, it is common to calculate the negative of the angle of the GH product, which is the

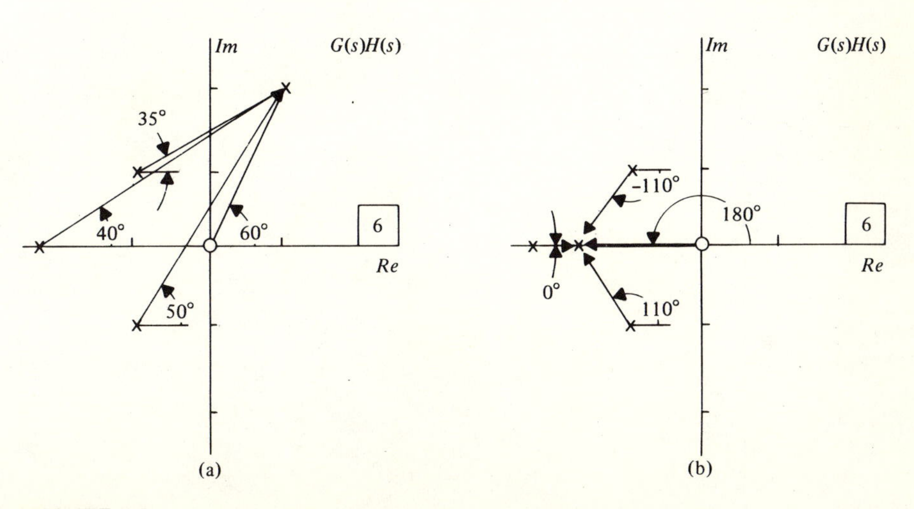

FIGURE 5-6
Testing the angle of $G(s)H(s)$ to determine if a point is on the root locus.

sum of the pole angles minus the zero angles. Since 180° and −180° are the same angle, the angle criterion becomes

$$\text{Sum of pole angles to } s_0 - \text{sum of zero angles to } s_0 = -\angle G(s_0)H(s_0) = 180^\circ$$

We will use this "reverse angle evaluation" in the development to follow.

DRILL PROBLEMS

D5-4. Graphically find the angle of $G(s)H(s)$ at each of the indicated points. Do not calculate angles; rather, estimate them from the plot.

ans. (a) 180°
(b) 77°
(c) −117°

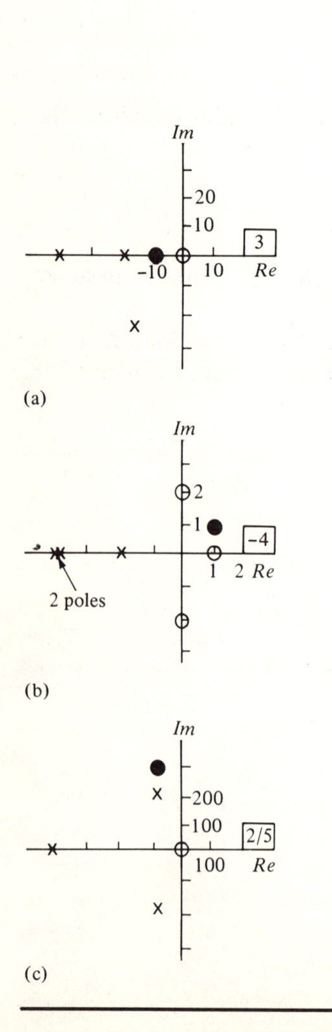

5.4 INTRODUCTION TO ROOT LOCUS CONSTRUCTION

One can, of course, test a whole collection of points on the complex plane to see where the root loci lie. If the angle of GH for a point is near 180°, then that point is near the root locus. In this manner, the various segments of the root locus can be traced. Fortunately, there are some simple principles which allow approximate root locus sketches to be made quite easily. If greater accuracy is needed in one region or another, points near the approximate locus may be tested to determine its actual location more accurately.

Root locus construction principles are summarized in the following sections as a set of six rules.

5.4.1 Loci Branches

RULE 1. Continuous curves, which comprise the branches of the locus, start at each of the n *poles of* GH, *for which* $K=0$. *As* K *approaches* $+\infty$, *the branches of the locus approach the* m *zeros of* GH. *Locus branches for excess poles extend to infinity; for excess zeros, locus segments extend from infinity.*

If $G(s)H(s)$ has more poles than zeros, some of the segments of the root locus, which start on poles (for $K\rightarrow 0$) do not have a zero to end upon (for $K\rightarrow\infty$). These segments of the locus extend from the poles, infinitely far from the origin of the complex plane. It is said that these loci extend "to infinity." If $G(s)H(s)$ has more poles n than zeros m, m segments of the locus extend from a pole to a zero, and $(n-m)$ excess segments each start at a pole and extend infinitely far from the origin. Segments never extend from a pole to infinity and then back from infinity to a zero. If $G(s)H(s)$ has more zeros than poles, the situation is similar, with n segments extending from a pole to a zero and $(m-n)$ excess segments coming from infinity to a zero.

Some examples of root locus plots are given in Table 5-1.

5.4.2 Real axis Segments

RULE 2. The locus includes all points along the real axis to the left of an odd number of poles plus zeros of GH.

The easiest points on the complex plane to test to see if they are on the root locus are points on the real axis. For these points, the angle contribution of each real axis pole or zero is either 0° or 180°, depending upon whether the root is to the right or to the left of the real axis point being tested, as indicated in Fig. 5-7a and b. A set of complex conjugate roots contributes angles to real axis points which are negatives of one another, so the net contribution to angle of a complex set of roots is zero, as indicated in Fig. 5-7c.

A point on the real axis is thus on the root locus if and only if it is to the left

TABLE 5-1
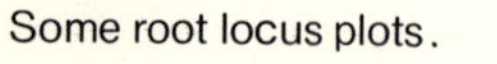
Some root locus plots.

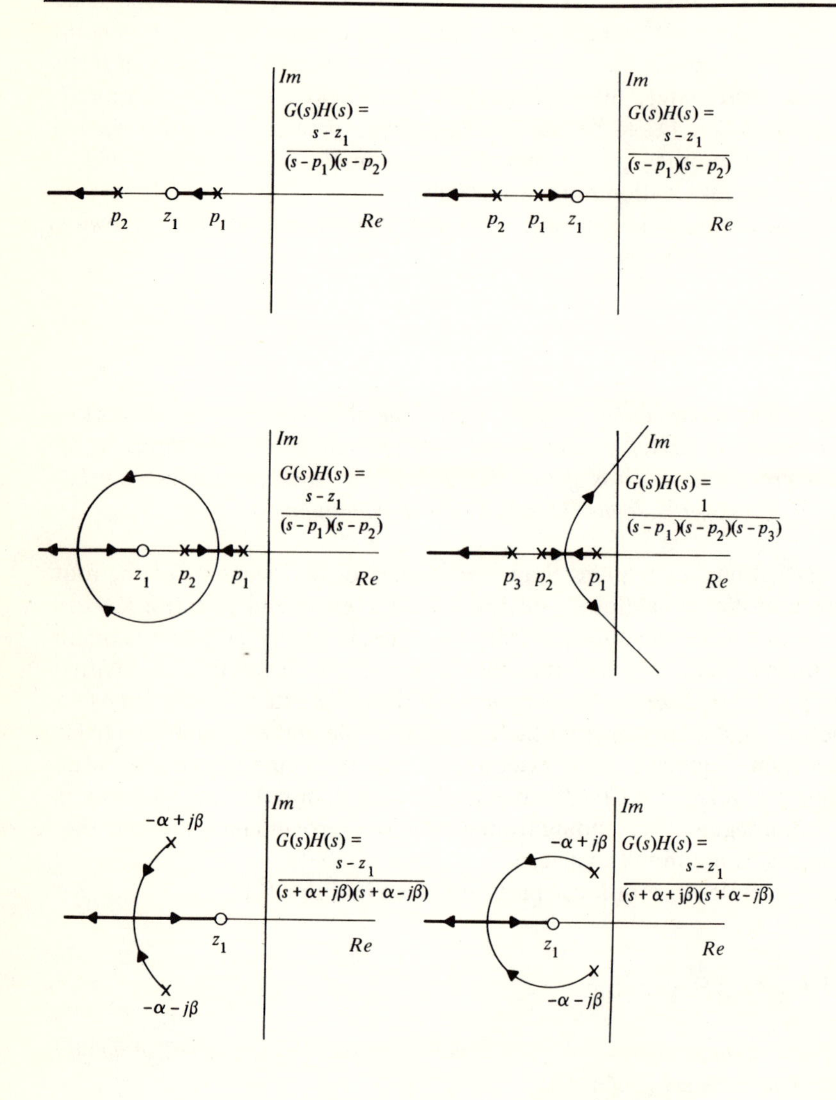

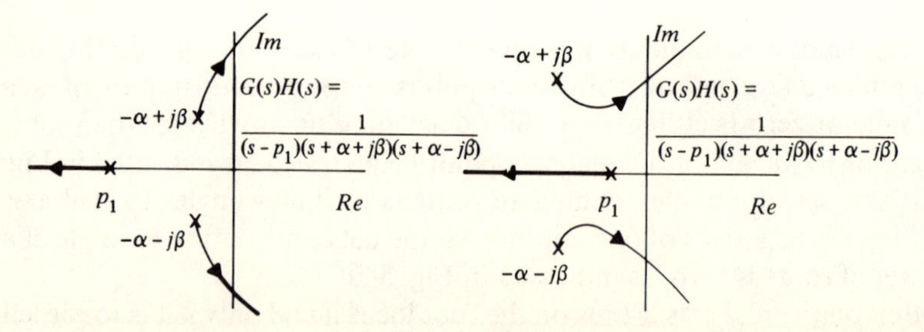

TABLE 5-1 (Continued).

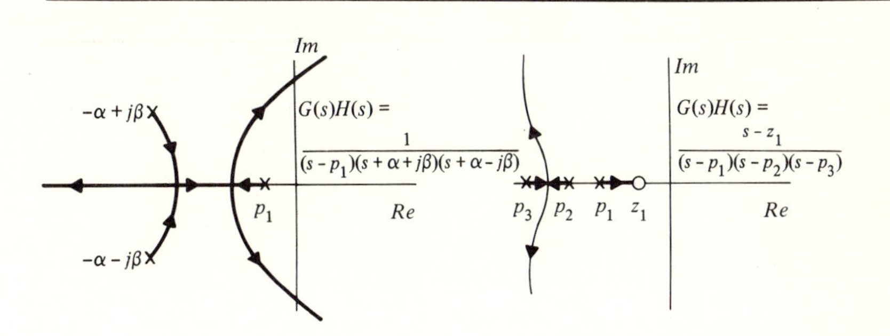

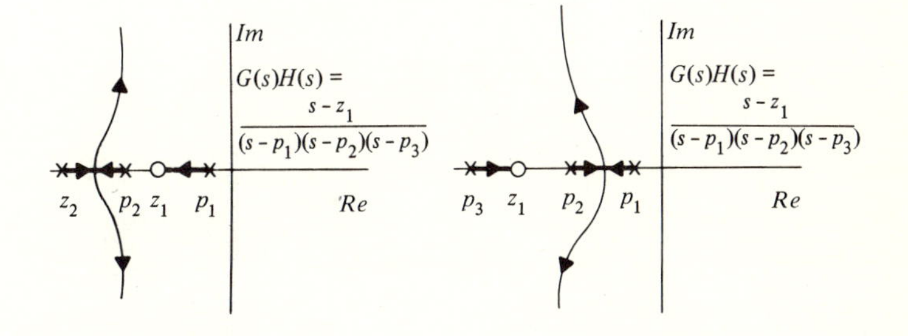

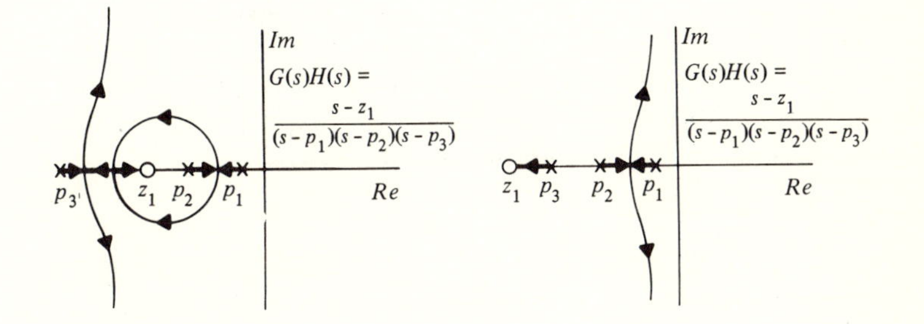

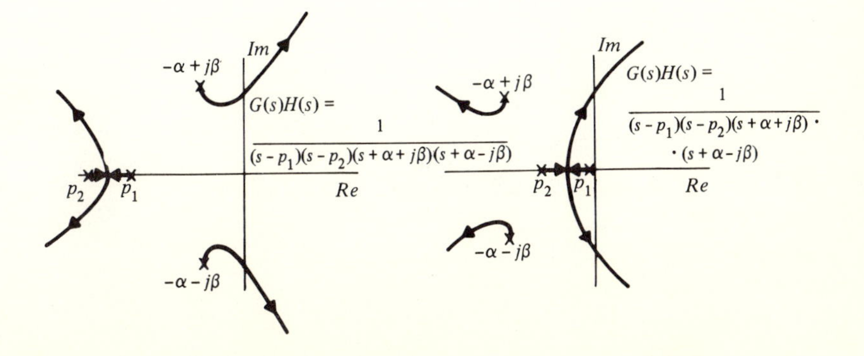

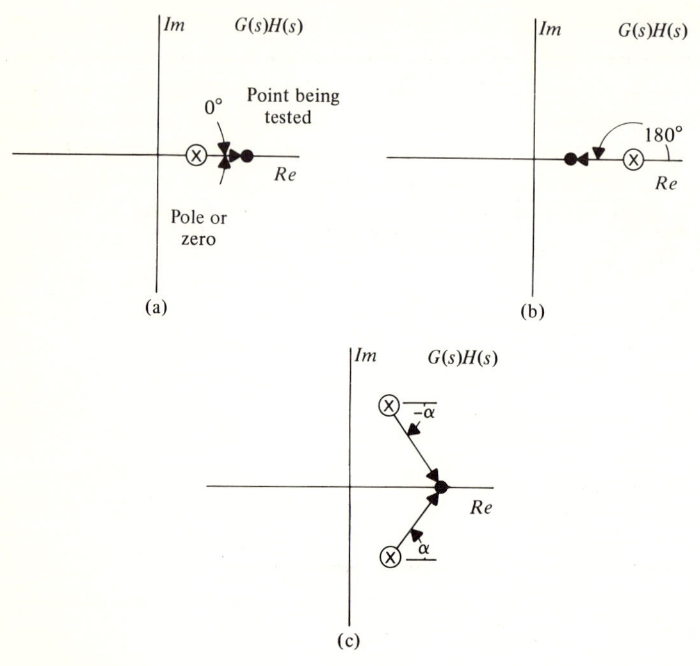

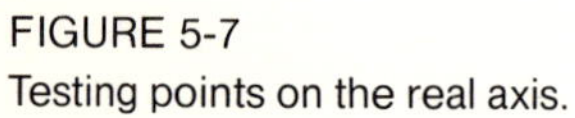
FIGURE 5-7
Testing points on the real axis.

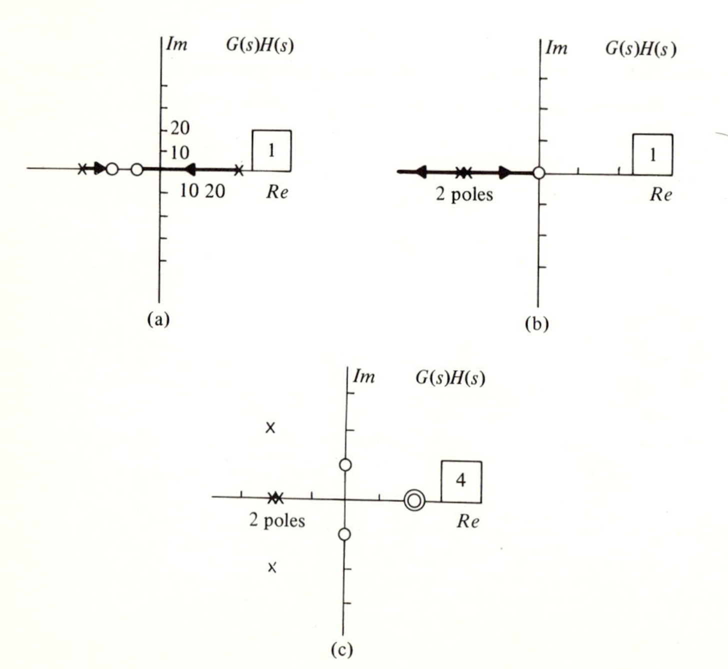

FIGURE 5-8
Examples of real-axis locus segments.

of an odd number of roots (poles and zeros), so that the angle of GH at that point is an odd multiple of 180°. The real axis root locus segments of several systems are sketched in Fig. 5-8. In Fig. 5-8a, the locus is entirely along the real axis. In Fig. 5-8b, one segment extends from the double pole to the zero at $s=0$, and one extends from the double pole to infinity. Two other root locus segments will extend from the complex poles. The sketching of these segments will be discussed later. In Fig. 5-8c, there are no real axis locus segments.

5.4.3. Asymptotic Angles

RULE 3. As K approaches $+\infty$, the branches of the locus become asymptotic to straight lines with angles

$$\theta=\frac{180^\circ+i360^\circ}{n-m}\quad \textit{for } i=0, \pm 1, \pm 2, \textit{ until all } (n-m) \textit{ angles not differing by multiples of } 360^\circ \textit{ are obtained}$$

where n *is the number of poles and* m *is the number of zeros of* GH.

At points on the complex plane very far from all of the poles and zeros of $G(s)H(s)$, the angle of GH is virtually

$$(\text{Number of poles} - \text{number of zeros})\theta$$

where θ is the angle of the point itself, as indicated in Fig. 5-9.

At large distances from the cluster of poles and zeros, root loci extending to or from infinity approach straight-line asymptotes at angles given by

$$(\text{Number of poles} - \text{number of zeros})\theta = 180^\circ \pm i360^\circ$$

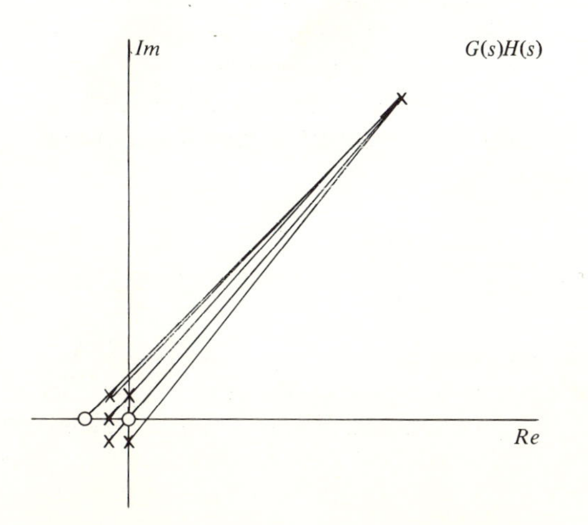

FIGURE 5-9
Angle contributions far from the poles and zeros of GH.

It is important to include the multiples of 360° in this formulation because it is from this term that multiple solutions, where there is more than one asymptotic angle, arise.

If, for example, the GH product has two zeros and six poles, the asymptotic angles are given by

$$(6-2)\theta = 180° \pm i360°$$

$$\theta = \frac{180° \pm i360°}{4}.$$

Substituting various integer values of i, say 0, 1, -1, 2, . . . , the four different angles

$$\theta = -45°, \ -135°, \ +45°, \ -225°$$

result. Substitution of additional integers simply give repetitions of the same angles. There are $n-m=4$ different asymptotic angles in this example.

5.4.4. Centroid of the Asymptotes

RULE 4. The starting point on the real axis from which the asymptotic lines radiate is given by

$$\sigma = \frac{\Sigma \text{ open-loop pole values} - \Sigma \text{ open-loop zero values}}{n-m}$$

This point is termed the centroid of the asymptotes.

Far from all the poles and zeros of GH, the asymptotes would be virtually the same, whether they are drawn to the origin or to some point near the origin around the poles and zeros. Nearer to the pole-zero cluster, much better approximations to the loci are obtained by centering the asymptotes at their centroid rather than the origin. The centroid is a sort of "center of gravity" of the pole-zero plot, and is given by

$$\sigma = \frac{\text{sum of pole locations} - \text{sum of zero locations}}{\text{number of poles} - \text{number of zeros}}$$

Consider the system with pole-zero plot given in Fig. 5-10a. A real axis segment of the locus extends from the real axis pole to the zero at $s=0$. The other two locus segments extend from the imaginary axis poles to infinity. Their asymptotic angles are given by

$$(3-1)\theta = 180° \pm i360°$$

$$\theta = 90°, \ -90°$$

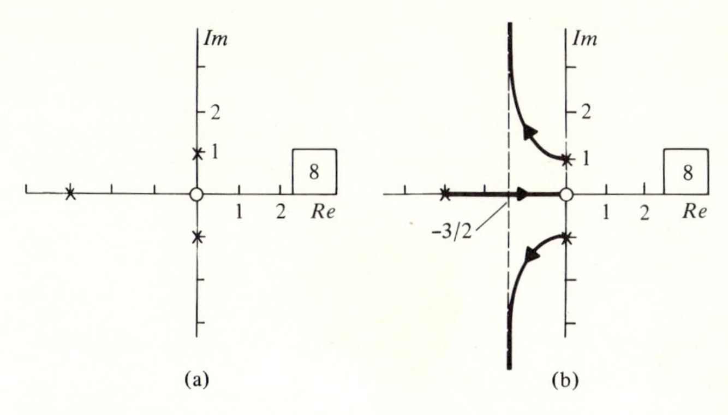

FIGURE 5-10
Example root locus plot with asymptotes and centroid.

The centroid of the asymptotes is

$$\sigma = \frac{(0+j+0-j-3)-0}{3-1} = -\frac{3}{2}$$

The imaginary part contributions of conjugate sets of roots always cancel one another, so only the real parts of the root locations need to be included in the centroid calculation. The complete root locus is shown in Fig. 5-10b.

DRILL PROBLEM

D5-5. Sketch root locus plots, for an adjustable constant K between 0 and $+\infty$, for systems with the following GH products. Find the asymptotic angles and centroid if applicable.

(a) $$G(s)H(s) = \frac{3s}{(s+2)(s^2+6s+18)}$$

ans. $\pm 90°$; -4

(b) $$G(s)H(s) = \frac{10}{(s+6)(s^2+8s+41)^2}$$

ans. $\pm 36°$, $\pm 108°$, $-180°$; -4.4

(c) $$G(s)H(s) = \frac{1}{(s^2+8s+41)(s^2+2s+5)}$$

ans. $\pm 45°$, $\pm 135°$; -2.5

(d) $$G(s)H(s) = \frac{7}{(s+1)(s^2+10s+26)}$$

ans. $\pm 60°$, $-180°$; $\frac{11}{3}$

(e) $G(s)H(s)=\dfrac{2(s^2+4s+3)}{(s+2)^2(s+4)}$

ans. -180°; centroid not applicable

5.5 MORE ABOUT ROOT LOCUS

5.5.1 Root Locus Calibration

The values of the adjustable constant K corresponding to various points on a root locus may be found by applying the relation

$$|G(s)H(s)|=\frac{1}{|K|}$$

for points on the locus. The magnitude $|G(s)H(s)|$ may be found graphically for a point on the locus using

$$|G(s)H(s)|=\frac{\begin{pmatrix}\text{magnitude of the}\\ \text{multiplying constant}\end{pmatrix}\begin{pmatrix}\text{product of}\\ \text{zero distances}\end{pmatrix}}{\text{product of pole distances}}$$

For the system of Fig. 5-11, the values of K corresponding to the imaginary axis points on the locus, where $T(s)$ is marginally stable, is given approximately by

$$|G(s)H(s)|=\frac{10}{(3.3)(4.1)(5.0)}=\left|\frac{1}{K}\right|$$

$$K=6.5$$

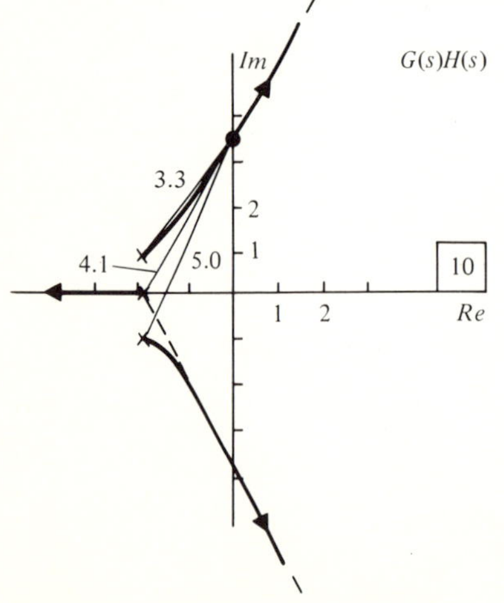

FIGURE 5-11
Calibrating a point on the root locus.

Thus for K greater than about this value, the overall system will be unstable. If the root locus is only approximate, this solution for K is approximate, too. More accuracy may easily be obtained by testing the angles of some points near the approximate locus and refining it.

Finding the values of K corresponding to specific points on the root locus is rather simple, while the reverse problem of solving for the locus points corresponding to a given value of K is generally much more difficult, since this is equivalent to factoring the characteristic equation. To find locus points for a given K, it is easiest to find the values of K for several test points along the loci, then interpolate.

5.5.2 Breakaway Points

RULE 5. Roots leave the real axis at a gain K which is the maximum possible value of K in that region of the real axis. Roots enter the real axis at a gain K which is the minimum possible value of K in that region of the real axis. Two roots leave or strike the axis at the breakaway point at angles of $\pm 90^\circ$.

For the system with $G(s)H(s)$ as in Fig. 5-12, there is a real axis root locus segment, but it is not complete because it does not extend from an open-loop pole to an open-loop zero. For this system, there are asymptotes with angles

$$\theta = \frac{180^\circ + i360^\circ}{2} = 90^\circ,\ -90^\circ$$

and centroid

$$\sigma = \frac{-3-1}{2} = -2$$

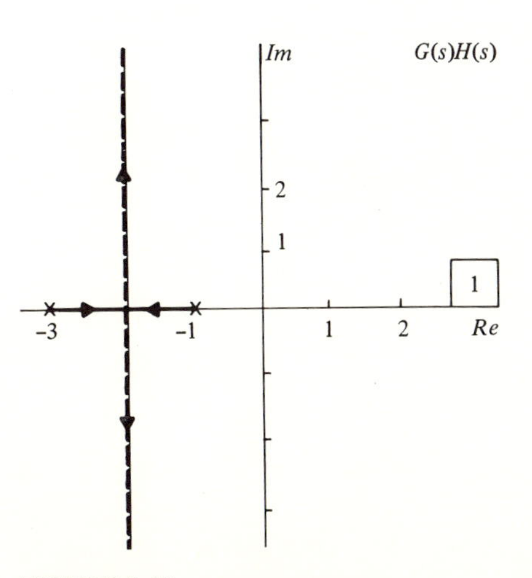

FIGURE 5-12
Breakaway of the locus from the real axis.

As K varied from zero to infinity, the poles of the overall transfer function begin as the poles of GH. For larger K, the overall transfer function has two real axis poles which become closer and closer together with increasing K. At some gain K there are two repeated poles in $T(s)$ at $s=-2$. For still larger K, the poles are complex conjugate with larger and larger imaginary parts. In this simple case, the locus breaks away from the real axis at the centroid, $s=-2$, as can be verified by examining the angle of $G(s)H(s)$ at points slightly above and below $s=-2$.

In more involved systems, a breakaway of two root locus segments from the real axis is also at 90° angles, but the breakaway point is not necessarily midway between the real axis GH roots. For a segment extending from poles, as in the example of Fig. 5-13a, the real axis point at which the loci break away will correspond to the largest value of K for that segment. Calculations of K for several real axis test points between 0 and -5 are summarized in Table 5-2.

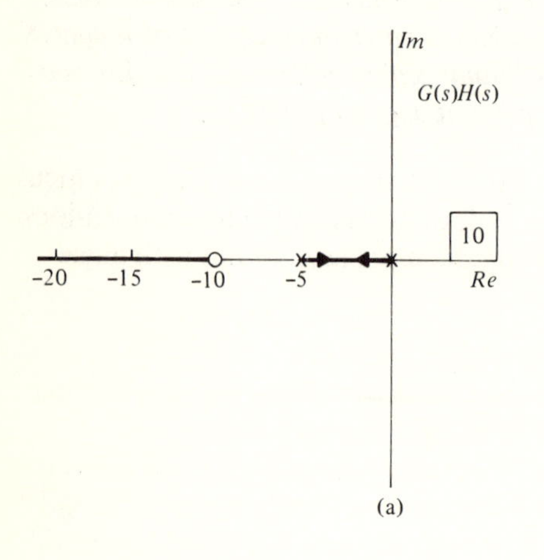

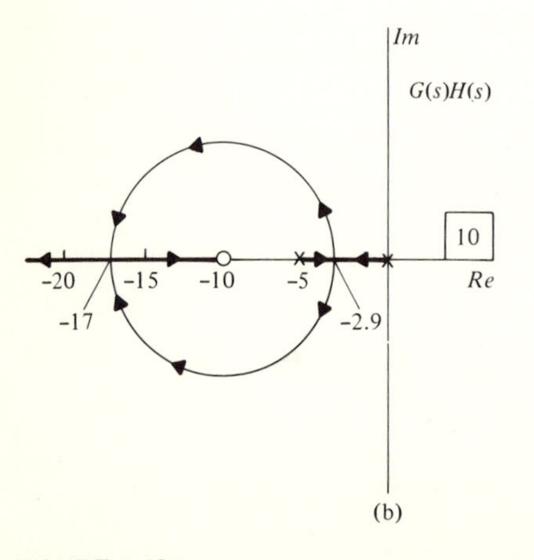

FIGURE 5-13
Calculating breakaway and entry points of root locus.

TABLE 5-2
Breakaway point calculations for the example of Fig. 5-13.

s_0	$\lvert G(s_0)H(s_0)\rvert = 1/K$	K
-4	$\frac{10(6)}{(1)(4)}$	0.067
-3	$\frac{10(7)}{(2)(3)}$	0.0857
-2	$\frac{10(8)}{(3)(2)}$	0.075
-3.5	$\frac{10(6.5)}{(1.5)(3.5)}$	0.081
-2.5	$\frac{10(7.5)}{(2.5)(2.5)}$	0.083
-2.9	$\frac{10(7.1)}{(2.1)(2.9)}$	0.0858
-2.8	$\frac{10(7.2)}{(2.2)(2.8)}$	0.0855

From these, it is seen that the largest value of K on that real axis segment is in the vicinity of $s = -2.9$, which must be the location of the breakaway.

In the example, Fig. 5-13b, there is also a point to the left of $s = -10$ on the real axis where the loci enter the real axis. The real axis entry point of the loci will correspond to the smallest value of K on that segment of the axis, since for larger K, the loci approach the zero and $-\infty$. Calculations of K for real axis test points to the left of $s = -10$ are given in Table 5-3, and it is seen that the entry point of the locus is near $s = -17$.

5.5.3 Angles of Departure and Approach

RULE 6. The angle of departure ϕ of a locus branch from a complex pole is found according to

Sum of all other GH pole angles to the pole under consideration
$+\phi -$ sum of GH zero angles to the pole $= 180°$

The angle of approach ϕ' of the locus to a complex zero is given by

Sum of GH pole angles to the zero under consideration
$-$ sum of all other GH zero angles to the zero $- \phi' = 180°$

TABLE 5-3
Entry point calculations for the example of Fig. 5-13.

s_0	$\lvert G(s_0)H(s_0)\rvert = 1/K$	K
-25	$\frac{10(15)}{(20)(25)}$	3.33
-20	$\frac{10(10)}{(15)(20)}$	3.00
-15	$\frac{10(5)}{(10)(15)}$	3.00
-17	$\frac{10(7)}{(12)(17)}$	2.91
-18	$\frac{10(8)}{(13)(18)}$	2.925
-17.5	$\frac{10(7.5)}{(12.5)(17.5)}$	2.917

Multiple angles of departure and approach at multiple complex roots of GH are found similarly, using multiple contributions of ϕ or ϕ' and equating to $180° \pm i360°$.

For a set of complex conjugate poles of $G(s)H(s)$, the angle at which the root locus leaves one of the poles may be found by considering points very close to that pole, as in the example of Fig. 5-14. Since the point is very close to the pole under consideration, the angles to the point are virtually the angles to the pole

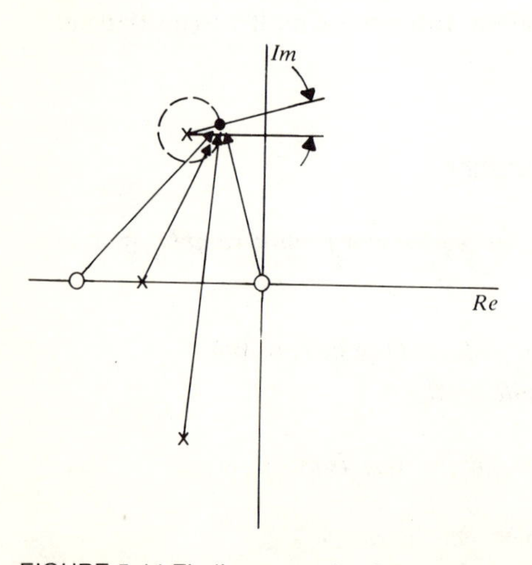

FIGURE 5-14 Finding an angle of departure.

itself. Solving

$$\text{Sum of other pole angles to the pole under consideration} + \phi \\ - \text{sum of zero angles to the pole} = 180^\circ$$

will give the angle of departure, ϕ. Of course, the angle of departure from the lower pole of the conjugate set is the negative of the upper pole's angle of departure.

Angles of approach to complex zeros may be found similarly:

$$\text{Sum of pole angles to the zero under consideration} \\ - \text{sum of other zero angles to the zero} - \phi' = 180^\circ$$

Consider the system with GH product given in Fig. 5-15a. For a point on the root locus near the top complex pole (approximately),

$$90^\circ + 50^\circ + \phi - 130^\circ = 180^\circ$$
$$\phi = 180^\circ$$

where ϕ is the angle of departure of the locus from the pole.

There is a complete real axis segment of the locus between the real axis pole and zero. The other two locus segments extend from the complex poles to infinity, with asymptotic angles given by

$$(3-1)\theta = 180^\circ \pm i360^\circ$$
$$\theta = \pm 90^\circ$$

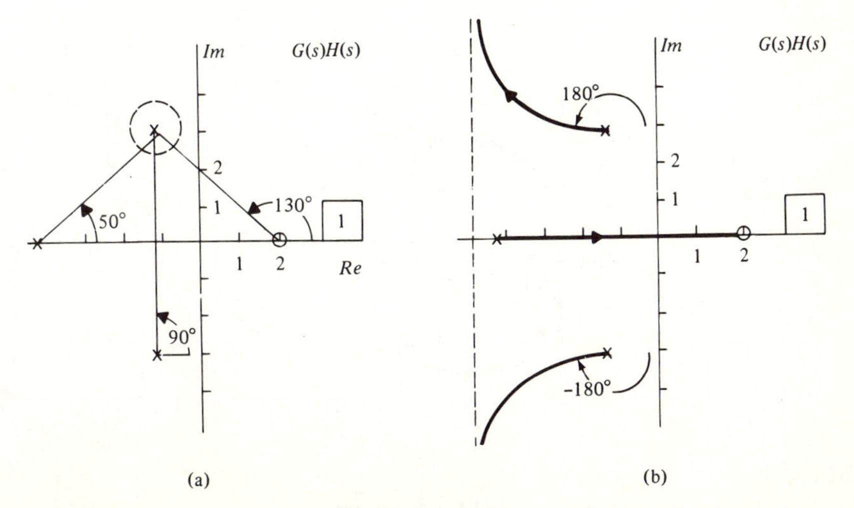

FIGURE 5-15
Root locus construction using angles of departure.

and centroid

$$\sigma=\frac{(-4-\frac{3}{2}-\frac{3}{2})-2}{3-1}=-\frac{9}{2}$$

A completed root locus sketch is given in Fig. 5-15b.

Consider the system of Fig. 5-16a. For a point on the root locus near the top complex pole, approximately

$$90^\circ+\phi-200^\circ-135^\circ=180^\circ$$
$$\phi=65^\circ$$

where ϕ is the angle of departure of the locus from the top pole.

For a point near the top complex zero, Fig. 5-16b, the angle ϕ' of arrival of

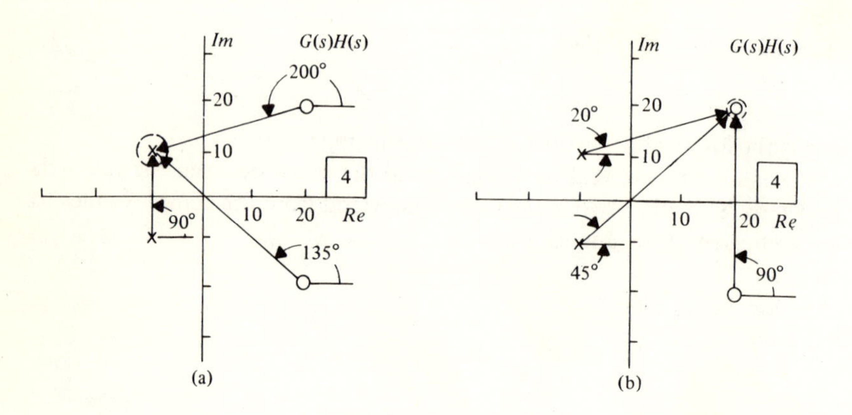

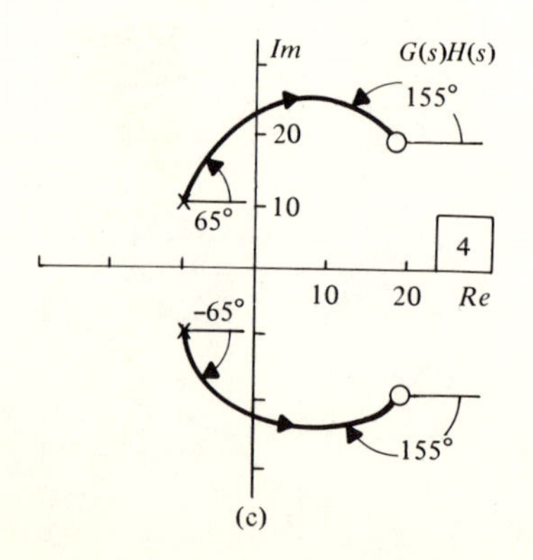

FIGURE 5-16
Root locus construction involving angles of departure and of approach.

the locus is given approximately by

$$20^\circ + 45^\circ - 90^\circ - \phi' = 180^\circ$$
$$\phi' = 155^\circ$$

A complete root locus sketch is shown in Fig. 5-16c.

At repeated complex roots of GH, more than one locus segment begins or ends at the root location, so more than one angle of departure or approach will be found, a different angle for each locus segment. For the system of Fig. 5-17a, the angle contribution to a point near the top double pole is given approximately by

$$90^\circ + 90^\circ + 108^\circ + \phi + \phi - 124^\circ = 180^\circ \pm i360^\circ$$
$$\phi = 8^\circ,\ 188^\circ$$

Inclusion of the multiples of 360° is important here because it is this term that gives the multiple solutions. It is easy to see that multiple departure (or arrival) angles will be evenly spaced around the multiple roots.

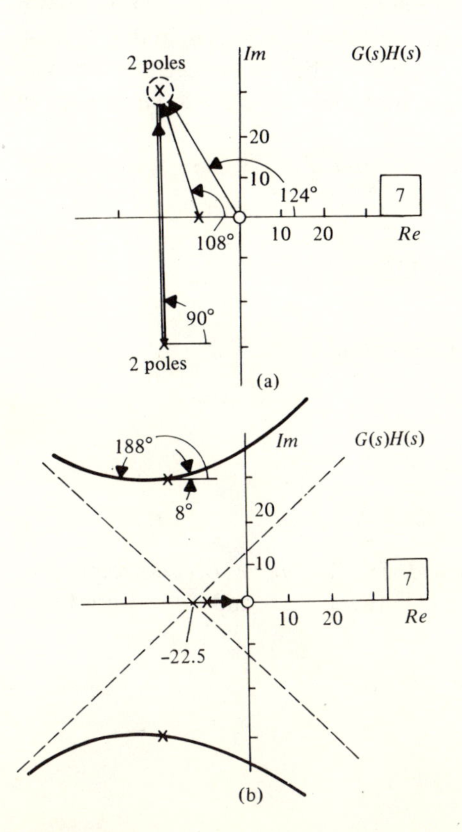

FIGURE 5-17
Angles of departure from multiple poles.

To complete the root locus sketch as in Fig. 5-17b, the real axis locus segment is drawn and the asymptotic angles and centroid are found:

$$(5-1)\theta = 180^\circ \pm i360^\circ$$

$$\theta = \pm 45^\circ,\ \pm 135^\circ$$

$$\sigma = \frac{-20-20-20-20-10}{5-1} = -22.5$$

5.5.4 Another Example

As an example of application of the six root locus construction rules, consider the open-loop transfer function

$$G(s)H(s) = \frac{1}{s(s+3)(s^2+6s+64)}$$

A step-by-step procedure is as follows:

Locate the open-loop poles and zeros and plot them (*rule 1*). There are no zeros of GH. The poles of GH are located at 0, -3, $-3+j7.4$, and $-3-j7.4$.

Locate real axis portions of the locus (*rule 2*). The real axis segment between $s=0$ and $s=-3$ is on the root locus. The root locus diagram so far is given in Fig. 5-18a.

Determine the angles of the asymptotes (*rule 3*). The asymptotic angles are given by

$$\theta = \frac{180^\circ \pm i360^\circ}{4-0} = 45^\circ,\ 135^\circ,\ -45^\circ,\ -135^\circ$$

Determine the centroid of the asymptotes (*rule 4*).

$$\sigma = \frac{0-3-3+j7.4-3-j7.4}{4-0}$$

$$= \frac{-9}{4} = -2.25$$

Find the real axis breakaway point (*rule 5*). The values of K corresponding to various real axis points on the locus between $s=0$ and $s=-3$ are found from graphical measurements as follows:

Value of s	*Approximate value of K*
-1	118
-1.5	129
-2	112

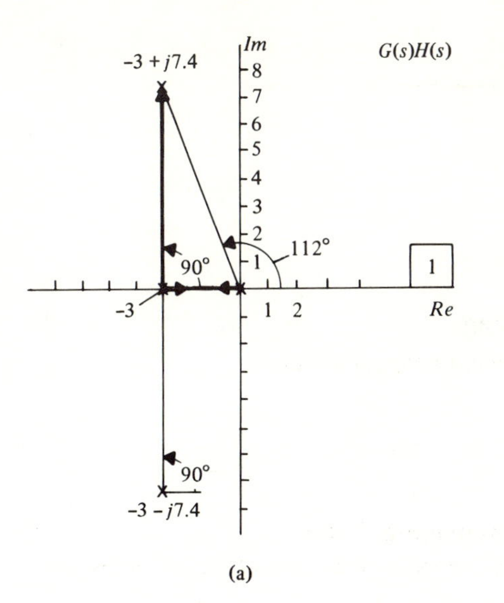

(a)

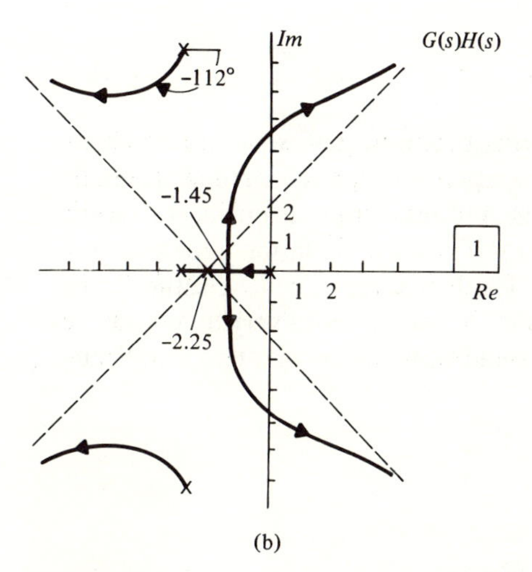

(b)

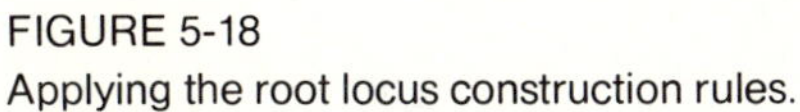

FIGURE 5-18
Applying the root locus construction rules.

A value of -1.5 for the breakaway point can be taken. Alternatively, numbers may be substituted into

$$K = \left|\frac{1}{G(s)H(s)}\right| = \frac{1}{|(s)(s+3)(s^2+6s+64)|}$$

With the aid of a pocket calculator, very accurate results may be obtained if desired:

Value of s	*Value of K*
-1.3	127.93
-1.4	128.93
-1.45	129.01
-1.5	128.81
-1.6	127.59

A value of 1.45 is obtained quickly.

The real axis breakaways are at $\pm 90^\circ$.

Determine the angle of departure from the top pole (rule 6).

$$112^\circ + 90^\circ + 90^\circ + \phi = 180^\circ$$
$$\phi = -112^\circ$$

The completed root locus diagram is given in Fig. 5-18b.

Of course, in many situations, only certain of the construction rules are applicable.

5.5.5 Other Root Locus Properties

There are several other root locus properties of a more general nature which are of help in sketching approximate plots. Except in very special circumstances, the locus segments do not touch or cross one another except at breakaway points on the real axis and points of symmetry. Thus, the situation shown in Fig. 5-19a should not be expected to occur. The loci seem to behave rather lazily' doing nothing terribly involved unless the GH product is of very high order, so that the situations of Fig. 5-19b and c could only occur for very complicated systems.

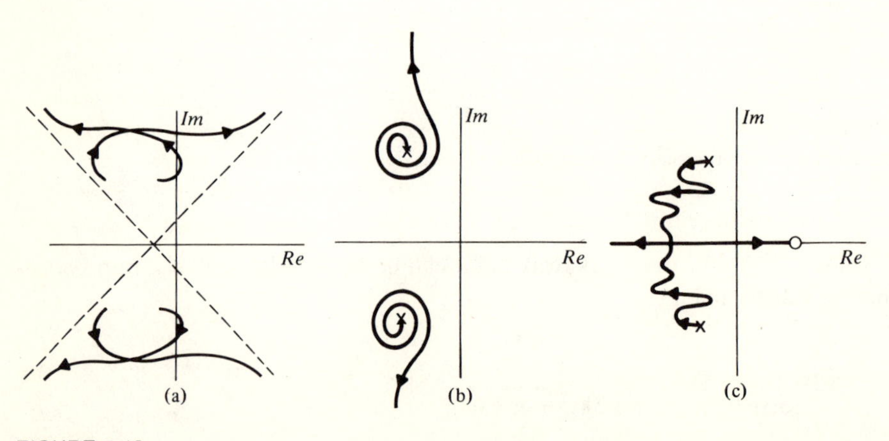

FIGURE 5-19
Incorrect root loci.

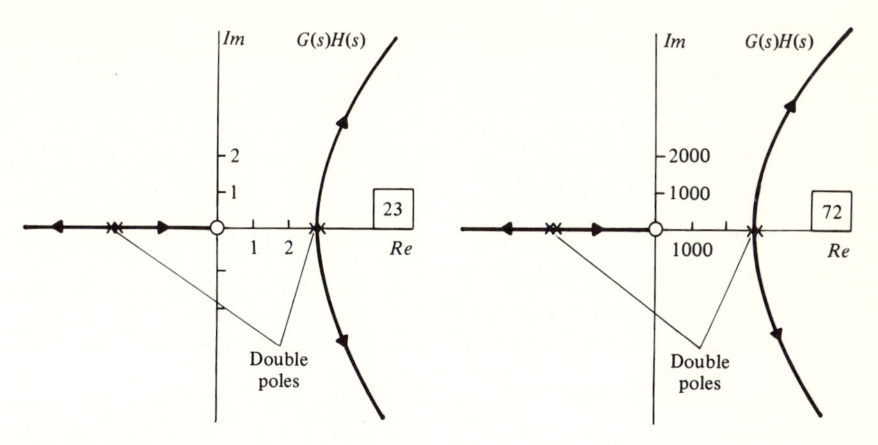

FIGURE 5-20
Root locus plots to different scales.

Two otherwise identical pole-zero plots with different scales, Fig. 5-20, have the same loci because the angle criterion is a geometric one. So long as it is positive, the multiplying constant of $G(s)H(s)$ does not affect the shape of the locus, either. Interchanging all poles and zeros in $G(s)H(s)$ gives the same root locus, except with the senses of the segments reversed.

5.5.6 Computer-Aided Root Locus Plotting

Now that the basics of root locus construction have been discussed, it is appropriate to consider the use of digital computers for root locus calculations. One way of doing root locus plotting with a computer is to repeatedly factor

$$1+KG(s)H(s)=0$$

for various values of K. As numerical polynomial factoring is quite involved in general, one would hope to make use of a good computer library routine for factoring. Figure 5-21 shows a programming flow diagram for performing root locus in this way. This program stores all of the points to be plotted, then performs the plot, which is expedient if the plot is made with a line printer.

Numerical polynomial-factoring methods have been studied extensively in the past because of the importance of the factoring problem to a variety of applications. A value of $s=x+jy$ is a root of

$$F(s)=1+KG(s)H(s)$$

for a fixed value of K only if both

$$\begin{cases} \text{Re}[F(s=x+jy)]=0 \\ \text{Im}[F(s=x+jy)]=0 \end{cases} \tag{5-1}$$

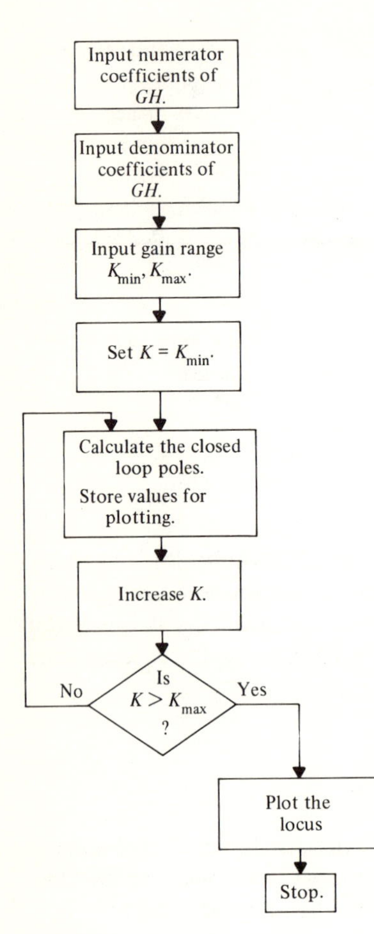

FIGURE 5-21
Computer flow diagram for factoring and root locus plotting.

or alternatively if

$$|F(x+jy)|=0 \tag{5-2}$$

Numerically locating polynomial roots using condition (5-2) is a lengthy two-dimensional minimization problem. Because of the polynomial nature of $F(s)$, conditions (5-1) are generally easier to apply. A popular method is Bairstow's,* in which x and y are incremented to trace (or follow) a curve $\text{Re}[F(s=x+jy)]=0$ until an $\text{Im}[F(s=x+jy)]=0$ curve is crossed, after which the intersection is iteratively located with suitable precision.

*In general, Bairstow's method refers to a search process along curves for which one parameter at a time is varied. One can, of course, apply Bairstow's method with parameters other than x and y—for example, $r=\sqrt{x^2+y^2}$ and $\phi=\tan^{-1}(y/x)$ or the two independent coefficients of a quadratic divisor.

Other numerical polynomial-factoring methods include synthetic division, Routh-Hurwitz testing with shifted axes, repeated root locus techniques, and countless variations on these. Because of the fundamental two-dimensional search nature of factoring, all of these techniques require more or less comparable computation time.

To plot root loci by repeated polynomial factoring is very inefficient compared to angle testing. Points $s=x+jy$ on the root loci are those for which

$$\angle GH(s=x+jy)=180^\circ \tag{5-3}$$

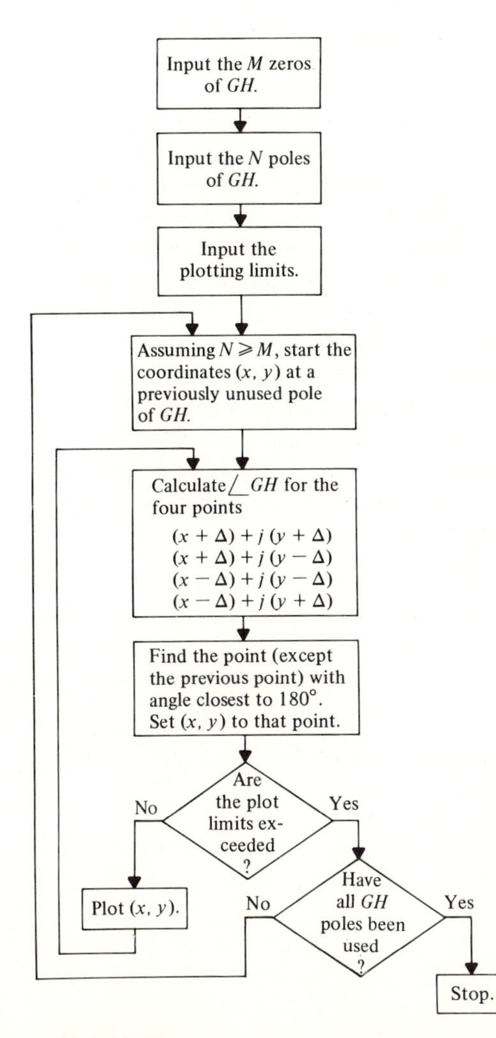

FIGURE 5-22
Computer flow diagram for root locus plotting using the angle criterion.

or

$$\begin{cases} \mathrm{Re}[GH(s=x+jy)]<0 \\ \mathrm{Im}[GH(s=x+jy)]=0 \end{cases} \tag{5-4}$$

Conditions (5-3) and (5-4) are also two-dimensional searches, equivalent to that for a factoring. However, *one* tracking of the curves $\angle GH(s=x+jy)=180^\circ$ or $\mathrm{Im}[GH(s=x+jy)]=0$ (which in the comparable factoring problem would yield the roots for one value of K) completes the entire root locus. Nevertheless, factoring (with a library routine) to plot root loci may be cost-effective for relatively low-order systems when one considers the high costs of program development. If the factoring routine has provision for initial guesses of the root locations, it is worthwhile to use the previous roots as starting points for root calculation with the next value of K.

An angle-testing root locus program is outlined in the flowchart of Fig. 5-22. It does what a very industrious person might do with ruler and protractor. Starting at an open-loop pole, it calculates the angle of the GH product at four test points, arranged in a diamond shape about the pole. The point with angle nearest 180° is selected and the locus is drawn to that point. Then new points about the new location are tested, and so on. This arrangement works well for sufficiently small step size when multiple roots are not involved. For a general routine, it is helpful to test more than just four points on each iteration and to keep track of previously encountered loci paths from multiple roots.

In using computers effectively, an important principle is to back up computer output with pencil-and-paper analysis. Hand calculations are done to examine alternatives and to give rough solutions. More refinement is obtained, when needed, with pocket calculator aid. Finally, digital computer routines offer precision, polish, and a check on previous results.

DRILL PROBLEMS

D5-6. Sketch root locus plots for the following systems. Find asymptotic angles, centroid, approximate breakaway points, angles of departure, and angles of approach where applicable.

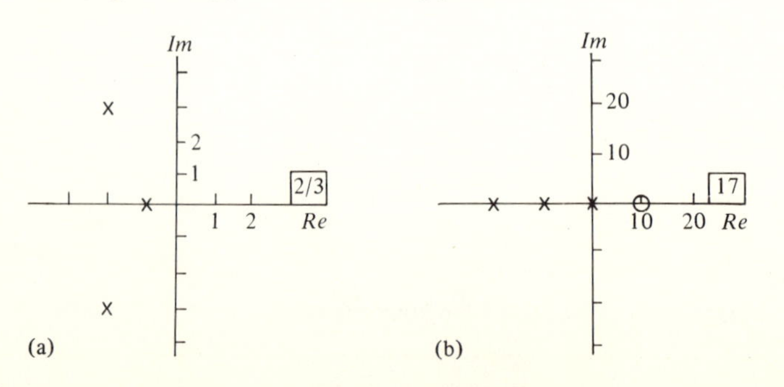

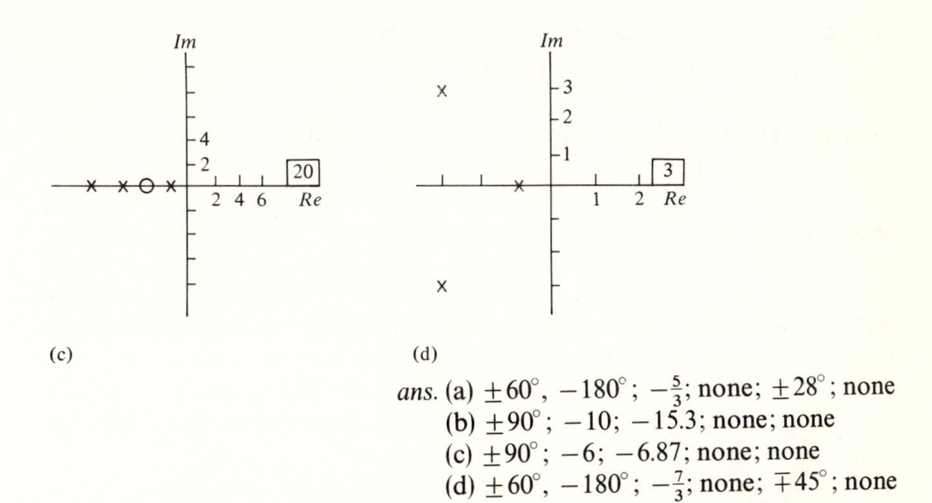

ans. (a) $\pm 60^\circ$, -180°; $-\frac{5}{3}$; none; $\pm 28^\circ$; none
(b) $\pm 90^\circ$; -10; -15.3; none; none
(c) $\pm 90^\circ$; -6; -6.87; none; none
(d) $\pm 60^\circ$, -180°; $-\frac{7}{3}$; none; $\mp 45^\circ$; none

D5-7. For systems with the following root locus plots, using graphical evaluation, find the value of the adjustable constant K for which the overall transfer function has a pole at the location indicated by the dot.

ans. (a) $K = 1$
(b) $K = \frac{26}{7}$
(c) $K = 1000$

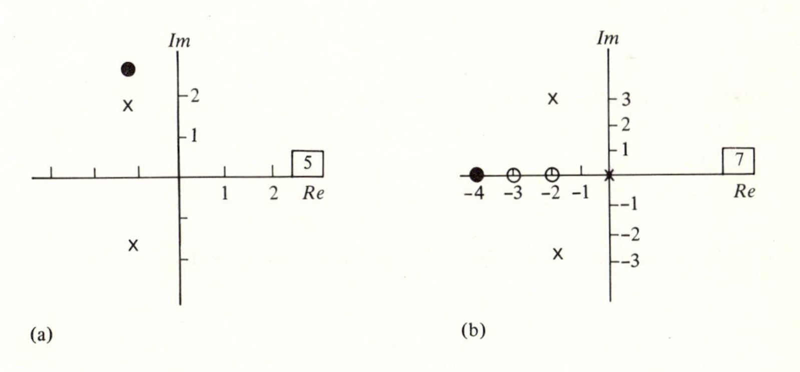

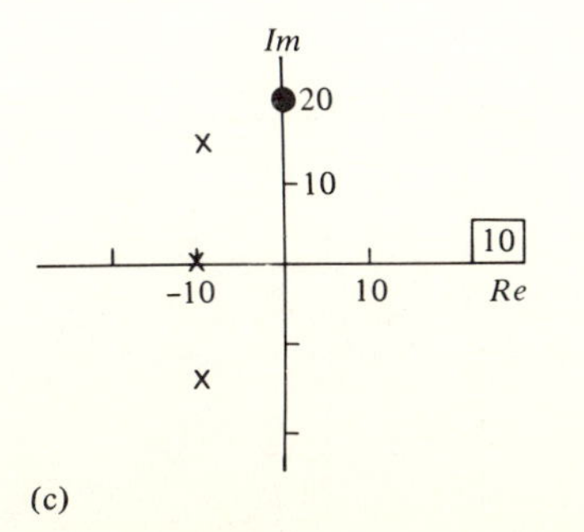

5.6 ROOT LOCI FOR OTHER SYSTEMS
5.6.1 Systems with Other Forms

Problems other than the basic one, which has been considered exclusively here to this point, may be cast in the root locus form,

$$1+K\frac{a(s)}{b(s)}=0$$

where $a(s)$ and $b(s)$ are known polynomials. For example, the adjustable system of Fig. 5-23a, where K is adjustable, has an overall transfer function

$$T(s)=\frac{\dfrac{s}{s+2}}{1+\left(\dfrac{s}{s+2}\right)\left(\dfrac{K}{s+K}\right)}$$

$$=\frac{s(s+K)}{s^2+2s+K(2s+2)}$$

$$=\frac{\dfrac{s(s+K)}{s^2+2s}}{1+K\dfrac{2s+2}{s^2+2s}}$$

Here the equivalent GH product may be taken to be

$$G(s)H(s)=\frac{2(s+1)}{s^2+2s}$$

The root locus for K ranging from zero to infinity is shown in Fig. 5-23b.

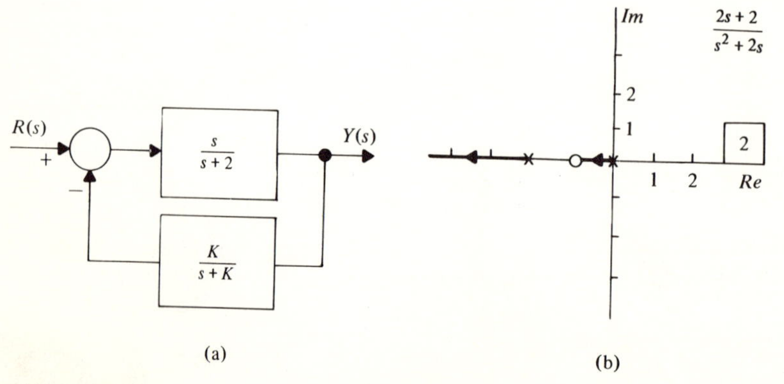

FIGURE 5-23
Root locus of a system with adjustable time constant.

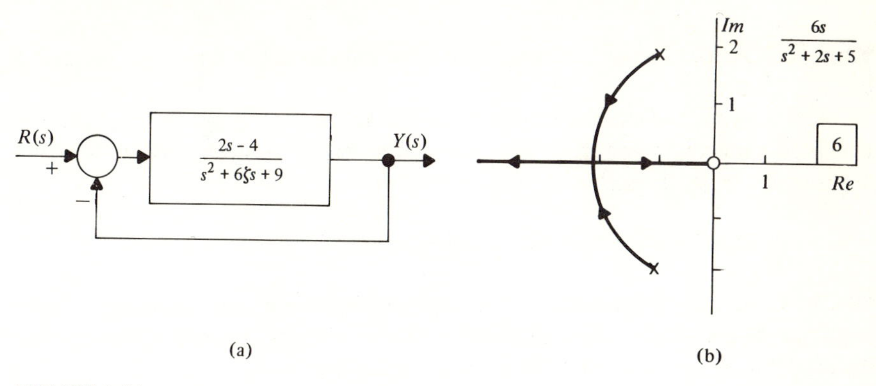

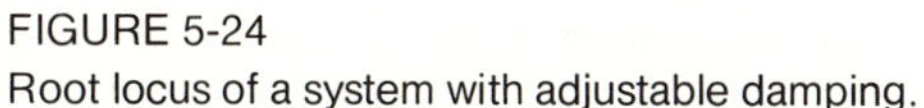
FIGURE 5-24
Root locus of a system with adjustable damping.

For the unity feedback system with adjustable damping in Fig. 5-24a,

$$T(s)=\frac{\dfrac{2s-4}{s^2+6\xi s+9}}{1+\dfrac{2s-4}{s^2+6\xi s+9}}$$

$$=\frac{2s-4}{s^2+2s+5+6\xi s}$$

$$=\frac{\dfrac{2s-4}{s^2+2s+5}}{1+\xi\,\dfrac{6s}{s^2+2s+5}}$$

The equivalent GH product may be taken to be

$$G(s)H(s)=\frac{6s}{s^2+2s+5}$$

Root locus for ξ in the range from zero to infinity is shown in Fig. 5-24b.

5.6.2 Negative Parameter Ranges

When it is desired to determine the root locus for a parameter K which ranges from zero to *minus* infinity, the root locus relations are

$$\angle G(s)H(s)=0^\circ \pm i360^\circ$$

$$|G(s)H(s)|=\frac{1}{|K|}$$

Points on the root locus are values of s for which the angle of the GH product is 0°.

Construction of the root locus is similar to the 180° angle procedure, with real axis segments being anywhere *not* to the left of an odd number of roots, asymptotic angles given by

$$(\text{Number of poles} - \text{number of zeros})\theta = 0^\circ \pm i360^\circ$$

and with similar expressions for angles of departure and approach. Locus segments begin on the poles of $G(s)H(s)$ (for $K \to 0$) and approach the zeros of $G(s)H(s)$ (for $K \to -\infty$), just as with the ordinary locus.

For example, the system with GH product given in Fig. 5-25a has the real axis loci shown for negative K. The asymptotic angles are given by

$$(4-1)\theta = 0^\circ \pm i360^\circ$$
$$\theta = 0^\circ,\ 120^\circ,\ -120^\circ$$

The centroid is

$$\sigma = \frac{-1+j2-1-j2+2-(-3)}{4-1} = 1$$

Angle of departure ϕ from the top pole is approximately given by

$$90^\circ + 145^\circ + 120^\circ + \phi - 45^\circ = 0^\circ$$
$$\phi = -310^\circ = 50^\circ$$

The completed root locus plot is shown in Fig. 5-25b.

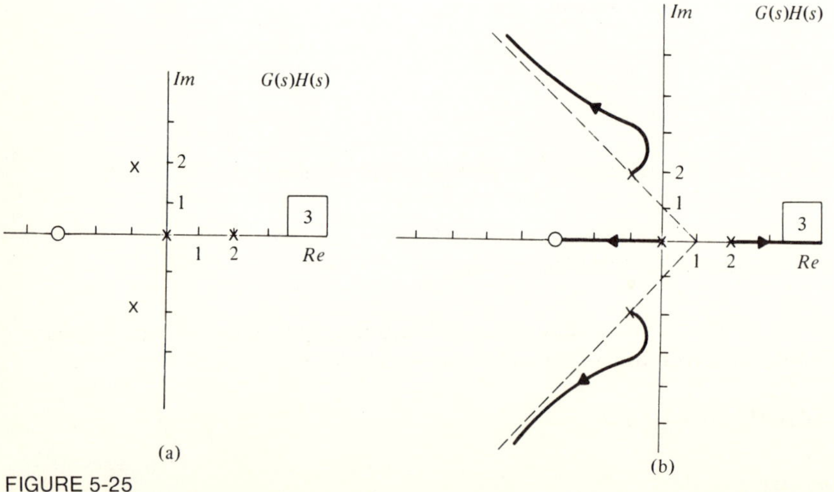

FIGURE 5-25
Root locus construction for negative K.

The same 0° locus situation occurs, too, for positive K if the multiplying constant of $G(s)H(s)$ is negative. Then, although

$$\angle G(s)H(s) = 180^\circ$$

is required, the angle contributions from the poles and zeros of GH must total 0° since 180° is contributed to the angle by the negative multiplying constant.

DRILL PROBLEMS

D5-8. Develop root locus plots for the following systems, for K ranging from 0 to $+\infty$. Find asymptotic angles, centroid, approximate breakaway points, angles of departure, and angles of approach where applicable.

ans. (a) $G(s)H(s) = \dfrac{(s+2+j2)(s+2-j2)}{(s+2+j1)(s+2-j1)}$;
none; none; none; $\pm 90^\circ$; $\mp 90^\circ$
(b) $G(s)H(s) = 1/s(s^2+2s+4)$; $\pm 60^\circ$, 180°; $-\frac{2}{3}$; none; $\mp 30^\circ$; none
(c) $G(s)H(s) = 2/s(s+4)(s^2+s+4)$; $\pm 45^\circ$, $\pm 135^\circ$; -1.25; -2.8; $\mp 55^\circ$; none
(d) $G(s)H(s) = \dfrac{s+4}{s(s^2+5s+6)}$; 180°; none; -2.59, -4.9; none; none

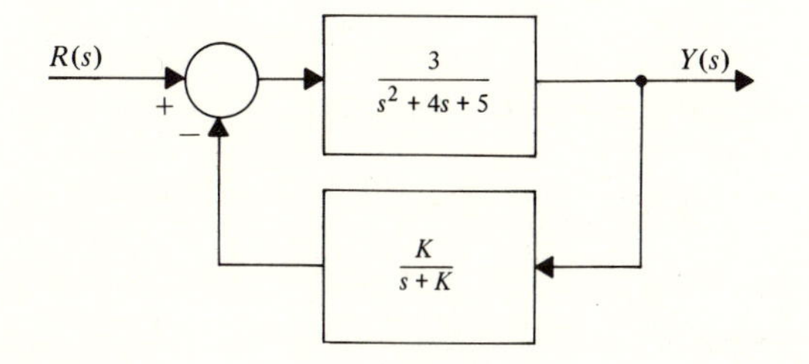

(a)

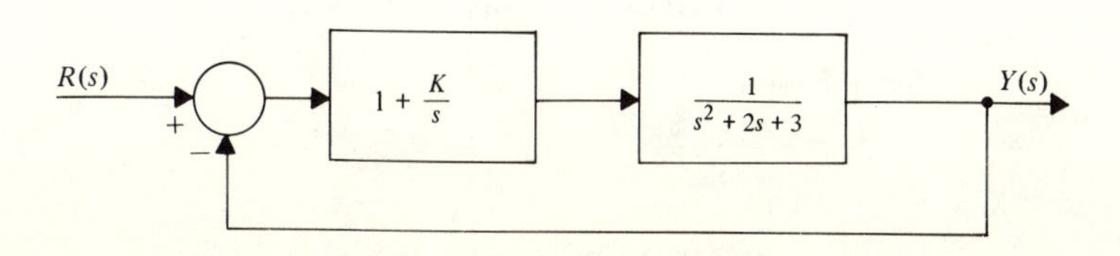

(b)

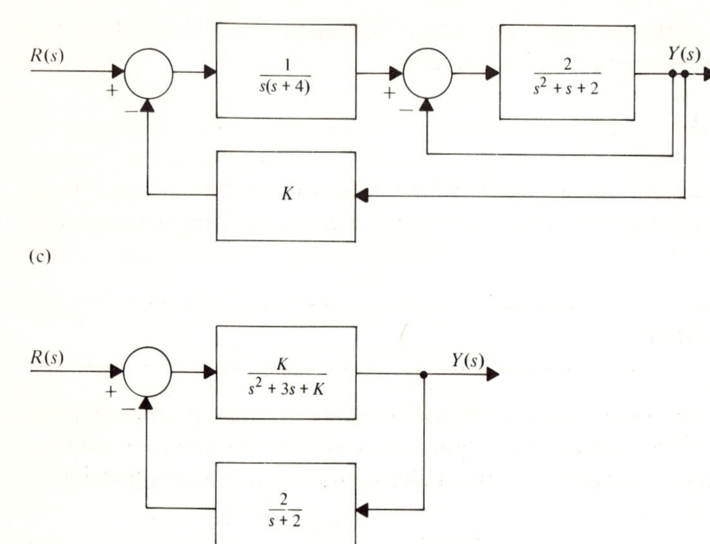

D5-9. Develop root locus plots for systems with the following overall transfer functions, for the indicated range of the adjustable parameter K:

(a) $$T(s)=\frac{6Ks+7}{s^3+Ks^2+(2K+9)s+K}$$

$0<K<\infty$

ans. $G(s)H(s)=(s+1)^2/s(s^2+9)$

(b) $$T(s)=\frac{10}{(1-K)s^2+3s+2}$$

$0<K<\infty$

ans. $G(s)H(s)=-[s^2/(s+1)(s+2)]$

(c) $$T(s)=\frac{K\dfrac{s}{s+4}}{1+K\dfrac{s(s-3)}{(s+2)^2(s+4)}}$$

$-\infty<K<0$

ans. $G(s)H(s)=s(s-3)/(s+2)^2(s+4)$

(d) $$T(s)=\frac{6s^2+Ks+2}{s^3+2s^2+(5+2K)s+2K}$$

$-\infty<K<\infty$

ans. $G(s)H(s)=2(s+1)/s(s^2+2s+5)$

5.7 INTRODUCTION TO COMPENSATION

5.7.1 Types of Compensators

The simple tracking control system configuration of Fig. 5-26 is created by providing unity feedback around the plant $G_p(s)$ and adding an error signal amplifier with gain K. The amplifier gain is adjusted to provide acceptable performance, if possible. Typically, the designer is interested in the closed-loop system's relative stability and its steady state error performance. Other considerations such as step response overshoot may also be of concern.

If adequate performance cannot be obtained with output feedback alone, additional transmittances termed *compensators* may be added to the system. Figure 5-27 shows several simple compensator configurations for single-input, single-output plants. In Fig. 5-27a, the compensator $G_c(s)$ is inserted into the system's forward path. It is termed a *cascade* compensator. In Fig. 5-27b, the compensator $H_c(s)$ is placed in the feedback path around the plant, forming a *feedback* compensator. A system involving both feedback and cascade compensation is given in Fig. 5-27c.

Compensators are further classified as to the forms of their transmittances. A number of common arrangements are given in Table 5-4, together with typical effects upon steady state errors and relative stability of the resulting system. In following sections of this chapter, the use of each of these compensators in systems with an example plant transmittance

$$G_p(s) = \frac{1}{s(s+2+j)(2+2-j)} = \frac{1}{s(s^2+4s+5)}$$

will be illustrated, along with design strategies and methods. Table 5-5 summarizes parameter values and other numerical data for the resulting designs.

5.7.2 The Uncompensated System with Feedback

For the simple feedback configuration of Fig. 5-27, the root locus in terms of the adjustable error signal gain K is given in Fig. 5-28. The asymptotic angles are $\pm 60°$ and $180°$, and the centroid of the asymptotes is at $s = -\frac{4}{3}$. If $K = 3$, the closed-loop poles are at $s = -0.77 \pm j0.8$ and -2.47, as indicated.

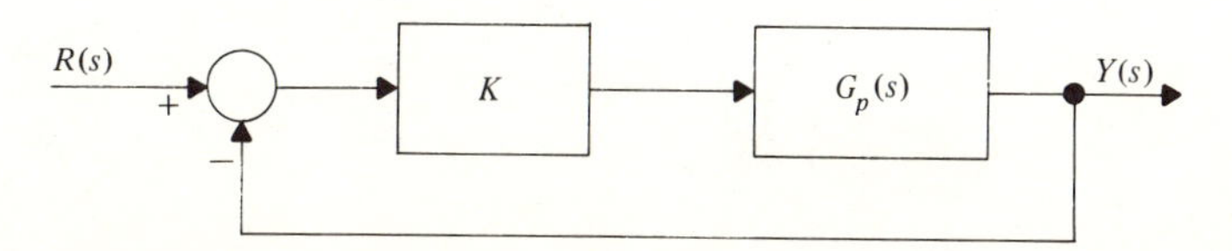

FIGURE 5-26
Unity feedback control system.

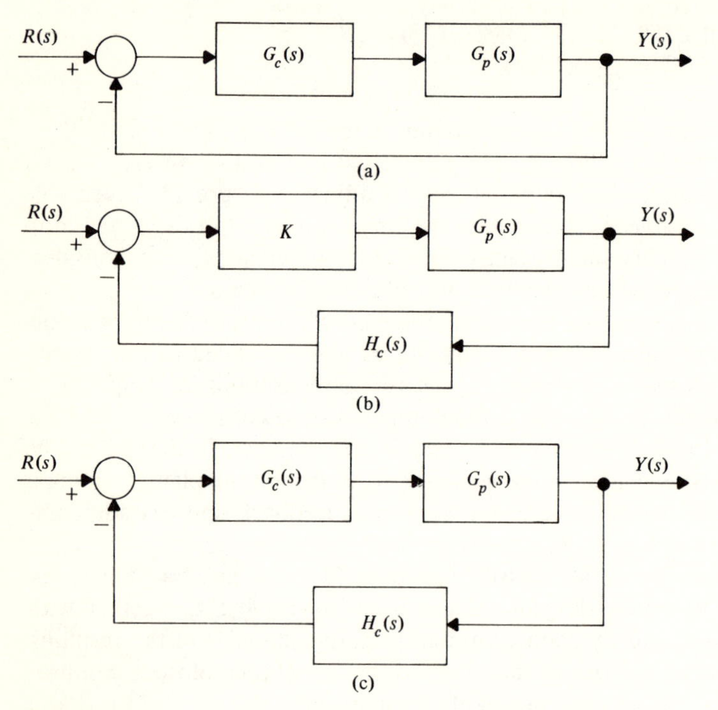

FIGURE 5-27
Compensator configurations. (a) Cascade compensated system. (b) Feedback compensated system. (c) System with feedback and cascade compensation.

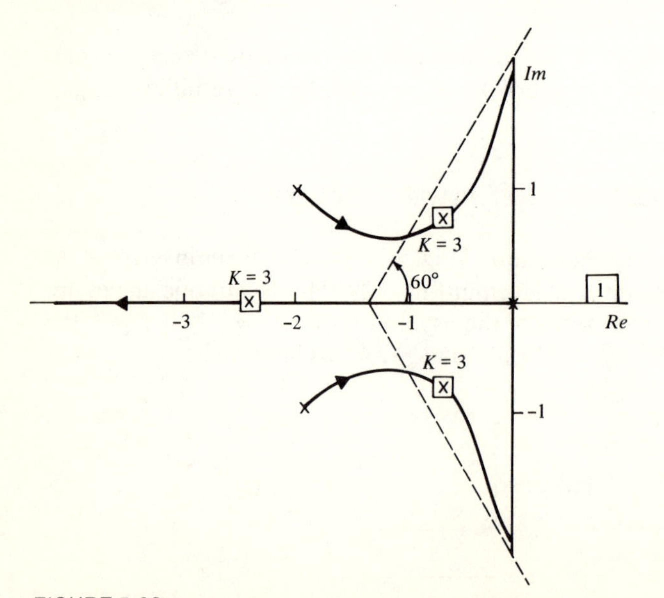

FIGURE 5-28
Root locus for the example uncompensated feedback system.

TABLE 5-4
Common types of compensators.

Compensator	*Transmittance*	*Typical Effect on Steady State Errors*	*Typical Effect on Relative Stability*
Cascade integral	$G_c(s)=\dfrac{K}{s}$	Greatly improved	Greatly reduced
Cascade integral plus proportional	$G_c(s)=\dfrac{K(s+a)}{s}$	Improved	Reduced
Cascade lag	$G_c(s)=\dfrac{K(s+a)}{s+b}$ $b<a$	Improved	Reduced
Cascade lead	$G_c(s)=\dfrac{K(s+a)}{s+b}$ $a<b$	Somewhat improved or somewhat worse	Increased
Feedback rate	$H_c(s)=1+As$	Improved	Increased

The ramp error coefficient of the system is

$$K_1=\lim_{s\to 0} sKG_p(s)=\frac{K}{5}=0.6$$

For a unit ramp input, the steady state error between output and input is thus given by

$$\text{Normalized steady state error to a ramp input}=\frac{1}{K_1}=1.67$$

If this steady state error ratio is too large, K could be increased; however, the closed-loop poles would then be further to the right on the complex plane, giving less relative stability. If greater relative stability is required, the gain K could be reduced, but the steady state ramp error would increase.

DRILL PROBLEM

D5-10. An uncompensated feedback system of the form of Fig. 5-26 has

$$G_p(s)=\frac{1}{(s+4)^2}$$

Select K so that the closed-loop poles are at $-4\pm j4$. What is K_0, the step error coefficient? Where are the closed-loop zeros?

ans. $K=16$; $K_0=1$; no closed-loop zeros

TABLE 5-5
Summary of designs.

Compensator	*Transmittance*	*Ramp Error Coefficient*	*Closed-Loop Zeros*	*Closed-Loop Poles*	*Relative Stability*
Uncompensated	$K=3$	0.6	None	$-0.77 \pm j0.80,\ -2.47$	0.77
Cascade integral	$G_c(s)=\dfrac{K}{s}$			Unstable	
Cascade integral plus proportional	$G_c(s)=\dfrac{3(s+0.1)}{s}$	∞	-0.1	$-0.123,$ $-0.724 \pm j0.695,\ -2.43$	0.123
Cascade lag	$G_c(s)=\dfrac{3(s+0.1)}{s+0.01}$	6	-0.1	$-0.120,$ $-0.729 \pm j0.705,\ -2.43$	0.120
Cascade lead	$G_c(s)=\dfrac{60(s+1.6)}{s+16}$	1.2	-1.6	$-1.106,$ $-1.315 \pm j1.9,\ -16.265$	1.106
Feedback rate	$H_c(s)=1+4s$ $K=6$	0.92	None	$-0.85 \pm j1.37,\ -2.31$	0.85

5.8 SIMPLE CASCADE COMPENSATION

5.8.1 Design Considerations

For the cascade compensation configuration of Fig. 5-27, the open-loop system transmittance is

$$G(s) = G_c(s)G_p(s)$$

Assuming that the overall system is stable, the steady state error coefficients are

$$K_n = \lim_{s\to 0} s^n G(s) = \lim_{s\to 0} s^n G_c(s)G_p(s)$$

Steady state performance may be improved by adding one or more poles, in the compensator, at or near $s=0$. If a pole is added at $s=0$, the system type number is increased by 1. Adding a pole very close to $s=0$ for a type n system does not change the system type number but increases K_n and thus reduces the normalized steady state error to a t^n input.

The addition of open-loop poles at or near $s=0$, however, tends to reduce the relative stability of the closed-loop system or to make it unstable. Thus, open-loop zeros are often also placed in the compensator transmittance to "draw away" the root locus into the LHP. The usual design strategy is then to place compensator poles near $s=0$ to improve steady state performance and to place compensator zeros so that the system root locus gives sufficient relative stability for a suitable value of the compensator multiplying constant K.

5.8.2 Cascade Integral Compensation

The cascade integral compensator

$$G_c(s) = \frac{K}{s}$$

adds a single pole at $s=0$ to the system's open-loop transmittance. It thus raises the system type number, improving steady state error performance provided that the resulting compensated system is stable. The addition of an open-loop pole at $s=0$ tends to reduce the relative stability because the additional pole in the open-loop transmittance means that the locus will have one more asymptote. The asymptotic angles of the root locus being evenly spaced, this increases the likelihood of closed-loop poles being near or in the RHP.

For the example plant $G_p(s)$, the compensated system transfer function is

$$T(s) = \frac{\frac{K}{s}\left[\frac{1}{s(s^2+4s+5)}\right]}{1+\frac{K}{s}\left[\frac{1}{s(s^2+4s+5)}\right]} = \frac{K}{s^2(s^2+4s+5)+K}$$

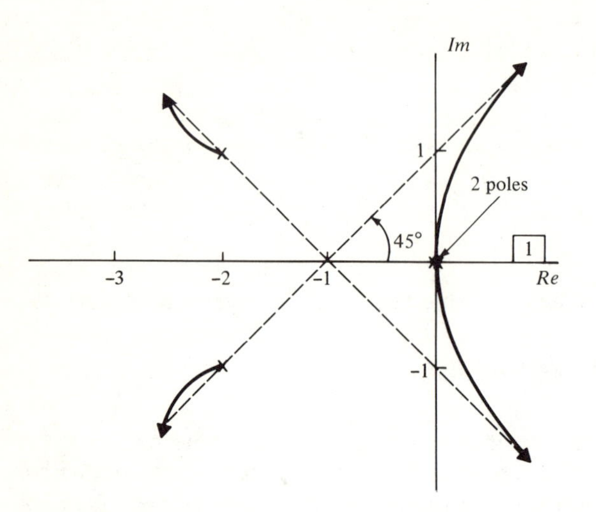

FIGURE 5-29 Root locus of the example system with cascade integral compensation.

A root locus plot, in terms of the adjustable constant K, is given in Fig. 5-29.

Unfortunately, this compensated system is unstable for all positive K. If it were stable, the ramp error coefficient would have meaning. It would be

$$K_1 = \lim_{s \to 0} sG_c(s)G_p(s)$$

$$= \lim_{s \to 0} \frac{K}{s(s^2 + 4s + 5)} = \infty$$

indicating zero steady state error to a ramp input because the compensator has raised the system type from type 1 to type 2. However, the detrimental aspect of this compensator, that of reducing the system's relative stability, makes this approach unusable for the example plant.

5.8.3 Cascade Integral Plus Proportional Compensation

The cascade integral plus proportional compensator has transmittance of the form

$$G_c(s) = A_p + \frac{A_i}{s} = \frac{K(s+a)}{s}$$

A pole at $s=0$ is added to the open-loop system transmittance $G_c(s)G_p(s)$ while a zero is added at $s=-a$.

System type increases by 1 because of the added open-loop pole at the origin. Hence, for a stable design, steady state error performance is improved. This compensator, having both a pole and a zero, does not change the difference between the number of open-loop poles and open-loop zeros. Thus the compensated system has the same root locus asymptotic angles as does the uncompensated one, typically meaning a greater relative stability than with integral compensation.

The centroid of the asymptotes does change, however. For the compensated system with open-loop transmittance $G_c(s)G_p(s)$,

$$\sigma=\frac{\Sigma \text{ poles of } G_cG_p-\Sigma \text{ zeros of } G_cG_p}{\text{number of poles of } G_cG_p-\text{number of zeros of } G_cG_p}$$

As the compensator contributes a pole at $s=0$ and a zero at $s=-a$,

$$\sigma=\frac{\Sigma \text{ poles of } G_p-\Sigma \text{ zeros of } G_p+a}{\text{number of poles of } G_p-\text{number of zeros of } G_p}$$

For a positive constant a (an LHP compensator zero), σ is moved to the right, tending to reduce relative stability from that of the uncompensated feedback system by an amount proportional to a.

5.8.4 Multiparameter Root Locus Design

For the cascade integral plus proportional compensator, there are two adjustable parameters, K and a, and the designer may alternate between adjusting one parameter then the other until a satisfactory design results. A reasonable starting point for such an iterative design procedure is with $a=0$, for which the compensator transmittance is

$$G_c(s)=\frac{K(s+0)}{s}=K$$

which is that for the uncompensated feedback system. The previous root locus plot of Fig. 5-28 then shows the effect of adjustable K when $a=0$. Choosing $K=3$, as was done for the uncompensated system, this compensator transmittance is, in terms of a,

$$G_c(s)=\frac{3(s+a)}{s}$$

giving an overall system transfer function

$$T(s)=\frac{G_c(s)G_p(s)}{1+G_c(s)G_p(s)}$$

$$=\frac{\dfrac{3(s+a)}{s}\left[\dfrac{1}{s(s^2+4s+5)}\right]}{1+\dfrac{3(s+a)}{s}\left[\dfrac{1}{s(s^2+4s+5)}\right]}$$

$$=\frac{3(s+a)}{s^2(s^2+4s+5)+3s+3a}$$

$$=\frac{\text{numerator}}{1+\dfrac{3a}{s^2(s^2+4s+5)+3s}}$$

$$=\frac{\text{numerator}}{1+\dfrac{3a}{s(s^3+4s^2+5s+3)}}$$

$$=\frac{\text{numerator}}{1+\dfrac{3a}{s(s+0.77+j0.8)(s+0.77-j0.8)(s+2.47)}}$$

With $K=3$ and a adjustable, the root locus plot for this system is given in Fig. 5-30a. Here, $a=0.1$ is chosen, giving closed-loop system poles at the indicated locations. Then, with $a=0.1$ and letting K be adjustable again, the root locus plot becomes that of Fig. 5-30b. A slight decrease in K from $K=3$ would increase the system's relative stability slightly, but we will stop the iterative design here, with

$$G_c(s)=\frac{3(s+0.1)}{s}$$

and

$$T(s)=\frac{G_c(s)G_p(s)}{1+G_c(s)G_p(s)}=\frac{\dfrac{3(s+0.1)}{s^2(s^2+4s+5)}}{1+\dfrac{3(s+0.1)}{s^2(s^2+4s+5)}}$$

$$=\frac{3(s+0.1)}{s^4+4s^3+5s^2+3s+0.3}$$

The compensated closed-loop system poles are at $s=-0.123$, $-0.724\pm j0.695$, and -2.43. Because of the compensator's added pole at the origin, the steady state error to a ramp input is zero.

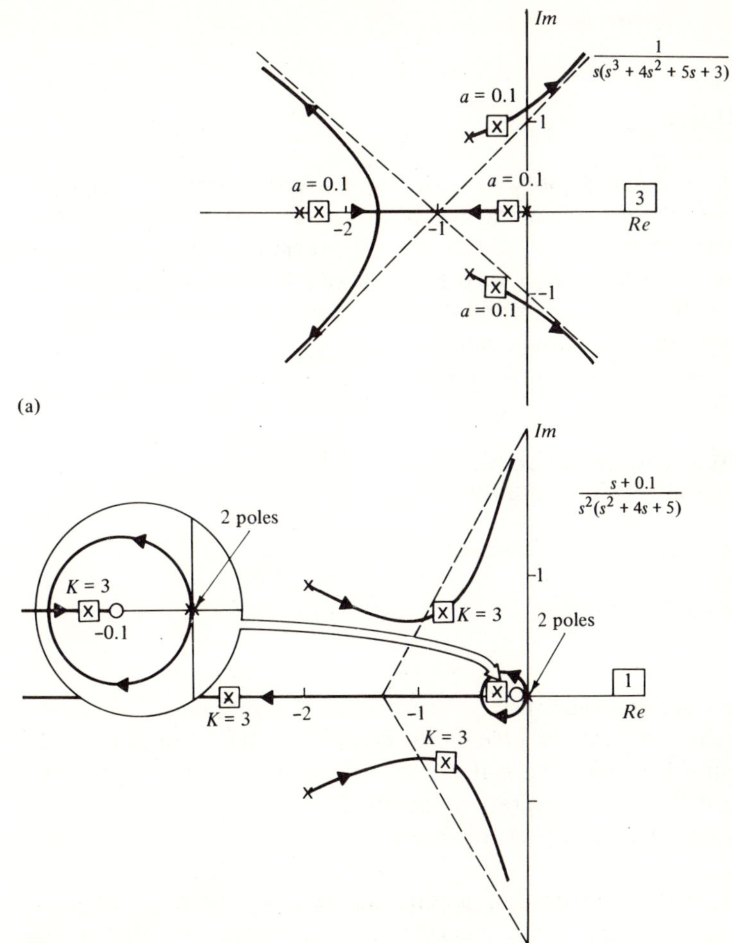

FIGURE 5-30
Root loci of the example system with cascade integral plus proportional compensation. (a) $K = 3$, a variable. (b) K variable, $a = 0.1$.

DRILL PROBLEMS

D5-11. Design cascade integral compensators for systems with the following plant transmittances. Select compensator gains K using root locus plots. For the values of K selected, find the steady state error coefficients and the relative stability of each system.

(a) $G_p(s) = \dfrac{1}{(s+4)^2}$

(b) $G_p(s)=\dfrac{s+1}{s(s+4)}$

(c) $G_p(s)=\dfrac{s+1}{s^2+4s+5}$

D5-12. Sketch root locus plots, in terms of the compensator multiplying constant K, for cascade integral plus proportional compensated systems with the plant transmittances of the previous problem. If possible, choose the compensator zero location and K, by trial and error, to achieve better relative stability and steady state error performance than with only integral compensation.

5.9 CASCADE LAG AND LEAD COMPENSATION

5.9.1 Cascade Lag Compensation

A cascade lag compensator has transmittance of the form

$$G_c(s)=\frac{K(s+a)}{s+b}$$

where K, a, and b are positive constants and $a>b$ so that the compensator zero is to the left of the compensator pole on the complex plane. The proportional plus integral compensator is a special case of lag compensation, for which the constant b is zero. The pure integration operation required when $b=0$ is often difficult to achieve in practice, so this more general form is of considerable practical interest.

For nonzero b, this compensator does not increase the system type number. However, steady state error performance can be improved over that of the uncompensated feedback system. For the example plant,

$$G_p(s)=\frac{1}{s(s^2+4s+5)}$$

the steady state ramp error coefficient of the compensated system, assuming it is stable, is given by

$$K_1=\lim_{s\to 0} sG_c(s)G_p(s)$$

$$=\lim_{s\to 0}\frac{K(s+a)}{(s+b)(s^2+4s+5)}=\frac{Ka}{5b}$$

The corresponding error ratio for a ramp input is

$$\text{Normalized steady state error to a ramp input}=\frac{1}{K_1}=\frac{5b}{Ka}$$

With this compensator, there are three adjustable parameters to be selected by the designer: a, b, and K. In this design, we will approximate the previous integral plus proportional controller, for which $b=0$, placing the controller pole a distance from the origin that is small compared to other root distances. Choosing $b=0.01$ and letting $a=0.1$ as in the previous case,

$$G_c(s)=\frac{K(s+0.1)}{s+0.01}$$

and the system root locus, in terms of adjustable gain K, is given in Fig. 5-31. The choice of $K=3$ gives the indicated four closed-loop system pole locations: -0.12, $-0.729\pm j0.705$, and -2.43. The corresponding ramp error coefficient is

$$K_1=\frac{3(0.1)}{5(0.01)}=6$$

so that the steady state error ratio for a ramp input is

$$\frac{1}{K_1}=0.167$$

a 10-fold improvement over the uncompensated feedback system. The relative stability has dropped from 0.77 to 0.12 unit, however.

Further adjustment of the three parameters is likely to increase both the ramp error coefficient and the relative stability slightly or to improve one to the detriment of the other.

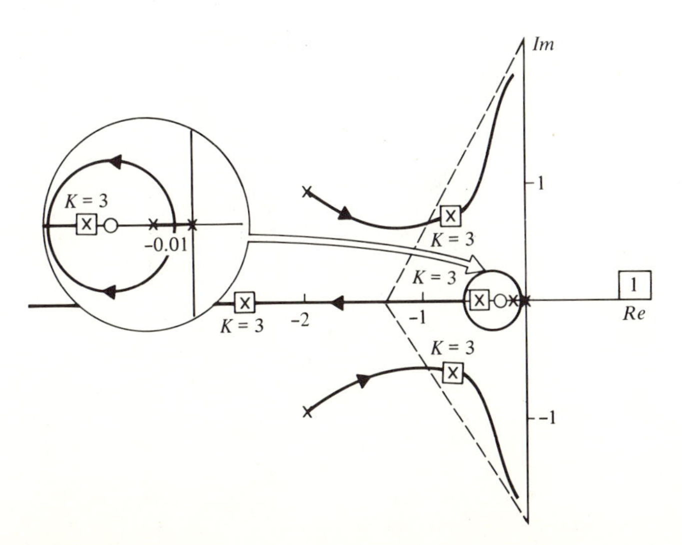

FIGURE 5-31
Root locus for the example system with cascade lag compensation.

5.9.2 Cascade Lead Compensation

A cascade lead compensator has transmittance of the form

$$G_c(s) = \frac{K(s+a)}{s+b}$$

where K, a, and b are positive constants and $b > a$ so that the compensator zero is to the right of the compensator pole on the complex plane. Lead compensation is often used to increase a system's relative stability, possibly at the expense of steady state error performance.

For cascade lead compensation of the example plant $G_c(s)$, the compensator zero is used to draw the locus segment from the plant pole at the origin into the LHP to improve the relative stability as in Fig. 5-32. The compensator zero and pole each pull the centroid of the asymptotes to the left, which also tends to improve the relative stability. Compared to the uncompensated feedback system, the addition of the compensator reduces the ramp error coefficient (increasing the steady state ramp error) unless the gain K is reduced. Hence there is a trade-off between improving relative stability and keeping the gain K sufficiently small for adequate steady state error performance. A compromise is chosen here for which the cascade lead compensator has the form

$$G_c(s) = \frac{K(s+1.6)}{s+16}$$

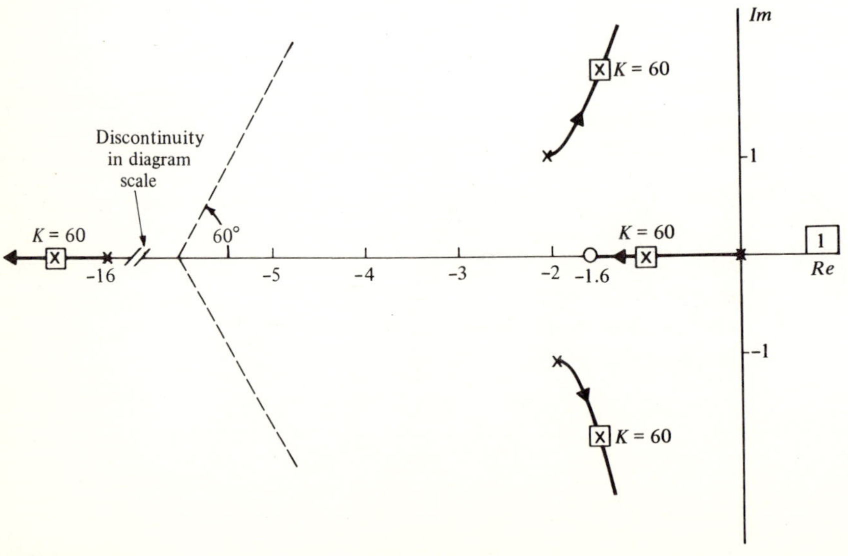

FIGURE 5-32
Root locus for the example system with cascade lead compensation.

In terms of K, the ramp error coefficient is

$$K_1 = \lim_{s \to 0} sG_c(s)G_p(s)$$

$$= \lim_{s \to 0} \frac{K(s+1.6)}{(s+16)(s^2+4s+5)} = \frac{K}{50}$$

A choice of $K = 60$ gives ramp error coefficient

$$K_1 = 1.2$$

The corresponding overall system transfer function is

$$T(s) = \frac{\dfrac{60(s+1.6)}{s+16}\left[\dfrac{1}{s(s^2+4s+5)}\right]}{1 + \dfrac{60(s+1.6)}{s+16}\left[\dfrac{1}{s(s^2+4s+5)}\right]}$$

$$= \frac{60(s+1.6)}{s^4+4s^3+21s^3+64s^2+140s+96}$$

which has poles at $s = -1.106$, $-1.315 \pm j1.9$, and -16.265 as indicated on the root locus plot. Both the error performance and the relative stability have been improved over the uncompensated feedback system. With more specific design objectives given, iterative improvement in the solution may be made as in the multiparameter design example of Sec. 5.8.4.

DRILL PROBLEMS

D5-13. Sketch root locus plots, in terms of the compensator multiplying constant K, for cascade lag compensated systems with the following plant transmittances. If possible, choose the compensator pole and zero locations and K, by trial and error, to achieve better relative stability and better steady state error performance than with the uncompensated feedback system.

(a) $G_p(s) = \dfrac{1}{(s+4)^2}$

(b) $G_p(s) = \dfrac{1}{s(s+4)}$

(c) $G_p(s) = \dfrac{s+1}{s(s+4)}$

D5-14. Sketch root locus plots, in terms of the compensator multiplying constant K, for cascade lead compensated systems with the following plant transmittances. If possible, choose compensator pole and zero locations and K, by trial and error, to achieve better relative stability and better steady state error performance than with the uncompensated feedback system.

(a) $G_p(s)=\dfrac{1}{(s+4)^2}$

(b) $G_p(s)=\dfrac{1}{s(s+2)}$

(c) $G_p(s)=\dfrac{s^2+4s+5}{s^2(s+4)}$

5.10 FEEDBACK COMPENSATION

5.10.1 Design Considerations

Feedback compensation, in whole as in Fig. 5-27b or in part as in Fig. 5-27c, is attractive in system design for several reasons. First, if the feedback transmittance is not unity, the steady state error performance of the system is generally different from that with a cascade compensator providing the same loop transmittance.

For unity feedback, the steady state error coefficients are given by

$$K_n=\lim_{s\to 0} s^n G(s)$$

provided the resulting system is stable. Then

$$\text{Normalized steady state error to a step input}=\frac{1}{1+K_0}$$

and, for ramp and higher power-of-t input,

$$\begin{pmatrix}\text{Normalized steady}\\ \text{state error to an}\\ \text{input of the form}\\ R(s)=A/s^{n+1}\end{pmatrix}=\frac{1}{K_n} \qquad n=1, 2, 3, \ldots$$

For nonunity feedback, the equivalent unity feedback system forward transmittance must be used for $G(s)$ above or, alternatively, the fundamental relation

$$\begin{pmatrix} \text{Normalized steady} \\ \text{state error to an} \\ \text{input of the form} \\ R(s)=A/s^{n+1} \end{pmatrix} = \lim_{s\to 0} s^n[1-T(s)]$$

applies, where $T(s)$ is overall system transfer function.

A second reason for placing compensation in the feedback path is the accuracy with which needed signals may be developed. For example, in a positioning system where the output is a mechanical shaft position, a tachometer will easily provide a rate of change of shaft position signal, while developing that signal from position sensing may be difficult.

A third consideration for feedback compensation is the possibility that the open-loop system may continue to operate with a reduced level of performance if the feedback path is broken. In aircraft and spacecraft applications, for example, the ability of the system as a whole to sustain operation in the face of control system damage and component failures is very important.

5.10.2 Feedback Rate Compensation

A feedback compensated system, Fig. 5-27b, with feedback transmittance of the form

$$H_c = 1 + As$$

where A is a constant, is said to be *feedback rate compensated*, since feedback proportional to the output derivative (or rate of change) is provided in addition to unity output feedback.

When feedback rate compensation is used for the example plant, the overall system transfer function is

$$T(s) = \frac{KG_p(s)}{1+KG_p(s)H_c(s)} = \frac{K/s(s^2+4s+5)}{1+K(1+As)/s(s^2+4s+5)}$$

$$= \frac{K}{s(s^2+4s+5)+K(1+As)}$$

The compensator is seen to place an additional zero, at location $s=-1/A$, onto the root locus. The steady state error to a unit ramp input, assuming stability, is

$$\begin{pmatrix} \text{Steady state error} \\ \text{to a unit ramp} \\ \text{input} \end{pmatrix} = \lim_{s\to 0} \frac{1}{s}[1-T(s)]$$

$$= \lim_{s\to 0} \frac{1}{s}\left[1 - \frac{K}{s^3+4s^2+5s+KAs+K}\right]$$

$$= \lim_{s\to 0} \frac{1}{s}\left[\frac{s^3+4s^2+5s+KAs}{s^3+4s^2+5s+KAs+K}\right] = A + \frac{5}{K}$$

which means that the ramp error coefficient is

$$K_1 = \frac{1}{A + 5/K}$$

Choosing $A = 1/4$ and $K = 6$ so that

$$G_p(s)H_c(s) = \frac{1/4(s+4)}{s(s^2+4s+5)}$$

the ramp error coefficient is

$$K_i = 0.92$$

and the system root locus plot is as given in Fig. 5-33. For error signal gain $K = 6.0$ as shown,

$$T(s) = \frac{6}{s^3 + 4s^2 + 6.5s + 6}$$

which has poles at $s = 0.85 \pm j1.37$ and -2.31.

Had the open-loop compensator zero been instead placed in the forward path, with a cascade compensator of the form

$$G_c(s) = K(1 + As)$$

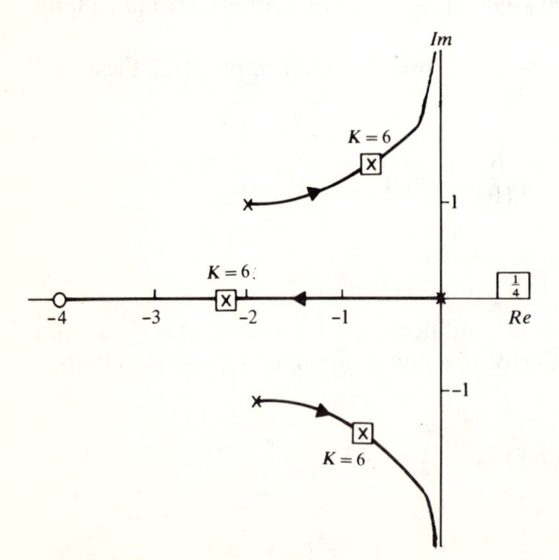

FIGURE 5-33
Root locus of the example system with feedback rate compensation.

the system transfer function would have been

$$T(s)=\frac{G_c(s)G_p(s)}{1+G_c(s)G_p(s)}$$

resulting in a ramp error coefficient

$$\begin{aligned}K_1&=\lim_{s\to 0} sG_c(s)G_p(s)\\&=\lim_{s\to 0}\frac{K(1+As)}{s^2+4s+5}\\&=\frac{K}{5}\end{aligned}$$

5.10.3 More Involved Cascade and Feedback Compensation

Generally speaking, the greater the compensator complexity, the more a system's performance may be improved. For example, a compensator for the example system with 10 poles and zeros at locations selected by the designer should be capable of outperforming the simple compensators discussed here. There are, however, some important practical considerations.

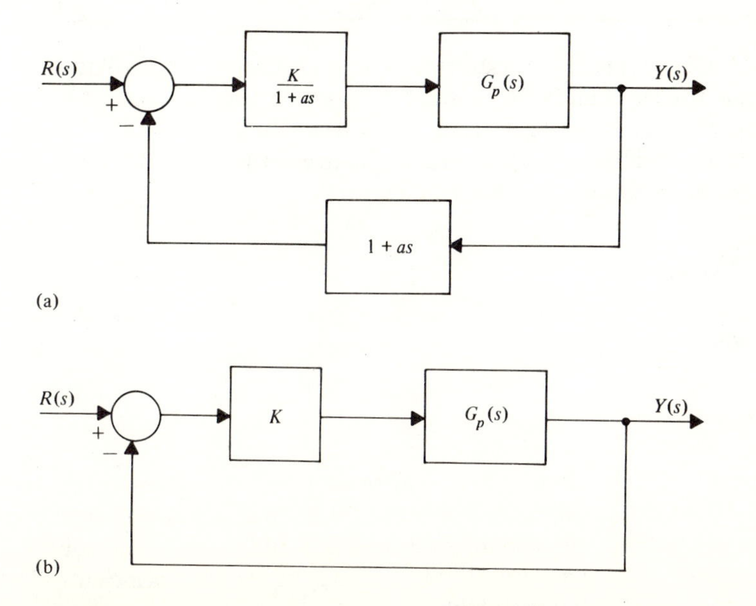

FIGURE 5-34
A compensation arrangement requiring special care. (a) Feedback system with compensator pole-zero cancellations. (b) A feedback system with the same open-loop transmittance.

Doubling the number of compensator poles and zeros is likely to improve performance by a much smaller percent. There is a diminishing return to increasing compensator order because

1. the accuracy of the plant model is limited; and
2. the accuracy of the signals used for compensation—for example, the measured system output—is limited.

Repeated signal differentiation should be avoided because differentiation is a noise-emphasizing operation. Directly monitoring rate-of-change signals from the plant when possible is highly preferable to electronically or otherwise forming a signal derivative by differentiation within the compensator. One tends to find, for example, good motion control systems with separate position, velocity, and acceleration sensors rather than only a position sensor and derivative operations to form signals proportional to velocity and acceleration.

Special care should be taken with designs such as that of Fig. 5-34a which assume open-loop pole-zero cancellations. Analytically, this system may appear to be very desirable. The closed-loop poles are those of the feedback system without compensation, Fig. 5-34b, but the error performance is affected by the open-loop zero at $s=-1/a$. An inexact open-loop pole-zero cancellation, however, could easily result in unexpected behavior.

DRILL PROBLEMS

D5-15. Design feedback rate compensators for systems with each of the following plant transmittances. If possible, choose the compensator zero location and the error signal gain K, by trial and error, to achieve better relative stability and better steady state error performance than with the uncompensated feedback system.

(a) $G_p(s)=\dfrac{1}{(s+4)^2}$

(b) $G_p(s)=\dfrac{1}{s(s+4)}$

(c) $G_p(s)=\dfrac{1}{s(s+4)^2}$

D5-16. Use root locus methods to design compensators for systems with each of the following plant transmittances. Each compensator should involve a combination of both rate feedback and integral cascade compensation. For each design, find the steady state error coefficients and the amount of relative stability.

(a) $G_p(s)=\dfrac{1}{s(s+4)^2}$

(b) $G_p(s)=\dfrac{s+1}{(s+4)^2}$

(c) $G_p(s)=\dfrac{s+4}{s(s^2+2s+5)}$

5.11 PROPORTIONAL-INTEGRAL-DERIVATIVE COMPENSATION

A cascade proportional-integral-derivative (PID) compensator has a transmittance of the form

$$G_c(s)=A_p+\frac{A_i}{s}+A_d s=\frac{K(s^2+bs+c)}{s}$$

Controllers of this type are widely used in simple process control applications in industry.

Provided that the resulting feedback system is stable, the added open-loop pole at the origin increases the system type number by 1. Relative stability is apt to be improved because the net addition of a zero will result in one less root locus asymptote.

A root locus plot, in terms of variable error signal gain K, for cascade PID compensation of the example system is given in Fig. 5-35. In this case, the com-

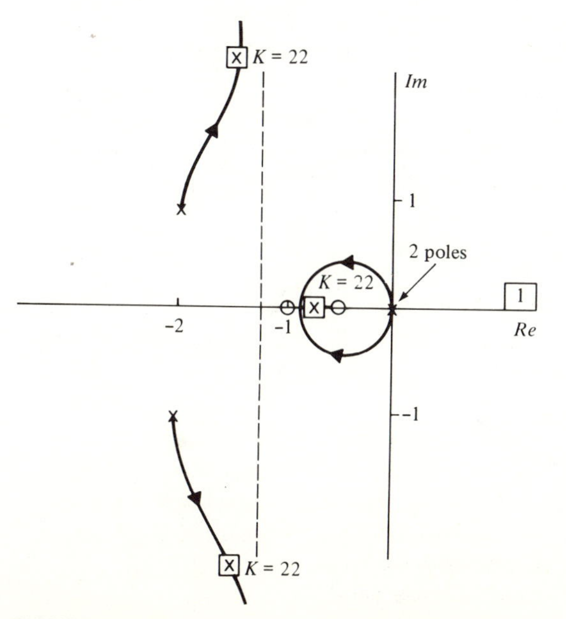

FIGURE 5-35
Root locus of the example system with PID compensation.

pensator zeros have been placed at $s = -0.5$ and $s = -1$ to draw the loci from the double pole at the origin into the LHP. For this design,

$$G_c(s) = \frac{K(s+0.5)(s+1)}{s}$$

The steady state ramp error is zero because of the added pole at the origin.

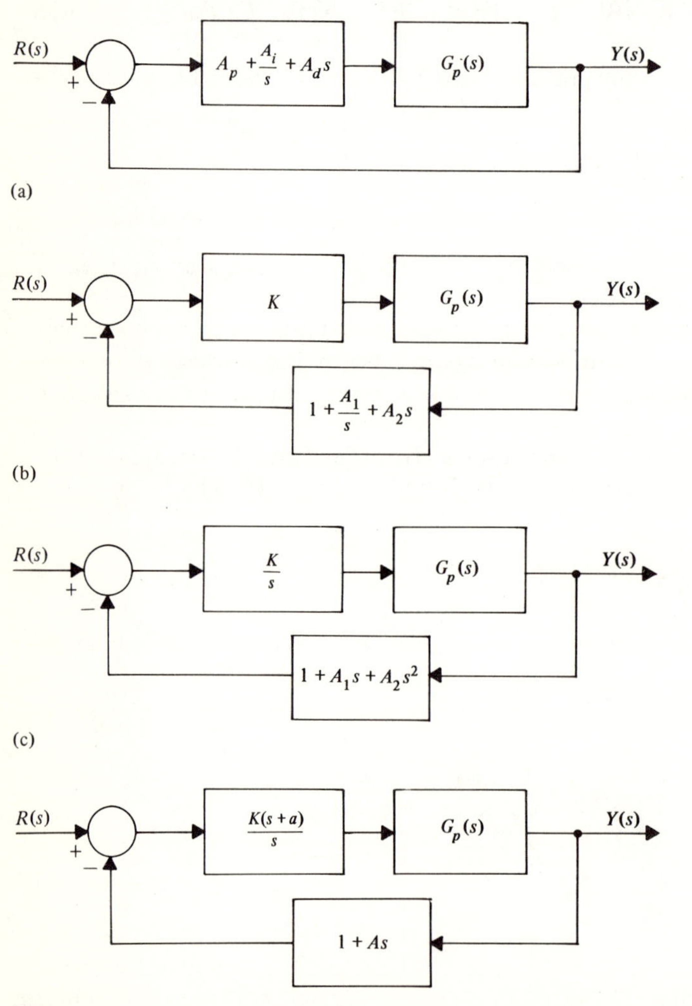

FIGURE 5-36
Some PID compensation arrangements. (a) Cascade compensation. (b) Feedback compensation. (c) A combination of cascade and feedback compensation. (d) Another combination of cascade and feedback compensation.

Choosing $K=22$, to give the closed-loop system poles locations indicated on the root locus diagram, the overall system transfer function is

$$T(s)=\frac{G_c(s)G_p(s)}{1+G_c(s)G_p(s)}$$

$$=\frac{22(s+0.5)(s+1)/s^2(s^2+4s+5)}{1+22(s+0.5)(s+1)/s^2(s^2+4s+5)}$$

$$=\frac{22(s+0.5)(s+1)}{s^4+4s^3+27s^2+33s+11}$$

Figure 5-36 shows a number of different feedback system configurations for which the open-loop system transmittance contains the same PID compensator terms. Each of these systems shares the same root locus plot, but their steady state error performances and other characteristics differ somewhat. A designer selects a system configuration based upon the ease of developing and generating needed signals and the required performance characteristics.

DRILL PROBLEM

D5-17. Use root locus methods to design cascade PID compensators for systems with the following plant transmittances. For each design, find the steady state error coefficients and the amount of relative stability of the closed-loop system.

(a) $G_p(s)=\dfrac{1}{(s+4)^2}$

(b) $G_p(s)=\dfrac{s+1}{s(s^2+4s+5)}$

(c) $G_p(s)=\dfrac{s^2+2s+5}{s(s+4)}$

5.12 SUMMARY

A pole-zero plot of a rational function consists of indications on the complex plane of the locations of the function's poles and zeros. The addition of the multiplying constant of the function to the diagram completely specifies the function.

To graphically evaluate a rational function $F(s)$ at a particular value $s=s_0$, directed line segments are drawn from each zero and each pole of $F(s)$ to the point s_0. The magnitude of the evaluated function is the product of the magnitude of the function's multiplying constant and the lengths of the directed line segments from the zeros, divided by the product of the lengths of the directed line segments from the poles. The angle of the evaluated function is the sum of the angles of the directed line segments from the zeros minus the sum of the angles of the directed line segments from the poles. An additional 180° must be added or subtracted from the angle of the function if the multiplying constant is negative.

A root locus plot consists of a pole-zero plot of the rational open-loop system transmittance $G(s)H(s)$, upon which is superimposed the locus of the roots

$$1+KG(s)H(s)=0$$

as K is varied from zero to $+\infty$. The root locus is symmetrical about the real axis. A value of s for which the angle of $G(s)H(s)$ is 180° (plus or minus any multiple of 360°, of course) is a point on the root locus, corresponding to some value of K.

A set of six rules for root locus construction were presented:

1. Continuous curves, which comprise the branches of the locus, start at each of the n poles of GH, for which $K=0$. As K approach $+\infty$, the branches of the locus approach the m zeros of GH. Locus branches for excess poles extend to infinity; for excess zeros, locus segments extend from infinity.
2. The locus includes all points along the real axis to the left of an odd number of poles plus zeros of GH.
3. As K approaches $+\infty$, the branches of the locus become asymptotic to straight lines with angles

   $$\theta=\frac{180^\circ+i360^\circ}{n-m}\qquad \text{for } i=0,\ \pm1,\ \pm2, \text{ until all angles not differing by multiples of } 360^\circ \text{ are obtained}$$

   where n is the number of poles and m is the number of zeros of GH.
4. The starting point on the real axis from which the asymptotic lines radiate is given by

   $$\sigma=\frac{\Sigma\ \text{open-loop pole values}-\Sigma\ \text{open-loop zero values}}{n-m}$$

   This point is termed the centroid of the asymptotes.
5. Roots leave the real axis at a gain K which is the maximum possible value of K in that region of the real axis. Roots enter the real axis at a gain K which is the minimum possible value of K in that region of the real axis. Two roots leave or strike the axis at the breakaway point at angles of $\pm 90^\circ$.
6. The angle of departure ϕ of a locus branch from a complex pole is found according to

Sum of all other GH pole angles to the poles under consideration $+\phi-$sum of GH zero angles to the pole$=180^\circ$

The angle of approach ϕ' of the locus to a complex zero is given by

Sum of GH pole angles to the zero under consideration $-$sum of all other GH zero angles to the zero $-\phi'=180^\circ$

Multiple angles of departure and approach at multiple complex roots of GH are found similarly, using multiple contributions of ϕ or ϕ' and equating to $180^\circ \pm i360^\circ$.

It was shown that root locus methods may be applied to other systems besides those of the standard form and that it is straightforward to extend the technique to systems with negative adjustable parameters.

In feedback tracking systems, compensators are used to improve relative stability, steady state error performance, and perhaps other characteristics such as damping ratios and overshoot. The root locus method is an especially powerful tool for compensator design because the effects of additional open-loop compensator poles and zeros upon the closed-loop pole locations is easily and quickly visualized.

A unity feedback system with a cascade compensator has type number equal to the number of $s=0$ poles in the forward transmittance $G(s)$. Provided the resulting design has sufficient relative stability for the application at hand, the system type number may be increased by introducing additional $s=0$ poles in the compensator. In cascade integral compensation, the compensator is an integrator having a single $s=0$ pole.

Unfortunately, the addition of $s=0$ poles to a system's open-loop transmittance tends to greatly reduce closed-loop relative stability, often resulting in unstable systems. Additional compensator zeros, placed to draw the root loci to the left on the complex plane, are then desirable. The cascade integral plus proportional compensator introduces a pole at $s=0$ and one zero, while the PID compensator has an $s=0$ pole and two zeros.

Accurate integrators are expensive or difficult to build for many applications, but steady state system performance can still be improved by placing compensator poles near $s=0$ in the LHP. The system type number is not increased, but normalized steady state error may be thereby reduced. Cascade lag compensation is often used for this purpose. Lead compensation, on the other hand, typically increases relative stability with little or no improvement in steady state error.

Placing compensation elements in the feedback path gives the designer additional flexibility. Relative stability depends only on loop transmittance, while steady state error performance depends upon how the compensation is distributed between the forward and feedback transmittances.

Increased compensator complexity will generally give improved system performance, although ultimately one should expect a diminishing return because of unavoidable inaccuracies.

REFERENCES

Root Locus

Clark, R. N. *Introduction to Automatic Control Systems.* New York: Wiley, 1962.

Evans, W. R. "Graphical Analysis of Control Systems." *Trans. AIEE* 67 (1948): 547–551.

_____. "Control System Synthesis by the Root Locus Method." *Trans. AIEE* 69 (1950): 67–69.

_____. *Control System Dynamics.* New York: McGraw-Hill, 1954.

Savant, C. J. Jr. *Basic Feedback Control System Design.* New York: McGraw-Hill, 1958.

Truxal, J. G. *Control System Synthesis.* New York: McGraw-Hill, 1955.

Numerical Methods

Ash, R. H., and Ash, G. R. "Numerical Computation of Root Loci Using the Newton-Raphson Technique." *IEEE Trans. Auto. Contr.* AC-13 (October 1968): 576–582.

Carnahan, B.; Luther, H. A.; and Wilkes, J. O. *Applied Numerical Methods.* New York: Wiley, 1969.

Hamming, R. W. *Numerical Methods for Scientists and Engineers,* 2nd ed. New York: McGraw-Hill, 1973.

James, M. L.; Smith, G. M.; and Wolford, J. C. *Applied Numerical Methods for Digital Computation,* 2nd ed. New York: Harper & Row, 1977.

Compensator Design

D'Azzo, J. J., and Houpis, C. H. *Linear Control System Analysis and Design.* New York: McGraw-Hill, 1975.

Dorf, R. C. *Modern Control Systems.* Reading, Mass.: Addison-Wesley, 1967.

Fallside, F. *Control System Design by Pole-Zero Assignment.* New York: Academic Press, 1977.

Garnell, P., and East, D. J. *Guided Weapon Control Systems.* New York: Pergamon Press, 1977.

Kuo, B. C. *Automatic Control Systems.* Englewood Cliffs, N.J.: Prentice-Hall, 1967.

Truxal, J. G. *Control System Synthesis.* New York: McGraw-Hill, 1955.

PROBLEMS

1. Draw pole-zero plots for the following rational functions:

(a) $F(s)=\dfrac{-3s^2+2}{(7s+1)(s+5)^2}$

(b) $F(s)=\dfrac{8(s^2-2s+10)(1+4s)}{s^3(s^2+2s+10)}$

(c) $F(s)=\dfrac{100s(s^2-4s+13)^2}{(s+3)^2(s^2+4s+13)^2}$

(d) $F(s)=\dfrac{3s^3+10s}{(s^2+2s+17)(s^2+2s+4)}$

2. Find the rational functions with the pole-zero plots of Fig. P5-1.

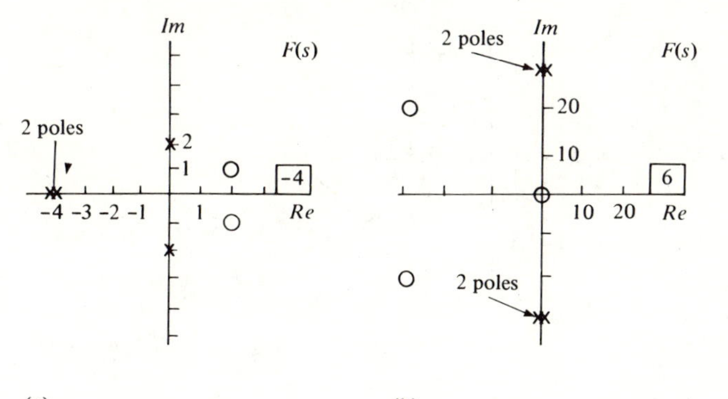

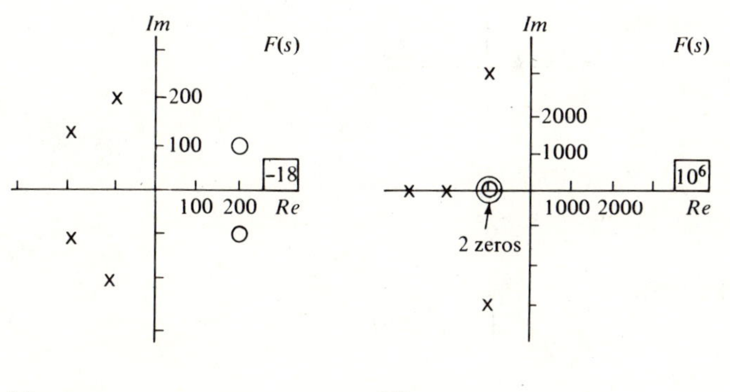

FIGURE P5-1

3. Evaluate the rational function of Fig. P5-2 at $s=j2$ using graphical methods.

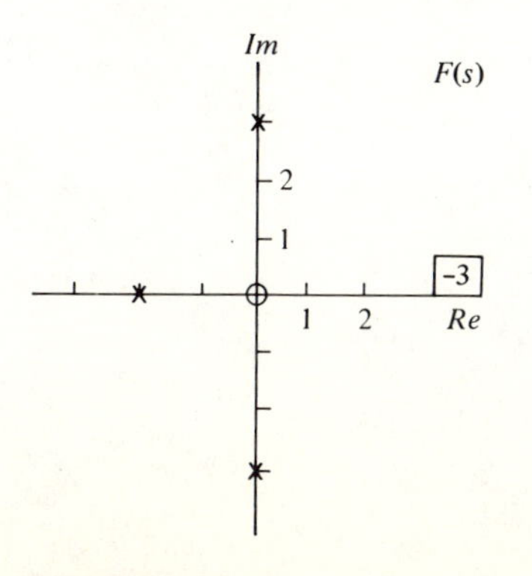

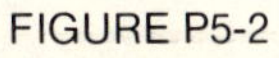

FIGURE P5-2

4. Sketch root locus plots for the systems with pole-zero plots given in Fig. P5-3. Find asymptotes, centroid, angles of departure, angles of approach, and approximate breakaway points where applicable.

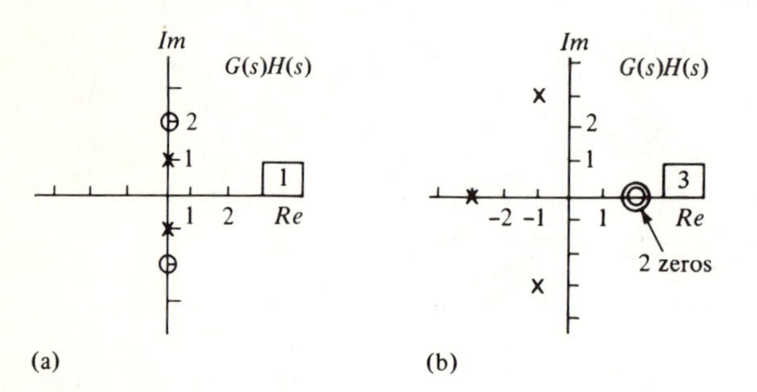

(a) (b)

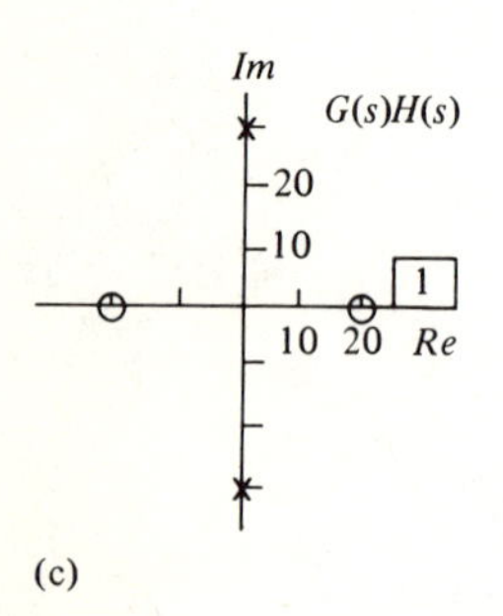

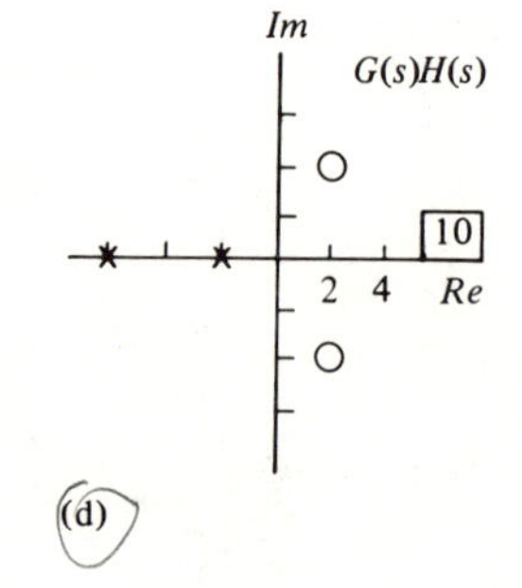

(c) (d)

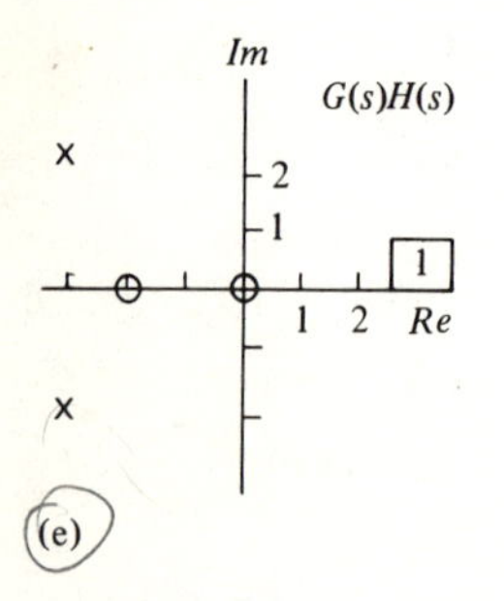

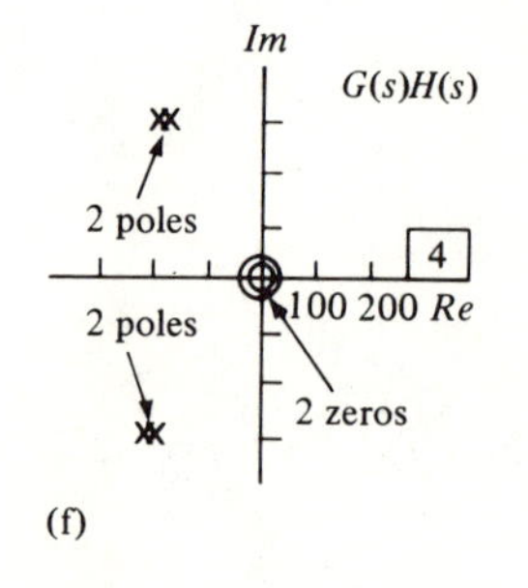

(e) (f)

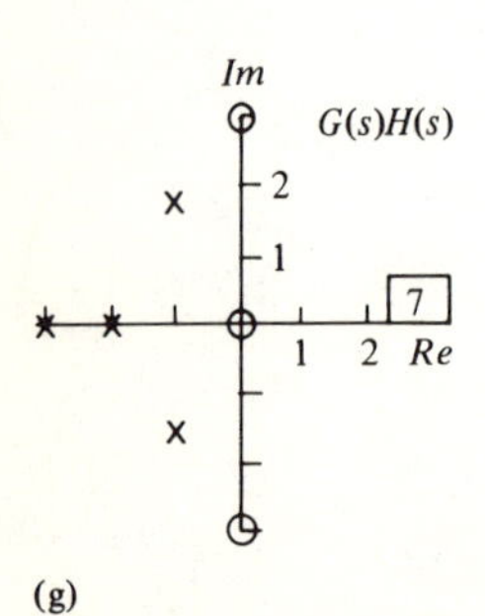

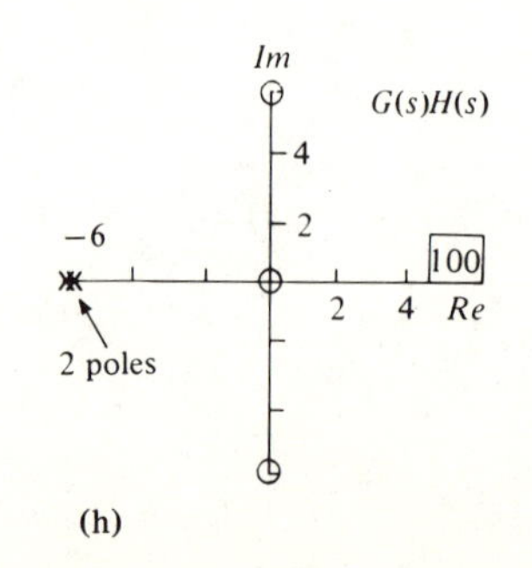

(g) (h)

FIGURE P5-3

5. For each of the following GH products, construct root locus sketches. Find the range of the positive adjustable gain K in

$$T(s)=\frac{G(s)}{1+KG(s)H(s)}$$

for which each overall system is stable.

(a) $G(s)H(s)=\dfrac{1}{s(2s+1)}$

(b) $G(s)H(s)=\dfrac{3s+1}{s(2s+1)}$

(c) $G(s)H(s)=\dfrac{s+1}{s(2s+1)}$

(d) $G(s)H(s)=\dfrac{1}{s(s+1)(2s+1)}$

(e) $G(s)H(s)=\dfrac{1}{s^2+10s+100}$

ans. stable for any K

(f) $G(s)H(s)=\dfrac{s+1}{s^2+s+10}$

ans. stable for any K

(g) $G(s)H(s)=\dfrac{s}{(s+1)(s+10)}$

ans. stable for any K

(h) $G(s)H(s)=\dfrac{1}{s(s+2)(s^2+2s+37)}$

ans. unstable for $K>416$

6. Sketch root locus plots for the systems of Fig. P5-4, for K between zero and $+\infty$. Find asymptotes, centroid, angles of departure, angles of approach, and approximate breakaway points where applicable.

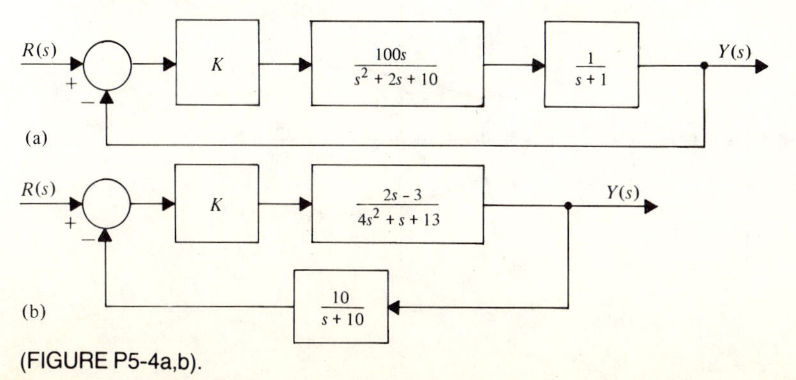

(FIGURE P5-4a,b).

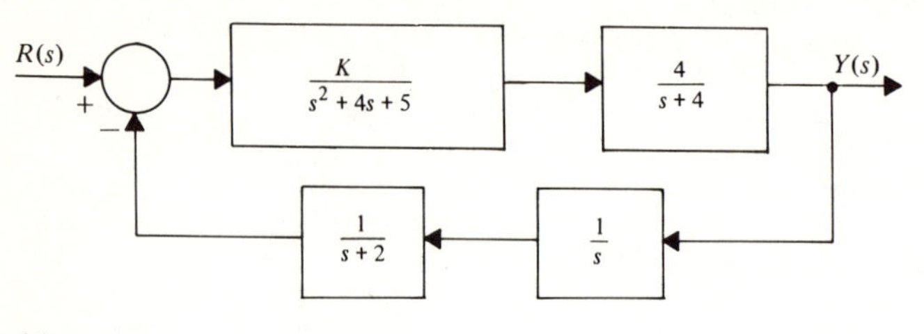

(c)

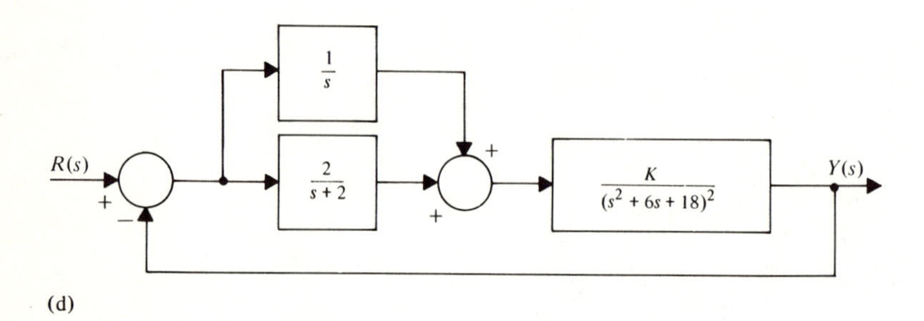

(d)

FIGURE P5-4

7. Are there any roots of $1+G(s)$ in the right half-plane for $G(s)$ given below?

$$G(s)=\frac{10}{s(0.1s+1)(0.5s+1)(s+1)}$$

8. For the root locus plots of Fig. P5-5, use graphical methods to find, approximately, the value of the adjustable constant K for which a root of

$$1+G(s)H(s)$$

is at the position indicated by the dot.

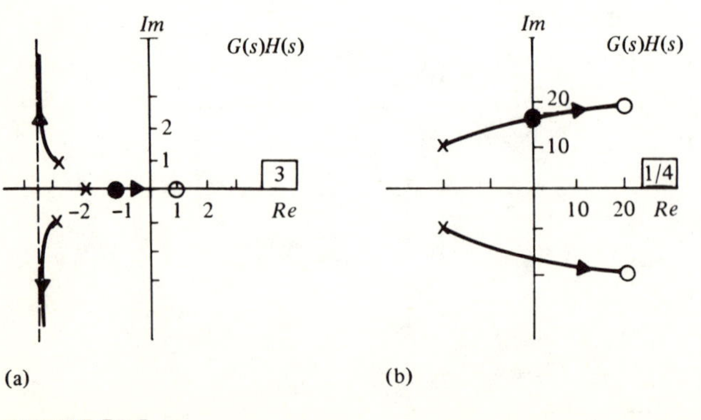

(a) (b)

FIGURE P5-5

9. For $K=3$, approximately locate the roots of

$$1+K\frac{s^2+4}{(s+2)(s+4)}$$

on a root locus plot, using graphical methods.

10. For the systems with the following GH products, find the value of adjustable gain K for which (if possible) there is a complex conjugate set of roots of the overall system with damping ratio $\xi=0.2$.

(a) $$\frac{1}{s(s+1)(0.2s+1)}$$

ans. $K=2.3$

(b) $$\frac{s+0.2}{s(s+1)(0.2s+1)(s+20)}$$

ans. $K=238$

(c) $$\frac{s+2}{s(s+1)(0.2s+1)(s+200)}$$

ans. $K=2600$

(d) $$\frac{s+5}{s(s+1)(0.2s+1)(s+500)}$$

ans. $K=625$

11. For each of the following GH products, plot the root locus diagram and approximately determine the roots of the overall system characteristic equation for values of K equal to 1, 10, 100, and 1000:

(a) $$\frac{1}{s(s+1)}$$

ans. $-\frac{1}{2}\pm j(\sqrt{3}/2)$, $-\frac{1}{2}\pm j(\sqrt{39}/2)$,
$-\frac{1}{2}\pm j(\sqrt{399}/2)$ $-\frac{1}{2}\pm j(\sqrt{3999}/2)$

(b) $$\frac{1}{s^2(s+1)^2}$$

12. On the complex plane, sketch the locus of the roots of the following functions, as the parameter K is varied from zero to $+\infty$:

(a) $$T(s)=\frac{4}{s^2+K}$$

(b) $$T(s)=\frac{s^2+3s-2}{Ks^2+3s+2}$$

(c) $$T(s)=\frac{100}{s^2+Ks+1}$$

(d) $$T(s)=\frac{-100}{s^2+(K-2)s+3+K}$$

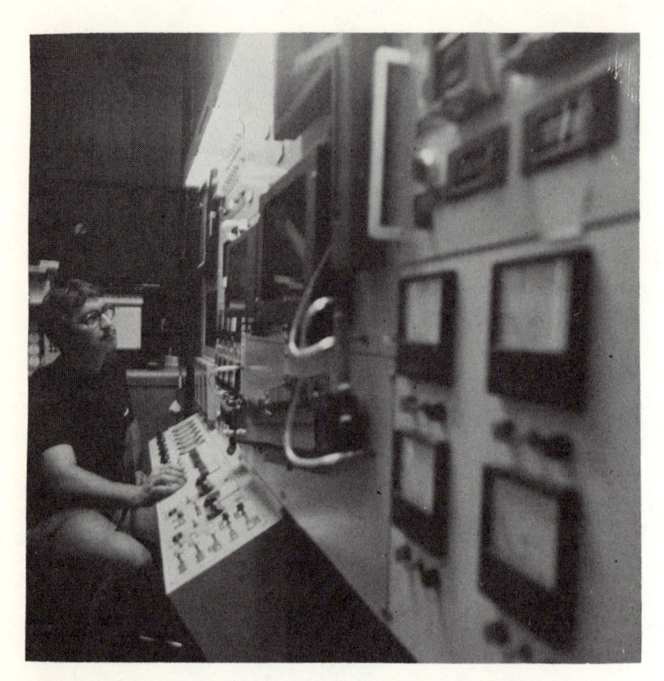

Adjustments are made at the master control panel for an industrial antipollution system. Pollutant measurements are used to control plant processes to minimize polution. (Photo courtesy of Barber Colman Company.)

13. On the complex plane, sketch the locus of the poles of the function

$$T(s)=\frac{-10s+8}{s^2+2s+Ks-3K}$$

as the parameter K is varied through the range $-\infty$ to $+\infty$.

14. Sketch the root locus, for K from zero to $+\infty$, for a system with

$$G(s)H(s)=-\frac{s+2}{(s+1)^2(s^2+4s+5)}$$

Note the negative algebraic sign.

15. Sketch the root locus for K in the range from zero to *minus* infinity for a system with

$$G(s)H(s)=\frac{2(s-2)^2}{(s+3)(s^2+2s+10)}$$

Using this root locus, approximately locate the poles of

$$T(s)=\frac{G(s)}{1+KG(s)H(s)}$$

for $K=-2$.

16. Each of the following systems consists of the plant driven by an error signal amplifier with gain K, as in Fig. 5-26. For each of the following plant transmittances and amplifier gains, use root locus methods to find, approximately, the closed-loop system poles. Then find the step error coefficient K_0.

(a) $G_p(s)=\dfrac{s+1}{s+10}$

$K=8$

(b) $G_p(s)=\dfrac{s+10}{s+1}$

$K=8$

ans. $-9;\ 80$

(c) $G_p(s)=\dfrac{s+4}{(s+2)^2}$

$K=1$

(d) $G_p(s)=\dfrac{1}{(s+10)(s+50)}$

$K=1300$

ans. $-30\pm j30;\ 2.6$

17. Design cascade integral compensators for unity feedback systems with the following plant transmittances. Use root locus plots to select compensator gains K. For each design, find the steady state error coefficients and the amount of relative stability.

(a) $G_p(s)=\dfrac{9}{(s+3)^2}$

(b) $G_p(s)=\dfrac{s+2}{(s+3)^3}$

(c) $G_p(s)=\dfrac{s+5}{(s^2+4s+20)}$

(d) $G_p(s)=\dfrac{s^2+4s+13}{(s+3)(s^2+2s+10)}$

18. Design cascade integral plus proportional compensators for unity feedback systems with the following plant transmittances. Use root locus plots to select the compensator parameters. For each design, find the steady state error coefficients and the amount of relative stability.

(a) $G_p(s)=\dfrac{10}{s^2+7s+10}$

(b) $G_p(s)=\dfrac{s+2}{s^2+4s+3}$

(c) $G_p(s)=\dfrac{s+4}{s^2+5s+6}$

(d) $G_p(s)=\dfrac{6s+30}{s^2+4s+13}$

19. A plant to be controlled has transmittance

$$G_p(s) = \frac{s+5}{s(s^2+4s+100)}$$

Use root locus methods to design tracking systems of each of the following types:
(a) Uncompensated feedback
(b) Lag compensated
(c) Rate feedback compensated with the compensator zero at $s=-8$
(d) Lead compensation with the compensator pole located 10 times further in the LHP than the compensator zero

For each design, find the normalized steady state errors to step, ramp, and parabolic inputs and the amount of relative stability.

20. Use root locus methods to design a PID compensator for a plant with transmittance

$$G_p(s) = \frac{500}{(s+2)(s^2+20s+500)}$$

The system should have zero steady state error to a ramp input and a relative stability of at least 5 units, if possible.

21. A simple block diagram of a control rod positioning system for a nuclear power plant is given in Fig. P5-6. Sketch a root locus plot for the system. Graphically test points along the imaginary axis to accurately determine where the loci cross into the RHP. Then determine graphically the value of K for which the system is marginally stable.

For a system of relatively low order such as this one, the same results are perhaps more easily obtained by Routh-Hurwitz testing in terms of K. With greater system complexity, a Routh-Hurwitz test becomes hopelessly involved, but root locus methods remain very suitable.

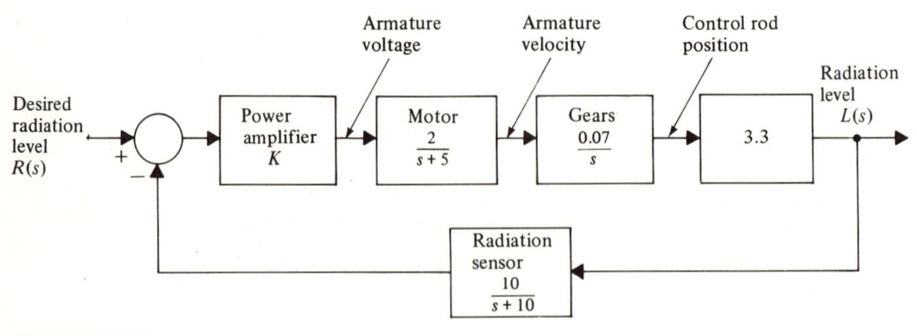

FIGURE P5-6

22. A fine positioning system for an elevator when it is in the vicinity of the correct position is modeled in Fig. P5-7. Use root locus methods to locate the overall system poles as a function of $K_1>0$, when $K_2=0$. Then, with a specific choice of K_1, use root locus methods to determine the overall system poles as a function of $K_2>0$. Using several iterations of choosing K_1, then choosing K_2, find values of K_1 and K_2, if possible, for which the system has a pair of complex poles with damping ratio approximately 0.7 and a relative stability in excess of 2 units.

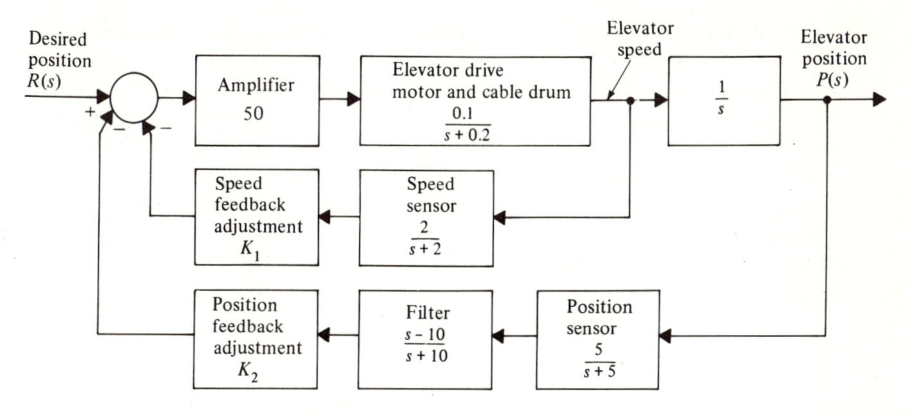

FIGURE P5-7

23. A linearized inventory control system model is given in Fig. P5-8. Based upon the difference between desired and actual inventory levels, management sets the production quota which in turn determines the production rate. Using graphical root locus methods, determine the production time constant

$$\tau=\frac{1}{K}$$

in the range

$$1<\tau<10$$

which gives greatest relative stability of the system.

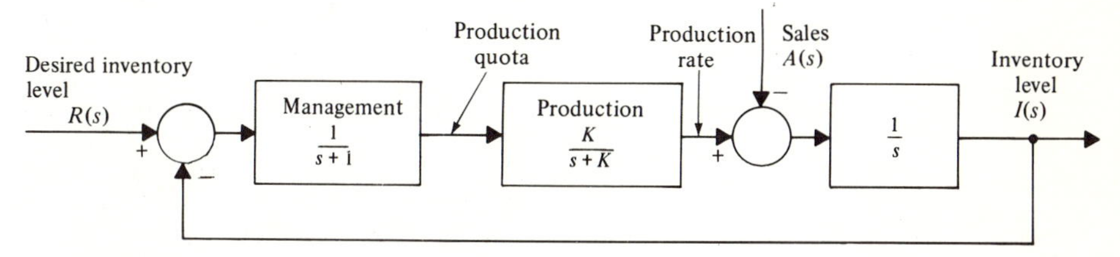

FIGURE P5-8

24. The pitch control system for a high-altitude aircraft is modeled in Fig. P5-9. Find the value of the adjustable constant $K>0$ for which the potentially complex pole pair of the closed-loop system has critical damping.

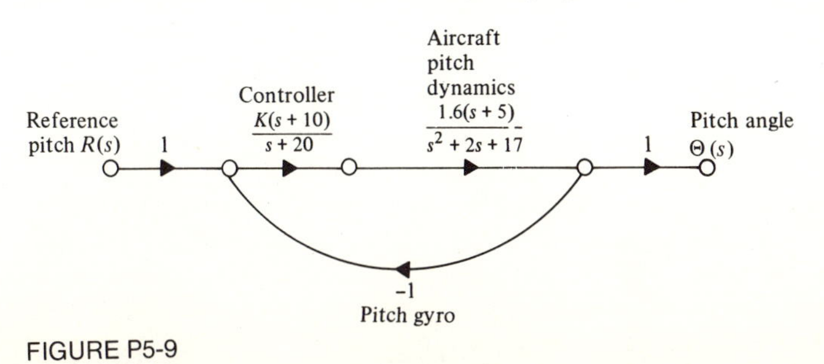

FIGURE P5-9

25. A simplified model of an automobile pollution control system is shown in Fig. P5-10. Sketch the system root locus for $K > 0$, paying particular attention to the manner in which the loci leave the repeated poles. Use the root locus to determine, approximately, the value of K, if any, for which the system's natural response will decay at least as rapidly as e^{-2t}.

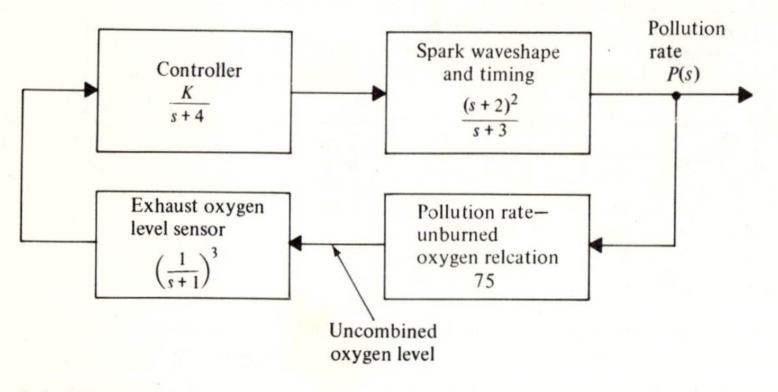

FIGURE P5-10

six

FREQUENCY RESPONSE ANALYSIS AND DESIGN

6.1 PREVIEW

It would be difficult to overstate the importance of frequency response methods to system analysis and design. Frequency response characterizations of systems have long been popular because of the relative ease and practicality of steady state sinusoidal measurements. In addition, these methods can cope with irrational transmittances such as time delays and can deal with system components so complicated that their properties can only be measured.

The forced response of a transmittance to a sinusoidal input signal is sinusoidal, of the same frequency but with generally different amplitude and phase angle than the input. The ratio of output to input amplitude and the input-output phase shift are termed the frequency response of the system and are related to the transmittance in Sec. 6.2.

Bode plots, the subject of Secs. 6.3 and 6.4, show frequency response graphically, in a form that allows easy construction for rational transmittances. A complicated transmittance is decomposed into simple first- and second-order terms, and the plots for the individual terms are simply added to obtain the overall frequency response plots.

When graphical pole-zero methods are combined with frequency response concepts in Sec. 6.5, the result is a powerful tool for visualizing the effects of poles and zeros upon frequency response. The incorporation of experimental frequency response data is then discussed and, in Sec. 6.6, the applicability of these methods to irrational transmittances such as time delays is covered.

To determine whether or not a feedback system is stable, one may imagine adjusting the amount of feedback from zero to the actual amount present. If the zero-feedback, open-loop system is stable and the closed-loop system is not, then there must be an intermediate amount of feedback gain, between zero and unity, for which the overall system is marginally stable, with poles on the imaginary axis. As the presence of imaginary axis roots is easy to determine from the frequency response, these methods of Sec. 6.7 provide an important stability test. For a feedback system which is stable, the added feedback gain and phase shift which would result in instability are important measures of relative system stability.

A definitive frequency response stability test, the Nyquist criterion, is introduced in Sec. 6.8. Nyquist methods apply to feedback systems which are unstable when connected open-loop, to systems with both rational and irrational transmittances, and to systems with components characterized by experimental data.

In the concluding section, the compensation of general-purpose operational amplifiers provides practical application examples for which frequency response methods are well suited.

6.2 FREQUENCY RESPONSE

6.2.1 Amplitude Ratio and Phase Shift

In the development to follow, a general symbol for a transmittance, $F(s)$, will be used. In later sections of this chapter, $F(s)$ is taken to represent the transmittance of a system component, the overall system transfer function $T(s)$, or the loop transmittance $G(s)H(s)$ of a feedback system, depending upon the application at hand.

The response of a transmittance $F(s)$ to a sinusoidal input signal, Fig. 6-1a,

$$r(t) = B\cos(\omega t + \beta)$$

generally consists of both forced and natural components. The forced part of the response is also sinusoidal, with the same frequency ω as the input. Generally, the amplitude and the phase angle of the forced sinusoidal response are different

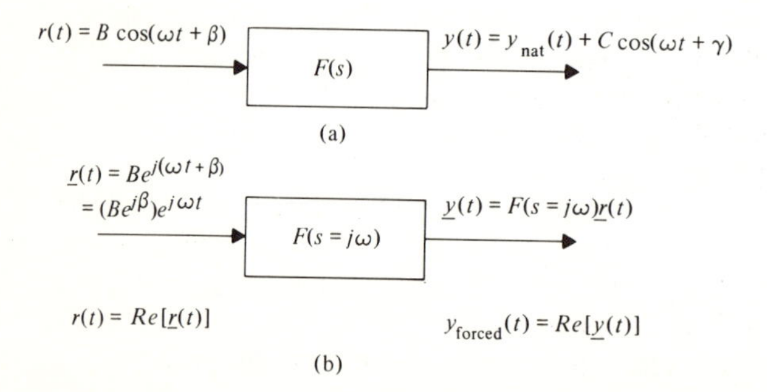

FIGURE 6 1
Response of a transmittance to a sinusoidal input. (a) Form of the solution. (b) Sinor solution for forced sinusoidal response.

than those of the input and they depend upon the input frequency:

$$y(t) = y_{\text{natural}}(t) + C\cos(\omega t + \gamma)$$

One way of calculating the forced sinusoidal response of a system is to solve, instead, the related sinor (or rotating phasor) problem of Fig. 6-1b, in which the forced sinusoidal signals are replaced by complex exponentials, the real parts of which are the actual signals of interest. Then

$$\frac{\underline{y}(t)}{\underline{r}(t)} = F(s = j\omega)$$

the transmittance being the ratio of exponential output to input as a function of the exponential constant s:

$$\begin{aligned}\underline{r}(t) &= Be^{j(\omega t + \beta)}\\ \underline{y}(t) &= Ce^{j(\omega t + \gamma)} = F(s = j\omega)\underline{r}(t)\\ &= F(s = j\omega)Be^{j(\omega t + \beta)}\end{aligned}$$

Letting

$$\begin{aligned}A(\omega) &= |F(s = j\omega)|\\ \Phi(\omega) &= \angle F(s = j\omega)\end{aligned}$$

then

$$\begin{aligned}\underline{y}(t) &= A(\omega)e^{j\Phi(\omega)}Be^{j(\omega t + \beta)}\\ &= A(\omega)Be^{j[\omega t + \beta + \Phi(\omega)]}\\ y_{\text{forced}}(t) &= C\cos(\omega t + \gamma) = Re[\underline{y}(t)]\\ &= A(\omega)B\cos[\omega t + \beta + \Phi(\omega)]\end{aligned}$$

The magnitude of the transmittance when evaluated at $s = j\omega$ is the ratio of output amplitude to input amplitude:

$$A(\omega) = |F(s = j\omega)| = \frac{C}{B}$$

The angle of the transmittance, for $s = j\omega$, is the difference in phase angles between output and input:

$$\Phi(\omega) = \angle F(s = j\omega) = \gamma - \beta$$

As a numerical example, consider the transmittance

$$F(s) = \frac{6}{s + 4}$$

and the input signal

$$r(t) = 3\cos(7t + 20^\circ)$$

For $s = j7$, the transmittance is

$$F(s = j7) = \frac{6}{j7 + 4} = 0.74e^{-j60^\circ}$$

The forced sinusoidal output has amplitude

$$C = (0.74)(3) = 2.22$$

and phase angle

$$\gamma = 20^\circ + (-60^\circ) = -40^\circ$$

giving

$$y_{\text{forced}}(t) = 2.22\cos(7t - 40^\circ)$$

6.2.2 Frequency Response Measurement

When a transmittance $F(s)$ is driven by a sinusoidal input signal, the ratio of the amplitude of the forced output to the amplitude of the input is

$$A(\omega) = \frac{\text{amplitude of sinusoidal output}}{\text{amplitude of sinusoidal input}} = |F(s = j\omega)|$$

which is a function of the radian frequency of the sinusoid, ω. The difference in the phase angles of the output and the input is

$$\begin{aligned} \Phi(\omega) &= \text{phase angle of sinusoidal output} - \text{phase angle of sinusoidal input} \\ &= \angle F(s = j\omega) \end{aligned}$$

To measure the frequency response of a stable component or system at some frequency, apply a sinusoidal input of that frequency. Choose any convenient amplitude which is not so large as to overload the system, yet not so small that the system signals are masked by noise. Wait until the system natural behavior, arising from the connection of the input or changes in its amplitude or frequency, dies out. Measure the amplitudes of the input and output sinusoids and form the ratio

$$A = \frac{\text{output sinusoid amplitude}}{\text{input sinusoid amplitude}}$$

The phase shift is the difference in phase between the forced-output sinusoid and the sinusoidal input signal. If the two signals can be plotted side by side, the phase shift may be found by comparing the plots. Stroboscopic light techniques are especially useful for comparing the phases of mechanical elements in rapid motion. If the system signals are electrical or can be monitored by electrical transducers, the elptical pattern formed by plotting the input versus the output signal on an oscilloscope or instruments known as phase meters or vector voltmeters may be used to display the amount of phase shift.

With a number of such measurements of amplitude ratio and phase shift at various frequencies, curves for $A(\omega)$ and $\phi(\omega)$ may be sketched.

6.2.3 Response at Low and High Frequencies

There are some considerations which make the determination of amplitude and phase shift curves from a limited number of measurements more accurate. If the transmittance is a rational function

$$F(s)=\frac{b_m s^m + b_{m-1}s^{m-1} + \cdots + b_1 s + b_0}{s^n + a_{n-1}s^{n-1} + \ldots + a_1 s + a_0}$$

then

$$F(j\omega)=\frac{b_m(j\omega)^m + b_{m-1}(j\omega)^{m-1} + \cdots + b_1(j\omega) + b_0}{(j\omega)^n + a_{n-1}(j\omega)^{n-1} + \cdots + a_1(j\omega) + a_0}$$

For large values of ω, all but the highest powers of ω in the numerator and denominator may be ignored:

$$F(j\omega)\cong\frac{b_m(j\omega)^m}{(j\omega)^n}=b_m j^{m-n}\omega^{m-n}$$

The amplitude curve approaches a power of ω, and the phase curve approaches a multiple of 90°.

For small ω, all but the lowest powers of ω in the transmittance numerator and denominator are negligible. If a_0 and b_0 are nonzero,

$$F(j\omega)\cong\frac{b_1(j\omega)+b_0}{a_1(j\omega)+a_0}\cong\frac{b_0}{a_0}$$

In general, at low frequencies, a rational transmittance's amplitude curve approaches a power of ω or a constant and the phase curve approaches a multiple of 90°.

For example, consider

$$F(s)=\frac{6s^3+2s^2+3s}{s^5+4s^4+2s^3+s^2+s+10}$$

At high frequencies,

$$F(j\omega)=\frac{6(j\omega)^3}{(j\omega)^5}=-6\omega^{-2}$$

$$A(\omega)=|F(j\omega)|\cong\frac{6}{\omega^2}=6\omega^{-2}$$

$$\Phi(\omega)=\angle F(j\omega)\cong 180^\circ$$

At low frequencies,

$$F(j\omega)\cong\frac{3(j\omega)}{10}$$

$$A(\omega)=|F(j\omega)|\cong\tfrac{3}{10}\omega$$

$$\Phi(\omega)=\angle F(j\omega)\cong 90^\circ$$

Once the experimenter is certain that measurements are being made in the small-ω or large-ω region, a few measurements will suffice for the entire region. It is most useful to make measurements at more closely spaced frequencies where the largest changes in amplitude or phase occur.

DRILL PROBLEMS

D6-1. Find the forced sinusoidal response of each of the following transmittances to the indicated input signals:

(a) $F(s)=\dfrac{s}{s+3}$

$r(t)=7\cos(3t-40^\circ)$

ans. $(7/\sqrt{2})\cos(3t+5^\circ)$

(b) $F(s)=\dfrac{4}{s+2}$

$r(t)=6\cos(5t+30^\circ)$

ans. $4.46\cos(5t-38^\circ)$

(c) $F(s)=\dfrac{10}{s^2+3s+10}$

$r(t)=8\cos(2t+70^\circ)$

ans. $(40/3\sqrt{2})\cos(2t+25^\circ)$

D6-2. For transmittances with the following amplitude ratios and phase shift functions, find the forced sinusoidal response to the given input signal:

(a) $A(\omega)=\dfrac{1}{\sqrt{\omega^2+100}}$

$\Phi(\omega)=\tan^{-1}\dfrac{\omega}{10}$

$r(t)=7\cos(6t+80^\circ)$

ans. $0.6\cos(6t+111^\circ)$

(b) $A(\omega)=\dfrac{4}{\sqrt{\omega^2+4}}$

$\Phi(\omega)=90^\circ-\tan^{-1}\dfrac{\omega}{2}$

$r(t)=3\cos 2t$

ans. $(3\sqrt{2})\cos(2t+45^\circ)$

(c) $A(\omega)=4$

$\Phi(\omega)=-3\omega$ (rad)

$r(t)=10\cos(5t-30^\circ)$

ans. $40\cos(5t-170^\circ)$

6.3 BODE PLOTS

6.3.1 Amplitude Plots in Decibels

The magnitude of the product of complex numbers is the product of the individual magnitudes, and the angle of a product is the sum of the individual angles. A transmittance which is the product of several simple terms thus has an amplitude curve which is the product of the amplitude curves for the individual terms. The overall phase shift curve is the sum of the individual phase shift curves.

If, instead of dealing with the amplitude curves directly, their logarithms are used, multiplication of individual amplitude curves is, in terms of logarithms, addition. Commonly, *decibels* (abbreviated dB) are used for the description of frequency response amplitude ratios:

$$\text{dB}=20\log_{10}A(\omega)=20\log_{10}|F(s=j\omega)|$$

If

$$F(s)=F_1(s)F_2(s)F_3(s)\cdots$$

then

$$F \text{ (in dB)} = F_1 \text{ (in dB)} + F_2 \text{ (in dB)} + F_3 \text{ (in dB)} + \cdots$$

There is little reason for this choice of units except common usage, dating from Alexander Graham Bell's study of the response of the human ear.

In electromagnetics and some other applications, the natural logarithm is used, and the definition does not appear to be so contrived. The units then are *nepers*.

For rational functions $F(s)$, it is only necessary to be able to plot amplitude and phase shift for the following types of terms:

Constants
Poles and zeros at the origin of the complex plane
Real axis poles and zeros
Complex conjugate pairs of poles and zeros

A rational function may be factored into terms of these types, and the individual dB and phase shift curves plotted. The complete dB curve is then the sum of the component dB curves, and the complete phase shift curve is the sum of the individual phase shift curves.

6.3.2 Constant Transmittances

A positive constant K has constant amplitude in dB

$$20 \log_{10} K$$

and angle 0°. A negative constant $-K$ has dB amplitude

$$20 \log_{10} |K|$$

and angle 180°. For example,

$$F = 100$$

has the frequency response curves of Fig. 6-2a and

$$F = -\tfrac{1}{10}$$

has the curves of Fig. 6-2b. Now 180° is the same angle as -180°, so the angle may be plotted either way. A table of dB is given as Table 6-1.

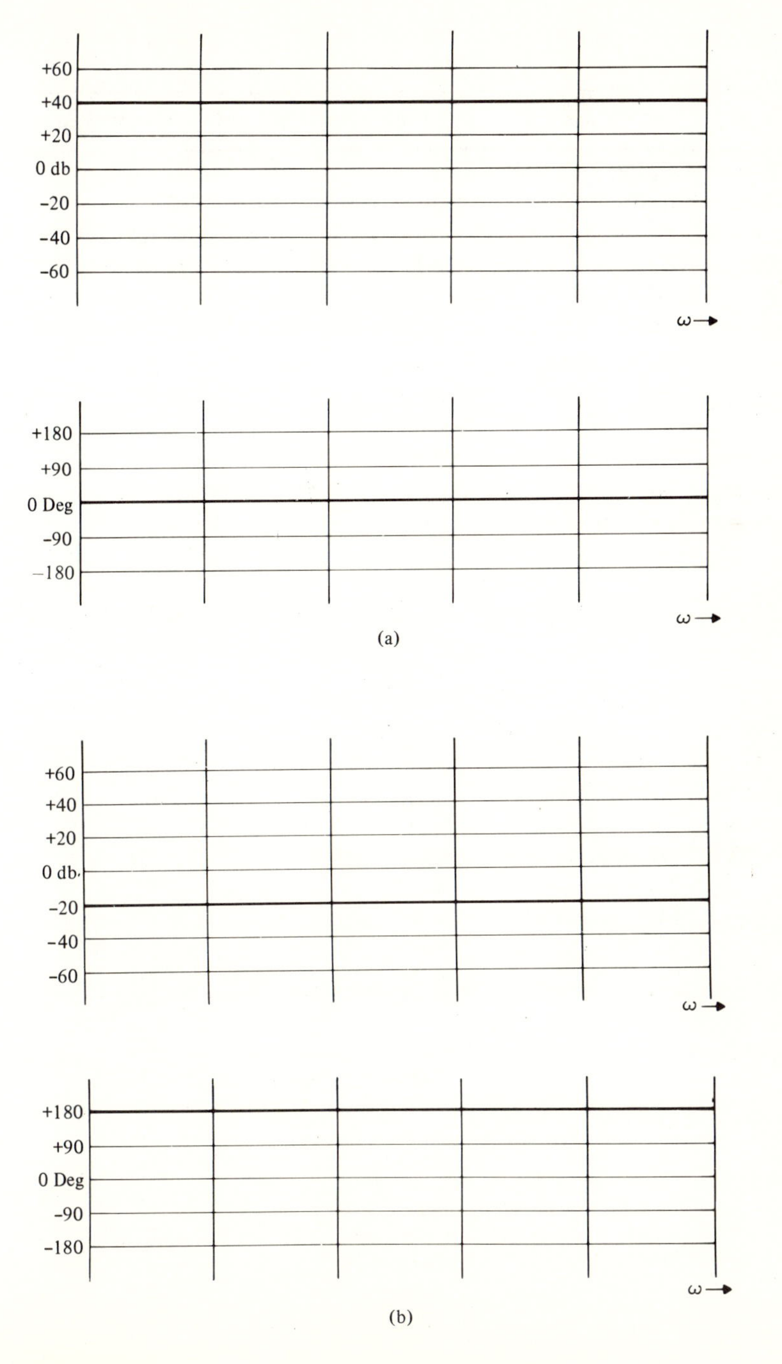

FIGURE 6-2
Frequency response curves for two constant transmittances.
(a) Frequency response for $F = 100$. (b) Frequency response for $F = -1/10$.

TABLE 6-1
Decibels.

Intensity Ratio A	$dB=20\ log_{10}\ A$	Intensity Ratio A	$dB=20\ log_{10}\ A$
1.000	0.0	1.000	− 0.0
1.012	0.1	0.989	− 0.1
1.023	0.2	0.977	− 0.2
1.035	0.3	0.966	− 0.3
1.047	0.4	0.955	− 0.4
1.059	0.5	0.944	− 0.5
1.072	0.6	0.933	− 0.6
1.084	0.7	0.923	− 0.7
1.096	0.8	0.912	− 0.8
1.109	0.9	0.902	− 0.9
1.122	1.0	0.891	− 1.0
1.189	1.5	0.841	− 1.5
1.259	2.0	0.794	− 2.0
1.334	2.5	0.750	− 2.5
1.413	3.0	0.703	− 3.0
1.496	3.5	0.668	− 3.5
1.585	4.0	0.631	− 4.0
1.679	4.5	0.596	− 4.5
1.778	5.0	0.562	− 5.0
1.884	5.5	0.531	− 5.5
1.995	6.0	0.501	− 6.0
2.113	6.5	0.473	− 6.5
2.239	7.0	0.447	− 7.0
2.371	7.5	0.422	− 7.5
2.512	8.0	0.398	− 8.0
2.661	8.5	0.376	− 8.5
2.818	9.0	0.355	− 9.0
2.985	9.5	0.335	− 9.5
3.162	10	0.316	− 10
3.548	11	0.282	− 11
3.981	12	0.251	− 12
4.467	13	0.224	− 13
5.012	14	0.200	− 14
5.62	15	0.178	− 15
6.31	16	0.159	− 16
7.03	17	0.141	− 17
7.94	18	0.126	− 18
8.91	19	0.112	− 19
10^1	20	10^{-1}	− 20
3.16×10^1	30	3.16×10^{-2}	− 30
10^2	40	10^{-2}	− 40
3.16×10^2	50	3.16×10^{-3}	− 50

TABLE 6-1 (Continued).

Intensity Ratio A	$dB = 20\ log_{10}\ A$	*Intensity Ratio A*	$dB = 20\ log_{10}\ A$
10^3	60	10^{-3}	−60
3.16×10^3	70	3.16×10^{-4}	−70
10^4	80	10^{-4}	−80
3.16×10^4	90	3.16×10^{-5}	−90
10^5	100	10^{-5}	−100
3.16×10^5	110	3.16×10^{-6}	−110
10^6	120	10^{-6}	−120

6.3.3 Roots at the Origin

The next simplest kind of transfer function is

$$F(s) = s$$

for which

$$A(\omega) = |F(s = j\omega)| = \omega$$
$$\text{dB} = 20\ \log_{10} A(\omega) = 20\ \log_{10} \omega$$

and

$$\Phi(\omega) = \angle F(s = j\omega) = 90^\circ$$

It is much easier to plot amplitude in dB versus *log* ω than versus ω. Versus $\log_{10} \omega$, the curve for $F(s) = s$ is the straight line, as in Fig. 6-3. Since it is really ω and not log ω which is of interest, the actual values of ω may be indicated along the log ω axis instead of the values of log ω. Or, if desired, the hertz (cycles per second) frequency $f = \omega/2\pi$ may be indicated.

At $\omega = 1$, $20\ \log_{10} \omega = 0$, so the straight-line dB plot for $F(s) = s$ goes through 0 dB at $\omega = 1$. Each increment of 1 unit in $\log_{10} \omega$ adds 20 dB to the dB value, so the slope of the amplitude curve is 20 dB/decade of ω.

A set of dB and phase shift plots for a transmittance, versus frequency on a logarithmic scale, are termed *Bode plots.*

The curves, which are shown in Fig. 6-4, for $F(s) = s^2$, are just the sums of two $F(s) = s$ curves, that is, double the plots for s. In fact, the nth power of any transfer function has curves which are n times the plots for the original transfer function. The frequency response plots for

$$F(s) = \frac{1}{s} = s^{-1}$$

Fig. 6-5, are then the negatives of the dB and phase angle plots for $F(s) = s$.

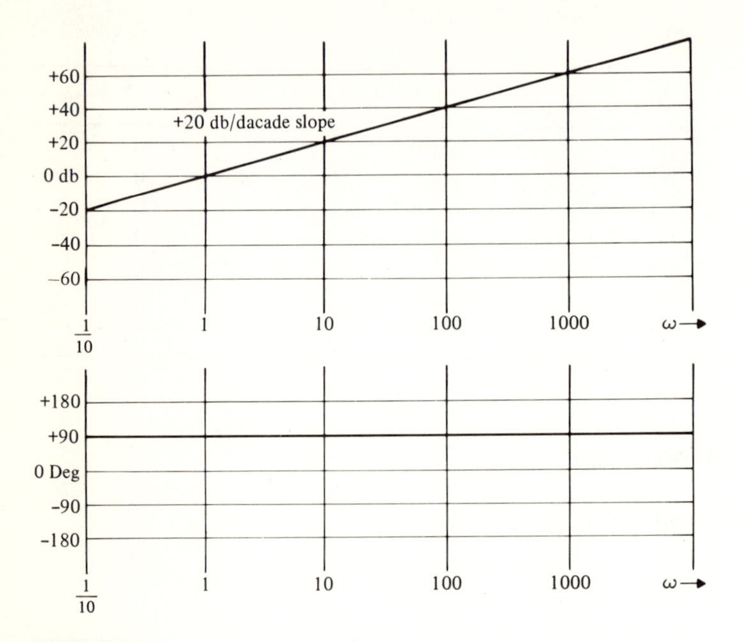

FIGURE 6-3
Frequency response plots for $F(s)=s$.

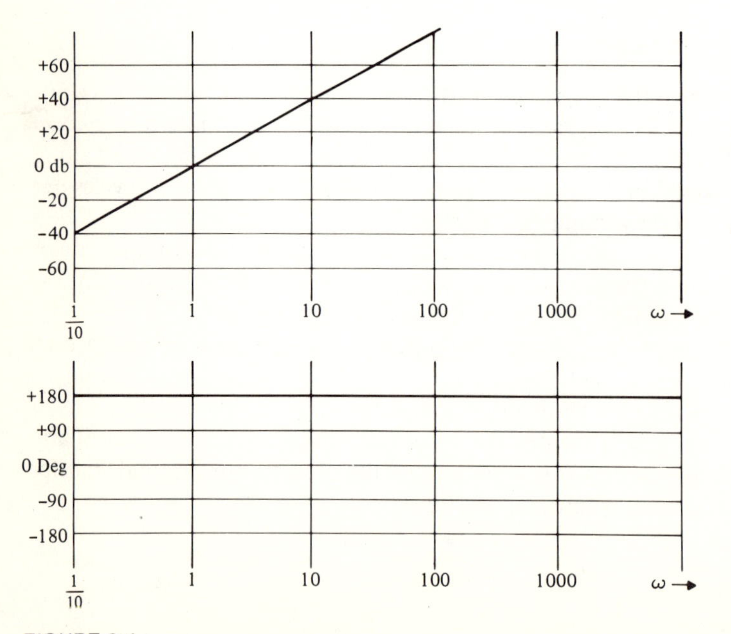

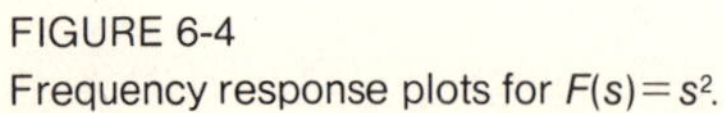
FIGURE 6-4
Frequency response plots for $F(s)=s^2$.

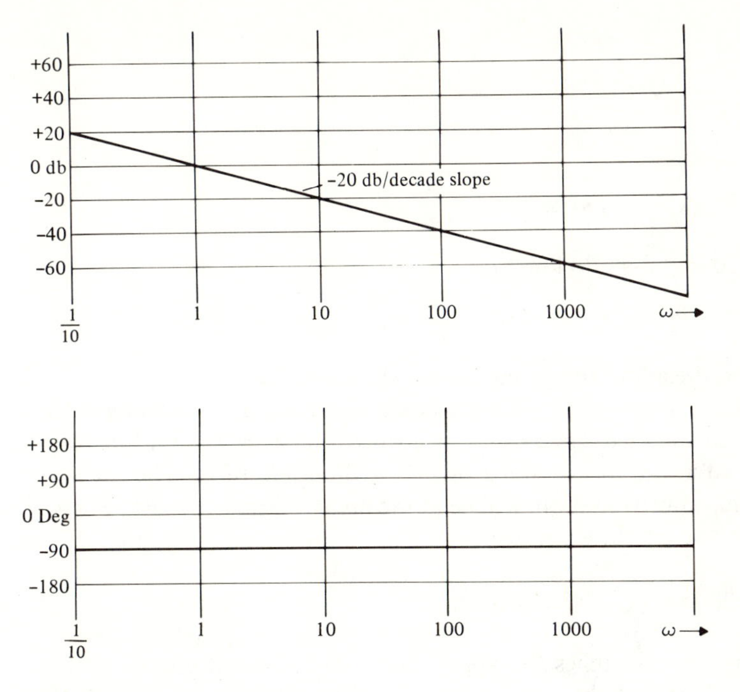

FIGURE 6-5
Frequency response plots for $F(s) = 1/s$.

6.3.4 Real Axis Roots

A more complicated transfer function is

$$F(s) = \frac{s+a}{a}$$

where a is a positive number. Instead of considering a single zero term, $s+a$, terms of this form are considered because the relations to be developed are simplest when

$$F(s=j\omega) = 1 + j\frac{\omega}{a}$$

For $\omega \ll a$,

$$F(s=j\omega) \cong 1$$

$$\text{dB} = 20 \log_{10} |F(s=j\omega)| \cong 0$$

$$\angle F(s=j\omega) \cong 0^\circ$$

For $\omega \gg a$,

$$F(s=j\omega) \cong j\frac{\omega}{a}$$

$$\text{dB} = 20\ \log_{10}|F(s=j\omega)| \cong 20\log_{10}\frac{\omega}{a}$$

$$= 20\ \log_{10}\omega - 20\ \log_{10} a$$

$$\angle F(s=j\omega) \cong 90^\circ$$

The actual and the approximate curves are shown in Fig. 6-6.

The dB curve may be approximated quite nicely by a curve along 0 dB up to $\omega = a$, termed the *break frequency*, or corner frequency, then a line sloping upward beyond $\omega = a$ with a slope of $+20$ dB/decade of ω. The maximum error, using this approximation, will be at the break frequency $\omega = a$, where the actual value of the amplitude is

$$|F(ja)| = 20\ \log_{10}\sqrt{2} = 3.01\ \text{dB}$$

In practice, one sketches the approximate curve and, if greater accuracy is desired, modifies it slightly so that it is smooth and goes through 3 dB at the break frequency. The straight-line approximation is sufficiently accurate for many purposes, however.

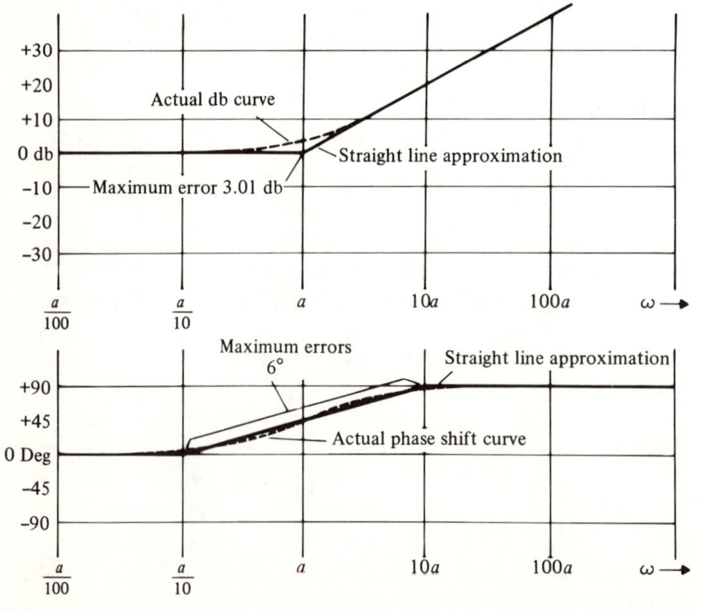

FIGURE 6-6
Frequency response for $F(s) = (s + a)/a$, for positive a.

The three-segment phase shift approximation indicated in Fig. 6-6 is accurate to about 6°. In the approximation, the angle is 0° up to one-tenth the break frequency, rises 45°/decade, through 45° at the break frequency. Beyond 10 times the break frequency, the approximate angle is 90°. At the break frequency a, the actual and approximate curves are equal, since

$$\angle F(ja)=\tan^{-1}\frac{1}{1}=45^\circ$$

The frequency response curves for

$$F(s)=\frac{a}{s+a}=\frac{1}{s/a+1}$$

are just the negatives of the curves for $(s+a)/a$ because the two transfer functions are inverses of one another. The approximate frequency response curves for

$$F_1(s)=\frac{s+10}{10}$$

and

$$F_2(s)=\frac{10}{s+10}$$

are shown, for comparison, in Fig. 6-7a and b.

Frequency response plots for terms of the form

$$F_3=\frac{s-a}{-a}$$

and

$$F_4(s)=\frac{-a}{s-a}$$

representing a zero and a pole, respectively, in the right half of the complex plane, are given in Fig. 6-7c and d. The RHP roots have the same dB curves as their LHP counterparts, but the negative of the corresponding LHP term's phase shift. Note that there are two different amplitude ratio curves and two different phase shift curves in Fig. 6-7. Each of the four different possible amplitude and phase curve combinations represents one of the four types of real axis root terms.

Of course, a transfer function with an RHP pole is unstable. There may be part of a system of interest which is, by itself, unstable, even though the system as a whole is stable.

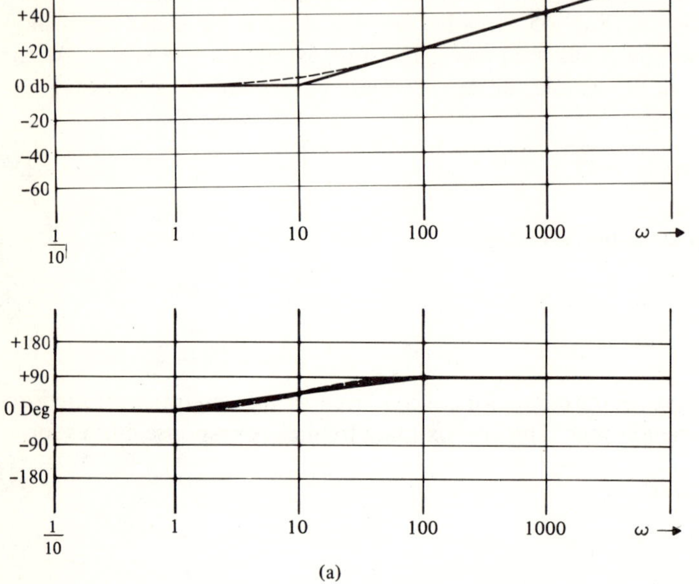

(a)

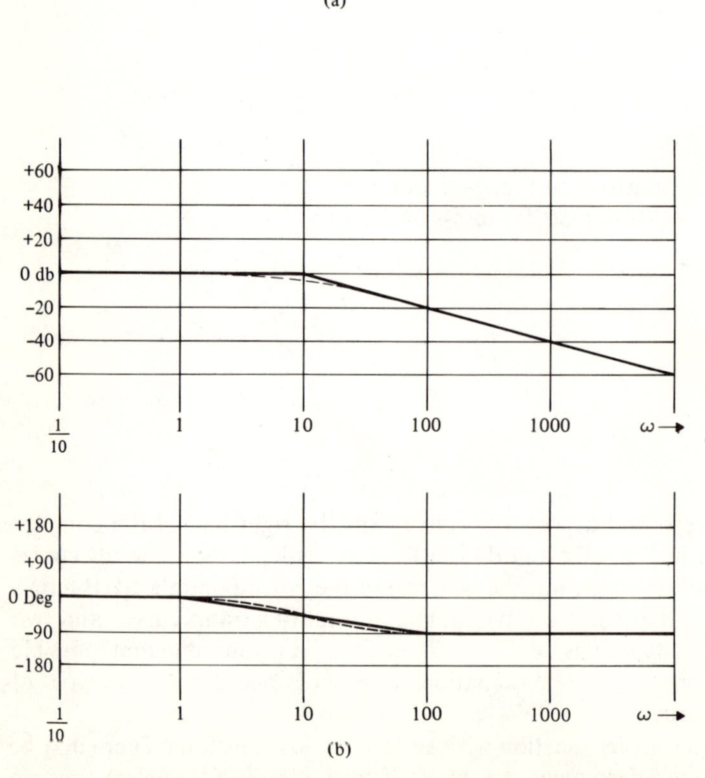

(b)

(FIGURE 6-7a,b).

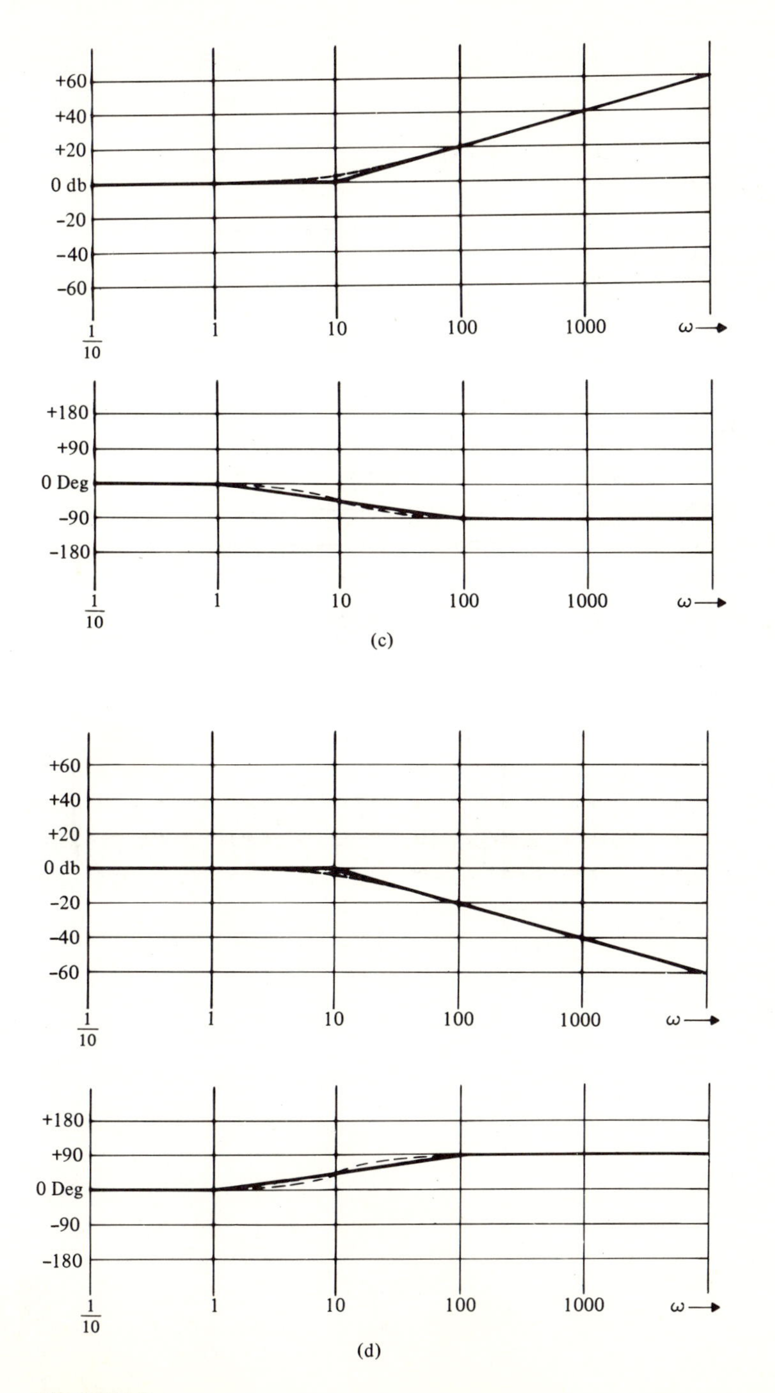

FIGURE 6-7
Frequency response curves for real axis poles and zeros.
(a) $F_1(s) = (s + 10)/10$. (b) $F_2(s) = 10/(s + 10)$. (c) $F_3(s) = (s + 10)/(-10)$.
(d) $F_4(s) = -10/(s - 10)$.

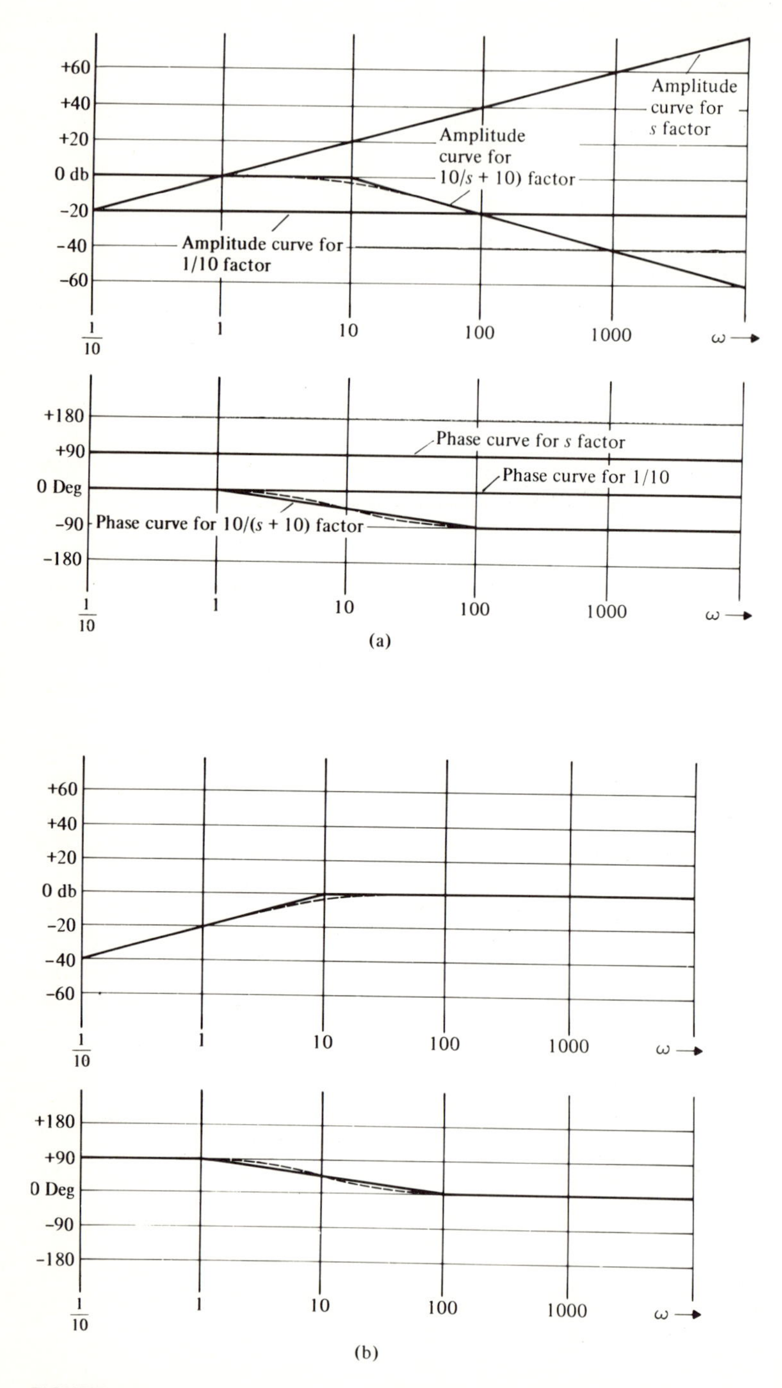

FIGURE 6-8
Frequency response for $s/(s+10)$. (a) Frequency response of the individual terms. (b) Overall frequency response.

6.3.5 Products of Terms

Consider plotting frequency response curves for

$$F(s)=\frac{s}{s+10}$$

First, decompose $F(s)$ into factors for which the frequency response is known and can be easily sketched:

$$F(s)=\left(\frac{1}{10}\right)(s)\left(\frac{10}{s+10}\right)$$

The curves for the individual terms are sketched in Fig. 6-8a. The sums of these curves are the curves for the overall transfer function, as shown in Fig. 6-8b.

To be complete, the frequency response curves should be drawn for a range of frequency which includes all the detail of the response. A minimum range of frequency for the above example is from below 1 rad/sec to above 100 rad/sec. Outside that range, the curves continue as straight lines. When convenient, phase shift is commonly plotted within the range $\pm 180^\circ$.

With some experience and practice, frequency response curves for transfer functions with real poles and zeros may be sketched by considering the various changes in slope that occur for each break frequency. To plot the dB curve in this manner for the example transmittance start at small ω, where the plot is that of $s/10$, a straight line with slope 20 dB/decade which passes through -20 dB at $\omega=1$. Beyond the break frequency $\omega=10$, the $10/(s+10)$ term contributes a downward slope of -20 dB/decade, giving a net zero slope.

The phase shift plot may be drawn similarly. For small ω, the frequency response is that for $s/10$, which has a 90° phase shift. Starting at one-tenth the break frequency for the $10/(s+10)$ term, the phase shift decreases by 45°/decade, reaching 45° at $\omega=10$ and 0° at $\omega=100$. Beyond 10 times the break frequency, $\omega=100$, the phase shift continues along 0°.

DRILL PROBLEM

D6-3. Sketch frequency response curves (both amplitude in dB and phase shift) for the following transmittances:

(a) $F(s)=-3s$

(b) $F(s)=\dfrac{1}{s+1000}$

(c) $F(s)=\dfrac{1}{(s+10)^3}$

(d) $F(s)=\dfrac{1000s}{(s+4)(s+10)}$

(e) $F(s)=\dfrac{s-10}{s+10}$

6.4 COMPLEX ROOTS

6.4.1 Decibel and Phase Curves for Conjugate Poles

Frequency response curves for complex conjugate pole or zero terms are plotted from straight-line asymptotes similar to simple real axis poles and zeros. Factor a set of complex conjugate poles or zeros to the form

$$F_1(s)=\frac{\omega_n^2}{s^2+2\xi\omega_n s+\omega_n^2}$$

for a set of complex conjugate poles, or

$$F_2(s)=\frac{s^2+2\xi\omega_n s+\omega_n^2}{\omega_n^2}$$

for a set of complex conjugate zeros.

First, consider complex conjugate set of poles, of the form of $F_1(s)$. For small values of ω compared to ω_n,

$$F_1(s=j\omega)\cong 1$$

For large values of ω compared to ω_n,

$$F_1(s=j\omega)\cong -\frac{\omega_n^2}{\omega^2}$$

which corresponds to a slope of -40 dB/decade and a phase shift of 180°. At $\omega=\omega_n$,

$$F_1(j\omega_n)=\frac{1}{j2\xi}$$

for which the dB curve has value

$$\text{dB}=20\log_{10}\frac{1}{2\xi}$$

and the phase shift is $-90°$.

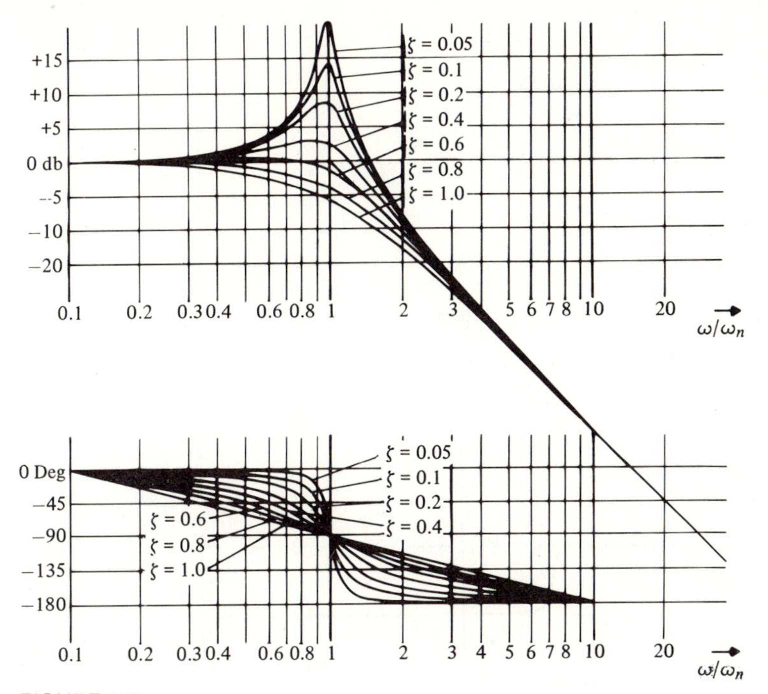

FIGURE 6-9
Frequency response of a complex conjugate pole pair term.

Plots of amplitude and phase shift for various damping ratios for complex conjugate pole terms are given in Fig. 6-9. For $\xi \geqslant 1$, the poles are real, not complex, and so can be handled by the previous real axis pole methods.

As an example, consider the transmittance

$$F(s) = \frac{1000}{s^2 + 20s + 40{,}000}$$

Decomposing into convenient terms,

$$F(s) = \left(\frac{1}{40}\right)\left(\frac{40{,}000}{s^2 + 20s + 40{,}000}\right)$$

For the complex conjugate pole term,

$$\omega_n = \sqrt{40{,}000} = 200$$

and

$$2\xi\omega_n = 400\xi = 20$$
$$\xi = \tfrac{1}{20}$$

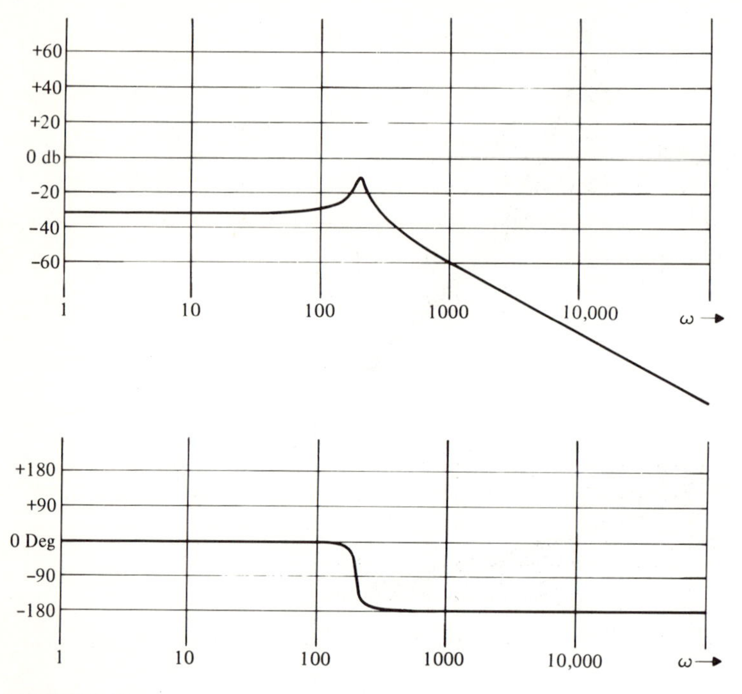

FIGURE 6-10
Frequency response plots for $1000/(s^2+20s+40{,}000)$.

The curves for this term may be traced from the given curves. The term $\frac{1}{40}$ has amplitude -32 dB and angle $0°$, and the results are as shown in Fig. 6-10.

6.4.2 Deviations from Approximations

Most experienced designers prefer to plot frequency response curves for complex conjugate pairs of roots by first approximating them with straight lines, then correcting the approximations. Figure 6-11 shows the amounts of correction necessary from the straight-line approximations for repeated pairs of real axis roots.

Consider the transmittance

$$F(s)=\frac{1000}{s^2+2s+100}=(10)\left(\frac{100}{s^2+2s+100}\right)$$

For the complex conjugate pole term

$$\omega_n=\sqrt{100}=10$$
$$2\xi\omega_n=20\xi=2 \qquad \xi=\frac{1}{10}$$

First the straight-line approximations shown in Fig. 6-12 are drawn. The

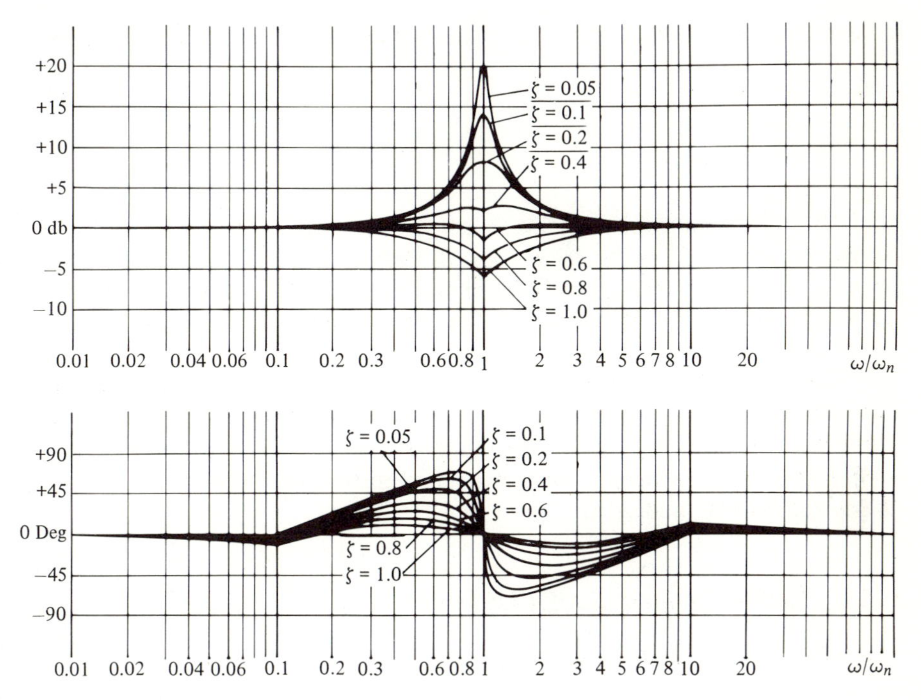

FIGURE 6-11
Correction curves for complex conjugate root frequency response plots.

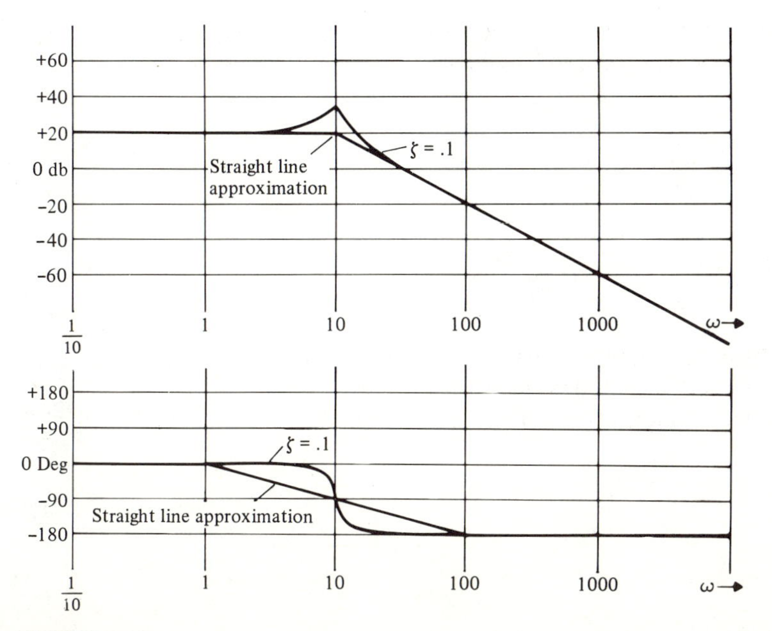

FIGURE 6-12
Frequency response plots for $F(s) = 1000/(s^2 + 2s + 100)$.

conjugate pair of roots are first approximated as if they were critically damped, of the form

$$\frac{100}{s^2+20s+100}=\left(\frac{10}{s+10}\right)^2$$

The dB approximation is 0 dB (+20 dB for the constant factor of 10) to the break frequency of $\omega=10$, then -40 dB/decade slope thereafter. The angle approximation is 0° to one-tenth the break frequency then $-90°$/decade slope, passing through $-90°$ at the break frequency and continuing until 10 times the break frequency.

The corrections of Fig. 6-11, for $\xi=\frac{1}{10}$, are then applied to the approximate curves to give the final result in Fig. 6-12. Transferring several points from each curve and then drawing a smooth curve through the result will usually suffice. As the error of the straight-line approximation is sizable, it is important that the corrections be made.

6.4.3 Other Complex Conjugate Roots

Complex conjugate LHP zeros of the form

$$F(s)=\frac{s^2+2\xi\omega_n s+\omega_n^2}{\omega_n^2}$$

have frequency response curves which are the negatives of the curves for the corresponding conjugate poles. Conjugate RHP poles and zeros,

$$F(s)=\frac{\omega_n^2}{s^2-2\xi\omega_n s+\omega_n^2}$$

and

$$F(s)=\frac{s^2-2\xi\omega_n s+\omega_n^2}{\omega_n^2}$$

differ from their LHP counterparts in algebraic sign of the phase, as indicated in Fig. 6-13; the dB amplitude remains the same.

As another example, consider the transmittance

$$F(s)=\frac{s^2+s+8}{s^2}$$

Decomposing,

$$F(s)=(8)\left(\frac{s^2+s+8}{8}\right)\left(\frac{1}{s}\right)^2$$

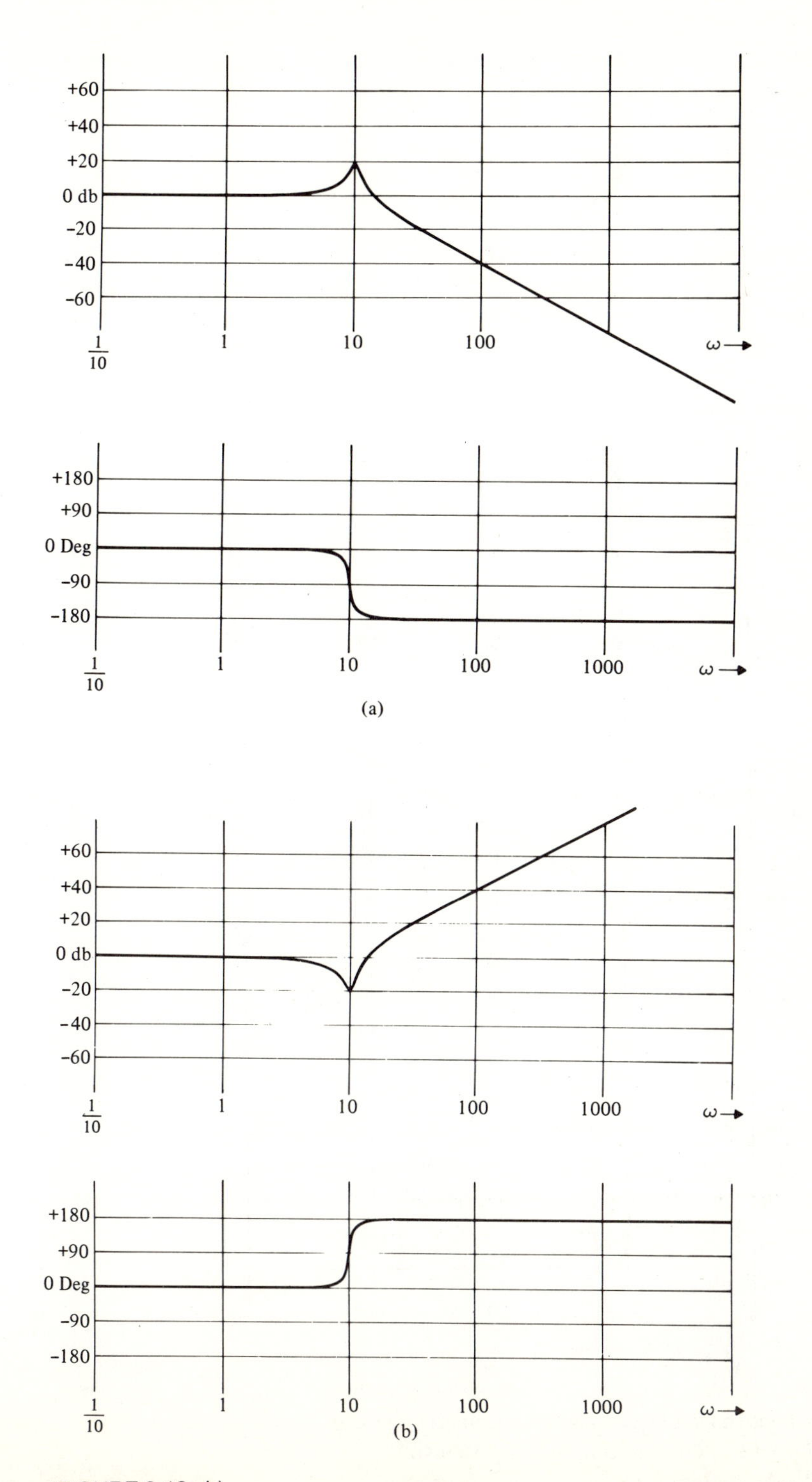

(FIGURE 6-13a,b).

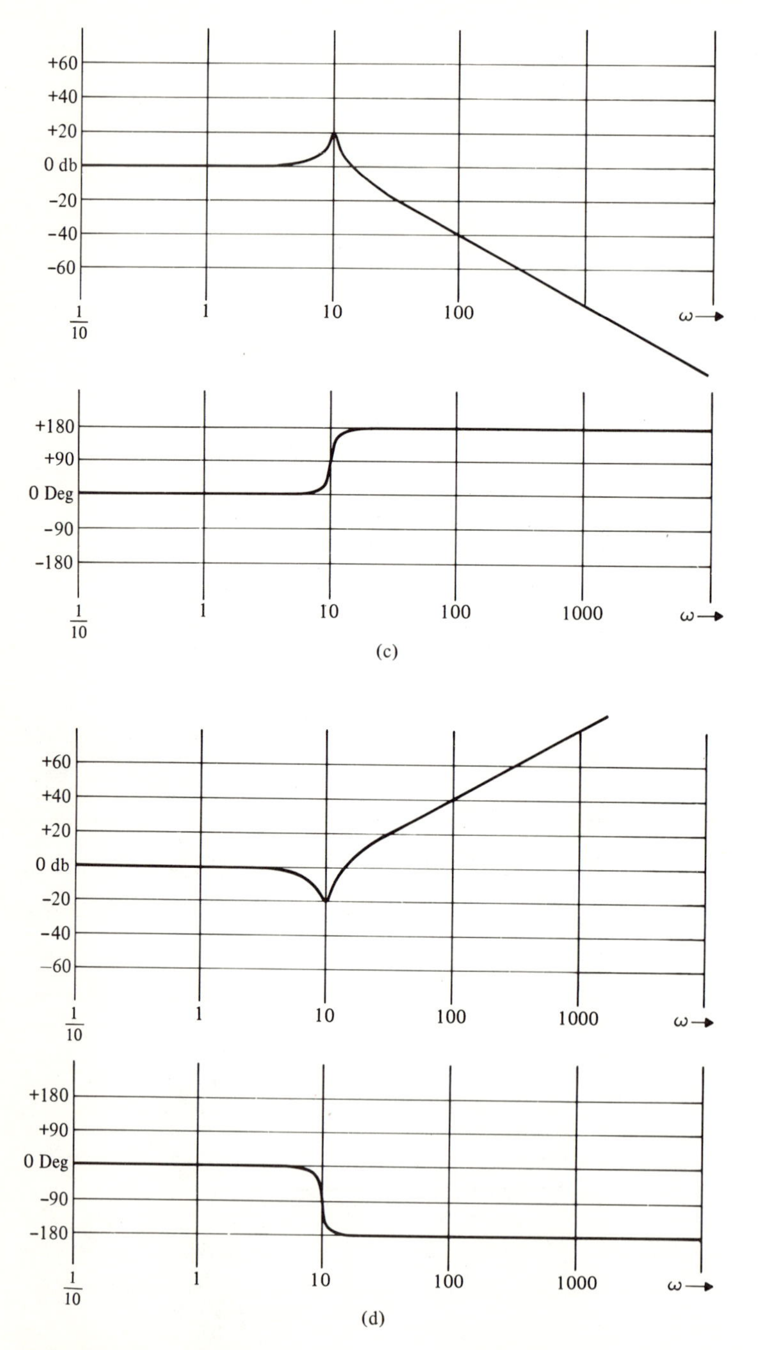

FIGURE 6-13
Frequency response curves for complex conjugate pairs of roots.
(a) $F_1(s) = 100/(s^2 + s + 100)$. (b) $F_2(s) = (s^2 + s + 100)/100$.
(c) $F_3(s) = 100/(s^2 - s + 100)$. (d) $F_4(s) = (s^2 - s + 100/100$.

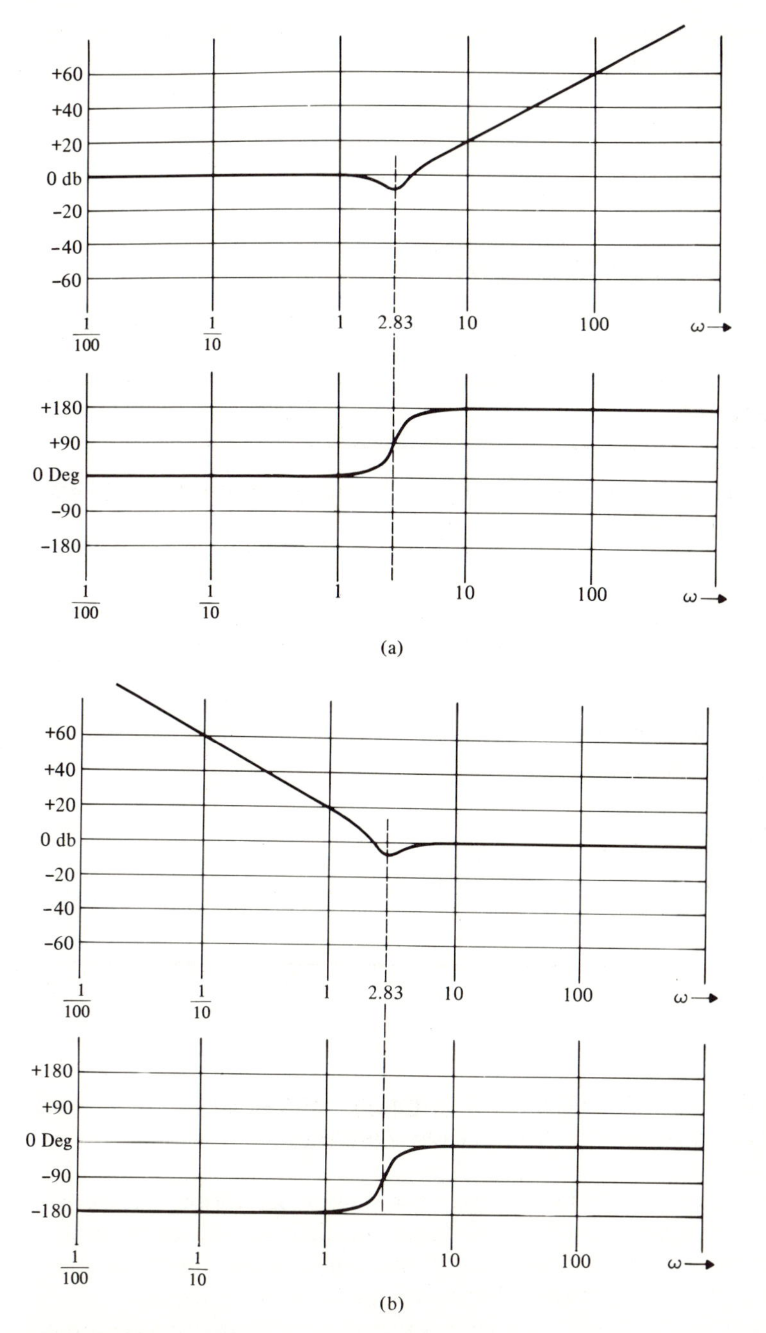

FIGURE 6-14

Frequency response of transmittances involving complex roots.
(a) Approximate frequency response curves for $F(s) = (s^2 + s + 8)/8$.
(b) Approximate frequency response curves for $F_2(s) = (s^2 + s + 8)/s^2$.

The set of complex conjugate zeros has curves which are the negatives of the given curves for a pole, with

$$\omega_n=\sqrt{8}=2\sqrt{2}=2.83$$
$$2\xi\omega_n=5.66\xi=1$$
$$\xi=0.177$$

from which the curves of Fig. 6-14 are sketched.

DRILL PROBLEM

D6-4. Sketch frequency response curves (both amplitude in dB and phase shift) for the following transmittances:

(a) $F(s)=\dfrac{100}{s^2+s+100}$

(b) $F(s)=\dfrac{10}{s^2+s+4}$

(c) $F(s)=\dfrac{s^2-4s+30}{(s+10)^2}$

(d) $F(s)=\dfrac{s}{s^2+20s+100}$

(e) $F(s)=\dfrac{s^2-2s+100}{s^2+100s+100}$

6.5 GRAPHICAL FREQUENCY RESPONSE METHODS

6.5.1 Frequency Response from a Pole-Zero Plot

Given a pole-zero plot for a transmittance, its evaluation for various values of $s=j\omega$ may be done graphically by drawing sets of directed line segments to points of evaluation on the imaginary axis. With a little practice, the amplitude and phase shift curves can be roughly sketched from just a few such evaluations. It is easiest to first sketch the curves directly rather than in dB and versus $\log \omega$. For example, the transfer function with pole-zero plot given in Fig. 6-15a has the approximate frequency response curves given in Fig. 6-15b.

A negative sign for the multiplying constant would contribute 180° to the phase shift curve.

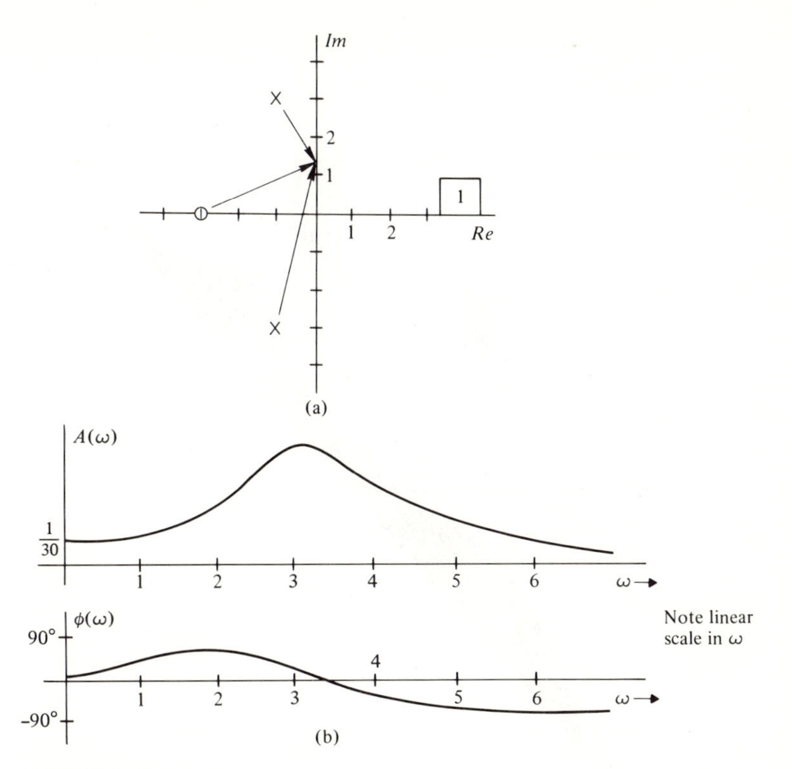

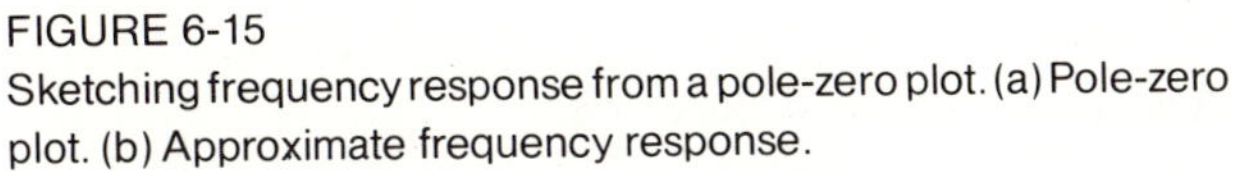

FIGURE 6-15
Sketching frequency response from a pole-zero plot. (a) Pole-zero plot. (b) Approximate frequency response.

6.5.2 Visualizing Amplitude Curves

The amplitude curve for a transmittance may be easily visualized from the pole-zero plot by imagining a very flexible rubber sheet suspended over the complex plane. The sheet is poked up by thin rods at each pole location and tacked down to the plane at each zero location. The height of the rubber sheet will represent $|F(s)|$ for each value of s. The rubber sheet analogy works because $|F(s)|$ and the displacement of an ideal rubber sheet satisfy the same partial differential equation. The frequency response amplitude curve is a cross section of the sheet displacement along the imaginary axis, as indicated in Fig. 6-16.

To use the rubber sheet analogy effectively, the height of the sheet at large distances from the origin, that is, as $|s| \to \infty$, must be determined. If

$$F(s) = \frac{b_m s^m + b_{m-1} s^{m-1} + \cdots + b_1 s + b_0}{a_n s^n + a_{n-1} s^{n-1} + \cdots + a_1 s + a_0}$$

$$\lim_{|s| \to \infty} |F(s)| = \left|\frac{b_m}{a_n}\right| \frac{|s|^m}{|s|^n}$$

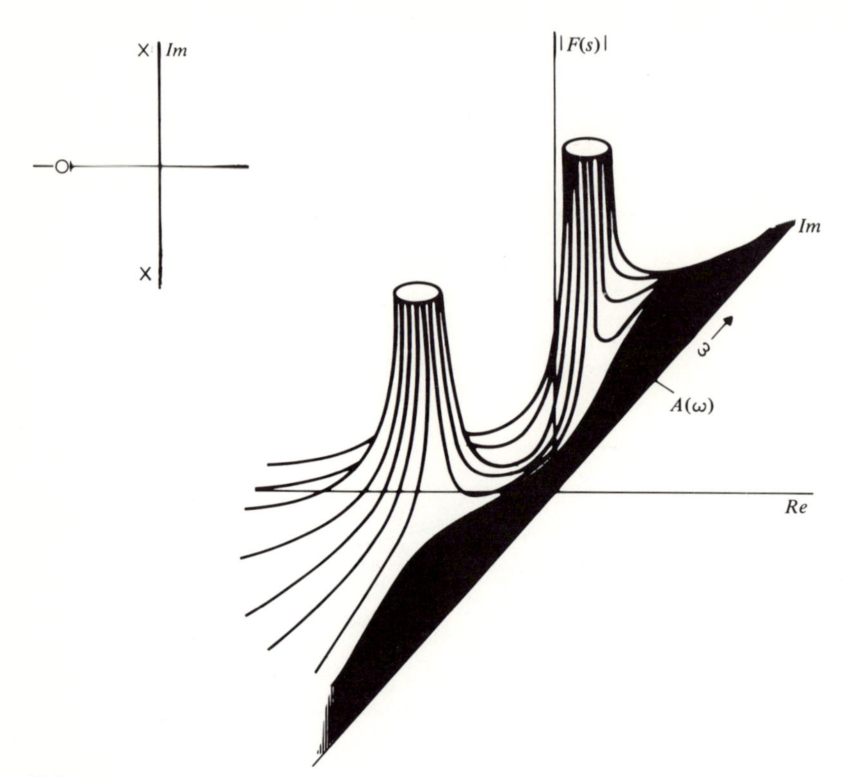

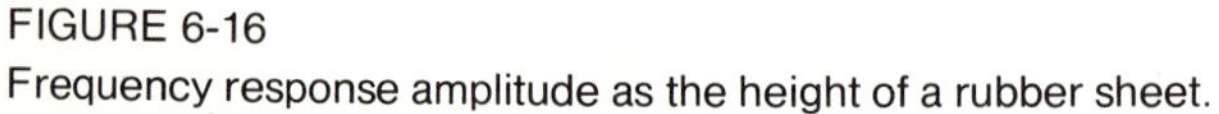
FIGURE 6-16
Frequency response amplitude as the height of a rubber sheet.

If the number of poles of $F(s)$ is greater than its number of zeros,

$$\lim_{|s| \to \infty} |F(s)| = 0$$

and the rubber sheet may be visualized as being tacked down to the complex plane at large distances from the origin, as in Fig. 6-17a.

If $F(s)$ has an equal number of poles and zeros,

$$\lim_{|s| \to \infty} |F(s)| = \left|\frac{b_m}{a_n}\right|$$

and the rubber sheet approaches a fixed height above the complex plane, as in Fig. 6-17b. If there are more zeros than poles in $F(s)$,

$$\lim_{|s| \to \infty} |F(s)| = \infty$$

This case is unusual in practical applications because it means that the amplitude ratio of the transmittance increases with frequency without bound.

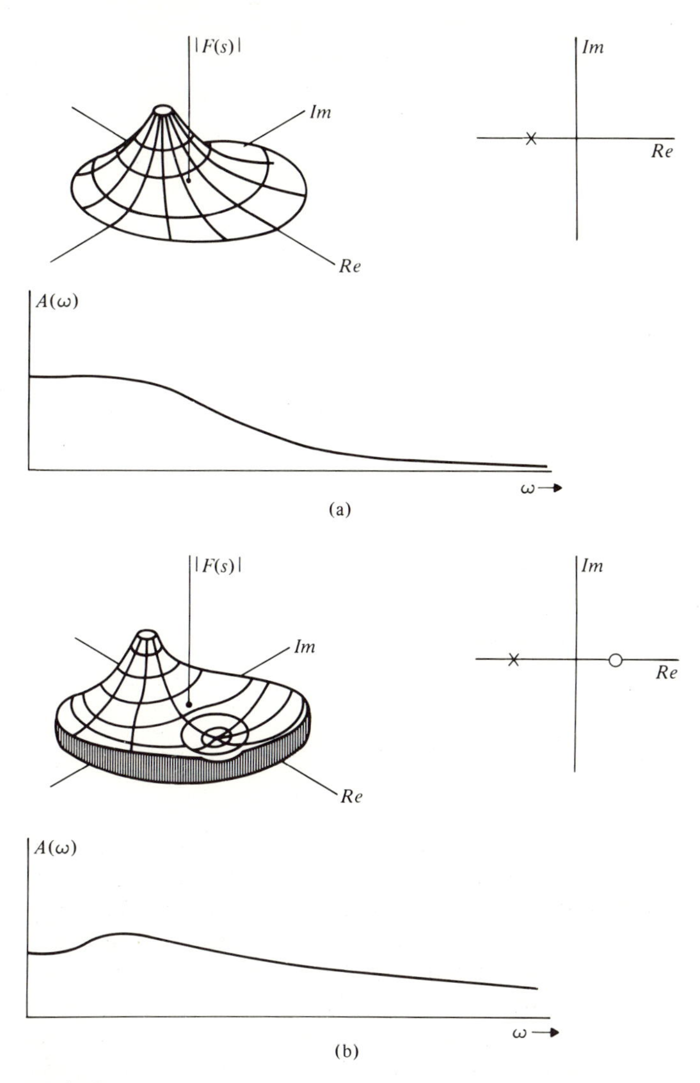

FIGURE 6-17
Using the rubber sheet analogy to visualize frequency response amplitude ratios. (a) More poles than zeros. (b) Equal number of poles and zeros.

6.5.3 Imaginary Axis Roots

Imaginary axis zeros and poles give zero or infinite amplitude for the values of ω where they occur. The phase shift is discontinuous at these values of ω, but limiting values of the phase shift are found by considering imaginary axis points slightly below and slightly above the zero or pole. An illustrative example is given in Fig. 6-18.

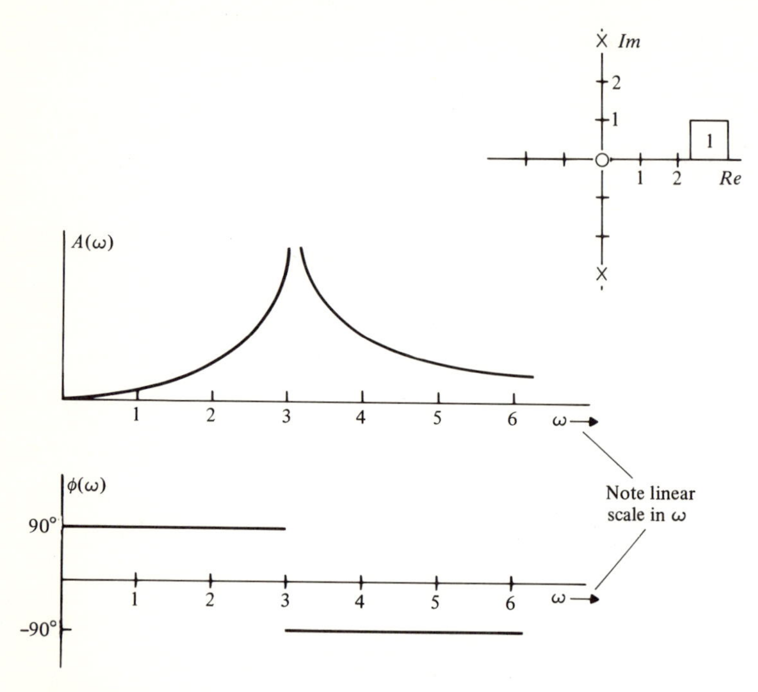

FIGURE 6-18
Pole-zero plot and approximate frequency response for a system with imaginary axis roots.

Complex conjugate poles on the imaginary axis mean that the corresponding system natural behavior is sinusoidal, neither decaying nor expanding with time. If such a system is *driven* with a sinusoidal signal of the same frequency as this natural behavior, the output becomes larger and larger. Of course, it cannot really become infinite in a practical system, since nonlinearities will eventually be reached which limit the output. Possibly, the system will be destroyed or it will not be possible to continue to supply the ever-increasing energy needed to maintain the input signal. A linear, time-invariant model does not take these practical constraints into account.

Complex conjugate poles in the LHP near the imaginary axis represent slowly decaying oscillatory natural behavior. Driving such a system with a sinusoidal signal of the same or nearly the same frequency as the natural behavior oscillations results in a relatively large, but not infinite, forced output. That is, there is typically a corresponding peak in the system's frequency response, termed a *resonance peak.*

6.5.4 Analyzing Experimental Data

One of the most important and powerful uses of the frequency response method of system design is in determining component transmittances. For many

TABLE 6-2
Experimental frequency data.

f	ω	Gain (dB)	Phase Shift (deg)
60	377	−7.75	−155
50	314	−4.3	−150
40	251	−0.2	−145
35	219	0.75	−140
25	157	5.16	−135
20	126	7.97	−120
16	100	10.5	−110
10	63	15.0	−100
7	44	16.9	−85
2.5	16	20.4	−45
1.3	8	21.6	−30
0.22	1.38	24.0	−5
0.16	1.0	24.1	0

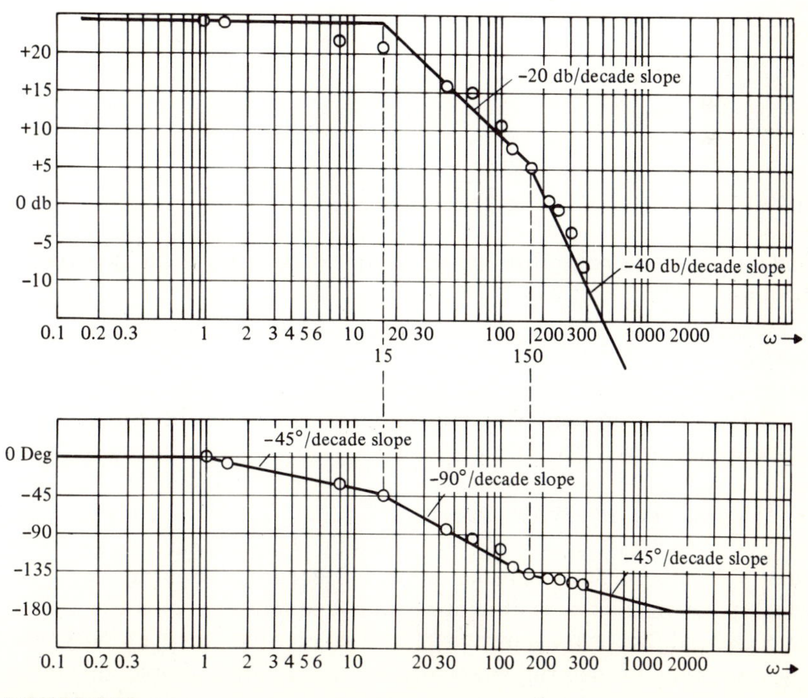

FIGURE 6-19
Incorporating experimental data for frequency response analysis.

practical system components, such as pneumatic valves and airframes, analytic expressions for transmittances are difficult to obtain from theory. If a frequency response test can be performed, however, the transmittance may be determined experimentally.

Consider an example transmittance for which the experimental amplitude and phase characteristics are as tabulated in Table 6-2. It is desired to obtain a transfer function which approximates these characteristics. If the characteristics are plotted, as in Fig. 6-19, a series of straight-line asymptotes may be fitted to these data for both amplitude and phase. By use of the slopes and corresponding break frequencies, a transfer function is obtained. For the example given, an approximate transmittance is

$$F(s)=16\left(\frac{15}{s+15}\right)\left(\frac{150}{s+150}\right)=\frac{16.0}{(0.05s+1)(0.007s+1)}$$

6.6 IRRATIONAL TRANSFER FUNCTIONS

Another advantage of frequency response methods is that it is not necessary to restrict the type of transfer function to rational polynomials. Frequency response methods may be brought to bear on such irrational transfer functions as

$$F(s)=\sqrt{s}$$

and

$$F(s)=\cos s$$

One type of irrational transmittance of considerable practical importance is of the form

$$F(s)=e^{-\tau s}$$

where τ is a positive constant. This transmittance represents the time delay of the incoming signal by τ sec. In the language of the Laplace transformation,

$$\mathscr{L}^{-1}[Y(s)e^{-\tau s}]=y(t-\tau)$$

One type of time delay scheme involves recording the input signal on magnetic tape. A time delay is obtained as the tape moves from the record to the playback head. This arrangement is often used on radio interview programs to allow censorship of the program before it is aired. Other simple time delay systems are transmission lines, digital shift registers, conveyor belts, and audio reverberation generators.

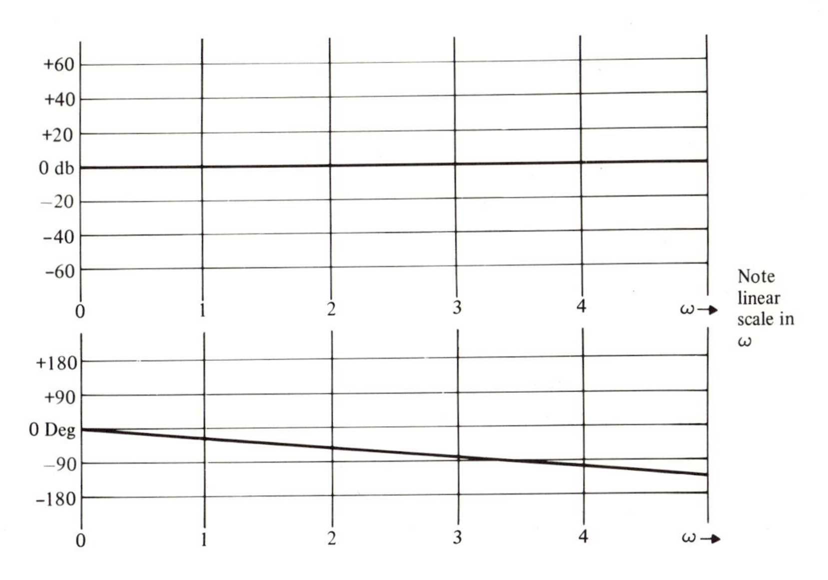

FIGURE 6-20
Frequency response plots for $F(s)=e^{-(1/2)s}$.

A steady state sinusoidal signal which is only delayed in time emerges with no change in amplitude, but it does undergo a change in phase. The higher the frequency of the sinusoid, the greater the phase shift for the same time delay. For

$$F(s)=e^{-s\tau}$$
$$F(j\omega)=e^{-j\omega\tau}$$

and

$$A(\omega)=|F(j\omega)|=1$$
$$\Phi(\omega)=\angle F(j\omega)=-\omega\tau \qquad \text{rad}$$

The frequency response of the time delay system

$$F(s)=e^{-(1/2)s}$$

is plotted on a linear scale of frequency, in Fig. 6-20. On a logarithmic frequency scale, the phase shift curve is more and more compressed for larger values of ω.

DRILL PROBLEMS

D6-5. Sketch approximate frequency response curves (both amplitude and phase shift) for functions with the following pole-zero plots:

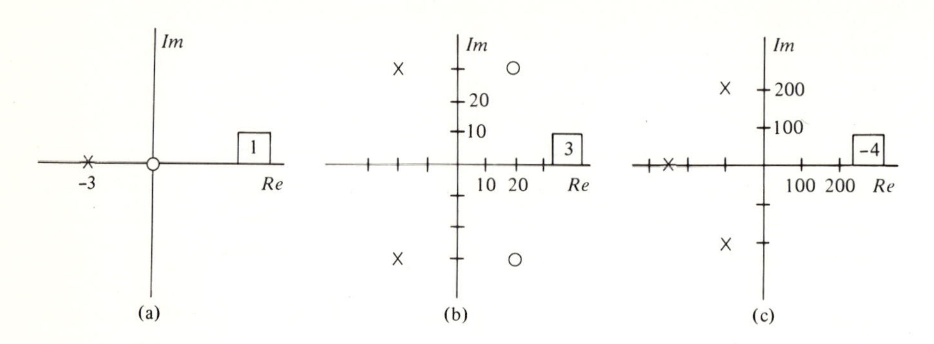

D6-6. Determine an approximate transmittance from the following experimental frequency response data:

ω	$F(s=j\omega)$	ω	$F(s=j\omega)$
0.1	$0\ -j20$	2.0	$-0.4\ -j0.6$
0.2	$-1.3-j10$	4.0	$-0.2\ -j0.2$
0.4	$-0.8-j5$	8.0	$-0.05\ -j0.008$
1.0	$-0.6-j1.7$	16.0	$-0.008+j0.0003$

D6-7. Find and sketch the frequency response (both amplitude and phase shift) for the following irrational functions:

(a) $F(s)=6e^{-0.2s}$

ans. $6,\ -0.2\omega$

(b) $F(s)=\dfrac{e^{-4s}}{s}$

ans. $1/\omega,\ -4\omega-\pi/2$

(c) $F(s)=\sqrt{s}$

ans. $\sqrt{\omega}$, 45° or 225°

6.7 GAIN AND PHASE MARGINS

6.7.1 Feedback System Stability

Many practical systems are of the feedback type shown in Fig. 6-21. The overall transfer function which relates the output $Y(s)$ to the input $R(s)$ is

$$T(s)=\frac{G(s)}{1+G(s)H(s)}$$

If the factors of $G(s)H(s)$ are known, frequency response plots of GH are relatively easy to construct, and they are of help in determining the properties of the overall

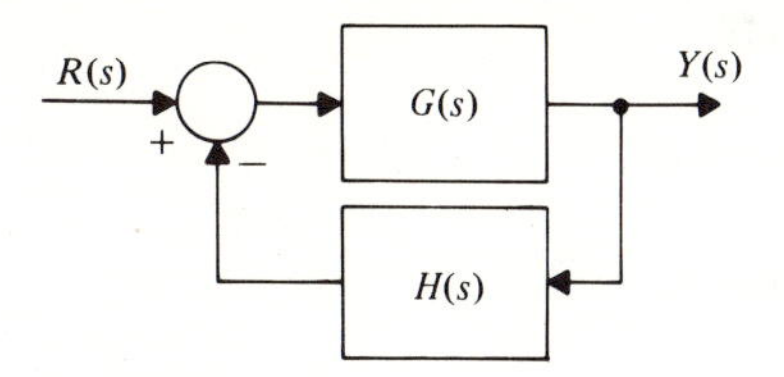

FIGURE 6-21
A feedback system.

system. Or the frequency response plots for GH may have been determined experimentally.

Whether or not the overall system is stable may be determined in the following way: Suppose the $H(s)$ block is replaced by

$$H'(s)=KH(s)$$

where K is a constant. Then

$$T'(s)=\frac{G(s)}{1+G(s)H'(s)}=\frac{G(s)}{1+KG(s)H(s)}$$

For $K=0$,

$$T'(s)=G(s)$$

and for $K=1$,

$$T'(s)=T(s),$$

where $T(s)$ is the actual system of interest. If $T'(s)=G(s)$, for $K=0$, is stable and if $T'(s)=T(s)$, for $K=1$, is unstable, there must be some intermediate value of K for which $T'(s)$ is marginally stable. That is, the root locus of $T'(s)$, as K is varied from 0 to 1, must cross the imaginary axis on the complex plane. There must be a value of K between zero and 1 for which there are poles of $T'(s)$ on the imaginary axis, for which

$$|KG(s=j\omega)H(s=j\omega)|=1$$

and

$$\angle KG(s=j\omega)H(s=j\omega)=180^\circ$$

A positive constant K only affects the amplitude ratio, so $T'(s)$ has poles on the imaginary axis for K between 0 and 1 if and only if

$$|G(s=j\omega)H(s=j\omega)|=\frac{1}{K}$$

is greater than unity (0 dB) at a frequency where

$$\angle KG(s=j\omega)H(s=j\omega) = \angle G(s=j\omega)H(s=j\omega) = 180^\circ$$

For example, the frequency response plots for

$$G(s)H(s) = \frac{100}{(s+1)^3}$$

in the system of Fig. 6-22a are given in Fig. 6-22b. To determine whether the overall system is stable, one imagines the system of Fig. 6-22c, for which

$$T'(s) = \frac{G(s)}{1+KG(s)H(s)}$$

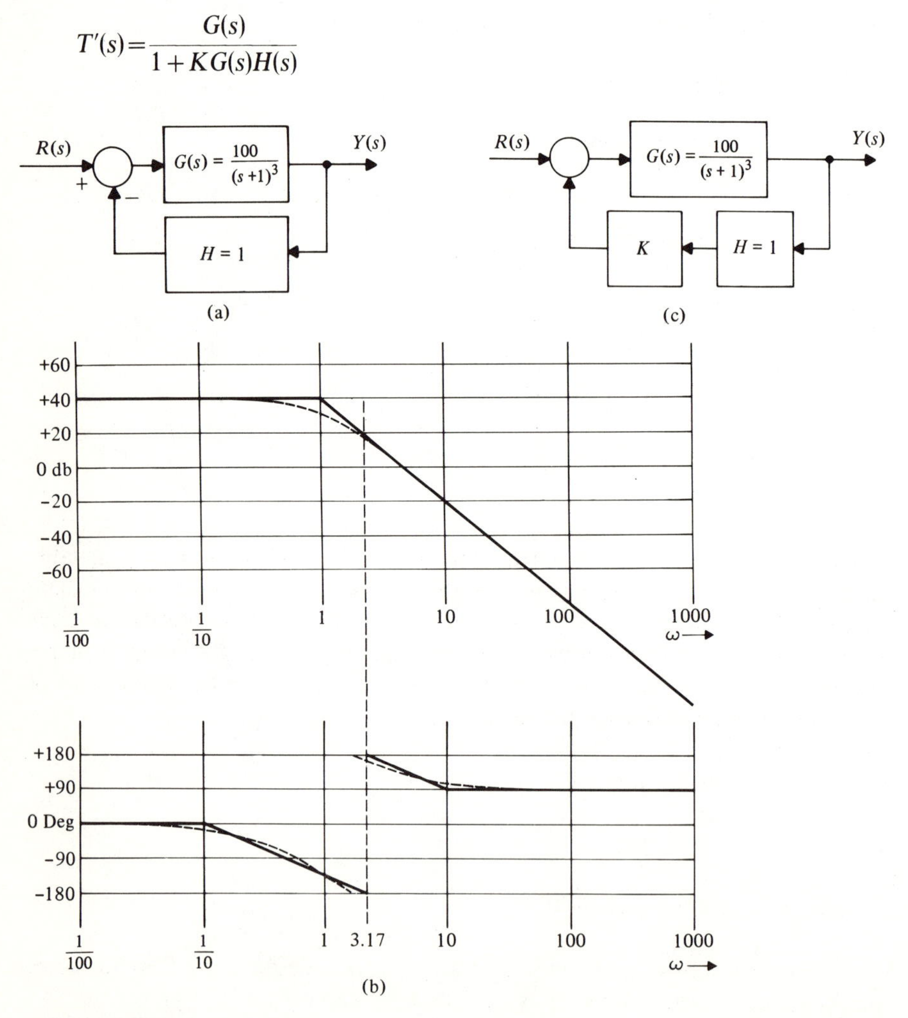

FIGURE 6-22
Feedback system example.

For $K=0$,

$$T'(s)=G(s)=\frac{100}{(s+1)^3}$$

is stable, its three poles being at $s=-1$.

Various values of K between 0 and 1 do not affect the phase shift curve for $KG(s)H(s)$ but result in an amplitude curve with the same shape as that for $G(s)H(s)$, but at various lower levels, as indicated in Fig. 6-23. For $K\cong\frac{1}{10}$,

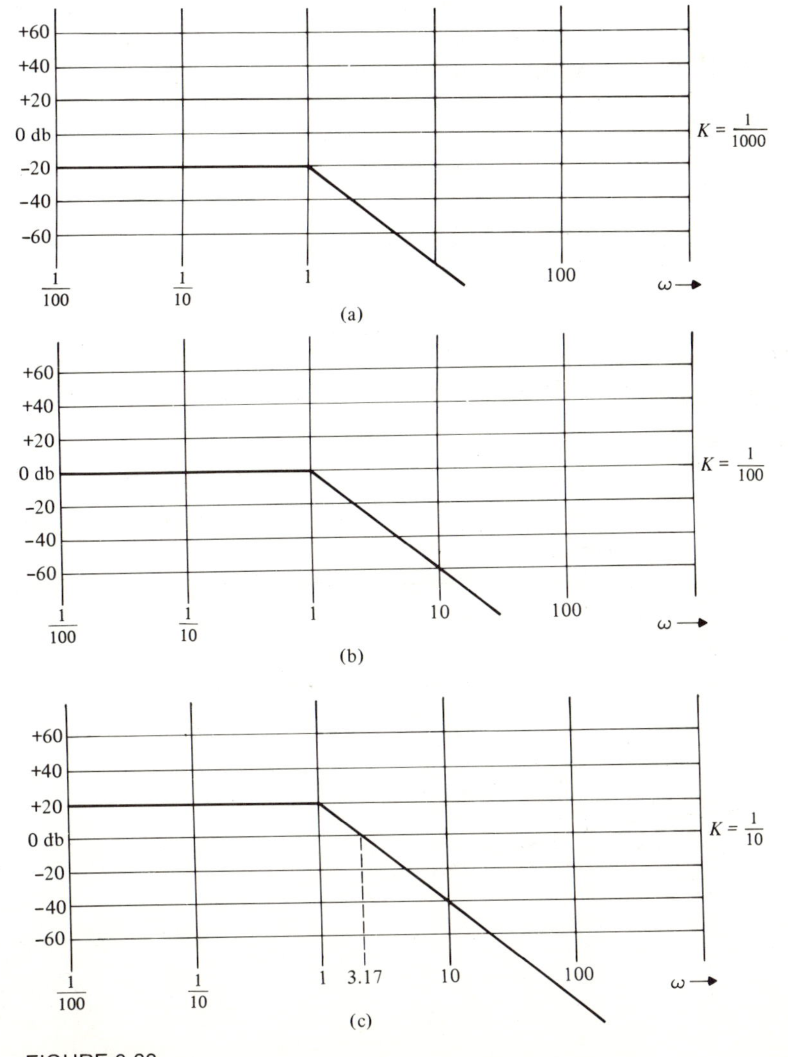

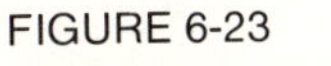
FIGURE 6-23
Amplitude curves of the example GH product for various values of the adjustable constant K.

the amplitude curve is at 0 dB at the same frequency as the phase shift curve is at 180°. For this value of K,

$$KG(s=j3.17)H(s=j3.17)=-1$$

and $T'(s)$ has a set of complex conjugate poles at $s=\pm j3.17$.

As K is varied from zero, where all the poles of $T'(s)$ are in the LHP, to unity, where $T'(s)=T(s)$, there is an intermediate value of K for which $T'(s)$ has imaginary axis poles. This means that the loci of two of the poles of $T'(s)$ must extend from the LHP, across the imaginary axis (at $\omega=\pm 3.17$), to the RHP as K goes from zero to 1. There is only one such set of crossings, so it is concluded that $T(s)$ has RHP poles and so is unstable.

There is the possibility that a segment of the root locus lies along the real axis, crossing the imaginary axis at the origin. This situation is easy to overlook if the frequency response plots are made versus log frequency because the response at $\omega=0$ is never explicitly plotted. If

$$G(s)H(s)=\frac{s-10}{s+1}=\left(\frac{s-10}{-s}\right)\left(\frac{1}{s+1}\right)(-10)$$

the loop transmittance frequency response is as plotted in Fig. 6-24. The phase shift is 180° at $\omega=0$. (It is nearly the asymptotic value below $\omega=\frac{1}{10}$, but reaches precisely 180° only at $\omega=0$.) As K is increased from zero to unity, the amplitude

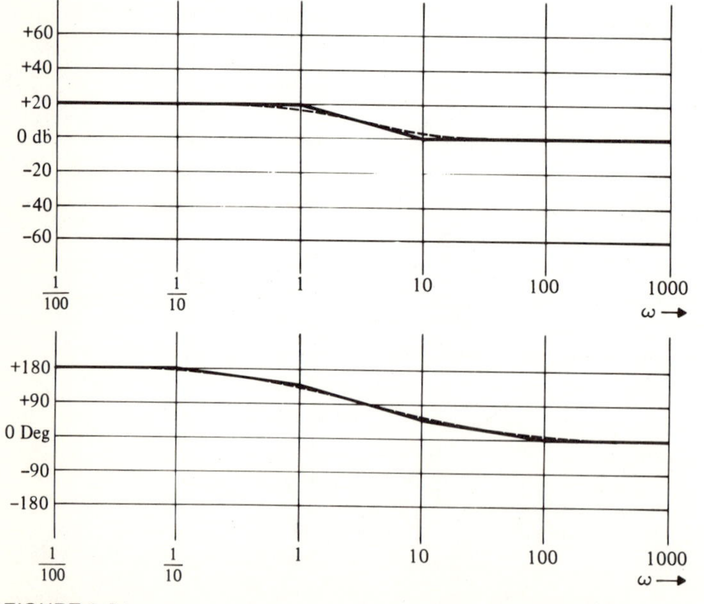

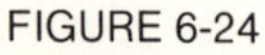

FIGURE 6-24
GH product for a system with an imaginary axis locus crossing at $\omega=0$.

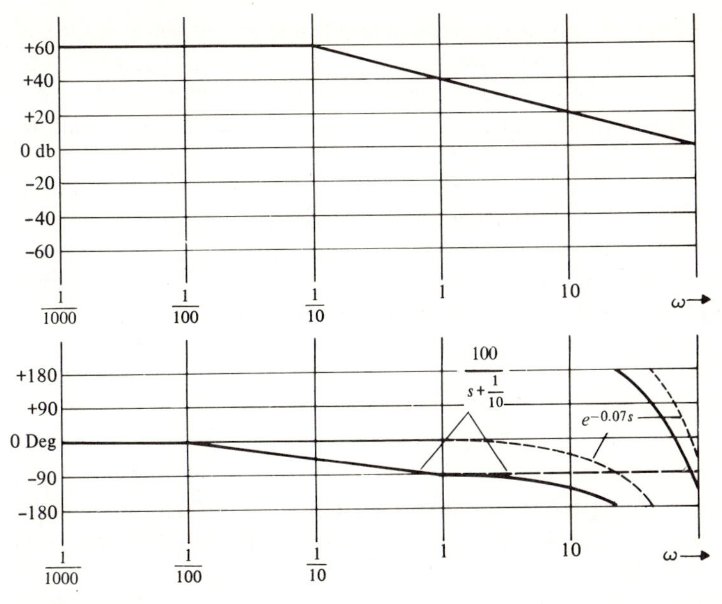

FIGURE 6-25
An irrational loop transmittance.

curve for $KG(s)H(s)$ has various positions below that of the amplitude curve for $G(s)H(s)$ in Fig. 6-24.

At the value $K=\frac{1}{10}$, the amplitude of KGH is at 0 dB at $\omega=0$, so there is a pole of $T'(s)$ at $s=j0$ for $K=\frac{1}{10}$. As K is further increased to unity, there are no other values of ω for which

$$KG(s=j\omega)H(s=j\omega)=-1$$

It is concluded that a root locus segment for $T'(s)$ must extend from the LHP, crossing the imaginary axis at $\omega=0$, into the RHP. The overall transfer function of this feedback system is thus unstable.

A major advantage of the frequency response method is that it applies to systems with irrational transfer functions. Consider the feedback system with

$$G(s)H(s)=\frac{100e^{-0{,}07s}}{s+\frac{1}{10}}$$

The frequency response for GH is given in Fig. 6-25. It is concluded that the overall system is unstable.

6.7.2 Gain Margin

Suppose it is known that a certain design results in a stable feedback system. The designed GH product will only be approximated in practice, owing to component tolerances. How much change in amplitude and in phase of the GH product can be tolerated before the overall system becomes unstable? The additional amplitude of the GH product and the additional phase angle of the product which result in imaginary axis roots of $T(s)$ are measures of the allowable tolerances in $G(s)H(s)$ for overall system stability.

A *phase crossover frequency* is any frequency at which the phase shift of the GH product is $\pm 180°$. The *gain margin* of a feedback system is the additional amplitude necessary to make the amplitude of GH unity at a phase crossover frequency. If the phase shift of GH crosses $\pm 180°$ at more than one frequency, the gain margin is the smallest value of the possibilities.

For example, frequency response plots for

$$G(s)H(s) = \frac{(s+1)^3}{s^3}$$

are given in Fig. 6-26. At the phase crossover frequency the amplitude curve is at -17 dB, giving a 17-dB gain margin. The loop gain could be increased by 17 dB before instability would result.

For the system with

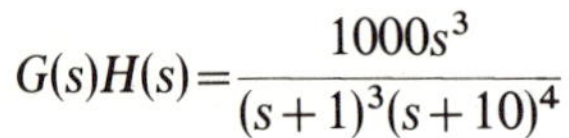

$$G(s)H(s) = \frac{1000s^3}{(s+1)^3(s+10)^4}$$

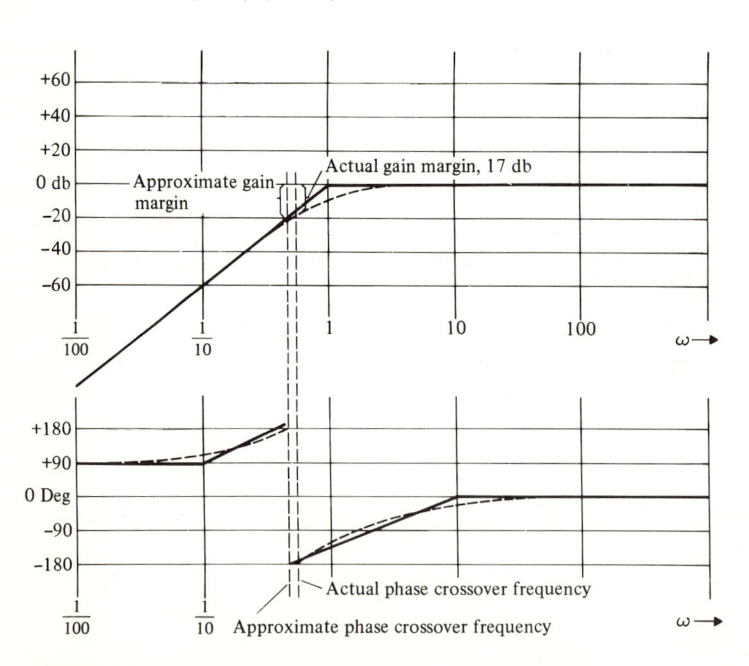

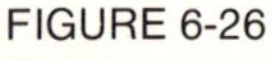

FIGURE 6-26
Gain margin of a simple feedback system. Frequency response plotted is for $G(s)H(s)$.

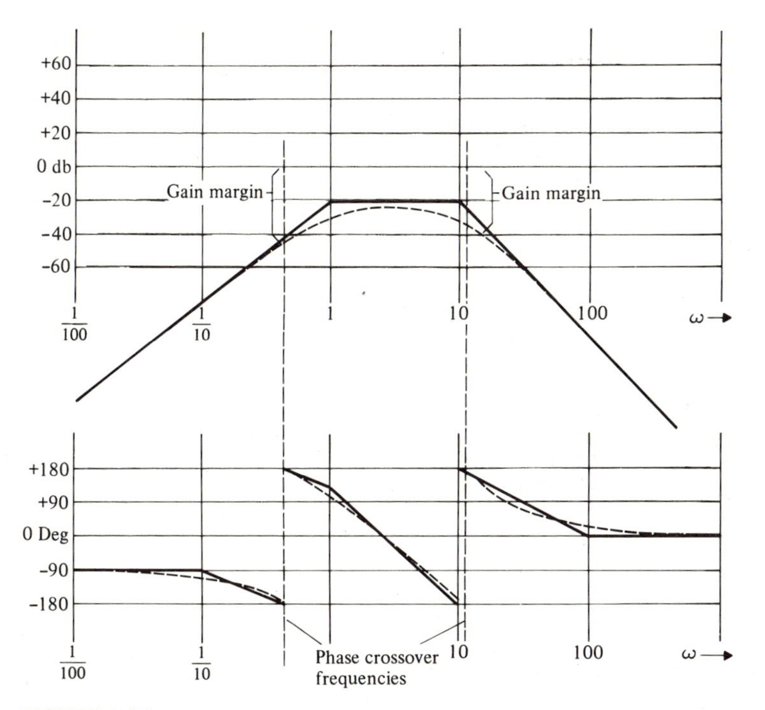

FIGURE 6-27
Gain margin when there is more than one phase crossover frequency.

there are two phase crossover frequencies, as shown in Fig. 6-27. The gain margin is the smallest of the two candidates. Negative gain margins are not defined in practice. If there is no phase crossover frequency, the gain margin may be said to be infinite.

6.7.3 Phase Margin

A *gain crossover frequency* is any frequency at which the amplitude ratio for GH is 0 dB. The *phase margin* is the additional *negative* phase shift necessary to make the phase of GH equal to $\pm 180^\circ$ at a gain crossover frequency.

The phase margin for an example feedback system with

$$G(s)H(s) = \frac{s^2}{s+1}$$

is indicated in Fig. 6-28.

If there is more than one gain crossover frequency, there is more than a

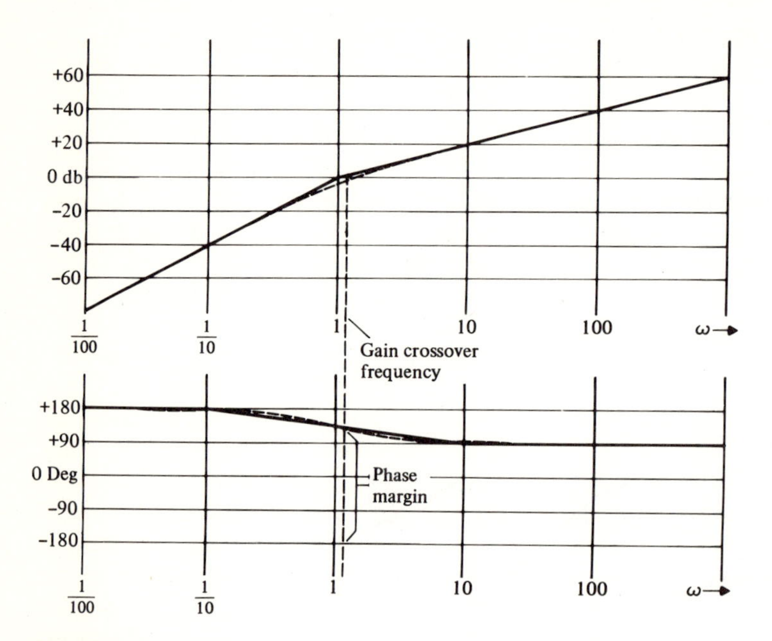

FIGURE 6-28
Phase margin for a simple feedback system. Frequency response plotted is for $G(s)H(s)$.

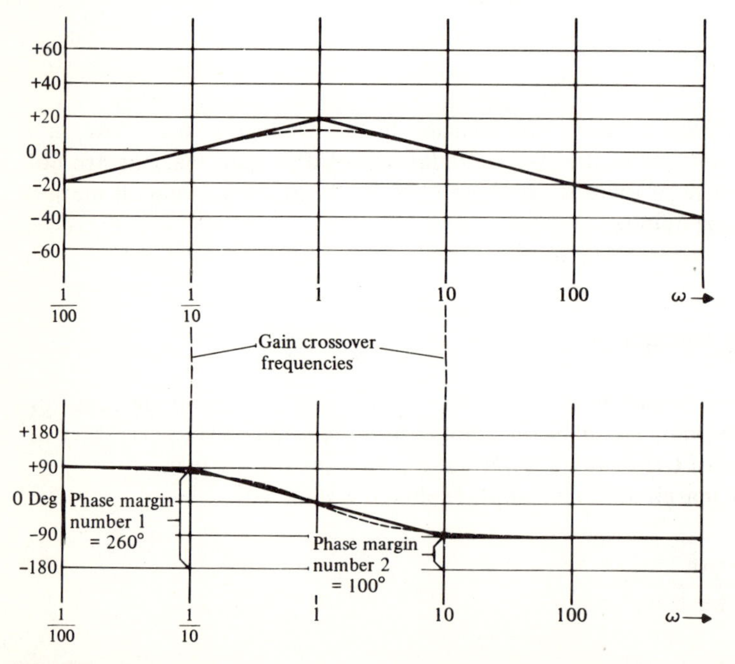

FIGURE 6-29
Phase margins when there are two gain crossover frequencies.

single phase margin. For

$$G(s)H(s)=\frac{10s}{(s+1)^2}$$

the two phase margins are shown in Fig. 6-29. A decrease of 100° or an increase of 360° − 260° = 100° in the loop transmittance phase shift would result in instability. Phase margin and gain margin are identified also.

6.7.4 Incorporating Experimental Data

The design of a closed-loop system can be accomplished by using experimental data directly, without approximating the transmittances of complicated components from their experimental data. The experimental data are plotted and the frequency responses of analytically known components are added directly to the experimental data. The resulting combined loop transmittance is used in the conventional fashion to determine stability and gain and phase margins.

As an example of this powerful technique, consider the following example: The yaw control system for a ground-effect vehicle is modeled in the block diagram of Fig. 6-30a. Experimental data for the vehicle are used in part to construct a frequency response plot for the loop gain $G_1(s)G_2(s)$. The experimental data are not approximated with asymptotes or an equation here. Instead, the experimental data for $G_2(s)$ are plotted and the analytical response for $G_1(s)$ is added in Fig. 6-30b. Assuming that there are no further 180° phase crossings or 0-dB amplitude crossings outside the range of frequencies plotted, the system gain margin is approximately 35 dB and the phase margin is about 35°.

When the closed-loop system is accurately modeled by a pair of complex conjugate poles near the imaginary axis, measured phase margin may be used to estimate the system damping ratio easily. A number of examples are given in Savant, *Control System Design*, 1964.

6.7.5 The Nichols Chart

The Nichols chart, Fig. 6-31a, provides a graphical conversion of loop transmittance amplitude and phase to overall amplitude and phase of a unity feedback system. When plotted in polar coordinates, the curves of constant amplitude ratio and constant phase shift for

$$F(j\omega)=\frac{G(j\omega)}{1+G(j\omega)}$$

are circles, as shown. For historical reasons these are often referred to as constant-M and constant-N circles, and are of the same form as the field and

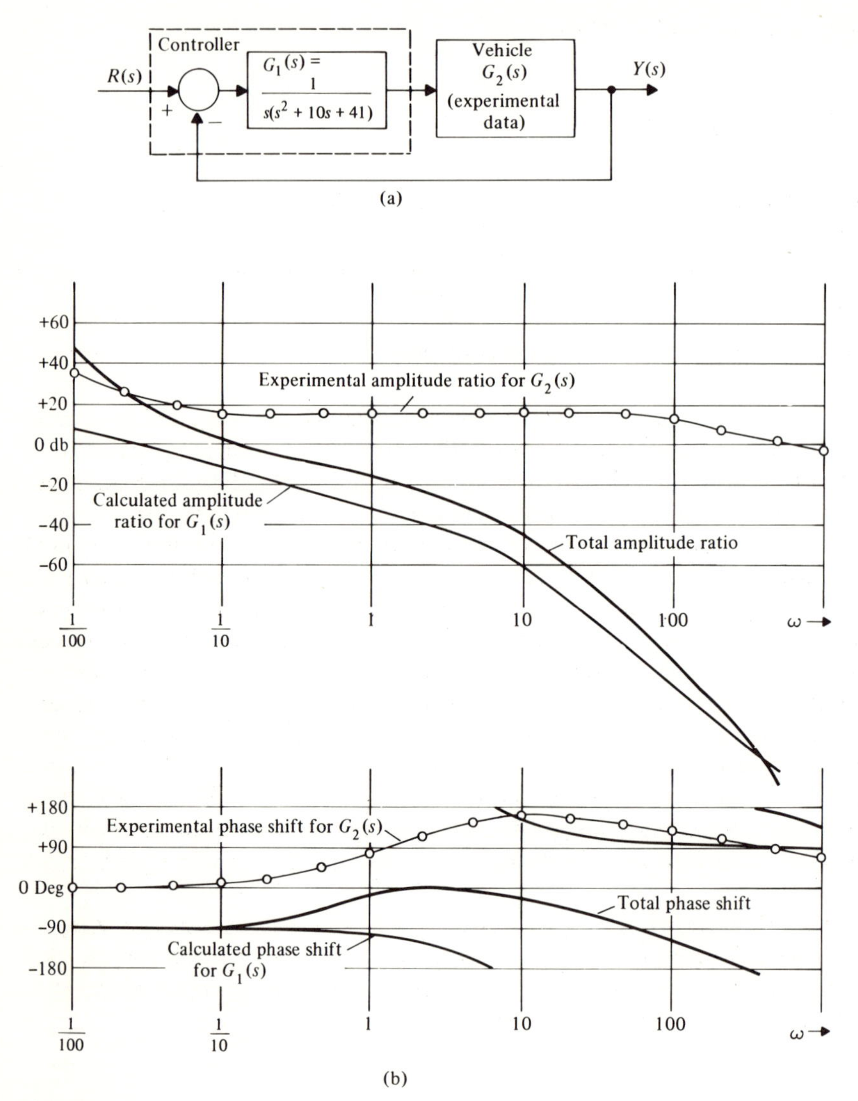

FIGURE 6-30
Incorporation experimental data into frequency response plots. (a) Model of yaw system. (b) Frequency response for the transmittance $G_1(s)\,G_2(s)$.

potential lines of a parallel-wire transmission line. An alternative form of the Nichols chart is given in Fig. 6-31b, where the curves are drawn with rectilinear dB and angle scales.

For example, the unity feedback system of Fig. 6-32a has

$$G(s)=\frac{10s}{(s+1)^2}$$

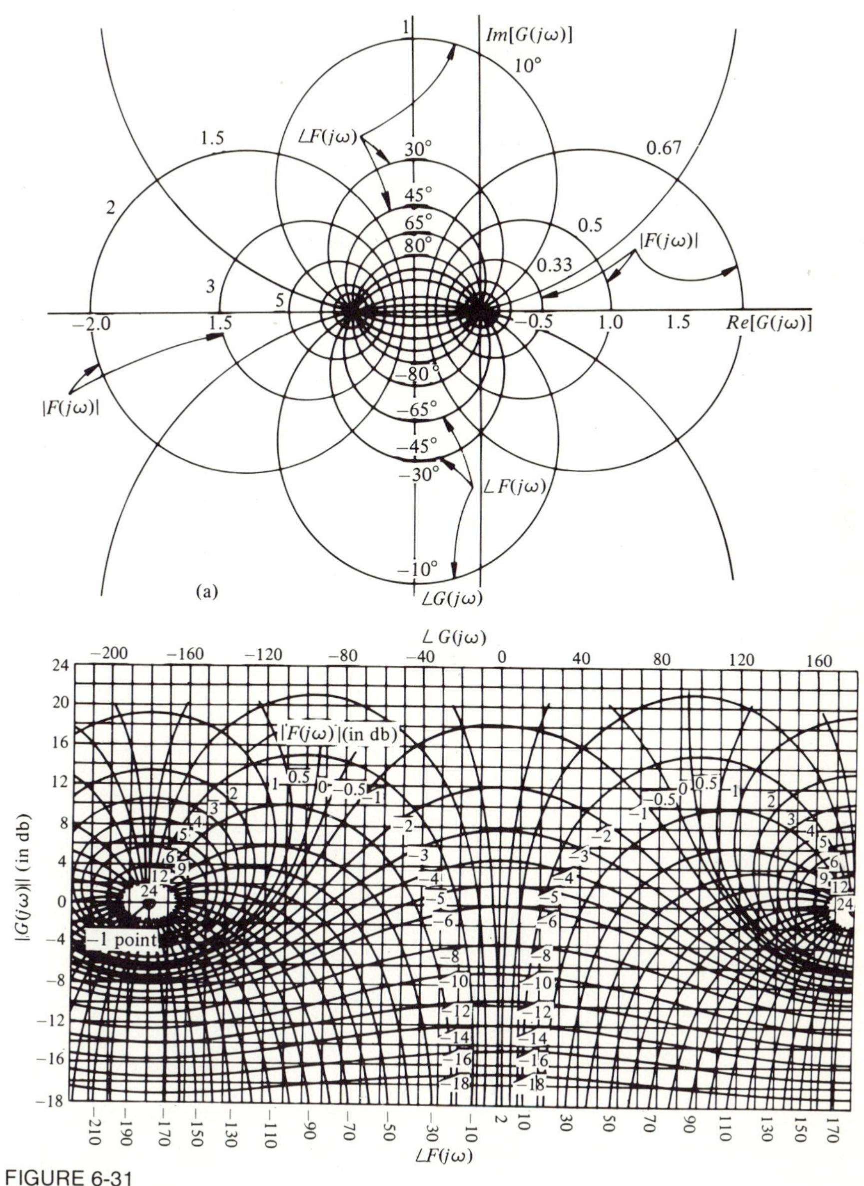

FIGURE 6-31
Nichols charts. (a) Polar chart. (b) Decibels versus angle.

Looking up the numbered points in Fig. 6-32b on the Nichols chart results in the corresponding overall system amplitude ratio and phase shift given in Fig. 6-32c. If desired, the points could be plotted directly on the Nichols chart.

Nichols charts were especially handy in the years before digital computers and sophisticated pocket calculators were so accessible to the designer. Nowadays, the simple routine outlined in Fig. 6-33 will easily perform these conversions.

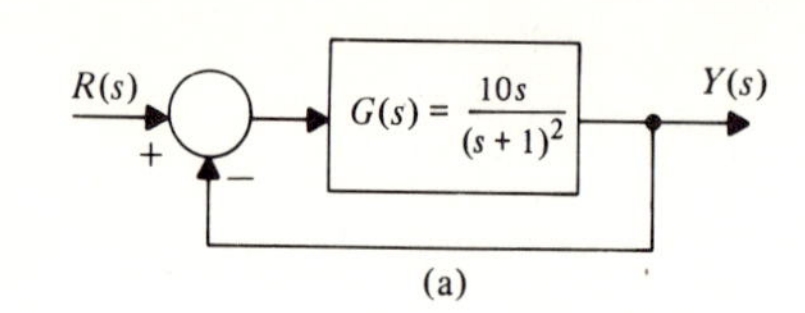

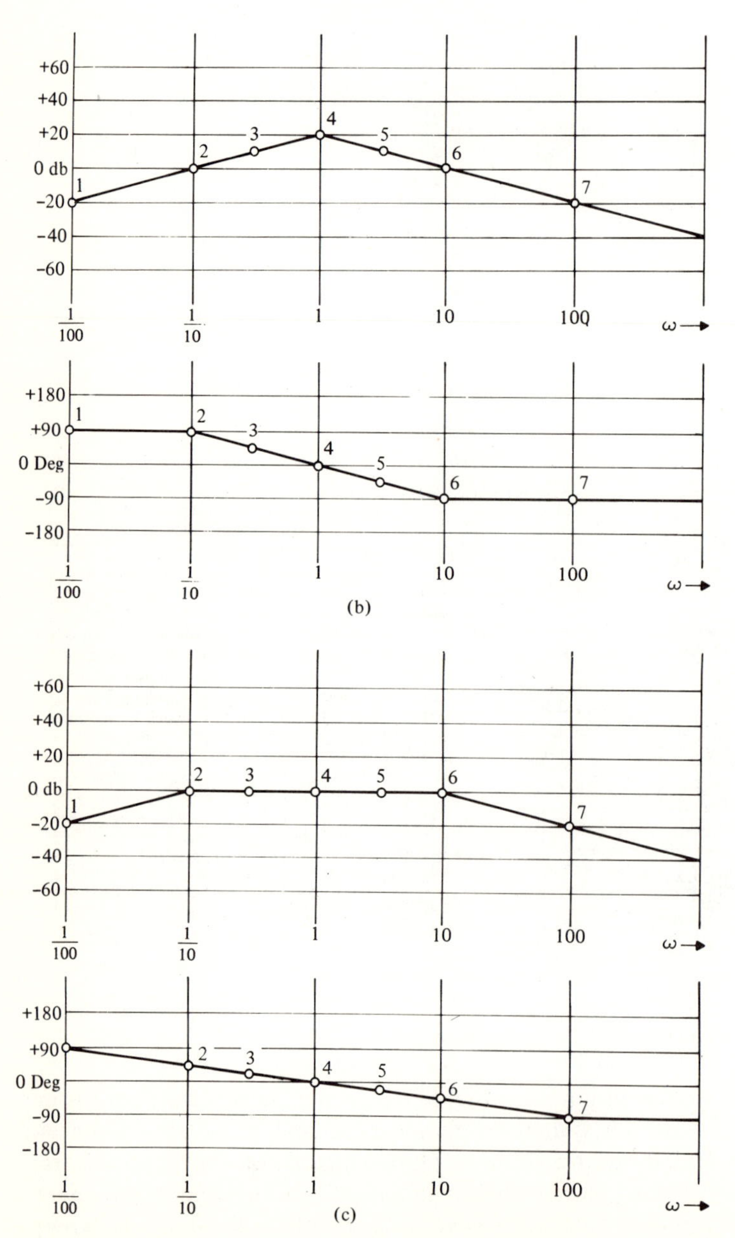

FIGURE 6-32
Using the Nichols chart to find overall system frequency response.

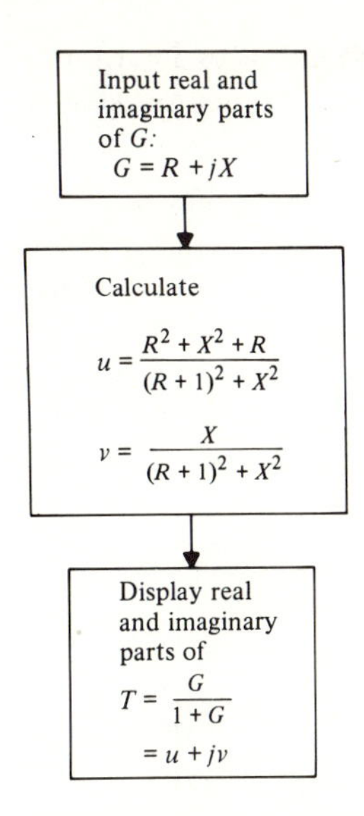

FIGURE 6-33
Nichols chart calculations.

DRILL PROBLEMS

D6-8. Find gain margins and phase margins (if they exist) for feedback systems with the following loop transmittances:

(a) $G(s)H(s)=\dfrac{2000}{(s+2)(s+7)(s+16)}$

ans. 5.4 dB, 20°

(b) $G(s)H(s)=\dfrac{20}{s(s^2+7s+140)}$

ans. 33.8 dB, 89°

(c) $G(s)H(s)=\dfrac{-s}{(s+100)^3}$

Note the negative algebraic sign.

ans. 88.5 dB, infinite phase margin

(d) $G(s)H(s)=\dfrac{e^{-0.1s}}{s}$

ans. 23.9 dB, 84°

D6-9. For the following systems, use frequency response methods to determine the range of the positive constant K for which the following systems are stable.

ans. (*a*) $K < 4.4$
(*b*) $K < 288$
(*c*) $K < 40{,}000$

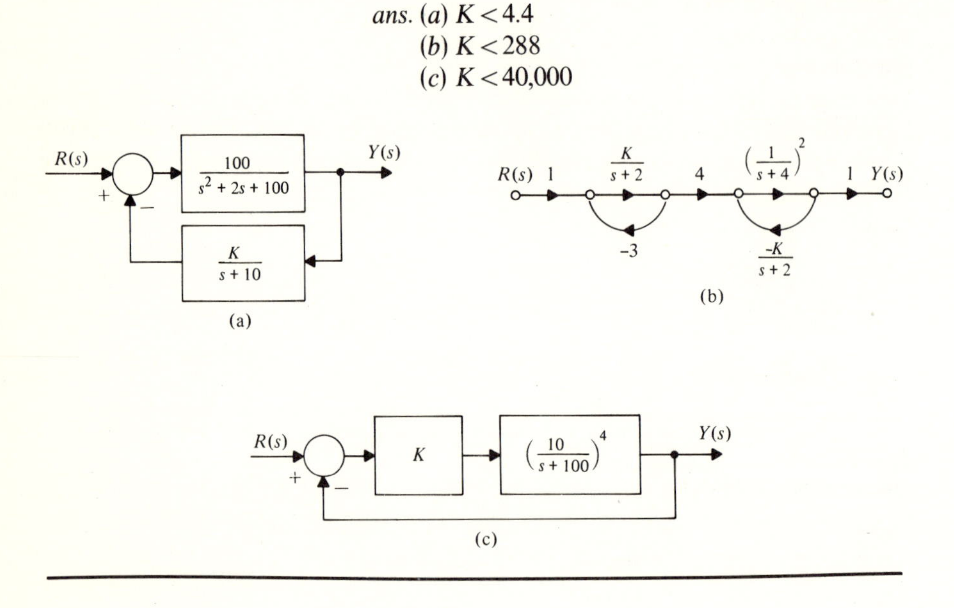

6.8 NYQUIST METHODS

The root locus frequency response procedure of the previous section does not give a definite answer as to the stability of a feedback system in all cases. If the transmittance $G(s)H(s)$ were to have both LHP and RHP poles, an imaginary axis crossing of the root locus could mean a RHP pole crossing into the LHP or a LHP pole crossing into the RHP. Or, if all poles of $G(s)H(s)$ are in the LHP, two imaginary axis crossings of the loci could mean two sets of loci extending from the LHP to the RHP or one set extending into the RHP and then crossing back to the LHP.

The Nyquist criterion, which is an extension of frequency response methods, offers an easy systematic method of determining stability in general.

6.8.1 Polar Frequency Response Plots

Another format for frequency response plots is the polar one, where the magnitude and angle of the transmittance are plotted on the complex plane. The following discussion will be concerned with frequency response of a loop transmittance $F(s) = G(s)H(s)$, but polar plotting is also applicable to other transmittances.

The loop transmittance

$$G(s)H(s)=\frac{s-1}{s+4}$$

for example, has the polar frequency response plot given in Fig. 6-34a. The function $G(s)H(s)$ may be evaluated for various values of $s=j\omega$ on the imaginary axis using directed line segments, as indicated in Fig. 6-34b. The process shown is termed a *mapping* of a *contour* (or curve) on the s-plane to the corresponding $G(s)H(s)$-curve on the GH-plane.

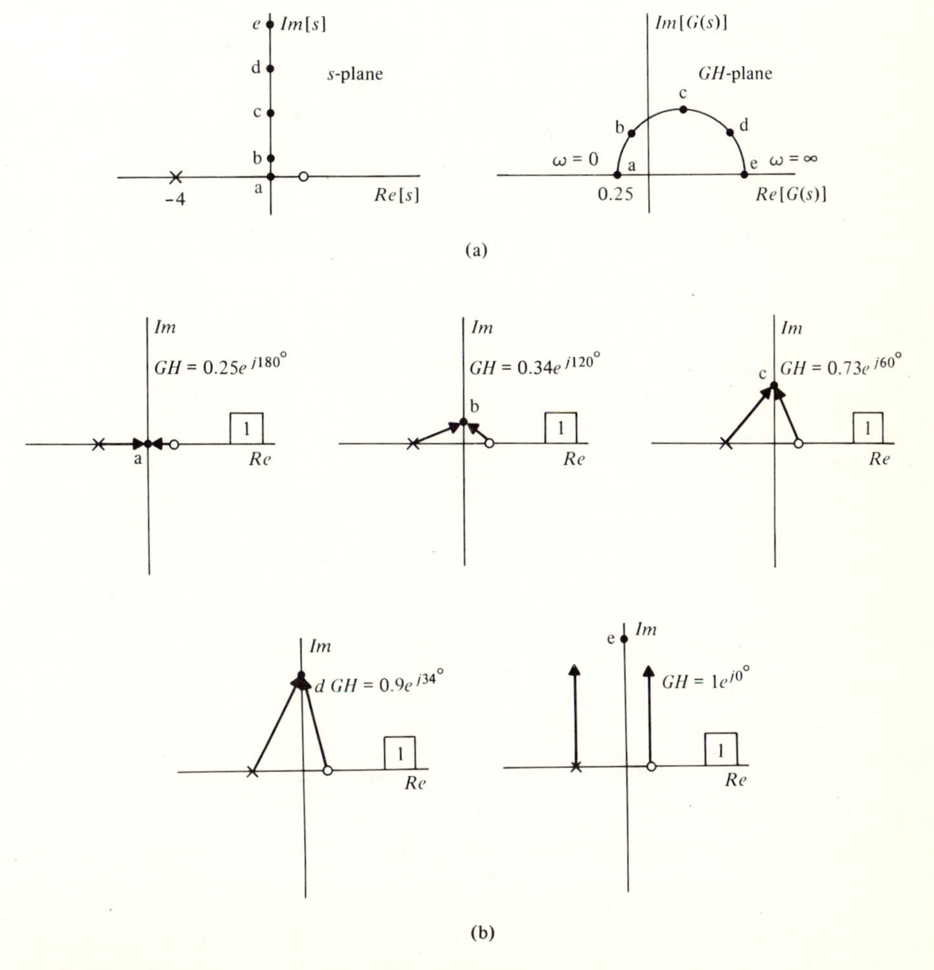

FIGURE 6-34
Polar frequency response plotting. (a) Mapping from the s-plane to the GH-plane. (b) Pole-zero calculations.

6.8.2 Nyquist Plot Construction

A polar plot of loop transmittance frequency response involves mapping the contour along the positive imaginary axis to the GH-plane. The closed contour used for a Nyquist plot involves the negative imaginary axis, the positive imaginary axis, and a half-circle of arbitrarily large radius in the RHP of the s-plane, as shown in Fig. 6-35a. A polar plot of $G(s)H(s)$ for values of s along the positive imaginary axis and the upper part of the arbitrarily large circular path is shown as a solid curve on the GH-plane in that figure. The magnitude

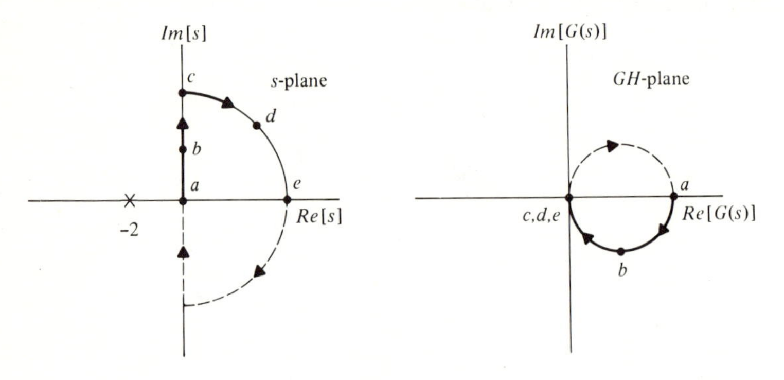

(a)

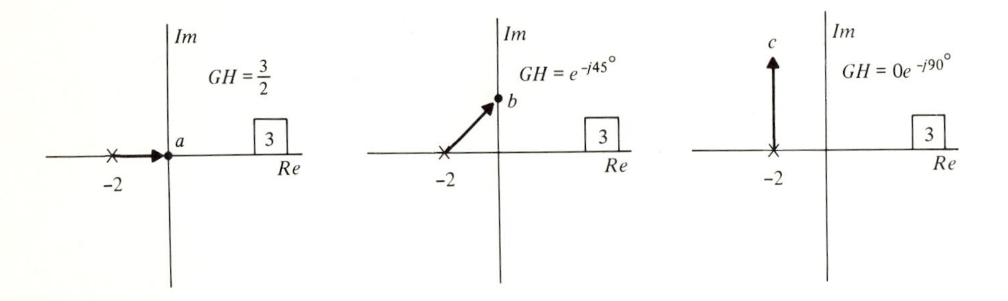

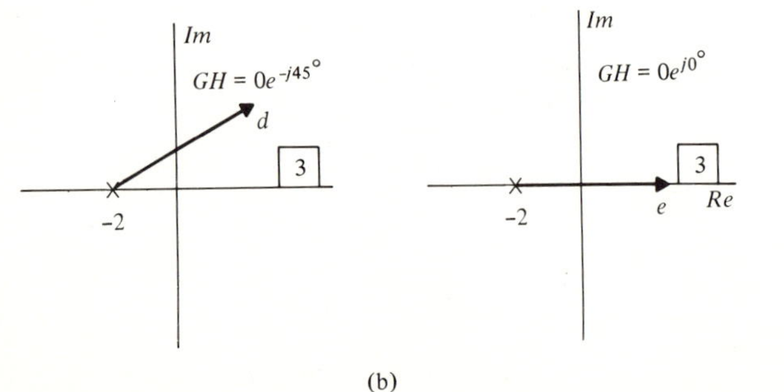

(b)

FIGURE 6-35
Nyquist plot construction for $G(s)H(s) = 3/(s+2)$. (a) Contour mapping. (b) Pole-zero calculations.

and angle of GH for selected points on the s-plane contour were calculated graphically, using directed line segments, in Fig. 6-35b. The dashed mirror image of this top portion of the closed Nyquist contour in the s-plane produces the dashed mirror image GH-plane curve, as in the figure, because of the symmetry of the pole-zero plot of $G(s)H(s)$ about the real axis.

One sketches the Nyquist plot by calculating, graphically or otherwise, the magnitude and angle of $G(s)H(s)$ for several representative points s along the upper portion of the semicircular closed contour in the s-plane. Then that portion of the plot is reflected across the real axis in the GH-plane to form the closed curve which is the Nyquist plot. It is helpful to indicate with arrows the sense of traversal of the Nyquist plot for clockwise traversal of the contour in the s-plane.

If $G(s)H(s)$ has a root on the imaginary axis, as in the example of Fig. 6-36,

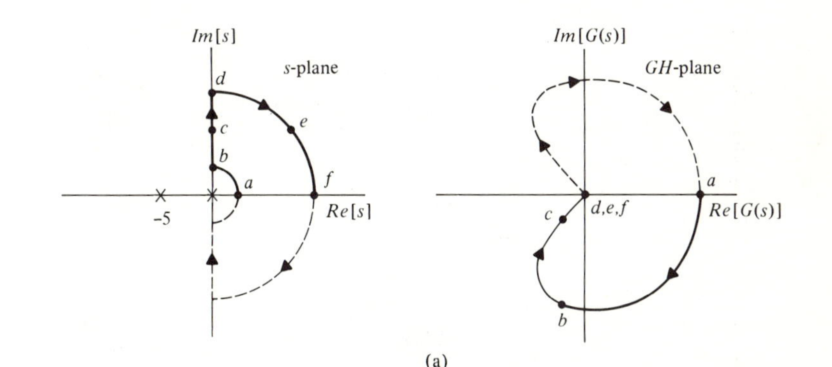

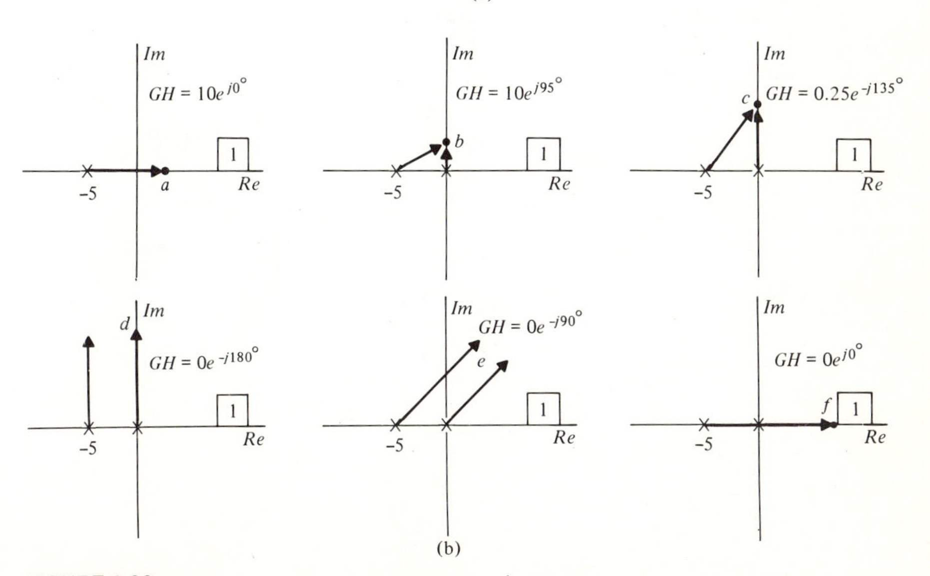

FIGURE 6-36
Nyquist plot construction for $G(s)H(s) = 1/s(s+5)$. (a) Contour mapping. (b) Pole-zero calculations.

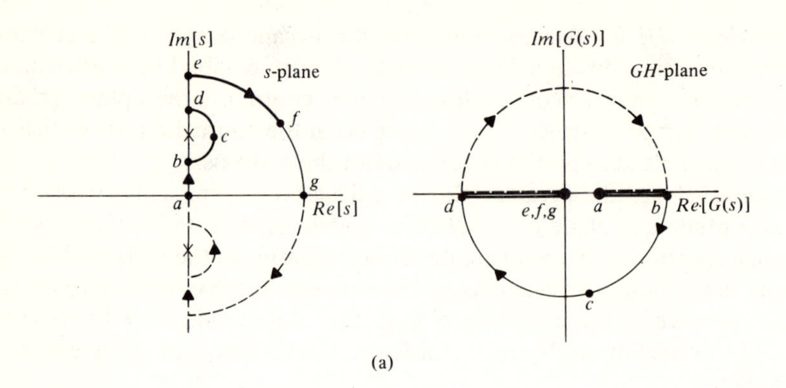

(a)

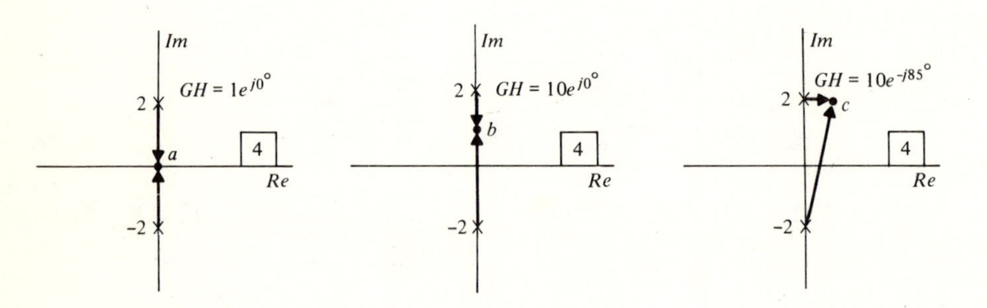

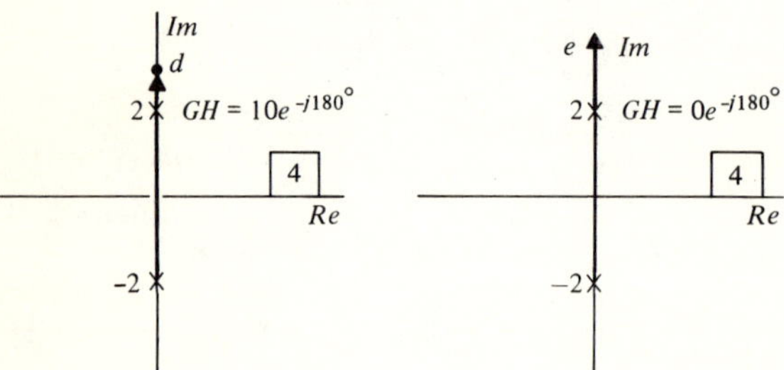

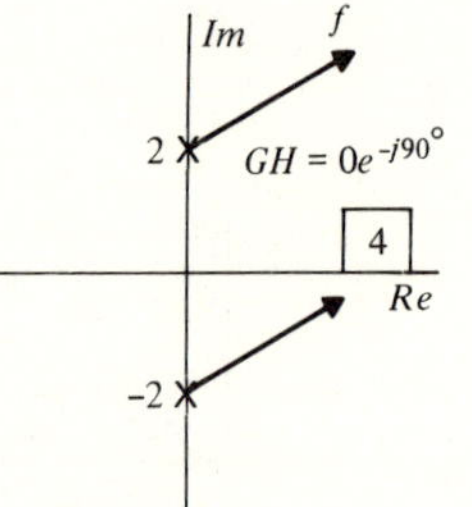

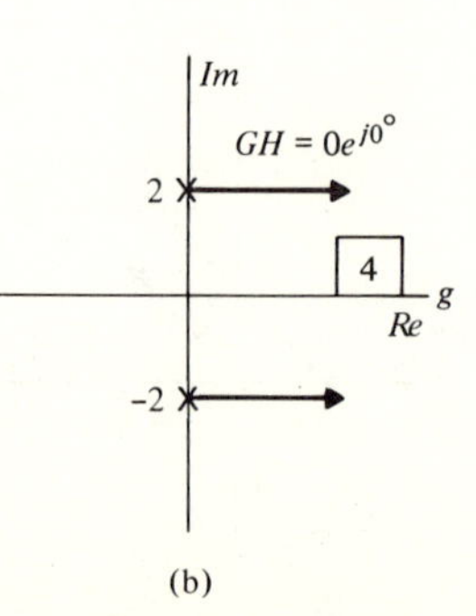

(b)

FIGURE 6-37
Nyquist plot construction for $G(s)H(s) = 4/(s^2 + 4)$. (a) Contour mapping. (b) Pole-zero calculations.

the path in the s-plane is modified to make circular detours of arbitrarily small radius into the RHP around the roots. Figure 6-37 shows another example Nyquist plot in which small detours are made around a complex conjugate set of poles. In this case, the Nyquist plot extends from point a to point b and from point d to point e, just below the real axis. The corresponding mirror image portions of the plot are just above the real axis.

Other example Nyquist plots are given in Figs. 6-38, 6-39, and 6-40. The closed path in the s-plane always results in a closed curve mapping in the GH-plane. It should perhaps be emphasized that very few people begin with a natural

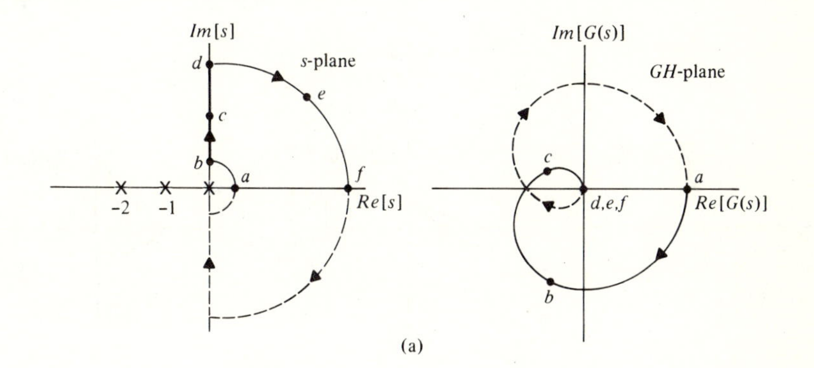

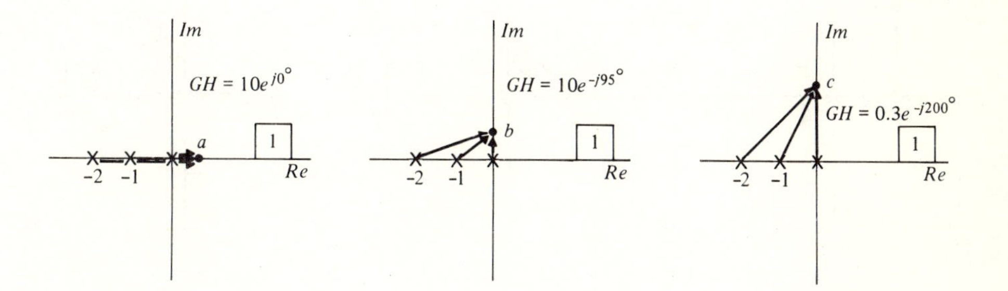

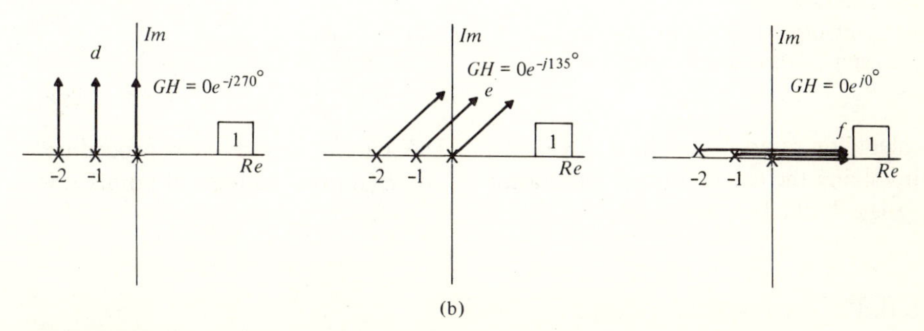

FIGURE 6-38
Nyquist plot construction for $G(s)H(s) = 1/s(s+1)(s+2)$.
(a) Contour mapping. (b) Pole-zero calculations.

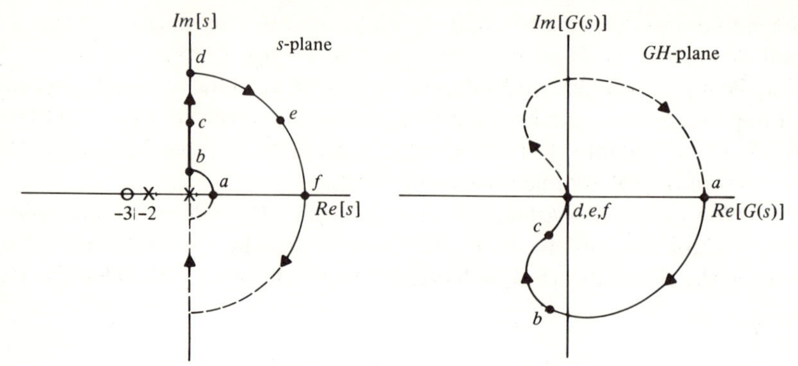

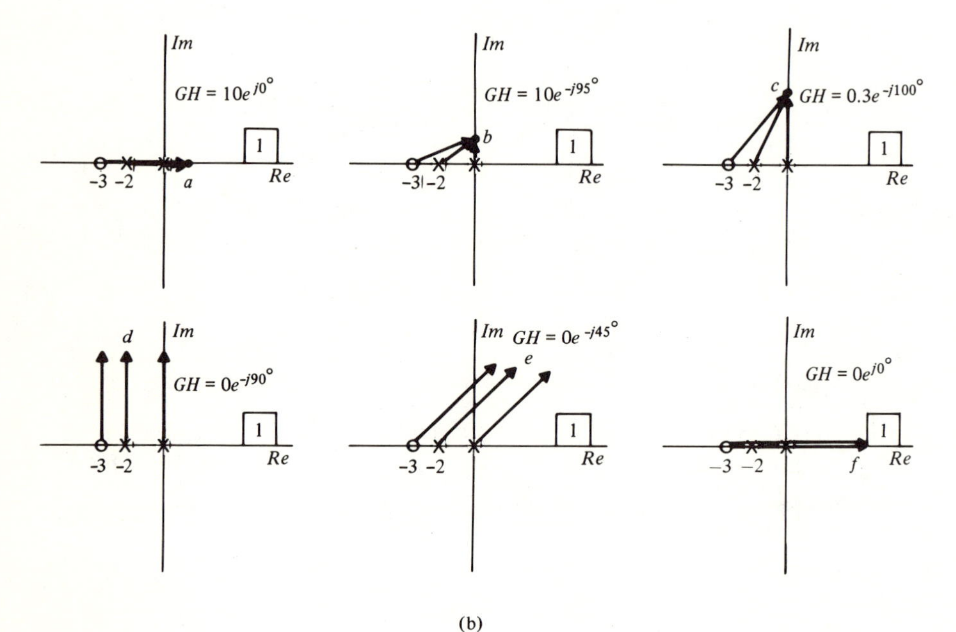

FIGURE 6-39
Nyquist plot construction for $G(s)H(s) = (s+3)/s(s+2)$. (a) Contour mapping. (b) Pole-zero calculations.

talent for rapidly sketching Nyquist plots. A careful step-by-step procedure, involving the calculation or estimation of $G(s)H(s)$ for a number of points s, is generally the best approach.

6.8.3 The Nyquist Criterion

The Nyquist criterion follows from a mathematical theorem known as the *principle of the argument*, which is illustrated in Fig. 6-41. If a function $P(s)$

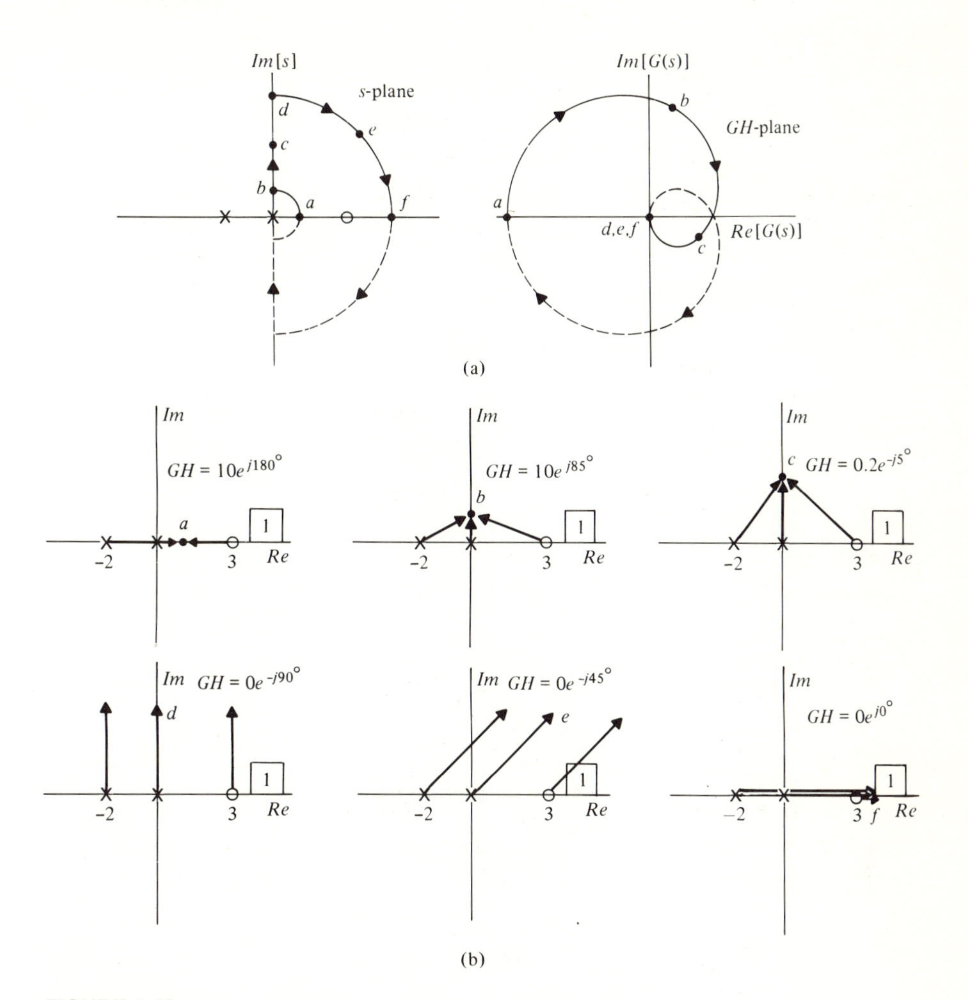

FIGURE 6-40
Nyquist plot construction for $G(s)H(s) = (s-3)/s(s+2)$. (a) Contour mapping. (b) Pole-zero calculations.

consists of a single zero, as in Fig. 6-41a, a circular contour about the zero maps to a circle about the origin of the $P(s)$-plane. As the circular contour about the zero is traversed in a clockwise (CW) sense, the angle of $P(s)$ rotates a full 360°, also clockwise. If, instead, $P(s)$ consists of a single pole as in Fig. 6-41b, CW traversal of a circular contour about the pole maps to a counterclockwise (CCW) circle on the $P(s)$ plane.

If the closed s-plane contour about a root of $P(s)$ is not a circle, as in Fig. 6-41c, the mapping is generally not a circle, but it does form a closed curve about the $P(s)$-plane's origin. The presence of other poles and zeros outside the s-plane contour affects the shape of the mapping but not its property of encirclement of the origin. If a CW s-plane contour encircles several poles and zeros of $P(s)$, then the mapping encircles the origin of the $P(s)$-plane in a CW

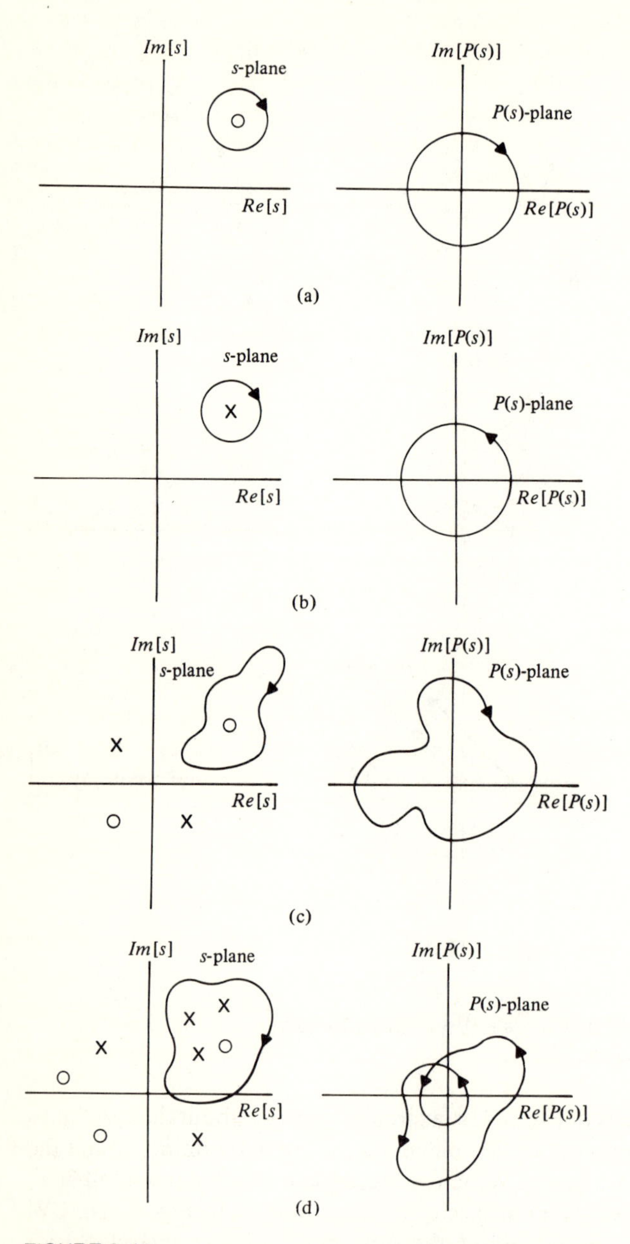

FIGURE 6-41.
Illustration of the principle of the argument. (a) Circular path about a zero: one circular CW encirclement of the origin of the $P(s)$-plane. (b) Circular path about a pole: one circular CCW encirclement of the origin of the $P(s)$-plane. (c) General closed path about a zero: one CW encirclement of the origin of the $P(s)$-plane. (d) Closed path about a collection of poles and zeros: $P(s)$-plane CW encirclements equal to the number of zeros, minus the number of poles inside the path in the s-plane.

sense a number of times equal to the number of zeros minus the number of poles within the contour. Figure 6-41d shows the s-plane contour surrounding one zero and three poles in a CW sense. The mapping of that contour onto the $P(s)$-plane encircles the origin -2 times in the CW sense, which is to say that it makes two CCW circuits about the origin.

It is the function

$$P(s)=1+G(s)H(s)$$

that is of concern in simple feedback systems, since the zeros of $P(s)$ are the poles of the overall system transfer function

$$T(s)=\frac{G(s)}{1+G(s)H(s)}$$

The functions $P(s)$ and $G(s)H(s)$ differ by unity, so the origin of the $P(s)$-plane is the point -1 on the $G(s)H(s)$-plane. Normally, the $P(s)$-plane origin is not redrawn; it is simply remembered that encirclements of the origin of the $P(s)$ plane are encirclements of the -1 point on the GH-plane.

For the Nyquist contour, each RHP pole of the overall system $T(s)$ is a zero of $P(s)$ and contributes one CW encirclement of the -1 point on the GH-plane. The poles of $P(s)$ are the same as the poles of $G(s)H(s)$, and each RHP pole of $G(s)H(s)$ contributes one CCW encirclement of the -1 point. The net number of CW encirclements of the -1 point on the GH-plane by the Nyquist plot is thus

$$\begin{pmatrix}\text{Number of CW}\\ \text{encirclements of } -1\end{pmatrix}=\begin{pmatrix}\text{number of RHP}\\ \text{poles of } T(s)\end{pmatrix}-\begin{pmatrix}\text{number of RHP}\\ \text{poles of } G(s)H(s)\end{pmatrix}$$

Or

$$\begin{pmatrix}\text{Number of RHP}\\ \text{poles of } T(s)\end{pmatrix}=\begin{pmatrix}\text{number of CW encirclements}\\ \text{of } -1\end{pmatrix}+\begin{pmatrix}\text{number of RHP}\\ \text{poles of } G(s)H(s)\end{pmatrix}$$

For example, a system with

$$G(s)H(s)=\frac{2(s+3)}{(s+2)^2(s-1)}$$

has the Nyquist plot given in Fig. 6-42. $G(s)H(s)$ has one RHP pole, but the Nyquist plot circles the -1 point once in a CCW sense, so

$$\begin{pmatrix}\text{Number of RHP}\\ \text{poles of } T(s)\end{pmatrix}=(-1)+(1)=0$$

and the overall system is stable.

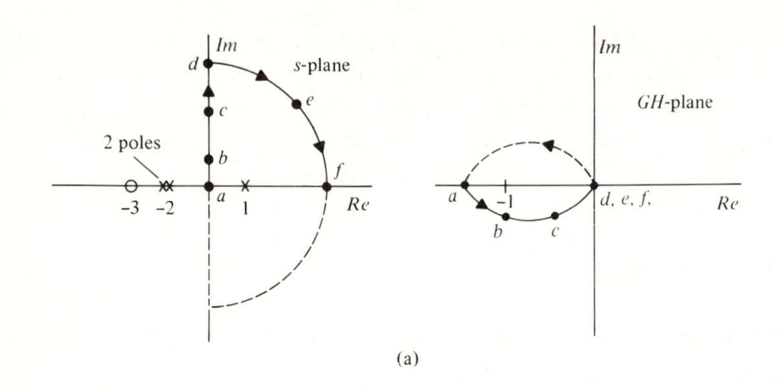

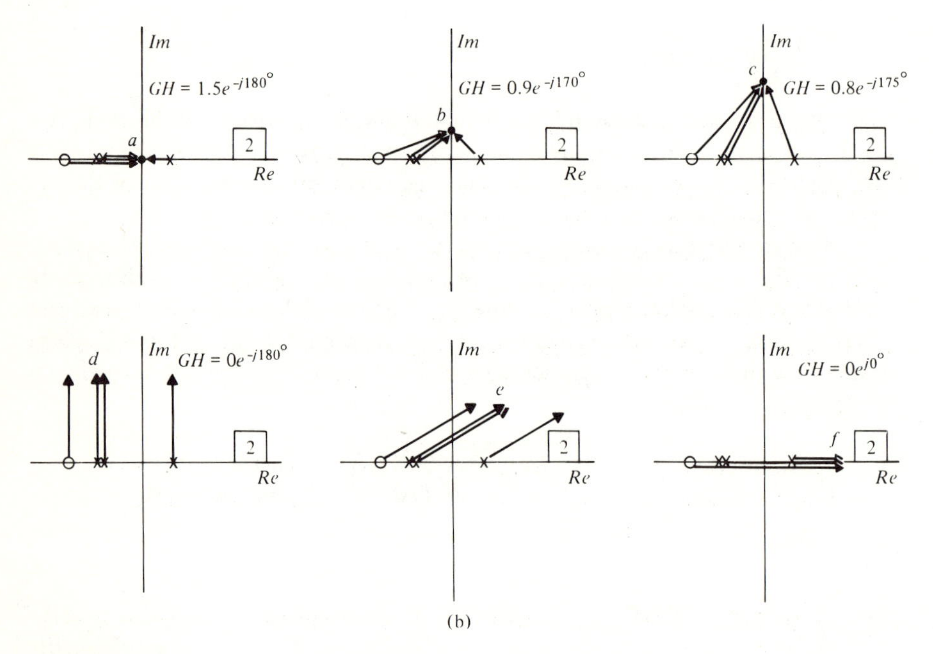

FIGURE 6-42
Applying the Nyquist criterion. (a) Contour mapping. (b) Pole-zero calculations.

Several other Nyquist plots are shown in Fig. 6-43 for systems of the form

$$G(s)H(s)=\frac{Ke^{-s}}{s}$$

for various values of the positive constant K. For sufficiently large K, this system's Nyquist plot circles the -1 point of the GH-plane several times, as in Fig. 6-43a, indicating the presence of several RHP poles in the overall system transfer function $T(s)$. For a smaller value of K, the Nyquist plot is given in

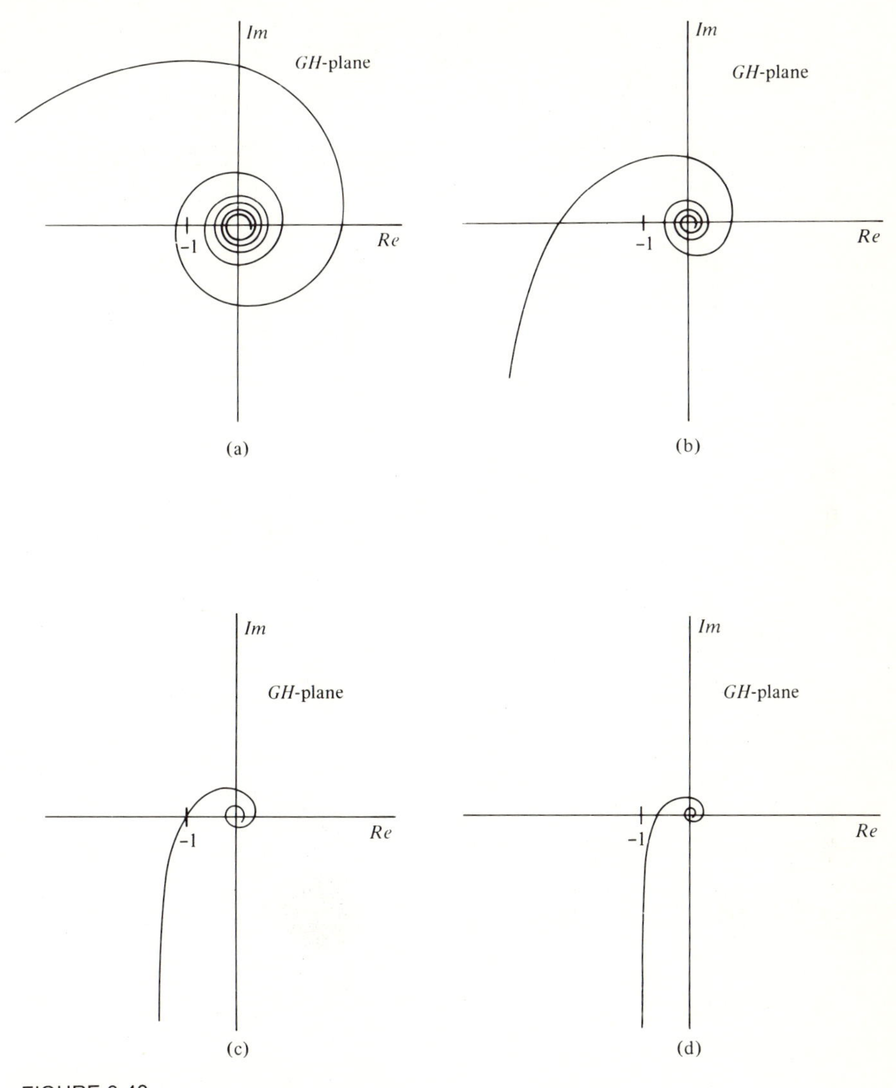

FIGURE 6-43
Nyquist plots for systems involving a time delay. For readability, only the positive frequency part of the plot is drawn. (a) $K=10$. (b) $K=4$. (c) $K=\pi/2$. (d) $K=1$.

Fig. 6-43b with a single CW encirclement of the -1 point, indicating one RHP pole in $T(s)$. For $K=\pi/2$, Fig. 6-43c, the Nyquist curve passes through the -1 point and $T(s)$ has imaginary axis poles. For smaller K, the Nyquist plot is as in Fig. 6-43d and $T(s)$ is stable.

In Table 6-3 are shown a number of Nyquist plots for systems with rational transmittances. In these, the GH-plane areas encircled by the plot are shaded, with double encirclement indicated by darker shading.

TABLE 6-3.
A collection of nyquist plots.

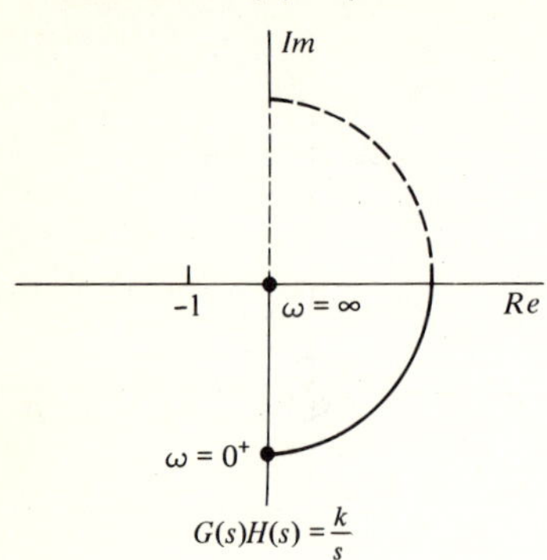

$G(s)H(s) = \frac{k}{s}$

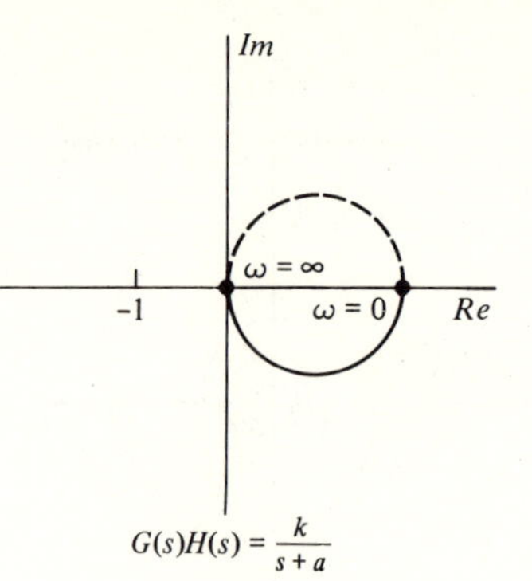

$G(s)H(s) = \frac{k}{s+a}$

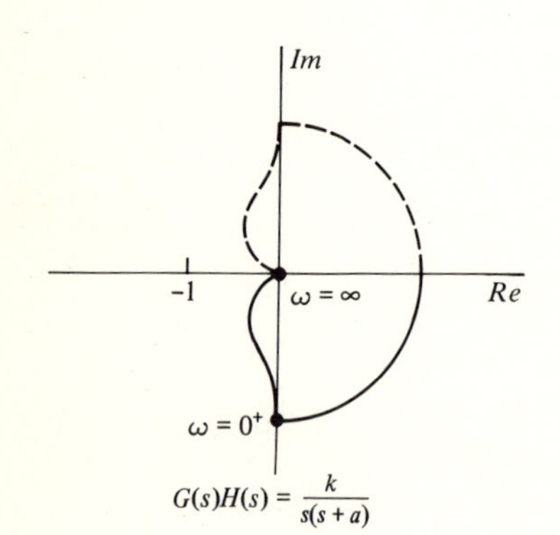

$G(s)H(s) = \frac{k}{s(s+a)}$

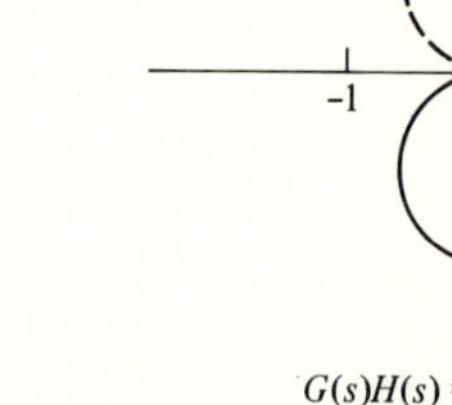

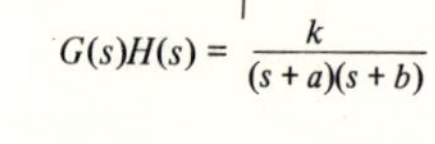

$G(s)H(s) = \frac{k}{(s+a)(s+b)}$

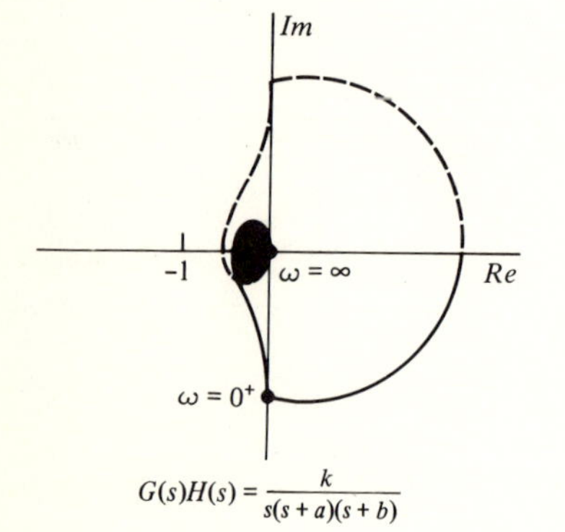

$G(s)H(s) = \frac{k}{s(s+a)(s+b)}$

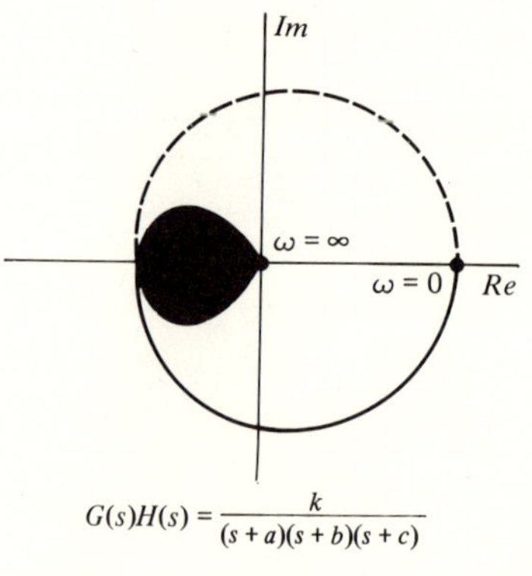

$G(s)H(s) = \frac{k}{(s+a)(s+b)(s+c)}$

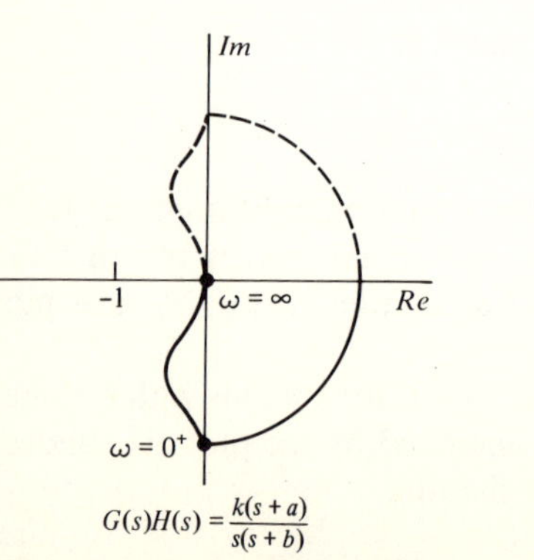

$G(s)H(s) = \frac{k(s+a)}{s(s+b)}$

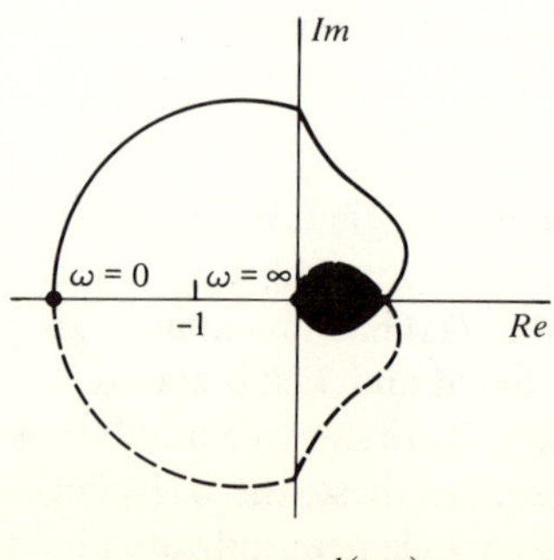

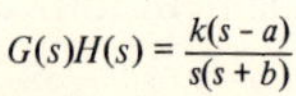

$G(s)H(s) = \frac{k(s-a)}{s(s+b)}$

TABLE 6-3 (Continued)

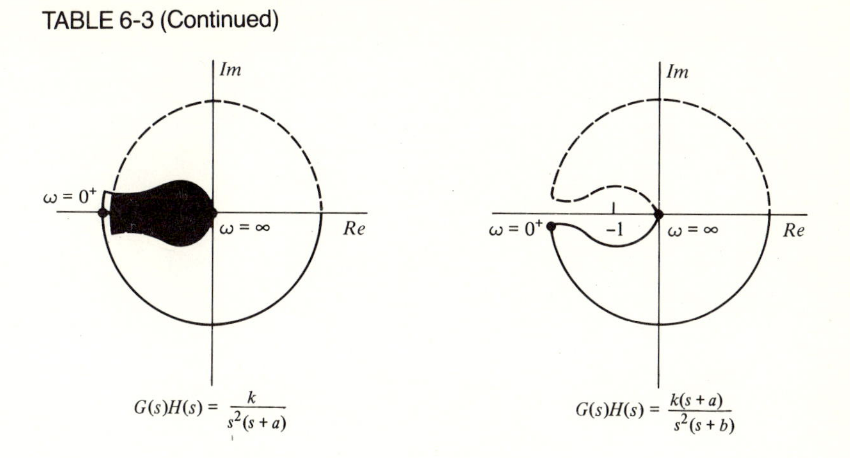

6.8.4 Closed-loop Frequency Response

The gain and phase margins of simple feedback systems may be easily determined from their Nyquist plots. The negative real axis of the GH-plane represents an angle of 180°, so any crossing of that line by the polar frequency response portion of the Nyquist curve is at a phase crossover frequency. As indicated in Fig. 6-44a, the magnitude of the loop gain is the polar distance to the crossing, 0.6 in the example. In decibels,

$$20 \log_{10} 0.6 = -4.44$$

a 4.44-dB gain margin.

The unit circle on the GH-plane represents unit magnitude, so any crossing of that circle by the Nyquist polar frequency response is a gain crossover

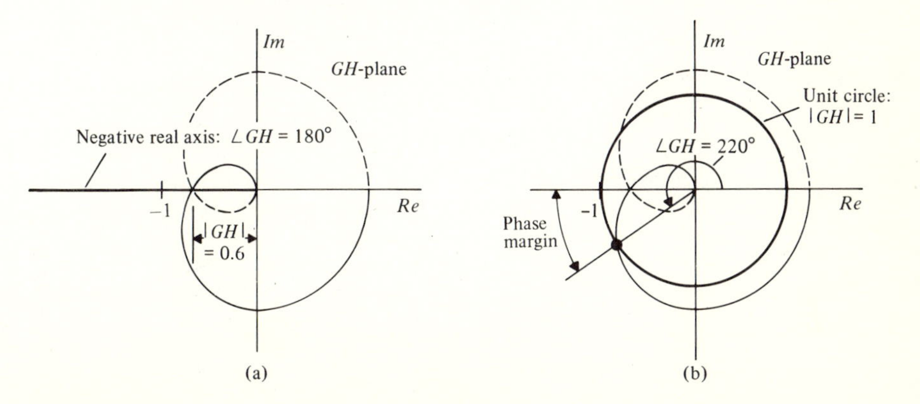

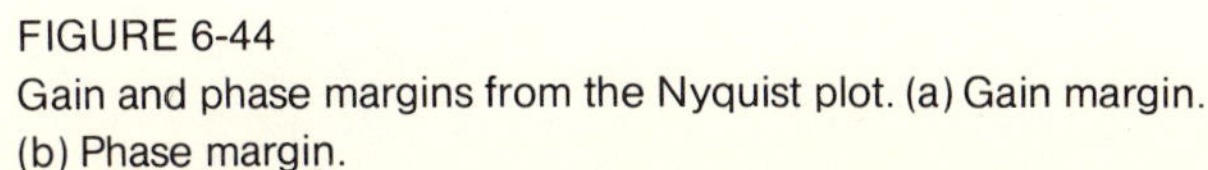
FIGURE 6-44
Gain and phase margins from the Nyquist plot. (a) Gain margin.
(b) Phase margin.

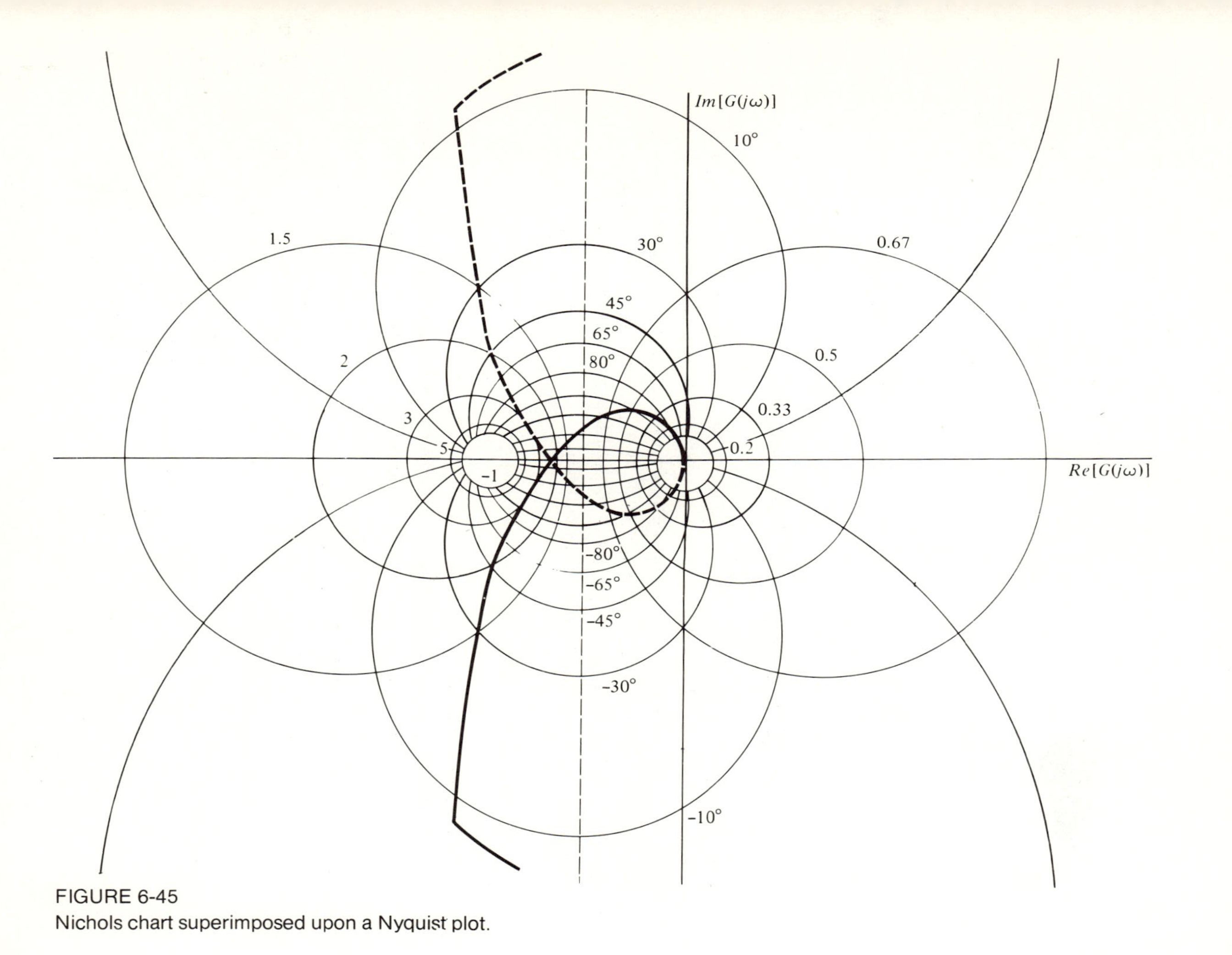

FIGURE 6-45
Nichols chart superimposed upon a Nyquist plot.

frequency. For the example of Fig. 6-44b, the angle of GH at the gain crossover frequency is 220° or −140°, corresponding to a phase margin of 40°.

When the Nichols chart is superimposed on the complex plane, the Nyquist plot for a unity feedback system not only shows stability measure, but the amplitude ratio and phase shift of the overall system as well. Figure 6-45a shows the Nyquist plot of the loop transmittance $G(s)$ for an example unity feedback system. Amplitude ratio and phase shift for the closed-loop system

$$T(s) = \frac{G(s)}{1+G(s)}$$

may be read from the chart. At a glance, it can be seen from the figure that the overall system has an amplitude ratio that begins at zero at low frequency, peaks at 3 units (a reasonable peak), then drops off at higher frequencies.

DRILL PROBLEMS

D6-10. Sketch Nyquist plots for feedback systems with the following loop transmittances, then use the plots to determine whether or not each system is stable:

(a) $G(s)H(s) = \frac{s}{s+4}$

ans. stable

(b) $G(s)H(s) = \frac{10}{(s+2)(s+6)}$

ans. stable

(c) $G(s)H(s) = \frac{1}{s(s^2+4)}$

ans. unstable

(d) $G(s)H(s) = \frac{s^2}{s^2+2s+10}$

ans. stable

(e) $G(s)H(s) = \frac{2}{s^2(s+3)}$

ans. unstable

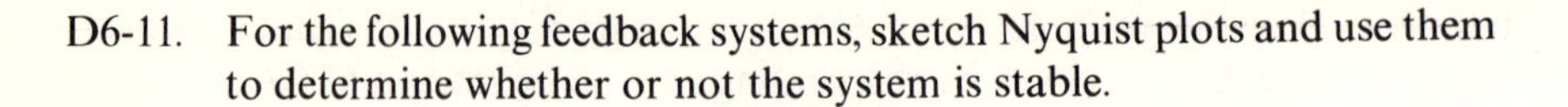

D6-11. For the following feedback systems, sketch Nyquist plots and use them to determine whether or not the system is stable.

ans. (a) stable
(b) unstable
(c) stable

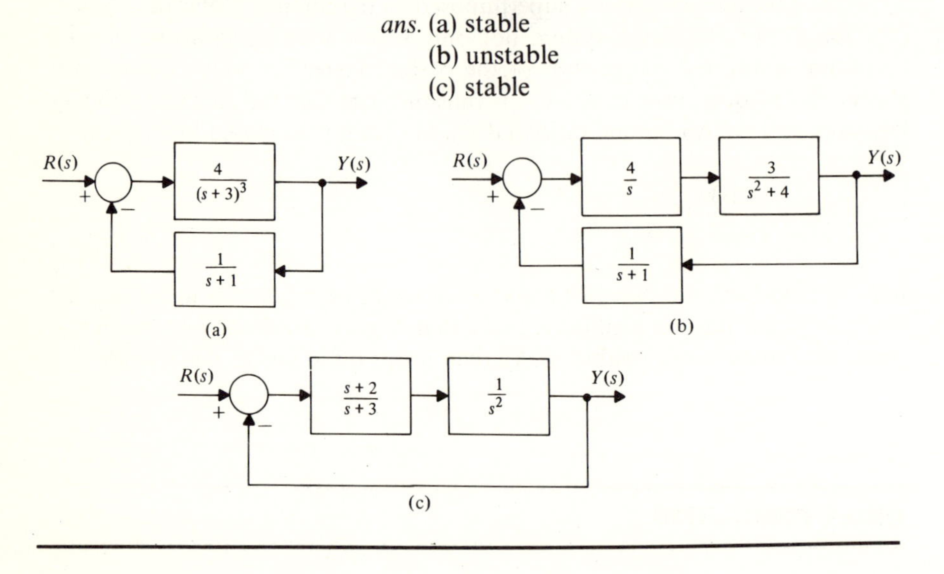

6.9 COMPENSATING OPERATIONAL AMPLIFIERS

Single-chip integrated circuit operational amplifiers (op amps) were first produced commercially in 1963. The type 101 uncompensated amplifier was introduced in 1967, followed by the internally compensated type 741 in 1968. In several years' time, the 101 and 741 became widely used by industry and are still in some sense industry standards to which other op amps are compared. Since then, many new type numbers have been marketed, each improved or optimized for one application or another. Because of their wide acceptance and application, the discussion to follow will concern the 101 and 741 types, although the methods and results apply to most other types as well.

When a general-purpose operational amplifier is connected as shown in Fig. 6-46a, it forms a noninverting amplifier of the input voltage V_{in}, with gain fixed primarily by the external resistors R_F and R_A. To a first approximation, the operational amplifier itself is a high-gain voltage-differencing amplifier with a model given in Fig. 6-46b. For this circuit,

$$\begin{cases} V_a = \dfrac{R_A}{R_A + R_F} V_{out} \\ V_b = V_{in} \\ V_{out} = A(V_b - V_a) \end{cases}$$

These relations are represented as a feedback system in Fig. 6-46c.

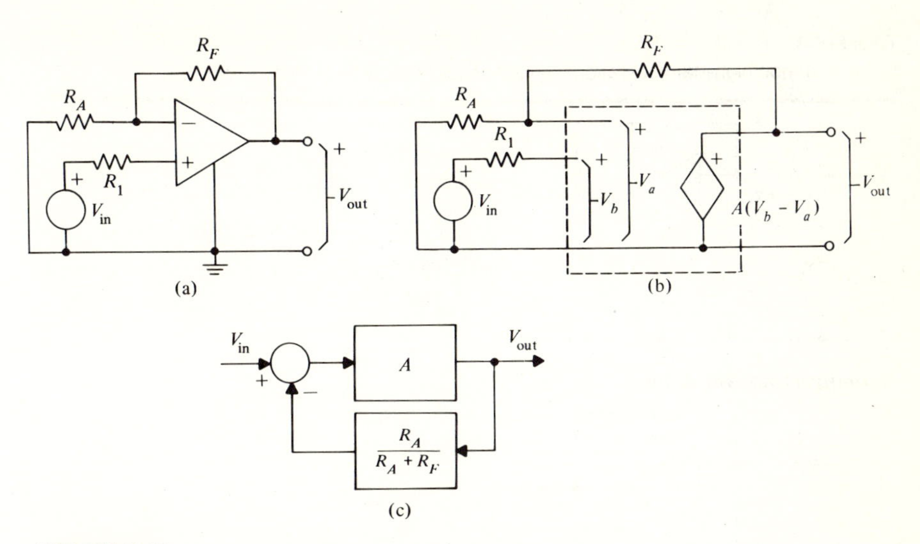

FIGURE 6-46
Analysis of an operational amplifier circuit. (a) Noninverting amplifier connection of an op amp. (b) A simple equivalent circuit. (c) The amplifier as a feedback system.

The transfer function of the noninverting op amp configuration is, applying the feedback relation to Fig. 6-46c,

$$T = \frac{A}{1 + AR_A/(R_A + R_F)} = \frac{R_A + R_F}{R_A + (R_A + R_F)/A}$$

For sufficiently large open-loop op amp gain A,

$$T \cong \frac{R_A + R_F}{R_A} = 1 + \frac{R_F}{R_A}$$

and

$$V_{out} \cong \left(1 + \frac{R_F}{R_A}\right) V_{in}$$

Typical general-purpose operational amplifiers have low-frequency open-loop gains A on the order of 10^5, so the approximation is a good one under those circumstances.

In a similar manner, approximate behavior of other operational amplifier configurations such as the ones listed in Table 6-4 may be calculated.*

*The operational amplifier symbol used here includes both the amplifier itself and its power supply; hence the explicit ground connection. Electronic designers commonly omit drawing the ground connection by considering the power supply to be externally connected and not of interest.

TABLE 6-4.
Some common high-gain operational amplifier connections.

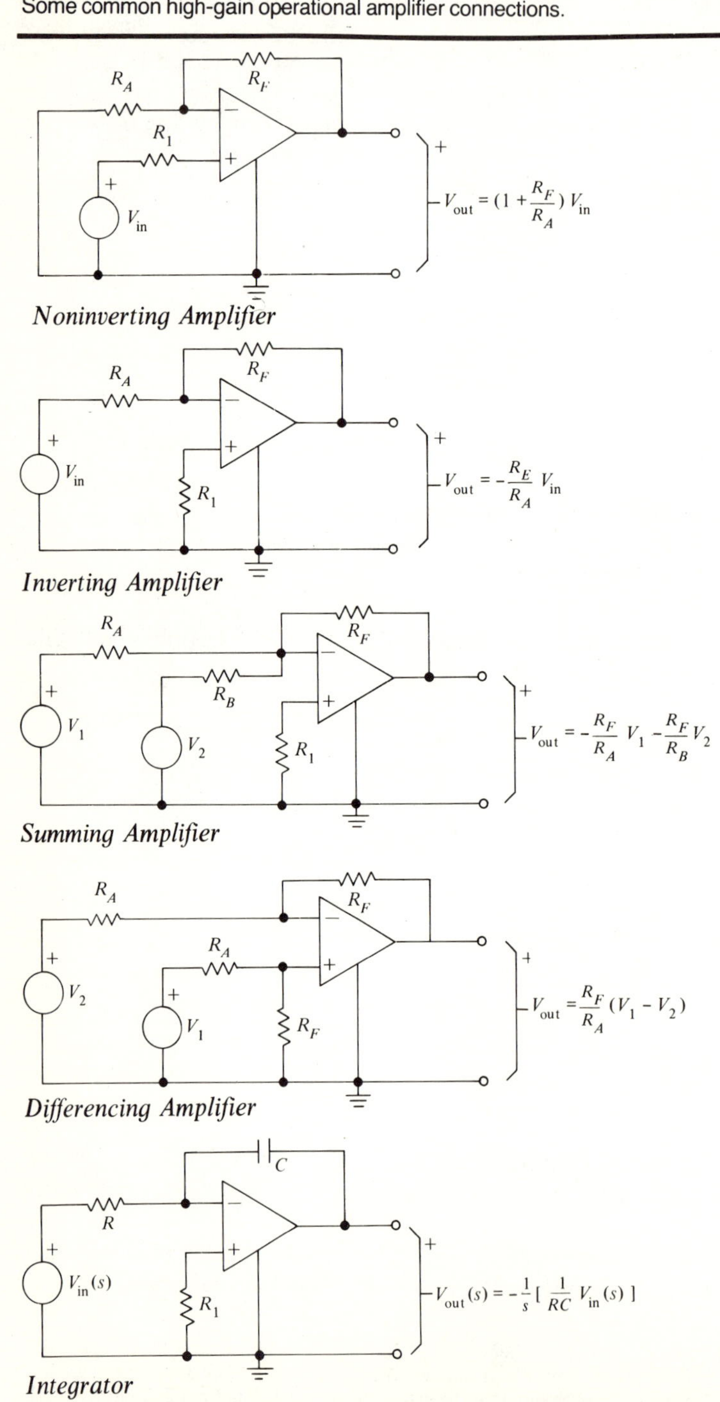

An operational amplifier's transmittance A is actually a function of s, and typical frequency response plots of $A(s)$ are shown in Fig. 6-47a. While the transmittance is large and nearly constant at low frequencies, there are three significant high-frequency breaks corresponding to three real poles of $A(s)$. One of these poles is associated with the transistor input circuitry, one with the transistor gains themselves, and one is due to capacitive effects at the output.

The loop transmittance of the feedback connection of Fig. 6-46 is

$$G(s)H(s) = A(s)\frac{R_A}{R_A + R_F}$$

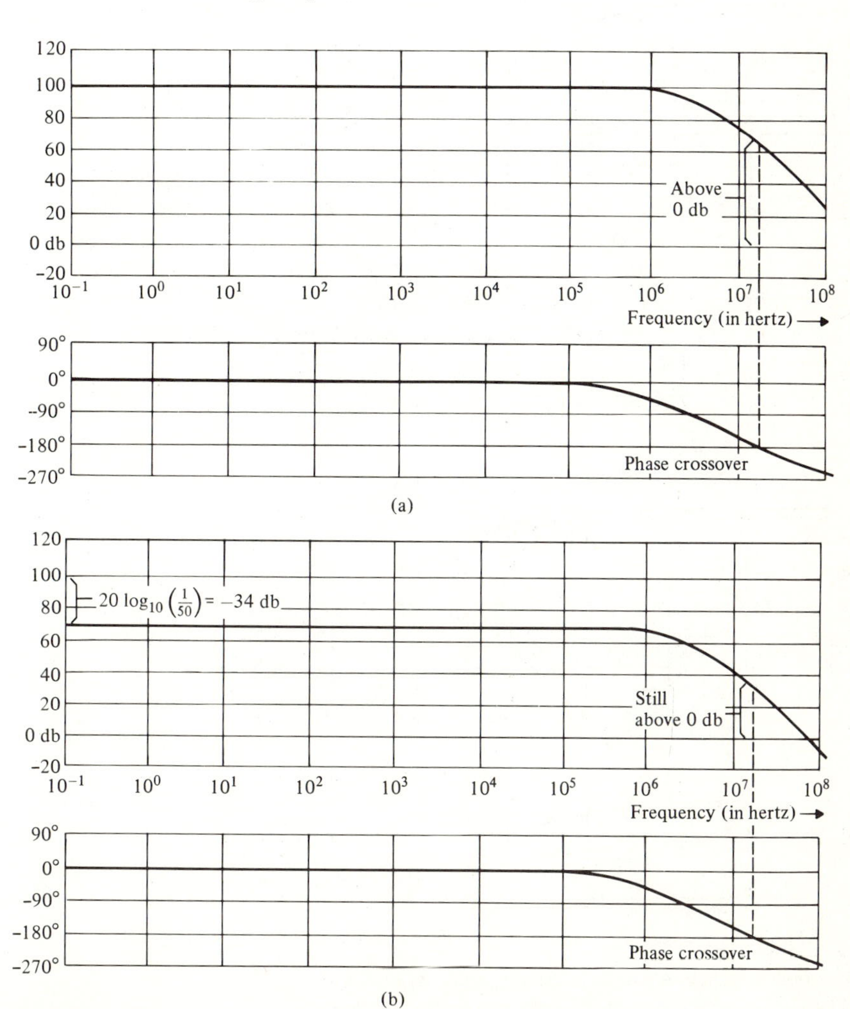

FIGURE 6-47
Operational amplifier frequency response. (a) Typical op amp transmittance $A(s)$. (b) Loop transmittance $A(s)R_A/(R_A + R_F)$ for a feedback factor $R_A/(R_A + R_F) = 1/50$.

which simply involves a downward shift of the amplitude curve for $A(s)$ by the dB represented by the factor $R_A/(R_A+R_F)$, as in Fig. 6-47b. As the feedback factor

$$K=\frac{R_A}{R_A+R_F}$$

is adjusted from a small value toward unity, there will be an amount of feed-

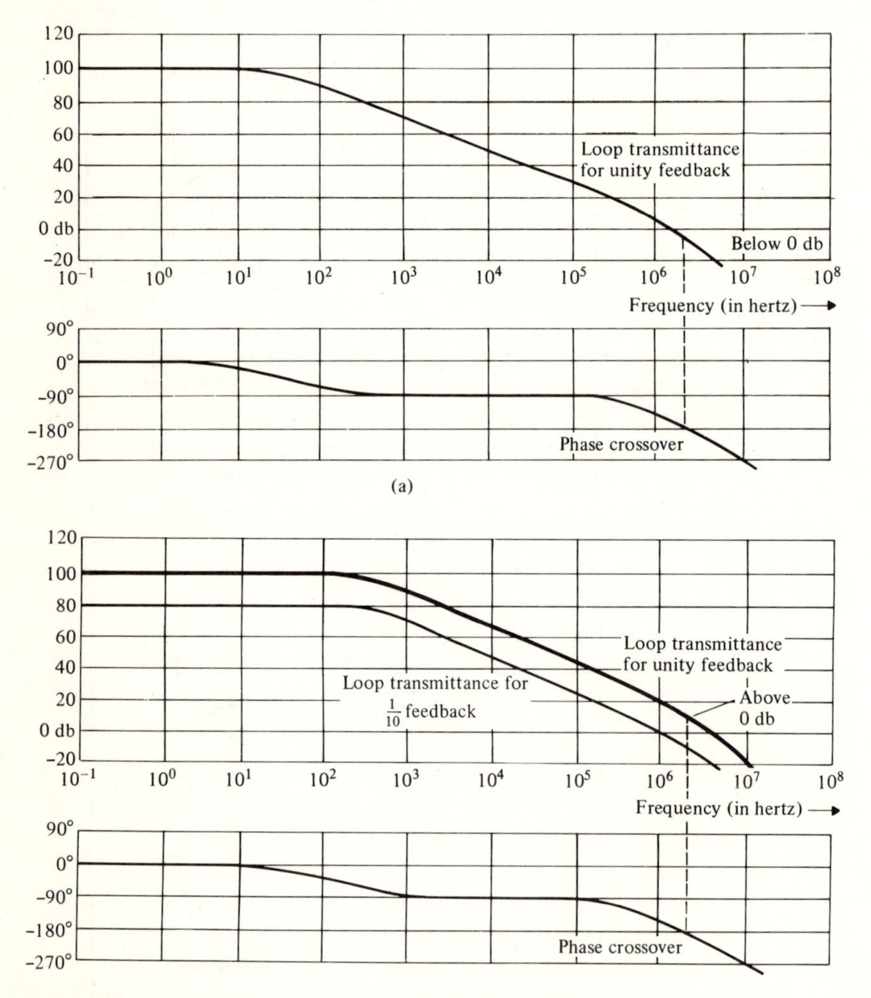

FIGURE 6-48
Open-loop frequency response of compensated operational amplifiers. (a) Open-loop frequency response with an additional pole at a sufficiently low break frequency. The closed-loop system is stable for any feedback ratio of unity or less. (b) An additional pole with a higher break frequency. The closed-loop system is stable only for a sufficiently small feedback ratio.

back for which the loop gain amplitude ratio is unity at the same frequency for which the phase shift is 180°. Hence for feedback factors greater than this amount, the op amp circuit will be unstable.

For general-purpose applications, it is desirable that the operational amplifier be stable for all feedback factors K from zero to unity. The worst case of smallest gain margin is for $K=1$, corresponding to $R_F=0$. For $K=1$, the noninverting configuration has unity overall gain at low frequencies. One approach to stabilization is to insert, with an RC network, an additional pole in $A(s)$ which will reduce the loop transmittance amplitude ratio to below 0 dB at the phase crossover frequency, as shown in Fig. 6-48a. To do so typically requires that the added break frequency be at a few hertz, which greatly reduces the high-frequency performance of the amplifier. The type 741 op amp contains such an internal RC network and is stable for all feedback factors including the worst case of unity feedback.

For feedback gain factors less than unity, the break frequency of the added pole may be increased as shown in Fig. 6-48b while still keeping the loop transmittance amplitude ratio below 0 dB at the phase crossover frequency. Stabilization of the type 101 op amp is done with an RC network for which the capacitor is connected externally. A 30-pF external capacitor will place the added break frequency in the same position as for the 741 op amp, making it unity-gain stable. A 3-pF capacitor will place the break frequency 10 times higher, giving better high-frequency response, but stability only for feedback factors of $\frac{1}{10}$ or less, corresponding to overall low-frequency gains of 10 or greater.

In more complicated compensation arrangements, both added zeros and poles are used to give sufficient gain margin while keeping the open-loop amplitude ratio large over a suitable range of frequency.

6.10 SUMMARY

When a transmittance $F(s)$ is driven with a sinusoidal signal, its forced response is also sinusoidal and of the same frequency. The ratio of sinusoidal output amplitude to input amplitude is the amplitude ratio

$$A(\omega)=|F(s=j\omega)|$$

as a function of radian frequency ω. The phase shift between input and output is

$$\Phi(\omega)=\angle F(s=j\omega)$$

At very small and at very large frequencies, the amplitude ratio of a rational transmittance is proportional to an integer power of ω and the phase shift is a multiple of 90°.

Bode plots are frequency response curves in a format which is especially convenient for rational transmittances. The amplitude ratio is plotted in decibels,

$$\text{dB} = 20 \log_{10} A(\omega)$$

and both amplitude ratio and phase shift are plotted on a logarithmic frequency scale. The frequency response contributions of real axis pole and zero terms are approximated well with straight-line segments, while the frequency response for complex conjugate pairs of roots may be constructed from standard, normalized curves.

A process for plotting dB and phase shift curves for complicated transmittances is the following:

1. Decompose the transmittance into simple factors.
2. Plot the dB and phase shift curves for each factor.
3. Add the individual dB curves to obtain the overall dB curve.
4. Add the individual phase shift curves to obtain the overall phase shift. Multiples of 360° may be added to or subtracted from the phase shift to keep that curve within a convenient range of angle.

Frequency response curves are sketched from the pole-zero plot of a transmittance by considering directed line segments from the poles and zeros to various points on the imaginary axis. The frequency response amplitude curve may be visualized as the height along the imaginary axis of a rubber sheet laid over the complex plane, pushed up by poles and tacked down by zeros.

Frequency response methods apply also to systems with irrational transmittances. One such system, of considerable practical importance, is the time delay

$$F(s) = e^{-s\tau}$$

where τ, the delay time, is a constant. For $F(s)$, the frequency response amplitude ratio is a uniform 0 dB. The phase shift is proportional to frequency:

$$\Phi(\omega) = -\tau\omega$$

A major advantage of frequency response methods, in addition to their applicability to systems with irrational transmittances, is that experimentally derived data is easily incorporated. Approximate transmittances may be determined from experimental data, or the data may be used directly in analysis and design.

If a simple feedback system has a loop transmittance $G(s)H(s)$ which is stable, and there are no frequencies (including $\omega = 0$) for which the frequency response for $KG(s)H(s)$ passes simultaneously through 0 dB and 180° for any K between 0 and 1, the overall system is stable. If there is a single such 0-dB, 180° frequency, the overall system is unstable.

For a stable system, a phase crossover frequency is any frequency at which the phase shift curve for $G(s)H(s)$ crosses 180°. The gain margin of a simple feedback system is the smallest additional amplitude of $G(s)H(s)$ necessary to give unity amplitude at a phase crossover frequency. A gain crossover frequency of a simple feedback system is any frequency at which the amplitude ratio for $G(s)H(s)$ crosses 0 dB. The phase margin of a stable system is the additional phase lag necessary to give 180° phase shift at a gain crossover frequency. If there is more than one gain crossover frequency, both the maximum and the minimum phase margins should be specified.

The Nichols chart is a graphical conversion between frequency response of the loop transmittance $G(s)$ and the transfer function

$$T(s)=\frac{G(s)}{1+G(s)}$$

for a unity feedback system.

A Nyquist plot consists of a curve on the complex plane representing the frequency response of the loop transmittance $G(s)H(s)$ of a simple feedback system. The Nyquist curve is a mapping of $G(s)H(s)$ for values of s along the closed curve from $-j\infty$ to $+j\infty$, then back to $-j\infty$ at large radius from the origin in the RHP. Small-radius RHP detours are made about any imaginary axis roots of $G(s)H(s)$. The number of RHP poles of the feedback system is equal to the algebraic number of clockwise encirclements of the point $s=-1$ by the Nyquist curve plus the number of RHP poles of $G(s)H(s)$. Overall system frequency response for a unity feedback system may be found from the Nyquist plot by superimposing the Nichols chart curves.

A simple compensation problem from the frequency response point of view was illustrated with the operational amplifier example. In this application, stability and wide bandwidth were desired. The gain margin approach fits this type of problem nicely because the design is directly linked to frequency response. For the simple single-pole compensation commonly employed, a great penalty in closed-loop bandwidth is traded for the required stability. If more complicated compensation is allowed, placing more poles and/or zeros in the open-loop transmittance, the open-loop (and thus the closed-loop) frequency response may be more carefully tailored, giving stability together with much wider bandwidth.

REFERENCES

Frequency Response Methods

Bode, H. W. *Network Analysis and Feedback Amplifier Design*. Princeton, N.J.: Van Nostrand, 1945.

James, H. M.; Nichols, N. B.; and Phillips, R. S. *Theory of Servomechanisms*. New York: McGraw-Hill, 1947.

Nyquist, H. "Regeneration Theory." *Bell Syst. Tech. J.*, January 1932, pp. 126–147.

Savant, C. J., Jr. *Control System Design*. New York: McGraw-Hill, 1964.

Operational Amplifiers

Giles, J. N. *Fairchild Semiconductor Linear Integrated Circuits Applications Handbook.* Mountain View, California: Fairchild Semiconductor Company, 1967.

Huelsman, L. P. *Theory and Design of Active R.C. Networks.* New York: McGraw-Hill, 1966.

Stout, D. F. and Kaufman, M. *Handbook of Operational Amplifiers.* 2nd ed. London: Butterworths, 1979.

Tobey, G. E.; Graeme, J. G.; and Huelsman, L. P. *Operational Amplifiers—Design and Applications.* New York: McGraw-Hill, 1971.

PROBLEMS

1. For the following transmittances, find the forced sinusoidal response to the indicated input signals:

(a) $F(s)=\dfrac{1}{s+10}$

$r(t)=\cos 5t$

ans. $0.09\cos(5t-27^\circ)$

(b) $F(s)=\dfrac{s^2}{s+2}$

$r(t)=100\cos(2t+40^\circ)$

ans. $(100/\sqrt{2})\cos(2t+175^\circ)$

(c) $F(s)=\dfrac{10}{s^2+2s+5}$

$r(t)=4\cos(2t-70^\circ)$

ans. $9.7\cos(2t-146^\circ)$

(d) $F(s)=\dfrac{s}{(s+3)(s+5)}$

$r(t)=10\cos(4t+120^\circ)$

ans. $1.25\cos(4t+118.2^\circ)$

2. Sketch frequency response curves (both amplitude in dB and phase shift) for the following transmittances:

(a) $F(s)=\dfrac{100}{s^4}$

(b) $F(s)=\dfrac{s}{s+10}$

(c) $F(s)=\dfrac{10(s+10)}{s+1}$

(d) $F(s)=\dfrac{-10}{s(s+1)}$

(e) $F(s)=\dfrac{s+1}{s(s+100)}$

3. Sketch frequency response curves (both amplitude in dB and phase shift) for the following transmittances:

(a) $F(s)=\dfrac{1}{s^2+20s+100}$

(b) $F(s)=\dfrac{1}{3s(s^2+s+4)}$

(c) $F(s)=\dfrac{s^2-s+10}{s^2+s+10}$

(d) $F(s)=\dfrac{s+100}{s^2+100s+100}$

(e) $F(s)=\dfrac{4s-1}{(s^2+2s+16)^2}$

4. Sketch approximate frequency response curves (both amplitude and phase shift) for functions with the pole-zero plots of Fig. P6-1.

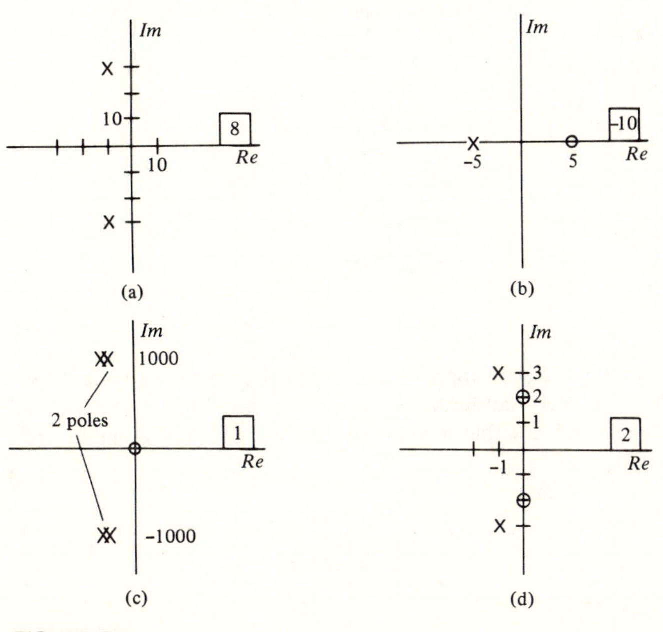

FIGURE P6-1

5. Draw Bode plots for the transmittances

$$T(s)=\frac{E_o(s)}{E_i(s)}$$

of the electrical networks of Fig. P6-2.

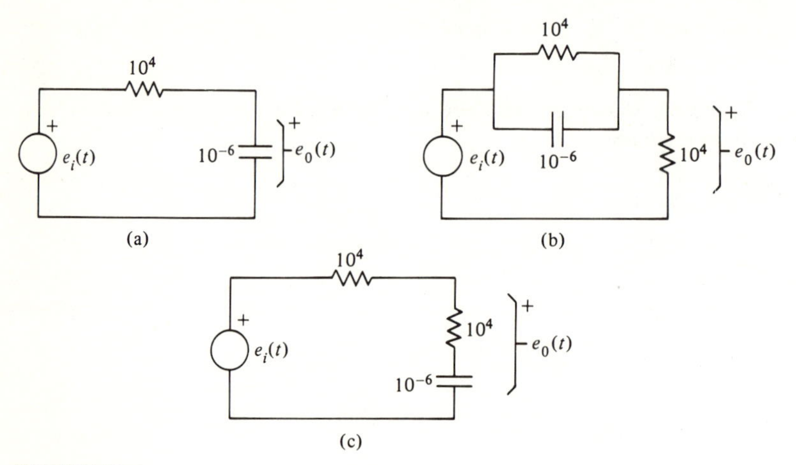

FIGURE P6-2

6. Draw Bode plots for the indicated transmittances of each of the mechanical networks of Fig. P6-3.

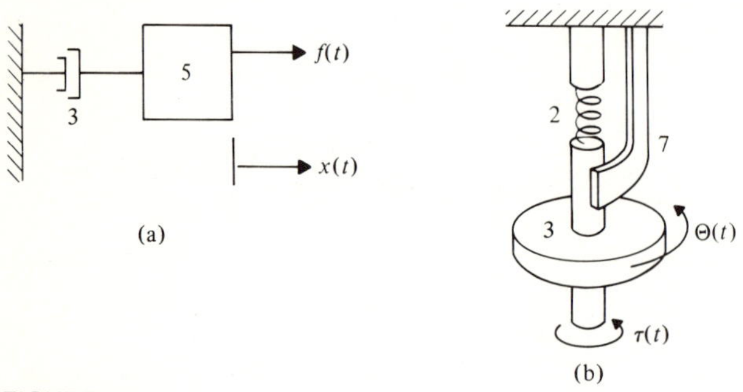

FIGURE P6-3

7. Find transmittances which have the approximate amplitude response curve as shown in Fig. P6-4. A *minimum-phase* transmittance has all poles and zeros in the left half of the complex plane. Find a stable minimum-phase solution and a stable non-minimum-phase solution.

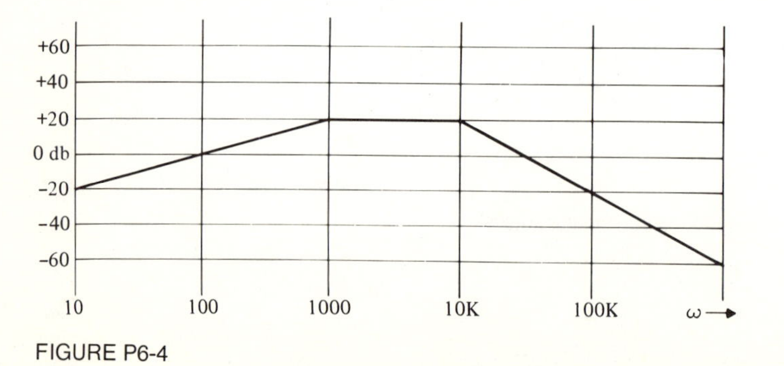

FIGURE P6-4

8. Find a stable transmittance which has the approximate frequency response phase shift shown in Fig. P6-5.

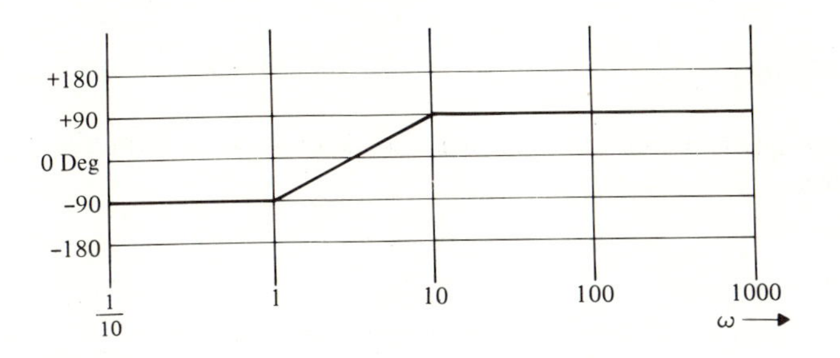

FIGURE P6-5

9. Experimental frequency response data for the forward transmittances $G(s)$ of unity feedback systems are given below. Find, approximately, the overall system transfer function $T(s)$.

(a)

Radian Frequency ω	$G(s=j\omega)$
0.1	$0-j20$
0.2	$-1.3-j10$
0.4	$-0.8-j5$
1.0	$-0.6-j1.7$
2.0	$-0.4-j0.6$
4.0	$-0.2-j0.2$
8.0	$-0.05-j0.008$
16.0	$-0.008+j0.003$

(b)

Frequency f (Hz)	*Amplitude ratio* $\lvert G\rvert$ *(dB)*	*Phase shift* $\angle G$ *(deg)*
120	−7.8	−165
100	−4.3	−160
80	−0.2	−160
70	0.75	−150
50	5.2	−140
40	8.0	−130
32	10.5	−125
20	15.0	−90
14	16.9	−55
5	20.4	−40
2.5	22.0	−30
0.5	24.0	−10
0	24.0	0

10. Find and sketch the frequency response (both amplitude ratio and phase shift) for the following irrational transmittances:

    (a) $F(s)=\sqrt[3]{s}$
    (b) $F(s)=se^{-3s}$
    (c) $F(s)=1+e^{-2s}$

11. Find gain margins and phase margins (if they exist) for feedback systems with the following GH functions:

    (a) $G(s)H(s)=\dfrac{100}{(s+10)^3}$

    ans. 38 dB; phase may be changed by any amount

    (b) $G(s)H(s)=\dfrac{1000}{s(s+10)^3}$

    (c) $G(s)H(s)=\dfrac{27s}{(s+3)^3}$

    (d) $G(s)H(s)=\dfrac{10^7(s+1)^2(s+100)^2}{(s+10)^4}$

    ans. gain margin is infinite

    (e) $G(s)H(s)=\dfrac{e^{-0.2s}}{(s+10)^2}$

12. For the system of Fig. P6-6, use frequency response methods to determine values of $K>0$, if any, which result in marginal stability of the overall system.

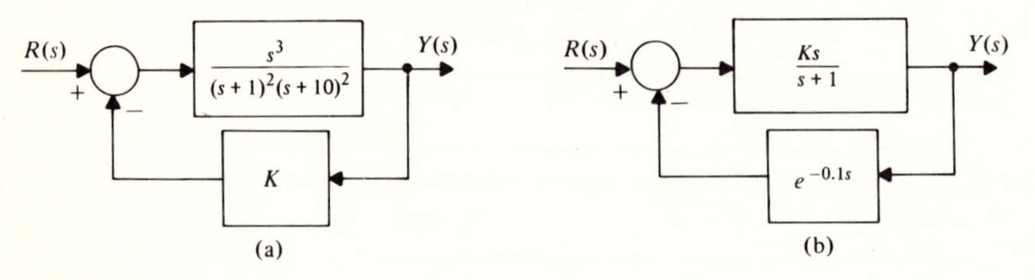

FIGURE P6-6

13. Use frequency response methods to find the ranges of the positive constant K (if any) for which each of the systems of Fig. P6-7 is stable.

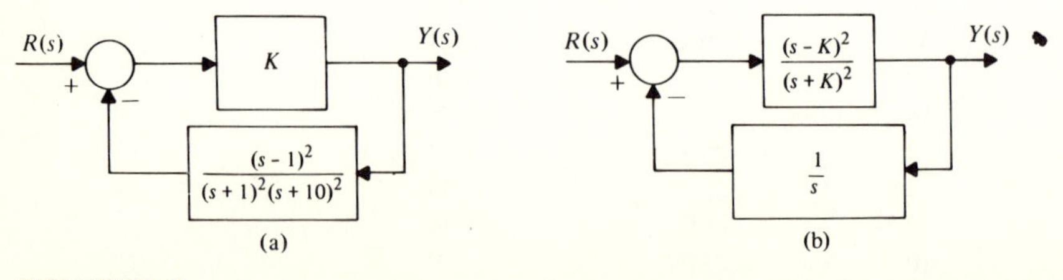

FIGURE P6-7

14. Unity feedback systems have the $G(s)$ functions with frequency response given in Fig. P6-8. Use the Nichols chart or program a calculator to develop plots of the frequency responses of the overall systems.

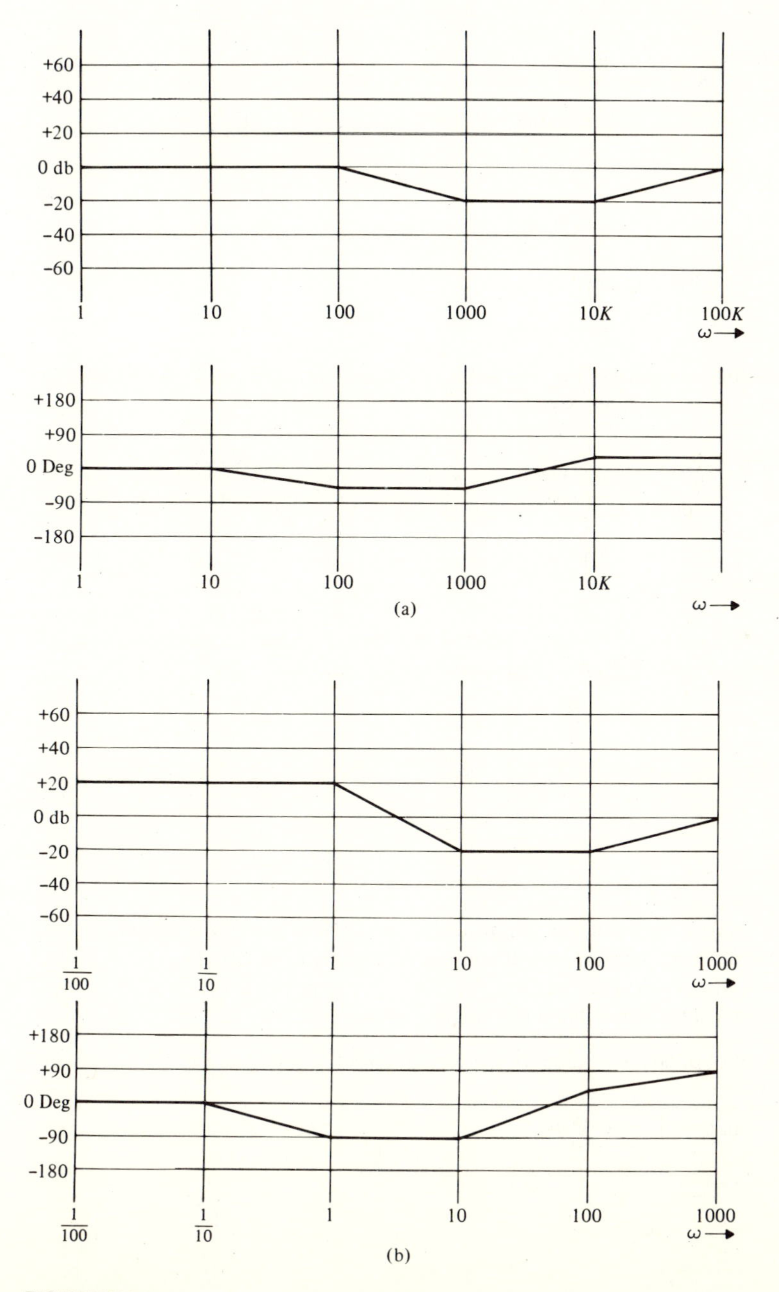

FIGURE P6-8

15. Sketch Nyquist plots for feedback systems with the following loop transmittances, and then use the plots to determine whether or not each system is stable.

(a) $G(s)H(s)=\dfrac{s+2}{(s+1)(s+3)}$

(b) $G(s)H(s)=\dfrac{10}{s(s+2)(s+5)}$

ans. stable

(c) $G(s)H(s)=\dfrac{4}{(s^2+1)(s+2)}$

(d) $G(s)H(s)=\dfrac{s^2+9}{s^2(s^2+4s+4)}$

ans. unstable

16. For the feedback systems of Fig. P6-9, sketch Nyquist plots and use them to determine whether or not each system is stable.

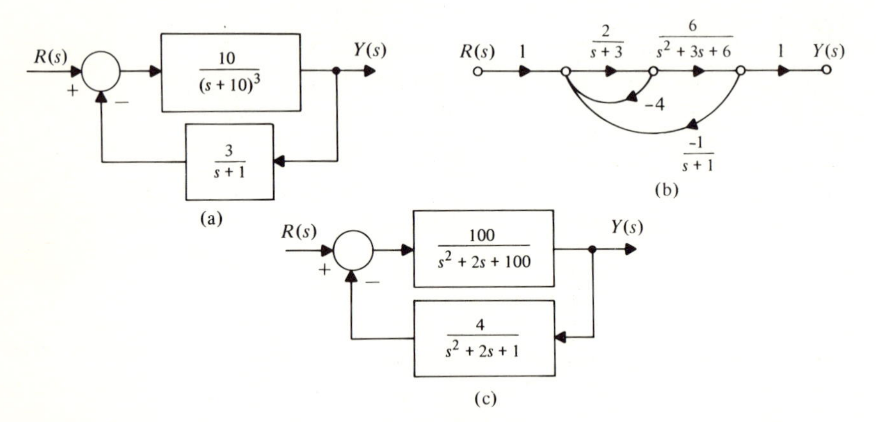

FIGURE P6-9

17. Write a complete digital computer program to tabulate points for Nyquist plots. After input of the numerator and denominator coefficients of $G(s)H(s)$, the step size, and finite radius of closure to be used, the program is to print a succession of complex number coordinates of points on the Nyquist curve.

18. For what range of positive values of the constant a is the system of Fig. P6-10 stable?

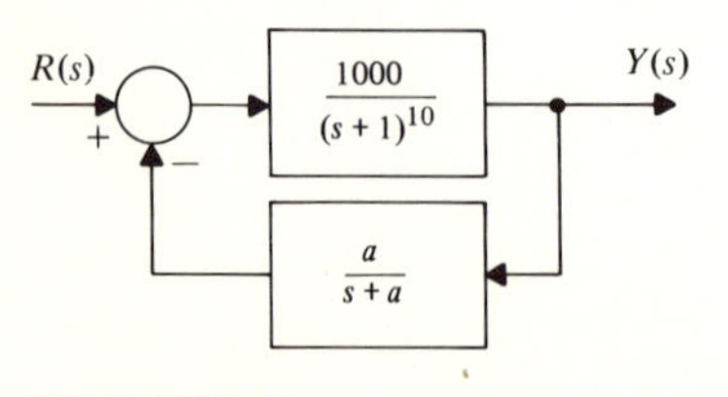

FIGURE P6-10

Sophisticated testing of blood samples is automated in this research laboratory. A digital computer guides the sequence of tests which is varied with the outcome of previous test results. (Photo courtesy of Tektronix, Incorporated.)

19. For the system of Fig. P6-11, how large can the constant K be for each positive integer value of n if the overall system is to be stable?

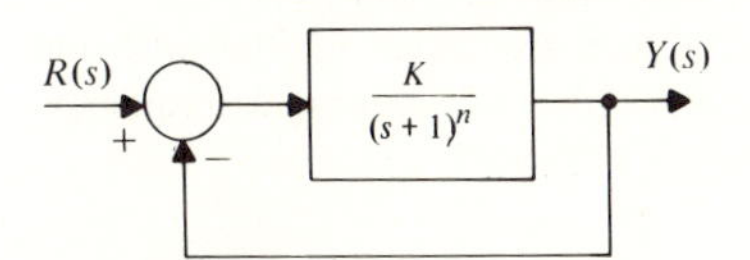

FIGURE P6-11

20. Carefully describe a method for measuring the frequency response of a control system component with an *unstable* transmittance by connecting it as part of a feedback system.

21. Find first-, second-, and third-order rational functions which approximate the desired "low-pass" amplitude function given in Fig. P6-12.

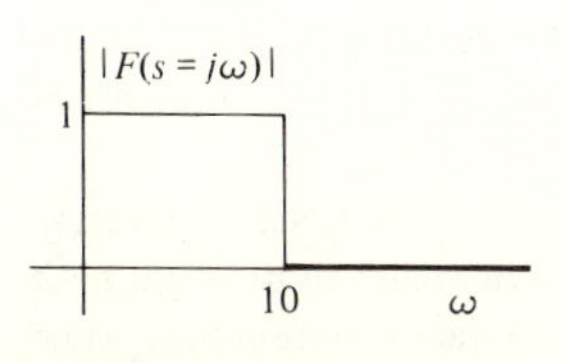

FIGURE P6-12

22. Find the poles and zeros of the function

$$F(s)=\frac{1-e^{-s}}{1+e^{-s}}$$

and sketch its frequency response. This function has the form of the impedance of a short-circuited transmission line.

23. Using the expansion

$$e^{-s}=\frac{1}{1+s+s^2/2!+s^3/3!+\ldots+s^n/n!+\ldots}$$

develop a *rational* approximation for the transmittance of a 1-sec time delay. Plot the frequency response of your approximation and compare with the frequency response for the time delay.

24. For a type 741 operational amplifier with open-loop frequency response given by Fig. 6-48a, find and draw the frequency response of the closed-loop noninverting configuration with feedback factor

$$\frac{R_A}{R_A+R_F}=\frac{1}{100}$$

25. Two improved operational amplifier compensation arrangements are termed "two-pole" and "feedforward." In two-pole compensation, two additional poles are placed in the open-loop transmittance $A(s)$, allowing higher break frequencies than the single-pole arrangement, thus giving high gain over a wider range of frequency.

For feedforward compensation, a pole and a zero are added to $A(s)$, with the zero placed to cancel or nearly cancel the pole in $A(s)$ with the lowest break frequency.

For an operational amplifier similar to the type 101 with

$$A(s)=\frac{10^{27}}{(s+10^7)(s+3\times 10^7)(s+10^8)}$$

design

(a) two-pole compensation

and

(b) feedforward compensation

such that the first open-loop break frequency is relatively high but the noninverting amplifier configuration is unity gain stable.

For each design, specify the compensator pole and zero locations and the frequency at which the open-loop transmittance is $\frac{1}{1000}$ its value of 10^5 at low frequencies.

26. A block diagram model of a chemical process temperature control system is given in Fig. P6-13. Because the temperature sensor is located downstream in the fluid path from the heater, an appreciable time delay is involved. Find the maximum delay time K for which this system is stable.

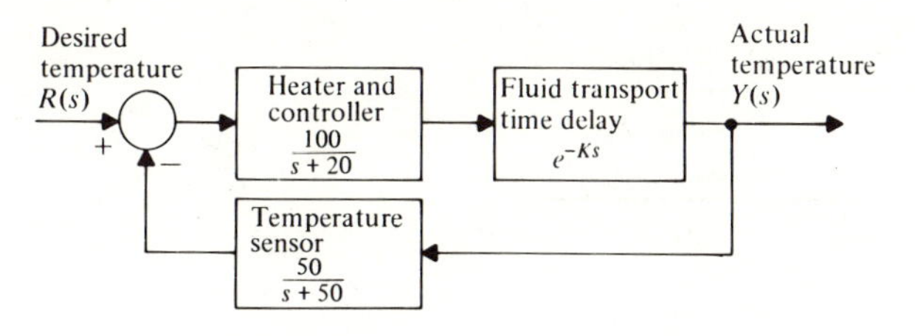

FIGURE P6-13

27. An aircraft heading control system is diagramed in Fig. P6-14. Use frequency response methods to determine a suitable value of "pilot gain" K.

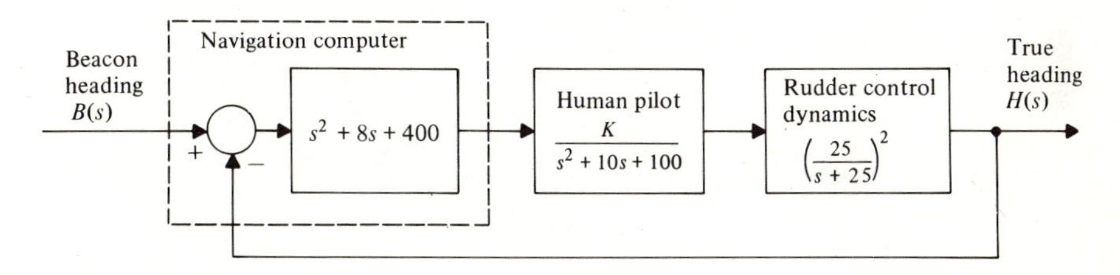

FIGURE P6-14

28. A commercial videotape transport positioning system is modeled in Fig. P6-15. Use a Nyquist plot to determine the maximum value of $K > 0$ for which the system is stable. Then choose a value of K for good system performance.

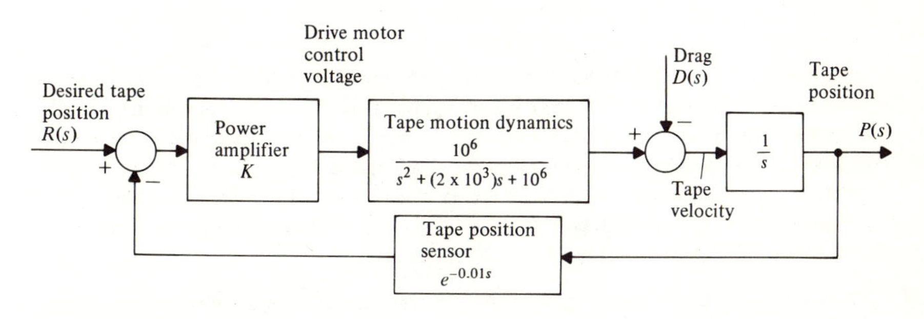

FIGURE P6-15

29. Experimental frequency response data for the near field of a loudspeaker in an auditorium are given in Table 6-5. One method of improving the frequency response (making the amplitude ratio more nearly constant) is to mount a velocity-sensing coil on the loudspeaker cone and feed back that signal through the amplifier as diagramed in Fig. P6-16. Plot the frequency response of this system both with and without the feedback connected.

TABLE 6-5.
Experimental frequency response data for a loudspeaker.

Frequency (*Hz*)	*Amplitude Ratio* (*dB*)	*Phase Shift* (*deg*)
40	36	−22
70	40	−20
100	45	−24
150	46	−32
200	42	−40
300	40	−45
400	46	−50
550	40	−40
900	34	−46
1200	39	−55
3000	35	−62
4000	30	−70
5000	27	−78
6000	20	−85

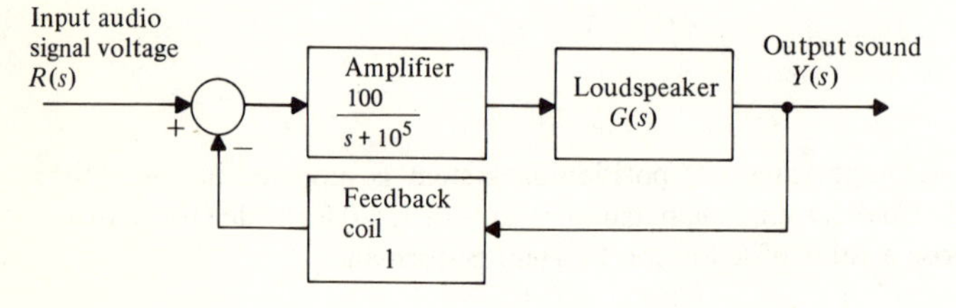

FIGURE P6-16

The open-loop system has a frequency response amplitude ratio which varies from 20 to 46 dB over the range of 40–6000 Hz. Its response is thus within ± 13 dB of being "flat" (uniform) in that frequency range. Such large variations in public address system frequency response are not at all unusual in motion picture theaters, sports arenas, and even home systems when the loudspeaker response and room acoustics are taken into account. Give a similar frequency response amplitude specification for the closed-loop system.

30. A numerical technique for approximating the derivative of a function of time is to take the difference between the function at time t and at time $t-\Delta t$, where Δt is small:

$$y(t)=\frac{r(t)-r(t-\Delta t)}{\Delta t}\cong\frac{dr}{dt}$$

This system has transfer function

$$T(s)=\left.\frac{Y(s)}{R(s)}\right|_{\substack{\text{zero initial}\\ \text{conditions}}}=\frac{1}{\Delta t}\left[1-e^{-\Delta ts}\right]$$

(a) Plot the frequency response of this system and compare with the frequency response of a true differentiator. For what class of signals does $T(s)$ give a good approximation to the time derivative of the incoming signal? For what class of signals is $T(s)$ poor as an approximate differentiator?

(b) Design a better approximate differentiator, of the form

$$y(t)=ar(t)+br(t-\Delta t)+cr(t-2\Delta t)$$

by choosing appropriate constants a, b, and c.

seven

STATE VARIABLE DESCRIPTIONS OF CONTINUOUS-TIME SYSTEMS

7.1 PREVIEW

In the early 1960s, the space program brought forth systems which were extremely complicated, and *state variable* methods, nowadays called "modern" control, received a great deal of emphasis. State variable (or *state space*) descriptions give a standard equation arrangement which offers economy of notation and ease of data entry into digital computer processing and is well suited for extensions to nonlinear and time-varying systems.

This chapter begins with the development of methods for synthesizing transfer functions. Simulation diagrams involving integrators are constructed and Mason's gain rule is employed to give solutions which are simply related to the transfer function coefficients.

In Sec. 7.3, the mathematical relations represented by simulation diagrams are expressed as a set of simultaneous state equations, in terms of state variables, and a set of output equations. These are written compactly in matrix notation, where the system input, outputs, and state variables are each arranged as vectors. The matrix operations involved in systematically calculating the system transfer functions from the state and output equations are developed. It is shown that all transfer functions of a system share a common characteristic denominator polynomial, the characteristic polynomial of the state coupling matrix.

A nonsingular change of state variables leaves the input-output relations of a system unchanged; it is its internal description which is different. Decoupling of the state equations from one another through such a change of variables is termed diagonalization of the system. The mathematical study of diagonalization is known as the characteristic value problem. Here we emphasize the easily visualized approach of partial fractions. Complex conjugate and repeated characteristic root terms are treated in detail.

Section 7.5 concerns the possibilities that certain systems may not be completely controlled from their inputs or that it may not be possible to determine their behavior from the available outputs. These topics, known as system controllability and observability, are very important to a wide range of applications, particularly those which are involved and of high order.

System time-domain response is discussed in Sec. 7.6, where the state transition matrix is introduced. Introductory computational considerations are given in Sec. 7.7, including sample Fortran and Basic programs for numerical response calculation.

Important aspects of system design using state variables are covered in Sec. 7.8 and illustrated with a control system design example, a magnetically levitated train, in Sec. 7.9.

7.2 SIMULATION DIAGRAMS

7.2.1 Phase-Variable Form

An important problem in control system design is the synthesis of specific transmittances through the interconnection of simple components, as is needed for many of the controllers (or compensators) of the previous chapters. Synthesis is important also in the simulation of systems, where system behavior is predicted from a model governed by equivalent equations. Above all, the viewpoint of synthesis leads to fundamental techniques for system description, analysis, and design. These methods are systematic, compact, and suitable for computer analysis. They are also extendable to nonlinear and time-varying systems.

A basic component for synthesis is the integrator, a block or branch having transmittance $1/s$. A block diagram or signal flow graph composed only of constant transmittances and integrators is termed a *simulation diagram*. The order of such a system is simply the number of integrators present. Signal flow graphs are especially convenient for representing simulation diagrams because in many cases, system transfer functions are evident by inspection, using Mason's gain rule.

A transfer function that is the ratio of two polynomials in s is termed rational. If the numerator degree is less than the denominator degree, the transfer function is said to be *proper*. Any proper rational transfer function may be synthesized with a simulation diagram, that is, using only integration, multiplication by a constant, and summation operations. One very useful synthesis arrangement, known as the *phase-variable* form, is described below. The development, which is in terms of a specific numerical example for clarity, is applicable to any proper rational transfer function.

For the transfer function

$$T(s)=\frac{-5s^2+4s-12}{s^3+6s^2+s+3}=\frac{-\dfrac{5}{s}+\dfrac{4}{s^2}+\dfrac{-12}{s^3}}{1+\dfrac{6}{s}+\dfrac{1}{s^2}+\dfrac{3}{s^3}}$$

dividing the numerator and denominator by the highest power of s term in the denominator places a 1 in the denominator and results in other numerator and denominator terms which are inverse powers of s, representing multiple integrations. In this form, the transfer function may be interpreted as a Mason's gain rule expression. The numerator terms

$$\frac{-5}{s}+\frac{4}{s^2}+\frac{-12}{s^3}$$

are each taken to be paths through integrators and the paths are intermingled as in Fig. 7-1a so as to require a minimum number of integrators—in this case,

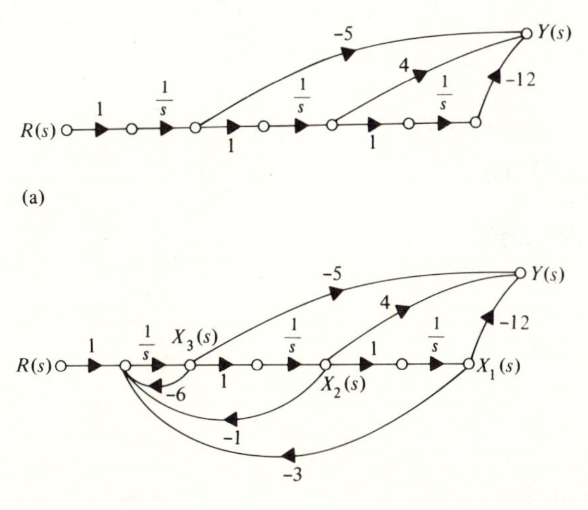

FIGURE 7-1
Phase-variable systhesis of a single-input, single-output system.
(a) Paths in the simulation diagram. (b) Complete simulation diagram.

three. The denominator terms

$$\frac{6}{s}+\frac{1}{s^2}+\frac{3}{s^3}$$

are taken to be loop gains. By placing each of these loops through the node to which $R(s)$ couples, all loops touch one another, so no product of loop gain terms is involved. All the loops touch each of the paths, so each path cofactor is unity.

In Fig. 7-1b, each integrator output signal has been labeled. These signals are termed the *state variables* of the system. This realization of the example transfer function is then described by the following Laplace-transformed equations:

$$X_1(s)=\frac{1}{s}X_2(s)$$

$$X_2(s)=\frac{1}{s}X_3(s)$$

$$X_3(s)=\frac{1}{s}[-3X_1(s)-X_2(s)-6X_3(s)+R(s)]$$

$$Y(s)=-12X_1(s)+4X_2(s)-5X_3(s)$$

Or

$$sX_1(s)=X_2(s)$$
$$sX_2(s)=X_3(s)$$
$$sX_3(s)=-3X_1(s)-X_2(s)-6X_3(s)+R(s)$$
$$Y(s)=-12X_1+4X_2(s)-5X_3(s)$$

As functions of time, the signals satisfy

$$\frac{dx_1}{dt}=x_2(t)$$

$$\frac{dx_2}{dt}=x_3(t)$$

$$\frac{dx_3}{dt}=-3x_1(t)-x_2(t)-6x_3(t)+r(t)$$

$$y(t)=-12x_1(t)+4x_2(t)-5x_3(t)$$

which is a set of coupled first-order differential equations.

7.2.2 Dual Phase-Variable Form

Another especially convenient way to synthesize a transfer function with integrators is to arrange the signal flow graph so that all of the paths and all of the loops touch an output node. For the previous transfer function

$$T(s)=\frac{-5s^2+4s-12}{s^3+6s^2+s+3}$$

$$=\frac{-\frac{5}{s}+\frac{4}{s^2}-\frac{12}{s^3}}{1+\frac{6}{s}+\frac{1}{s^2}+\frac{3}{s^3}}$$

for example, the diagram of Fig. 7-2 shows this *dual phase-variable* arrangement. The output signal is derived from a single node, while the input signal is coupled to each integrator.

The Laplace transform relations describing this system are, in terms of the indicated state variables,

$$sX_1(s)=-6X_1(s)+X_2(s)-5R(s)$$
$$sX_2(s)=-X_1(s)+X_3(s)+4R(s)$$
$$sX_3(s)=-3X_1-12R(s)$$
$$Y(s)=X_1(s)$$

As functions of time, the signals satisfy

$$\frac{dx_1}{dt}=-6x_1(t)+x_2(t)-5r(t)$$

$$\frac{dx_2}{dt}=-x_1(t)+x_3(t)+4r(t)$$

$$\frac{dx_3}{dt}=-3x_1(t)-12r(t)$$

$$y(t)=x_1(t)$$

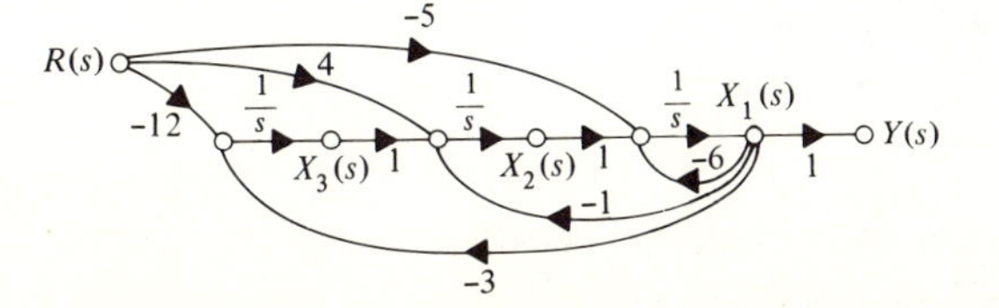

FIGURE 7-2
Synthesizing a transfer function in the dual phase-variable form.

7.2.3 Multiple Outputs and Inputs

Additional system outputs may be easily derived from the phase-variable arrangement. For example, the single-input, two-output system of Fig. 7-3a has the following transfer functions:

$$T_{11}(s) = \left.\frac{Y_1(s)}{R(s)}\right|_{\text{initial conditions}=0} = \frac{-5/s + 4/s^2 + (-12/s^3)}{1 + 6/s + 1/s^2 + 3/s^3}$$

$$= \frac{-5s^2 + 4s - 12}{s^3 + 6s^2 + s + 3}$$

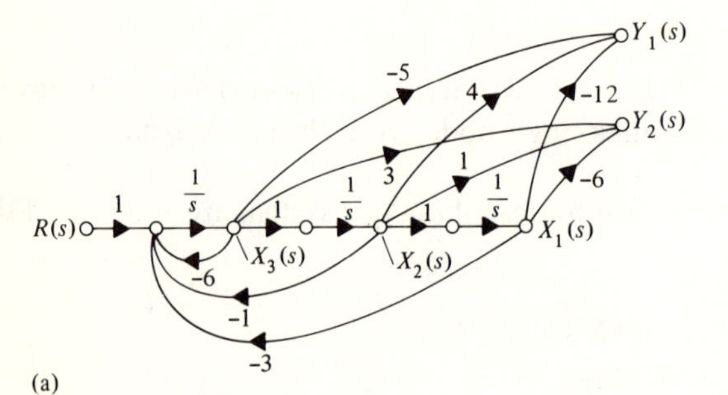

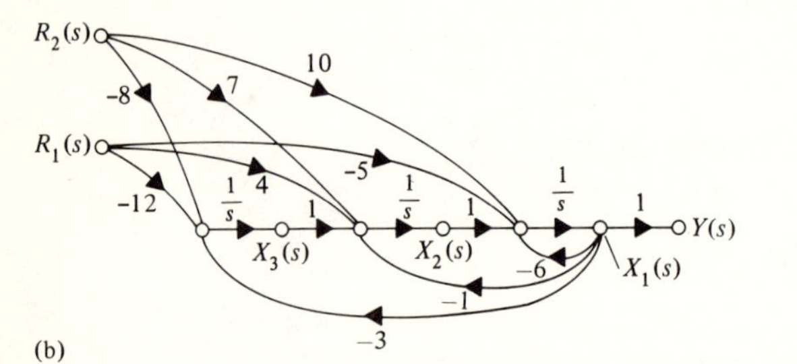

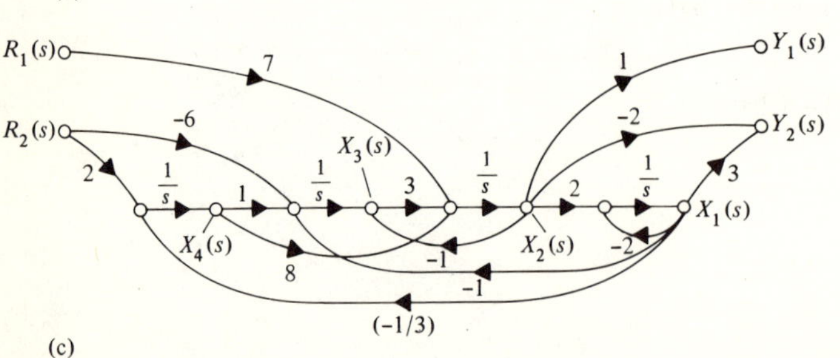

FIGURE 7-3
Achieving multiple outputs and multiple inputs. (a) Multiple outputs from the phase-variable arrangement. (b) Multiple inputs with the dual phase-variable arrangement. (c) System with both multiple inputs and multiple outputs.

$$T_{21}(s)=\left.\frac{Y_2(s)}{R(s)}\right|_{\text{initial conditions}=0}=\frac{3/s+1/s^2+(-6/s^3)}{1+6/s+1/s^2+3/s^3}$$
$$=\frac{3s^2+s-6}{s^3+6s^2+s+3}$$

Additional inputs are easily added to the dual phase-variable arrangement. The example two-input, single-output system of Fig. 7-3b has the following transfer functions:

$$T_{11}(s)=\left.\frac{Y(s)}{R_1(s)}\right|_{\substack{\text{initial}\\ \text{conditions}\\ \text{and } R_2=0}}=\frac{-5/s+4/s^2-12/s^3}{1+6/s+1/s^2+3/s^3}$$
$$=\frac{-5s^2+4s-12}{s^3+6s^2+s+3}$$

$$T_{12}(s)=\left.\frac{Y(s)}{R_2(s)}\right|_{\substack{\text{initial}\\ \text{conditions}\\ \text{and } R_1=0}}=\frac{10/s+7/s^2-8/s^3}{1+6/s+1/s^2+3/s^3}$$
$$=\frac{10s^2+7s-8}{s^3+6s^2+s+3}$$

The system described by the simulation diagram of Fig. 7-3c is neither in phase-variable nor dual phase-variable form. Its two inputs and two outputs are governed by the following Laplace-transformed equations:

$$sX_1(s)=-2X_1(s)+2X_2(s)$$
$$sX_2(s)=-3x_2(s)+3X_3(s)+8X_4(s)+7R_1(s)$$
$$sX_3(s)=-X_1(s)+X_4(s)-6R_2(s)$$
$$sX_4(s)=-\tfrac{1}{3}X_1(s)+2R_2(s)$$
$$Y_1(s)=X_2(s)$$
$$Y_2(s)=3X_1(s)-2X_2(s)$$

Applying Mason's gain rule to the simulation diagram, the four transfer functions describing this system are:

$$T_{11}(s)=\left.\frac{Y_1(s)}{R_1(s)}\right|_{\substack{\text{initial}\\ \text{conditions}\\ \text{and } R_2=0}}=\frac{(7/s)(1+2/s)}{1-(-2/s-3/s^2-6/s^3-2/s^4)+(-2/s)(-3/s)}$$
$$=\frac{7s^3+14s^2}{s^4+5s^3+6s^2+6s+2}$$

$$T_{12}(s)=\left.\frac{Y_1(s)}{R_2(s)}\right|_{\substack{\text{initial}\\ \text{conditions}\\ \text{and } R_1=0}}=\frac{(-2/s^2)(1+2/s)+(6/s^3)(1+2/s)}{1-(-2/s-3/s^2-6/s^3-2/s^4)+(-2/s)(-3/s)}$$
$$=\frac{-2s^2+2s+12}{s^4+5s^3+6s^2+6s+2}$$

$$T_{21}(s)=\left.\frac{Y_2(s)}{R_1(s)}\right|_{\substack{\text{initial}\\ \text{conditions}\\ \text{and } R_2=0}}=\frac{(-14/s)(1+2/s)+42/s^2}{1-(-2/s-3/s^2-6/s^3-2/s^4)+(-2/s)(-3/s)}$$

$$=\frac{-14s^3+14s^2}{s^4+5s^3+6s^2+6s+2}$$

$$T_{22}(s)=\left.\frac{Y_2(s)}{R_2(s)}\right|_{\substack{\text{initial}\\ \text{conditions}\\ \text{and } R_1=0}}$$

$$=\frac{(4/s^2)(1+2/s)-12/s^3(1+2/s)-12/s^3+36/s^4}{1-(-2/s-3/s^2-6/s^3-2/s^4)+(-2/s)(-3/s)}$$

$$=\frac{4s^2-16s+12}{s^4+5s^3+6s^2+6s+2}$$

DRILL PROBLEM

D7-1. Draw simulation diagrams in either the phase-variable or the dual phase-variable form for systems with the following transfer functions:

(a) $T(s)=\dfrac{-4s+3}{s^2+6s+2}$

(b) $T(s)=\dfrac{-s^2+5s+9}{3s^3+2s^2+4s+1}$

(c) $T_{11}(s)=\dfrac{0.4s^2+1.4s+0.8}{s^3+0.3s^2+1.7s+0.2}$

$T_{12}(s)=\dfrac{-0.5s^2+0.7s-1.9}{s^3+0.3s^2+1.7s+0.2}$

(d) $T_{11}(s)=\dfrac{4s^2-1}{s^3+6s^2+2s+5}$

$T_{21}(s)=\dfrac{3s+6}{s^3+6s^2+2s+5}$

7.3 STATE REPRESENTATIONS OF SYSTEMS

7.3.1 State Variable Equations

Phase-variable form is particularly convenient for the synthesis of single- and multiple-output systems, while in dual phase-variable form, single- and multiple-input systems are easily arranged. There are a whole spectrum of other ways of connecting integrators to achieve systems with desired transfer functions, including systems with both multiple inputs and multiple outputs. Moreover, the representation of systems in terms of integrators is useful not only for transfer function synthesis, but for the description of systems of all kinds, particularly those that are very complicated, for which a standard, compact notation is especially helpful.

A general state variable description of an nth-order system involves n integrators, the outputs of which are the state variables. The inputs of each of the integrators is driven with a linear combination of the state signals and the inputs:

$$\begin{aligned}
sX_1(s) &= a_{11}X_1(s) + a_{12}X_2(s) + \cdots + a_{1n}X_n(s) + b_{11}R_1(s) + \cdots + b_{1i}R_i(s) \\
sX_2(s) &= a_{21}X_1(s) + a_{22}X_2(s) + \cdots + a_{2n}X_n(s) + b_{21}R_1(s) + \cdots + b_{2i}R_i(s) \\
&\vdots \\
sX_n(s) &= a_{n1}X_1(s) + a_{n2}X_2(s) + \cdots + a_{nn}X_n(s) + b_{n1}R_1(s) + \cdots + b_{ni}R_i(s)
\end{aligned} \tag{7-1}$$

In the time domain, these are a set of n first-order differential equations in the n state variables and the inputs:

$$\begin{aligned}
\frac{dx_1}{dt} &= a_{11}x_1 + a_{12}x_2 + \cdots + a_{1n}x_n + b_{11}r_1 + \cdots + b_{1i}r_i \\
\frac{dx_2}{dt} &= a_{21}x_1 + a_{22}x_2 + \cdots + a_{2n}x_n + b_{21}r_1 + \cdots + b_{2i}r_i \\
&\vdots \\
\frac{dx_n}{dt} &= a_{n1}x_1 + a_{n2}x_2 + \cdots + a_{nn}x_n + b_{n1}r_1 + \cdots + b_{ni}r_i
\end{aligned}$$

These state equations are compactly written in matrix notation as

$$\frac{d}{dt}\begin{bmatrix} x_1 \\ x_2 \\ \vdots \\ x_n \end{bmatrix} = \begin{bmatrix} \dot{x}_1 \\ \dot{x}_2 \\ \vdots \\ \dot{x}_n \end{bmatrix} = \begin{bmatrix} a_{11} & a_{12} & \cdots & a_{1n} \\ a_{21} & a_{22} & \cdots & a_{2n} \\ \vdots & & & \\ a_{n1} & a_{n2} & \cdots & a_{nn} \end{bmatrix} \begin{bmatrix} x_1 \\ x_2 \\ \vdots \\ x_n \end{bmatrix} + \begin{bmatrix} b_{11} & \cdots & b_{1i} \\ b_{21} & \cdots & b_{2i} \\ \vdots & & \\ b_{n1} & \cdots & b_{ni} \end{bmatrix} \begin{bmatrix} r_1 \\ r_2 \\ \vdots \\ r_i \end{bmatrix}$$

or

$$\frac{d\mathbf{x}}{dt} = \dot{\mathbf{x}} = \mathbf{Ax} + \mathbf{Br}$$

The column matrix of state variables

$$\mathbf{x} = \begin{bmatrix} x_1 \\ x_2 \\ \vdots \\ x_n \end{bmatrix}$$

is called the *state vector*. The inputs are arranged to form the *input vector*:

$$\mathbf{r}=\begin{bmatrix} r_1 \\ \vdots \\ r_i \end{bmatrix}$$

The system outputs are similarly arranged in an *output vector*,

$$\mathbf{y}=\begin{bmatrix} y_1 \\ \vdots \\ y_m \end{bmatrix}$$

related linearly to the state variables through the output equations:

$$\begin{cases} y_1 = c_{11}x_1 + c_{12}x_2 + \cdots + c_{1n}x_n \\ \vdots \\ y_m = c_{m1}x_1 + c_{m2}x_2 + \cdots + c_{mn}x_n \end{cases}$$

Or

$$\begin{bmatrix} y_1 \\ \vdots \\ y_m \end{bmatrix} = \begin{bmatrix} c_{11} & c_{12} & \cdots & c_{1n} \\ \vdots & & & \\ c_{m1} & c_{m2} & \cdots & c_{mn} \end{bmatrix} \begin{bmatrix} x_1 \\ \vdots \\ x_n \end{bmatrix}$$

or

$$\mathbf{y}=\mathbf{C}\mathbf{x}$$

The state equations describe how the system state vector evolves in time. One may imagine the tip of the vector tracing a curve, the *state trajectory*, in an n-dimensional space. The output equations describe how the output signals are related to the state.

Consider the two-input, two-output system of the simulation diagram of Fig. 7-4. The indicated state variables are governed by

$$\begin{cases} \dot{x}_1 = -2x_1 + x_2 + 5x_3 + 10r_2 \\ \dot{x}_2 = -3x_2 + 2x_3 \\ \dot{x}_3 = -4x_2 + 3r_1 \end{cases}$$

In matrix form, these state equations are

$$\begin{bmatrix} \dot{x}_1 \\ \dot{x}_2 \\ \dot{x}_3 \end{bmatrix} = \begin{bmatrix} -2 & 1 & 5 \\ 0 & -3 & 2 \\ 0 & -4 & 3 \end{bmatrix} \begin{bmatrix} x_1 \\ x_2 \\ x_3 \end{bmatrix} + \begin{bmatrix} 0 & 10 \\ 0 & 0 \\ 3 & 0 \end{bmatrix} \begin{bmatrix} r_1 \\ r_2 \end{bmatrix}$$

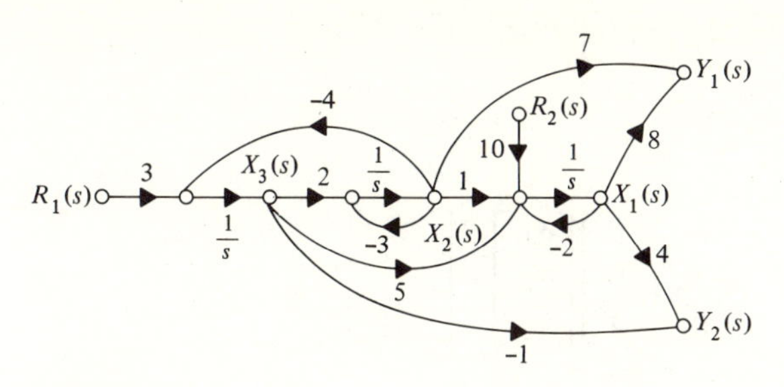

FIGURE 7-4
Simulation diagram for a certain two-input, two-output system.

The outputs of this system are related to the state variables by

$$\begin{cases} y_1 = 8x_1 + 7x_2 \\ y_2 = 4x_1 - x_3 \end{cases}$$

In matrix form, these output equations are

$$\begin{bmatrix} y_1 \\ y_2 \end{bmatrix} = \begin{bmatrix} 8 & 7 & 0 \\ 4 & 0 & -1 \end{bmatrix} \begin{bmatrix} x_1 \\ x_2 \\ x_3 \end{bmatrix}$$

For systems described by linear constant-coefficient integrodifferential equations, the state variable arrangement is simply a standard form for the equations describing a system. Instead of dealing with a mixed collection of simultaneous system equations, some of first order, some of second order, some involving running integrals, and so on, additional manipulation of the original equations is done to place them in the standard form. The advantages of a standard form are that systematic methods may be easily brought to bear upon very involved problems and that a degree of unification results.

DRILL PROBLEMS

D7-2. Write state variable equations in matrix form for the systems described by the following simulation diagrams.

ans. (a) $\begin{bmatrix} \dot{x}_1 \\ \dot{x}_2 \end{bmatrix} = \begin{bmatrix} -2 & 5 \\ -4 & -3 \end{bmatrix} \begin{bmatrix} x_1 \\ x_2 \end{bmatrix} + \begin{bmatrix} -7 \\ 6 \end{bmatrix} r$

$$y = [1 \quad 0] \begin{bmatrix} x_1 \\ x_2 \end{bmatrix}$$

(b) $$\begin{bmatrix} \dot{x}_1 \\ \dot{x}_2 \end{bmatrix} = \begin{bmatrix} -2 & 1 \\ -3 & -1 \end{bmatrix} \begin{bmatrix} x_1 \\ x_2 \end{bmatrix} + \begin{bmatrix} 0 \\ 1 \end{bmatrix} r$$

$$y = \begin{bmatrix} 1 & 1 \end{bmatrix} \begin{bmatrix} x_1 \\ x_2 \end{bmatrix}$$

(c) $$\begin{bmatrix} \dot{x}_1 \\ \dot{x}_2 \\ \dot{x}_3 \end{bmatrix} = \begin{bmatrix} -2 & 0 & 0 \\ 0 & -3 & 0 \\ 0 & 0 & -4 \end{bmatrix} \begin{bmatrix} x_1 \\ x_2 \\ x_3 \end{bmatrix} + \begin{bmatrix} 1 \\ -6 \\ 10 \end{bmatrix} r$$

$$\begin{bmatrix} y_1 \\ y_2 \end{bmatrix} = \begin{bmatrix} 1 & 1 & 0 \\ 0 & 1 & 1 \end{bmatrix} \begin{bmatrix} x_1 \\ x_2 \\ x_3 \end{bmatrix}$$

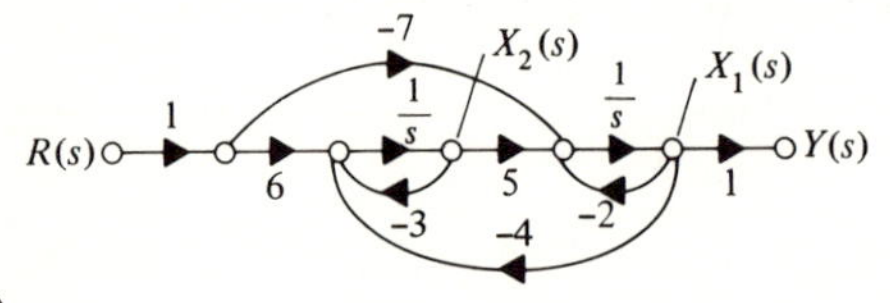

(a)

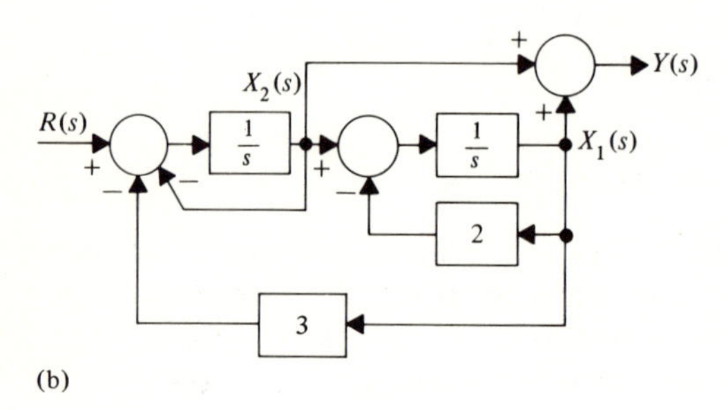

(b)

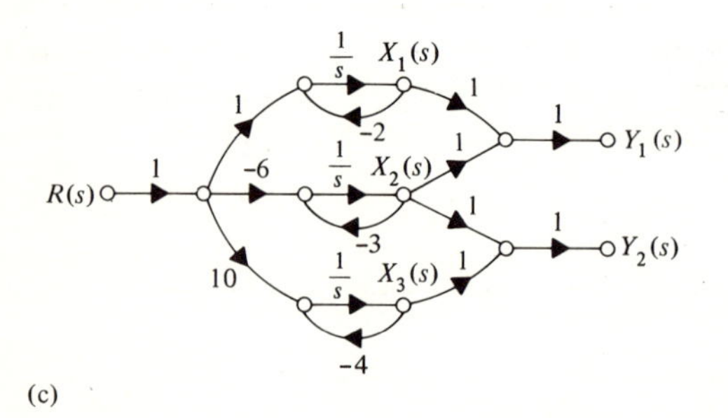

(c)

D7-3. Draw simulation diagrams to represent the following systems:

(a) $$\begin{bmatrix} \dot{x}_1 \\ \dot{x}_2 \end{bmatrix} = \begin{bmatrix} -2 & -3 \\ 4 & 1 \end{bmatrix} \begin{bmatrix} x_1 \\ x_2 \end{bmatrix} + \begin{bmatrix} 5 \\ -6 \end{bmatrix} r$$

$$y = \begin{bmatrix} 7 & -8 \end{bmatrix} \begin{bmatrix} x_1 \\ x_2 \end{bmatrix}$$

(b)
$$\begin{bmatrix} \dot{x}_1 \\ \dot{x}_2 \\ \dot{x}_3 \end{bmatrix} = \begin{bmatrix} 0 & 10 & 3 \\ 5 & 8 & 0 \\ -2 & -7 & -3 \end{bmatrix} \begin{bmatrix} x_1 \\ x_2 \\ x_3 \end{bmatrix} + \begin{bmatrix} 0 & 1 \\ 2 & 0 \\ -1 & 3 \end{bmatrix} \begin{bmatrix} r_1 \\ r_2 \end{bmatrix}$$

$$y = [0 \quad 4 \quad -3] \begin{bmatrix} x_1 \\ x_2 \\ x_3 \end{bmatrix}$$

(c)
$$\begin{bmatrix} \dot{x}_1 \\ \dot{x}_2 \\ \dot{x}_3 \end{bmatrix} = \begin{bmatrix} -2 & 0 & -6 \\ 3 & 5 & 0 \\ -4 & 0 & 7 \end{bmatrix} \begin{bmatrix} x_1 \\ x_2 \\ x_3 \end{bmatrix} + \begin{bmatrix} 8 \\ -2 \\ 0 \end{bmatrix} r$$

$$\begin{bmatrix} y_1 \\ y_2 \end{bmatrix} = \begin{bmatrix} 0 & -1 & 1 \\ -1 & 1 & 0 \end{bmatrix} \begin{bmatrix} x_1 \\ x_2 \\ x_3 \end{bmatrix}$$

7.3.2 Transfer Functions

The transfer functions of a system represented in state variable form may be found by Laplace-transforming the state equations with zero initial conditions. In general, these are the equations in (7-1). Collecting the terms involving **X**(s), there results

$$\begin{bmatrix} (s-a_{11}) & -a_{12} & \cdots & -a_{1n} \\ -a_{21} & (s-a_{22}) & \cdots & -a_{2n} \\ \vdots & & & \\ -a_{n1} & -a_{n2} & \cdots & (s-a_{nn}) \end{bmatrix} \begin{bmatrix} X_1(s) \\ X_2(s) \\ \vdots \\ X_n(s) \end{bmatrix} = \begin{bmatrix} b_{11} & b_{12} & \cdots & b_{1i} \\ b_{21} & b_{22} & \cdots & b_{2i} \\ \vdots & & & \\ b_{n1} & b_{n2} & \cdots & b_{ni} \end{bmatrix} \begin{bmatrix} R_1(s) \\ R_2(s) \\ \vdots \\ R_i(s) \end{bmatrix}$$

or

$$[s\mathbf{I} - \mathbf{A}]\mathbf{X}(s) = \mathbf{B}\mathbf{R}(s)$$

where **I** is the $n \times n$ identity matrix

$$\mathbf{I} = \begin{bmatrix} 1 & 0 & \cdots & 0 & 0 \\ 0 & 1 & \cdots & 0 & 0 \\ \vdots & & & & \\ 0 & 0 & \cdots & 0 & 1 \end{bmatrix}$$

Solving for the Laplace transform of the state vector,

$$\mathbf{X}(s) = [s\mathbf{I} - \mathbf{A}]^{-1}\mathbf{B}\mathbf{R}(s)$$

The output and state vectors are related by

$$\begin{bmatrix} Y_1(s) \\ Y_2(s) \\ \vdots \\ Y_m(s) \end{bmatrix} = \begin{bmatrix} c_{11} & c_{12} & \cdots & c_{1n} \\ c_{21} & c_{22} & \cdots & c_{2n} \\ \vdots & & & \\ c_{m1} & c_{m2} & \cdots & c_{mn} \end{bmatrix} \begin{bmatrix} X_1(s) \\ X_2(s) \\ \vdots \\ X_n(s) \end{bmatrix}$$

or

$$\mathbf{Y}(s) = \mathbf{C}\mathbf{X}(s) = \{\mathbf{C}[s\mathbf{I} - \mathbf{A}]^{-1}\mathbf{B}\}\mathbf{R}(s)$$

The $m \times i$ matrix in brackets $\{\}$ above consists of the input-output transfer functions of the system, arranged as a matrix:

$$C[s\mathbf{I} - \mathbf{A}]^{-1}\mathbf{B} = \begin{bmatrix} T_{11}(s) & T_{12}(s) & \cdots & T_{1i}(s) \\ T_{21}(s) & T_{22}(s) & \cdots & T_{2i}(s) \\ \vdots & & & \\ T_{m1}(s) & T_{m2}(s) & \cdots & T_{mi}(s) \end{bmatrix}$$

For example, a single-input, single-output system with state equations

$$\begin{bmatrix} \dot{x}_1 \\ \dot{x}_2 \end{bmatrix} = \begin{bmatrix} -3 & 1 \\ -2 & 0 \end{bmatrix} \begin{bmatrix} x_1 \\ x_2 \end{bmatrix} + \begin{bmatrix} 4 \\ -5 \end{bmatrix} r$$

$$y = [1 \quad -1] \begin{bmatrix} x_1 \\ x_2 \end{bmatrix}$$

has transfer function given by

$$T(s) = [1 \quad -1] \begin{bmatrix} s+3 & -1 \\ 2 & s \end{bmatrix}^{-1} \begin{bmatrix} 4 \\ -5 \end{bmatrix}$$

$$= \frac{[1 \quad -1] \begin{bmatrix} s & 1 \\ -2 & s+3 \end{bmatrix} \begin{bmatrix} 4 \\ -5 \end{bmatrix}}{s^2 + 3s + 2}$$

$$= \frac{[1 \quad -1] \begin{bmatrix} (4s-5) \\ (-5s-23) \end{bmatrix}}{s^2 + 3s + 2}$$

$$= \frac{9s + 18}{s^2 + 3s + 2}$$

The two-input, two-output system

$$\begin{bmatrix} \dot{x}_1 \\ \dot{x}_2 \end{bmatrix} = \begin{bmatrix} -3 & 1 \\ -2 & 0 \end{bmatrix} \begin{bmatrix} x_1 \\ x_2 \end{bmatrix} + \begin{bmatrix} 4 & 6 \\ -5 & 0 \end{bmatrix} \begin{bmatrix} r_1 \\ r_2 \end{bmatrix}$$

$$\begin{bmatrix} y_1 \\ y_2 \end{bmatrix} = \begin{bmatrix} 1 & -1 \\ 8 & 1 \end{bmatrix} \begin{bmatrix} x_1 \\ x_2 \end{bmatrix}$$

is described by the transfer function matrix given by

$$\mathbf{T}(s)=\begin{bmatrix}1 & -1\\ 8 & 1\end{bmatrix}\begin{bmatrix}s+3 & -1\\ 2 & s\end{bmatrix}^{-1}\begin{bmatrix}4 & 6\\ -5 & 0\end{bmatrix}$$

$$=\begin{bmatrix}1 & -1\\ 8 & 1\end{bmatrix}\frac{\begin{bmatrix}s & 1\\ -2 & s+3\end{bmatrix}}{s^2+3s+2}\begin{bmatrix}4 & 6\\ -5 & 0\end{bmatrix}$$

$$=\frac{\begin{bmatrix}1 & -1\\ 8 & 1\end{bmatrix}\begin{bmatrix}(4s-5) & 6s\\ (-5s-23) & -12\end{bmatrix}}{s^2+3s+2}$$

$$=\begin{bmatrix}\dfrac{9s+18}{s^2+3s+2} & \dfrac{6s+12}{s^2+3s+2}\\[2ex] \dfrac{27s-63}{s^2+3s+2} & \dfrac{48s-12}{s^2+3s+2}\end{bmatrix}=\begin{bmatrix}T_{11}(s) & T_{12}(s)\\ T_{21}(s) & T_{22}(s)\end{bmatrix}$$

where

$$T_{11}(s)=\frac{9s+18}{s^2+3s+2}=\left.\frac{Y_1(s)}{R_1(s)}\right|_{\substack{\text{initial}\\ \text{conditions}\\ \text{and } R_2=0}}$$

$$T_{12}(s)=\frac{6s+12}{s^2+3s+2}=\left.\frac{Y_1(s)}{R_2(s)}\right|_{\substack{\text{initial}\\ \text{conditions}\\ \text{and } R_1=0}}$$

$$T_{21}(s)=\frac{27s-63}{s^2+3s+2}=\left.\frac{Y_2(s)}{R_1(s)}\right|_{\substack{\text{initial}\\ \text{conditions}\\ \text{and } R_2=0}}$$

$$T_{22}(s)=\frac{48s-12}{s^2+3s+2}=\left.\frac{Y_2(s)}{R_2(s)}\right|_{\substack{\text{initial}\\ \text{conditions}\\ \text{and } R_1=0}}$$

All of the transfer functions of a system share the denominator polynomial

$$|s\mathbf{I}-\mathbf{A}|$$

where **A** is the state coupling matrix for the system, since

$$[s\mathbf{I}-\mathbf{A}]^{-1}=\frac{\text{adjoint }[s\mathbf{I}-\mathbf{A}]}{|s\mathbf{I}-\mathbf{A}|}$$

The nth-degree polynomial

$$|s\mathbf{I}-\mathbf{A}|=0$$

is termed the characteristic polynomial of an $n\times n$ matrix **A** and the n roots of that polynomial are the characteristic roots of the matrix. A system is stable if and only if the characteristic roots of the state coupling matrix are all in the left half of the complex plane.

DRILL PROBLEM

D7-4. Find the transfer function matrices of the following systems:

(a)
$$\begin{bmatrix}\dot{x}_1\\ \dot{x}_2\end{bmatrix}=\begin{bmatrix}-2 & 3\\ -1 & -1\end{bmatrix}\begin{bmatrix}x_1\\ x_2\end{bmatrix}+\begin{bmatrix}4 & 0\\ -5 & 6\end{bmatrix}\begin{bmatrix}r_1\\ r_2\end{bmatrix}$$

$$y=\begin{bmatrix}7 & 8\end{bmatrix}\begin{bmatrix}x_1\\ x_2\end{bmatrix}$$

$$ans.\ \begin{bmatrix}\dfrac{-12s-189}{s^2+3s+5} & \dfrac{48s+222}{s^2+3s+5}\end{bmatrix}$$

(b)
$$\begin{bmatrix}\dot{x}_1\\ \dot{x}_2\end{bmatrix}=\begin{bmatrix}-3 & 4\\ -2 & 0\end{bmatrix}\begin{bmatrix}x_1\\ x_2\end{bmatrix}+\begin{bmatrix}2\\ 1\end{bmatrix}r$$

$$\begin{bmatrix}y_1\\ y_2\end{bmatrix}=\begin{bmatrix}-4 & 6\\ 5 & -1\end{bmatrix}\begin{bmatrix}x_1\\ x_2\end{bmatrix}$$

$$ans.\ \begin{bmatrix}\dfrac{-2s-22}{s^2+3s+8}\\[2ex] \dfrac{9s+21}{s^2+3s+8}\end{bmatrix}$$

(c)
$$\begin{bmatrix}\dot{x}_1\\ \dot{x}_2\\ \dot{x}_3\end{bmatrix}=\begin{bmatrix}-4 & 0 & 2\\ -1 & -1 & 0\\ 3 & 0 & -3\end{bmatrix}\begin{bmatrix}x_1\\ x_2\\ x_3\end{bmatrix}+\begin{bmatrix}-3 & -2\\ 4 & 1\\ 0 & 0\end{bmatrix}\begin{bmatrix}r_1\\ r_2\end{bmatrix}$$

$$\begin{bmatrix}y_1\\ y_2\end{bmatrix}=\begin{bmatrix}1 & 1 & 1\\ -1 & 0 & 1\end{bmatrix}\begin{bmatrix}x_1\\ x_2\\ x_3\end{bmatrix}$$

$$ans.\ \begin{bmatrix}\dfrac{s^2+10s+15}{(s+1)(s^2+7s+6)} & \dfrac{-s(s+5)}{(s+1)(s^2+7s+6)}\\[2ex] \dfrac{3s}{s^2+7s+6} & \dfrac{2s}{s^2+7s+6}\end{bmatrix}$$

7.3.3 Change of State Variables

A nonsingular transformation of state variables results in a new system state representation with the same relation between inputs and outputs. A new set of n variables may be derived from the original n state variables through a constant transformation of the form

$$\begin{aligned}
x'_1 &= p_{11}x_1 + p_{12}x_2 + \cdots + p_{1n}x_n \\
x'_2 &= p_{21}x_1 + p_{22}x_2 + \cdots + p_{2n}x_n \\
&\vdots \\
x'_n &= p_{n1}x_1 + p_{n2}x_2 + \cdots + p_{nn}x_n
\end{aligned}$$

If the transformation is nonsingular, the original variables may be recovered from the new ones through the inverse transformation

$$\begin{aligned}
x_1 &= q_{11}x'_1 + q_{12}x'_2 + \cdots + q_{1n}x'_n \\
x_2 &= q_{21}x'_1 + q_{22}x'_2 + \cdots + q_{2n}x'_n \\
&\vdots \\
x_n &= q_{n1}x'_1 + q_{n2}x'_2 + \cdots + q_{nn}x'_n
\end{aligned}$$

In matrix notation,

$$\begin{bmatrix} x'_1 \\ x'_2 \\ \vdots \\ x'_n \end{bmatrix} = \begin{bmatrix} p_{11} & p_{12} & \cdots & p_{1n} \\ p_{21} & p_{22} & \cdots & p_{2n} \\ \vdots & & & \\ p_{n1} & p_{n2} & \cdots & p_{nn} \end{bmatrix} \begin{bmatrix} x_1 \\ x_2 \\ \vdots \\ x_n \end{bmatrix}$$

or

$$\mathbf{x}' = \mathbf{P}\mathbf{x}$$

and

$$\begin{bmatrix} x_1 \\ x_2 \\ \vdots \\ x_n \end{bmatrix} = \begin{bmatrix} q_{11} & q_{12} & \cdots & q_{1n} \\ q_{21} & q_{22} & \cdots & q_{2n} \\ \vdots & & & \\ q_{n1} & q_{n2} & \cdots & q_{nn} \end{bmatrix} \begin{bmatrix} x'_1 \\ x'_2 \\ \vdots \\ x'_n \end{bmatrix}$$

or

$$\mathbf{x} = \mathbf{Q}\mathbf{x}' = \mathbf{P}^{-1}\mathbf{x}'$$

A nonsingular change of state variables

$$\mathbf{x}' = \mathbf{P}\mathbf{x} \qquad \mathbf{x} = \mathbf{P}^{-1}\mathbf{x}'$$

in a state representation for a system

$$\begin{aligned}
\dot{\mathbf{x}} &= \mathbf{A}\mathbf{x} + \mathbf{B}\mathbf{r} \\
\mathbf{y} &= \mathbf{C}\mathbf{x}
\end{aligned}$$

gives a representation of the same form, in terms of an alternative set of state variables:

$$\mathbf{P}^{-1}\dot{\mathbf{x}}' = \mathbf{A}\mathbf{P}^{-1}\mathbf{x}' + \mathbf{B}\mathbf{r}$$
$$\mathbf{y} = \mathbf{C}\mathbf{P}^{-1}\mathbf{x}'$$

$$\dot{\mathbf{x}}' = (\mathbf{P}\mathbf{A}\mathbf{P}^{-1})\mathbf{x}' + (\mathbf{P}\mathbf{B})\mathbf{r}$$
$$\mathbf{y} = (\mathbf{C}\mathbf{P}^{-1})\mathbf{x}'$$

Defining

$$\mathbf{A}' = \mathbf{P}\mathbf{A}\mathbf{P}^{-1}$$
$$\mathbf{B}' = \mathbf{P}\mathbf{B}$$
$$\mathbf{C}' = \mathbf{C}\mathbf{P}^{-1}$$

the equations in terms of the new state variables $\mathbf{x}'$ are of the same form as the original equations:

$$\dot{\mathbf{x}}' = \mathbf{A}'\mathbf{x}' + \mathbf{B}'\mathbf{r}$$
$$\mathbf{y} = \mathbf{C}'\mathbf{x}'$$

The input-output relations of a system are unchanged by a nonsingular change of state variables; it is only the internal description, in terms of its state, that is changed. The matrix of system transfer functions for the prime system is given by

$$\mathbf{T}'(s) = \mathbf{C}'(s\mathbf{I} - \mathbf{A}')^{-1}\mathbf{B}'$$

Substituting the original system matrices, it is seen that the transfer functions are identical:

$$\begin{aligned}\mathbf{T}'(s) &= \mathbf{C}\mathbf{P}^{-1}(s\mathbf{I} - \mathbf{P}\mathbf{A}\mathbf{P}^{-1})^{-1}\mathbf{P}\mathbf{B}\\ &= \mathbf{C}\mathbf{P}^{-1}(s\mathbf{P}\mathbf{P}^{-1} - \mathbf{P}\mathbf{A}\mathbf{P}^{-1})^{-1}\mathbf{P}\mathbf{B}\\ &= \mathbf{C}\mathbf{P}^{-1}[\mathbf{P}(s\mathbf{I} - \mathbf{A})\mathbf{P}^{-1}]^{-1}\mathbf{P}\mathbf{B}\\ &= \mathbf{C}\mathbf{P}^{-1}\mathbf{P}(s\mathbf{I} - \mathbf{A})^{-1}\mathbf{P}^{-1}\mathbf{P}\mathbf{B}\\ &= \mathbf{C}(s\mathbf{I} - \mathbf{A})^{-1}\mathbf{B}\end{aligned}$$

which is the same transfer function matrix as for the system described in terms of the original state variables. The matrix relations

$$\mathbf{P}\mathbf{P}^{-1} = \mathbf{P}^{-1}\mathbf{P} = \mathbf{I}$$

and, for square matrices $\mathbf{X}$, $\mathbf{Y}$, and $\mathbf{Z}$,

$$(\mathbf{X}\mathbf{Y}\mathbf{Z})^{-1} = \mathbf{Z}^{-1}\mathbf{Y}^{-1}\mathbf{X}^{-1}$$

are used in this derivation.

As a numerical example, consider the system

$$\begin{bmatrix} \dot{x}_1 \\ \dot{x}_2 \\ \dot{x}_3 \end{bmatrix} = \begin{bmatrix} -2 & 1 & 1 \\ -3 & 0 & 1 \\ 0 & 0 & 0 \end{bmatrix} \begin{bmatrix} x_1 \\ x_2 \\ x_3 \end{bmatrix} + \begin{bmatrix} 1 & 2 \\ 0 & 0 \\ -1 & 0 \end{bmatrix} \begin{bmatrix} r_1 \\ r_2 \end{bmatrix} = \mathbf{A}\mathbf{x} + \mathbf{B}\mathbf{r}$$

$$\begin{bmatrix} y_1 \\ y_2 \end{bmatrix} = \begin{bmatrix} 1 & 0 & 1 \\ 0 & 0 & -2 \end{bmatrix} \begin{bmatrix} x_1 \\ x_2 \\ x_3 \end{bmatrix} = \mathbf{C}\mathbf{x}$$

The nonsingular transformation of state variables

$$\begin{bmatrix} x'_1 \\ x'_2 \\ x'_3 \end{bmatrix} = \begin{bmatrix} 1 & 0 & 1 \\ 0 & 1 & 2 \\ 0 & 0 & 4 \end{bmatrix} \begin{bmatrix} x_1 \\ x_2 \\ x_3 \end{bmatrix} = \mathbf{P}\mathbf{x}$$

$$\begin{bmatrix} x_1 \\ x_2 \\ x_3 \end{bmatrix} = \begin{bmatrix} 1 & 0 & -\frac{1}{4} \\ 0 & 1 & -\frac{1}{2} \\ 0 & 0 & \frac{1}{4} \end{bmatrix} \begin{bmatrix} x'_1 \\ x'_2 \\ x'_3 \end{bmatrix} = \mathbf{P}^{-1}\mathbf{x}'$$

give a new state representation as follows:

$$\mathbf{A}' = \mathbf{P}\mathbf{A}\mathbf{P}^{-1} = \begin{bmatrix} 1 & 0 & 1 \\ 0 & 1 & 2 \\ 0 & 0 & 4 \end{bmatrix} \begin{bmatrix} -2 & 1 & 1 \\ -3 & 0 & 1 \\ 0 & 0 & 0 \end{bmatrix} \begin{bmatrix} 1 & 0 & -\frac{1}{4} \\ 0 & 1 & -\frac{1}{2} \\ 0 & 0 & \frac{1}{4} \end{bmatrix}$$

$$= \begin{bmatrix} -2 & 1 & 1 \\ -3 & 0 & 1 \\ 0 & 0 & 0 \end{bmatrix} \begin{bmatrix} 1 & 0 & -\frac{1}{4} \\ 0 & 1 & -\frac{1}{2} \\ 0 & 0 & \frac{1}{4} \end{bmatrix} = \begin{bmatrix} -2 & 1 & \frac{1}{4} \\ -3 & 0 & 1 \\ 0 & 0 & 0 \end{bmatrix}$$

$$\mathbf{B}' = \mathbf{P}\mathbf{B} = \begin{bmatrix} 1 & 0 & 1 \\ 0 & 1 & 2 \\ 0 & 0 & 4 \end{bmatrix} \begin{bmatrix} 1 & 2 \\ 0 & 0 \\ -1 & 0 \end{bmatrix} = \begin{bmatrix} 0 & 2 \\ -2 & 0 \\ -4 & 0 \end{bmatrix}$$

$$\mathbf{C}' = \mathbf{C}\mathbf{P}^{-1} = \begin{bmatrix} 1 & 0 & 1 \\ 0 & 0 & -2 \end{bmatrix} \begin{bmatrix} 1 & 0 & -\frac{1}{4} \\ 0 & 1 & -\frac{1}{2} \\ 0 & 0 & \frac{1}{4} \end{bmatrix} = \begin{bmatrix} 1 & 0 & 0 \\ 0 & 0 & -\frac{1}{2} \end{bmatrix}$$

Simulation diagrams for the original and the transformed systems are given in Fig. 7-5. The two systems are indistinguishable from one another so far as input and output signals are concerned. In general, every system has an infinite number of different state representations, each involving a different choice of state variables. There is the possibility, then, of finding especially simple state variable system representations.

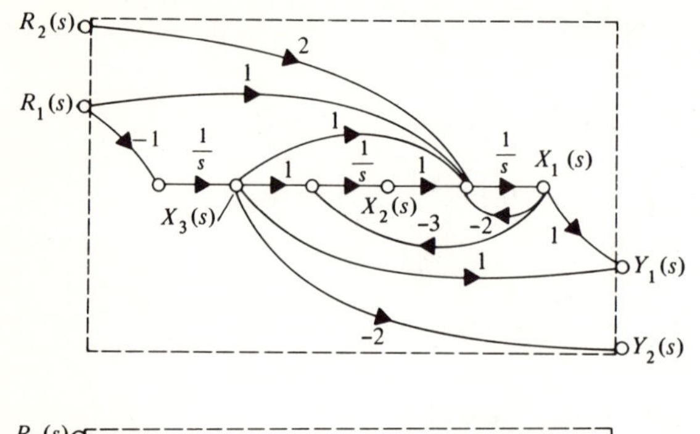

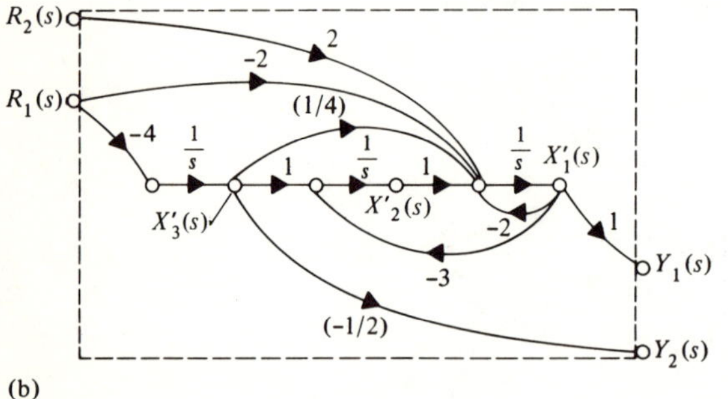

(b)

FIGURE 7-5
Simulation diagrams for two different representations of a system. (a) The system $\mathbf{x} = \mathbf{Ax} + \mathbf{Br}$, $\mathbf{y} = \mathbf{Cx}$. (b) The system $\mathbf{x}' = \mathbf{A}'\mathbf{x}' + \mathbf{B}'\mathbf{r}$, $\mathbf{y} = \mathbf{C}'\mathbf{x}'$.

DRILL PROBLEM

D7-5. Make the indicated change of state variables, finding the new set of state and output equations in terms of $\mathbf{x}'$.

(a) $$\begin{bmatrix} \dot{x}_1 \\ \dot{x}_2 \end{bmatrix} = \begin{bmatrix} -2 & 1 \\ -3 & 0 \end{bmatrix} \begin{bmatrix} x_1 \\ x_2 \end{bmatrix} + \begin{bmatrix} 4 \\ 5 \end{bmatrix} r$$

$$y = [1 \quad 0] \begin{bmatrix} x_1 \\ x_2 \end{bmatrix}$$

$$\begin{bmatrix} x'_1 \\ x'_2 \end{bmatrix} = \begin{bmatrix} 2 & 1 \\ 4 & 3 \end{bmatrix} \begin{bmatrix} x_1 \\ x_2 \end{bmatrix}$$

ans. $$\begin{bmatrix} \dot{x}'_1 \\ \dot{x}'_2 \end{bmatrix} = \begin{bmatrix} -\frac{29}{2} & \frac{11}{2} \\ -\frac{67}{2} & \frac{25}{2} \end{bmatrix} \begin{bmatrix} x'_1 \\ x'_2 \end{bmatrix} + \begin{bmatrix} 13 \\ 31 \end{bmatrix} r$$

$$y = [\tfrac{3}{2} \quad -\tfrac{1}{2}] \begin{bmatrix} x'_1 \\ x'_2 \end{bmatrix}$$

(b) $$\begin{bmatrix}\dot{x}_1\\ \dot{x}_2\end{bmatrix}=\begin{bmatrix}2 & -6\\ 12 & 16\end{bmatrix}\begin{bmatrix}x_1\\ x_2\end{bmatrix}+\begin{bmatrix}0 & -1\\ 2 & 1\end{bmatrix}\begin{bmatrix}r_1\\ r_2\end{bmatrix}$$

$$y=[1 \quad 1]\begin{bmatrix}x_1\\ x_2\end{bmatrix}$$

$$\begin{bmatrix}x'_1\\ x'_2\end{bmatrix}=\begin{bmatrix}1 & 2\\ 1 & 4\end{bmatrix}\begin{bmatrix}x_1\\ x_2\end{bmatrix}$$

ans. $$\begin{bmatrix}\dot{x}'_1\\ \dot{x}'_2\end{bmatrix}=\begin{bmatrix}39 & -13\\ 71 & -21\end{bmatrix}\begin{bmatrix}x'_1\\ x'_2\end{bmatrix}+\begin{bmatrix}4 & 1\\ 8 & 3\end{bmatrix}\begin{bmatrix}r_1\\ r_2\end{bmatrix}$$

$$y=[\tfrac{3}{2} \quad -\tfrac{1}{2}]\begin{bmatrix}x'_1\\ x'_2\end{bmatrix}$$

(c) $$\begin{bmatrix}\dot{x}_1\\ \dot{x}_2\\ \dot{x}_3\end{bmatrix}=\begin{bmatrix}-2 & 1 & 1\\ -3 & 0 & 0\\ 0 & 0 & 0\end{bmatrix}\begin{bmatrix}x_1\\ x_2\\ x_3\end{bmatrix}+\begin{bmatrix}1\\ 1\\ 1\end{bmatrix}r$$

$$\begin{bmatrix}y_1\\ y_2\end{bmatrix}=\begin{bmatrix}2 & -2 & 1\\ 0 & -1 & 1\end{bmatrix}\begin{bmatrix}x_1\\ x_2\\ x_3\end{bmatrix}$$

$$\begin{bmatrix}x'_1\\ x'_2\\ x'_3\end{bmatrix}=\begin{bmatrix}1 & 0 & 1\\ 0 & 1 & 3\\ 0 & 0 & 4\end{bmatrix}\begin{bmatrix}x_1\\ x_2\\ x_3\end{bmatrix}$$

ans. $$\begin{bmatrix}\dot{x}_1\\ \dot{x}_2\\ \dot{x}_3\end{bmatrix}=\begin{bmatrix}-2 & 1 & 0\\ -3 & 0 & \tfrac{3}{4}\\ 0 & 0 & 0\end{bmatrix}\begin{bmatrix}x_1\\ x_2\\ x_3\end{bmatrix}+\begin{bmatrix}2\\ 4\\ 4\end{bmatrix}r$$

$$y=\begin{bmatrix}2 & -2 & \tfrac{5}{4}\\ 0 & -1 & 1\end{bmatrix}\begin{bmatrix}x'_1\\ x'_2\\ x'_3\end{bmatrix}$$

7.4 DECOUPLING STATE EQUATIONS

7.4.1 Diagonal Forms for the Equations

When a nonsingular change of state variables in a system representation is made,

$$\mathbf{x}'=\mathbf{P}\mathbf{x} \qquad \mathbf{x}=\mathbf{P}^{-1}\mathbf{x}'$$

the new state coupling matrix $\mathbf{A}'$ is related to the original one $\mathbf{A}$ by

$$\mathbf{A}'=\mathbf{P}\mathbf{A}\mathbf{P}^{-1}$$

Such an operation on a matrix is termed a *similarity transformation.* One of the most important results of matrix algebra is that, provided that a square matrix **A** has no repeated characteristic roots, a similarity transformation **P** may be found for which

$$\mathbf{A}' = \mathbf{P}\mathbf{A}\mathbf{P}^{-1}$$

is diagonal, with the characteristic roots as the diagonal elements.

For example, for the system

$$\begin{bmatrix} \dot{x}_1 \\ \dot{x}_2 \\ \dot{x}_3 \end{bmatrix} = \begin{bmatrix} -1 & -2 & 0 \\ 1 & 2 & 0 \\ -2 & -1 & -3 \end{bmatrix} \begin{bmatrix} x_1 \\ x_2 \\ x_3 \end{bmatrix} + \begin{bmatrix} 1 \\ 0 \\ 0 \end{bmatrix} r = \mathbf{A}\mathbf{x} + \mathbf{b}r$$

$$\begin{bmatrix} y_1 \\ y_2 \end{bmatrix} = \begin{bmatrix} 1 & 0 & 1 \\ 1 & -1 & 0 \end{bmatrix} \begin{bmatrix} x_1 \\ x_2 \\ x_3 \end{bmatrix} = \mathbf{C}\mathbf{x}$$

the transformation

$$\mathbf{P} = \begin{bmatrix} 1 & 1 & 0 \\ -\frac{1}{4} & -\frac{1}{2} & 0 \\ \frac{3}{4} & \frac{1}{2} & 1 \end{bmatrix} \qquad \mathbf{P}^{-1} = \begin{bmatrix} 2 & 4 & 0 \\ -1 & -4 & 0 \\ -1 & -1 & 1 \end{bmatrix}$$

gives a state variable representation for which the state coupling matrix is diagonal:

$$\mathbf{A}' = \mathbf{P}\mathbf{A}\mathbf{P}^{-1} = \begin{bmatrix} 1 & 1 & 0 \\ -\frac{1}{4} & -\frac{1}{2} & 0 \\ \frac{3}{4} & \frac{1}{2} & 1 \end{bmatrix} \begin{bmatrix} -1 & -2 & 0 \\ 1 & 2 & 0 \\ -2 & -1 & -3 \end{bmatrix} \begin{bmatrix} 2 & 4 & 0 \\ -1 & -4 & 0 \\ -1 & -1 & 1 \end{bmatrix}$$

$$= \begin{bmatrix} 1 & 1 & 0 \\ -\frac{1}{4} & -\frac{1}{2} & 0 \\ \frac{3}{4} & \frac{1}{2} & 1 \end{bmatrix} \begin{bmatrix} 0 & 4 & 0 \\ 0 & -4 & 0 \\ 0 & -1 & -3 \end{bmatrix} = \begin{bmatrix} 0 & 0 & 0 \\ 0 & 1 & 0 \\ 0 & 0 & -3 \end{bmatrix}$$

$$\mathbf{B}' = \mathbf{P}\mathbf{B} = \begin{bmatrix} 1 & 1 & 0 \\ -\frac{1}{4} & -\frac{1}{2} & 0 \\ \frac{3}{4} & \frac{1}{2} & 1 \end{bmatrix} \begin{bmatrix} 1 \\ 0 \\ 0 \end{bmatrix} = \begin{bmatrix} 1 \\ -\frac{1}{4} \\ \frac{3}{4} \end{bmatrix}$$

$$\mathbf{C}' = \mathbf{C}\mathbf{P}^{-1} = \begin{bmatrix} 1 & 0 & 1 \\ 1 & -1 & 0 \end{bmatrix} \begin{bmatrix} 2 & 4 & 0 \\ -1 & -4 & 0 \\ -1 & -1 & 1 \end{bmatrix} = \begin{bmatrix} 1 & 3 & 1 \\ 3 & 8 & 0 \end{bmatrix}$$

The system described by the primed state variables,

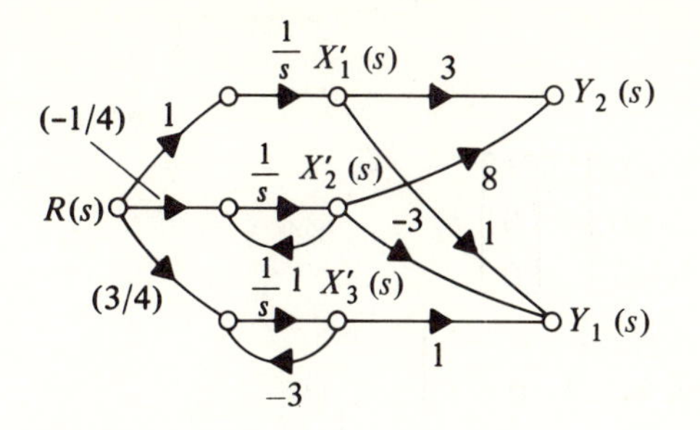

FIGURE 7-6
Simulation diagram for the diagonalized example system.

$$\begin{bmatrix} \dot{x}_1' \\ \dot{x}_2' \\ \dot{x}_3' \end{bmatrix} = \begin{bmatrix} 0 & 0 & 0 \\ 0 & 1 & 0 \\ 0 & 0 & -3 \end{bmatrix} \begin{bmatrix} x_1' \\ x_2' \\ x_3' \end{bmatrix} + \begin{bmatrix} 1 \\ -\frac{1}{4} \\ \frac{3}{4} \end{bmatrix} r$$

$$\begin{bmatrix} y_1 \\ y_2 \end{bmatrix} = \begin{bmatrix} 1 & 3 & 1 \\ 3 & 8 & 0 \end{bmatrix} \begin{bmatrix} x_1' \\ x_2' \\ x_3' \end{bmatrix}$$

has the same relation between r and $\mathbf{y}$. Because the state coupling matrix is diagonal, however, the state equations are decoupled from one another. The system is represented in the form of three separate first-order systems, as in the simulation diagram of Fig. 7-6.

Finding a transformation matrix that diagonalizes a square matrix $\mathbf{A}$ with distinct characteristic roots is a fundamental technique of linear algebra. It is termed the *characteristic value problem* and is discussed in detail in most texts on linear algebra, including those cited in the references at the end of this chapter.

7.4.2 Diagonalization Using Partial Fraction Expansion

Another method of determining a diagonal form for a system involves partial fraction expansion. For a single-input, single-output system such as

$$\begin{bmatrix} \dot{x}_1 \\ \dot{x}_2 \\ \dot{x}_3 \end{bmatrix} = \begin{bmatrix} -1 & -2 & 0 \\ 1 & 2 & 0 \\ -2 & -1 & -3 \end{bmatrix} \begin{bmatrix} x_1 \\ x_2 \\ x_3 \end{bmatrix} + \begin{bmatrix} 1 \\ 0 \\ 0 \end{bmatrix} r$$

$$y = [1 \quad 0 \quad 1] \begin{bmatrix} x_1 \\ x_2 \\ x_3 \end{bmatrix}$$

the transfer function is

$$T(s) = \mathbf{C}(s\mathbf{I} - \mathbf{A})^{-1}\mathbf{B}$$

$$= [1 \quad 0 \quad 1]\begin{bmatrix} s+1 & 2 & 0 \\ -1 & s-2 & 0 \\ 2 & 1 & s+3 \end{bmatrix}^{-1}\begin{bmatrix} 1 \\ 0 \\ 0 \end{bmatrix}$$

$$= [1 \quad 0 \quad 1]\frac{\begin{bmatrix} s^2+s-6 & -2s-6 & 0 \\ s+3 & s^2+4s+3 & 0 \\ -2s+3 & -s+3 & s^2-s \end{bmatrix}}{s^2+2s^2-3s}\begin{bmatrix} 1 \\ 0 \\ 0 \end{bmatrix}$$

$$= \frac{[1 \quad 0 \quad 1]}{s^3+2s^2-3s}\begin{bmatrix} (s^2+s-6) \\ (s+3) \\ (-2s+3) \end{bmatrix} = \frac{s^2-s-3}{s^3+2s^2-3s}$$

Expanding this transfer function in partial fractions, there results

$$T(s) = \frac{s^2-s-3}{s(s-1)(s+3)} = \frac{1}{s} + \frac{-\frac{3}{4}}{s-1} + \frac{\frac{3}{4}}{s+3}$$

which may be considered as the tandem (or parallel) connection of first-order systems shown in Fig. 7-7a. Each of these first-order subsystems is drawn in state variable form in Fig. 7-7b, where the three integrator output signals are labeled as state variables. The state variable equations for this alternate system repre-

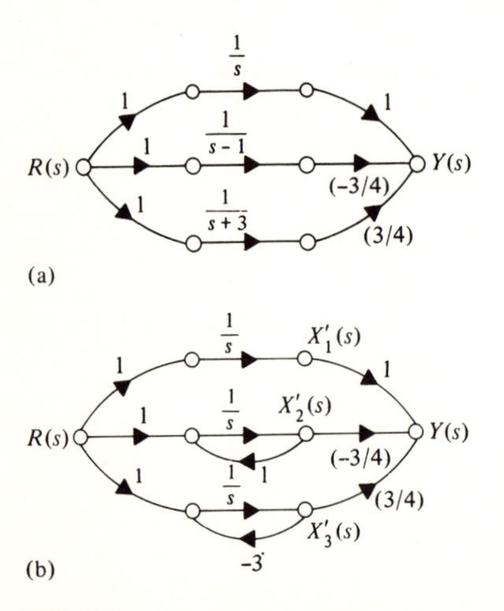

FIGURE 7-7
Diagonalizing a single-input, single-output system. (a) Tandem first-order subsystems from the partial fraction expansion of the transfer function. (b) Subsystems in simulation diagram form.

sentation, which has the same transfer function as the original system, are

$$\dot{x}_1' = r$$
$$\dot{x}_2' = x_2' + r$$
$$\dot{x}_3' = -3x_3' + r$$
$$y = x_1' - \tfrac{3}{4}x_2' + \tfrac{3}{4}x_3'$$

or

$$\begin{bmatrix} \dot{x}_1' \\ \dot{x}_2' \\ \dot{x}_3' \end{bmatrix} = \begin{bmatrix} 0 & 0 & 0 \\ 0 & 1 & 0 \\ 0 & 0 & -3 \end{bmatrix} \begin{bmatrix} x_1' \\ x_2' \\ x_3' \end{bmatrix} + \begin{bmatrix} 1 \\ 1 \\ 1 \end{bmatrix} r$$

$$y = \begin{bmatrix} 1 & -\tfrac{3}{4} & \tfrac{3}{4} \end{bmatrix} \begin{bmatrix} x_1' \\ x_2' \\ x_3' \end{bmatrix}$$

which is diagonal.

Additional outputs of the same system may be added by adding further couplings of the state variables. The related single-input, two-output system of Fig. 7-8a has the following transfer functions:

$$T_{11}(s) = \left.\frac{Y_1(s)}{R(s)}\right|_{\substack{\text{initial} \\ \text{conditions}=0}} = \frac{1}{s} + \frac{-\frac{3}{4}}{s-1} + \frac{\frac{3}{4}}{s+3}$$

$$= \frac{s^2 - s - 3}{s(s-1)(s+3)}$$

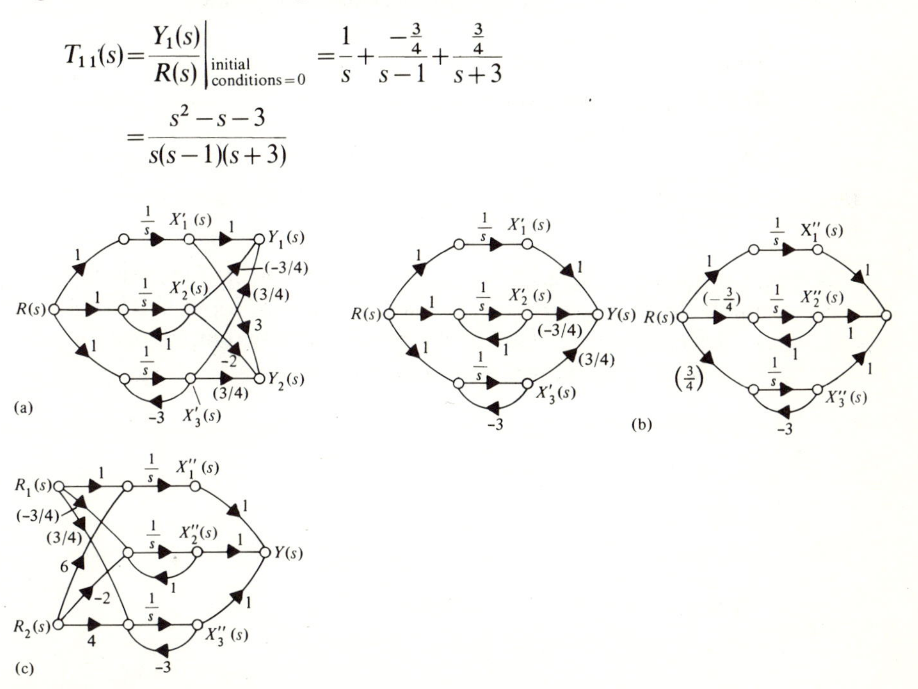

FIGURE 7-8
Diagonalizing multiple-output and multiple-input systems.
(a) Diagonalized one-input, multiple-output system. (b) Alternative diagonal forms for the single-input, single-output example system.
(c) Diagonalized multiple-input, one-output system.

$$T_{21}(s) = \left.\frac{Y_2(s)}{R(s)}\right|_{\substack{\text{initial}\\ \text{conditions}=0}} = \frac{3}{s} + \frac{-2}{s-1} + \frac{\frac{3}{4}}{s+3}$$

$$= \frac{\frac{7}{4}s^2 - \frac{3}{4}s - 9}{s(s-1)(s+3)}$$

An alternative diagonalized form for the original single-input, single-output system is given in Fig. 7-8b. In the new arrangement, the gains of 1, $-\frac{3}{4}$, and $\frac{3}{4}$ are placed on the input end of the diagram instead of the output end. The state variable equations for the modified system are slightly different,

$$\begin{bmatrix} \dot{x}_1'' \\ \dot{x}_2'' \\ \dot{x}_3'' \end{bmatrix} = \begin{bmatrix} 0 & 0 & 0 \\ 0 & 1 & 0 \\ 0 & 0 & -3 \end{bmatrix} \begin{bmatrix} x_1'' \\ x_2'' \\ x_3'' \end{bmatrix} + \begin{bmatrix} 1 \\ -\frac{3}{4} \\ \frac{3}{4} \end{bmatrix} r$$

$$y = \begin{bmatrix} 1 & 1 & 1 \end{bmatrix} \begin{bmatrix} x_1'' \\ x_2'' \\ x_3'' \end{bmatrix}$$

although the transfer function for the system remains the same:

$$T(s) = \left.\frac{Y(s)}{R(s)}\right|_{\substack{\text{initial}\\ \text{conditions}=0}} = \frac{1}{s} + \frac{-\frac{3}{4}}{s-1} + \frac{\frac{3}{4}}{s+3}$$

$$= \frac{s^2 - s - 3}{s(s-1)(s+3)}$$

Different input coupling gains will give different transfer function numerator polynomials. The numerator polynomials for multiple-input, single-output systems may then be chosen at will by this method. The two-input, one-output system of Fig. 7-8c has the following two transfer functions:

$$T_{11}(s) = \left.\frac{Y(s)}{R_1(s)}\right|_{\substack{\text{initial}\\ \text{conditions}\\ \text{and } R_2=0}} = \frac{1}{s} + \frac{-\frac{3}{4}}{s-1} + \frac{\frac{3}{4}}{s+3}$$

$$= \frac{s^2 - s - 3}{s(s-1)(s+3)}$$

$$T_{12}(s) = \left.\frac{Y(s)}{R_2(s)}\right|_{\substack{\text{initial}\\ \text{conditions}\\ \text{and } R_1=0}} = \frac{6}{s} + \frac{-2}{s-1} + \frac{4}{s+3}$$

$$= \frac{8s^2 + 2s - 18}{s(s-1)(s+3)}$$

For the case of both multiple inputs and multiple outputs, both the input coupling gains and the output coupling gains must be selected. Two inputs and

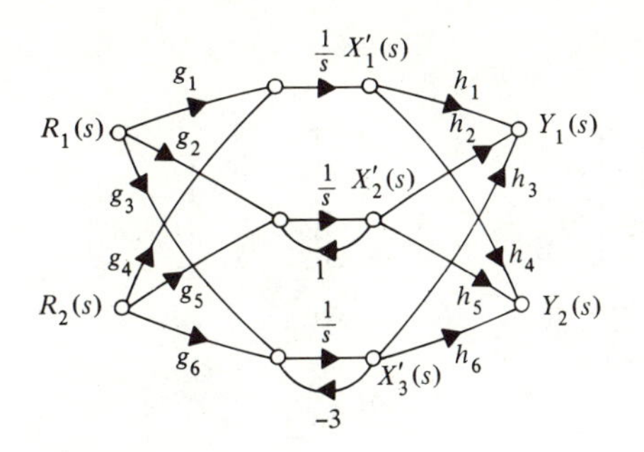

FIGURE 7-9
A diagonalized two-input, two-output system.

two outputs in the example system, as in Fig. 7-9, would result in four transfer functions, of the form

$$T_{11}(s) = \left.\frac{Y_1(s)}{R_1(s)}\right|_{\substack{\text{initial}\\ \text{conditions}\\ \text{and } R_2=0}} = \frac{k_1 s^2 + k_2 s + k_3}{s(s-1)(s+3)}$$

$$T_{12}(s) = \left.\frac{Y_1(s)}{R_2(s)}\right|_{\substack{\text{initial}\\ \text{conditions}\\ \text{and } R_1=0}} = \frac{k_4 s^2 + k_5 s + k_6}{s(s-1)(s+3)}$$

$$T_{21}(s) = \left.\frac{Y_2(s)}{R_1(s)}\right|_{\substack{\text{initial}\\ \text{conditions}\\ \text{and } R_2=0}} = \frac{k_7 s^2 + k_8 s + k_9}{s(s-1)(s+3)}$$

$$T_{22}(s) = \left.\frac{Y_2(s)}{R_2(s)}\right|_{\substack{\text{initial}\\ \text{conditions}\\ \text{and } R_1=0}} = \frac{k_{10} s^2 + k_{11} s + k_{12}}{s(s-1)(s+3)}$$

where the 12 k terms are determined by the 12 gains, $g_1, \ldots, g_6$ and $h_1, \ldots, h_6$.

DRILL PROBLEMS

D7-6. Use the partial fraction method to find diagonal state equations for single-input, single-output systems with the following transfer functions:

(a) $T(s) = \dfrac{-5s+7}{s^2+7s+12}$

ans. $$\begin{bmatrix} \dot{x}_1 \\ \dot{x}_2 \end{bmatrix} = \begin{bmatrix} -3 & 0 \\ 0 & -4 \end{bmatrix} \begin{bmatrix} x_1 \\ x_2 \end{bmatrix} + \begin{bmatrix} 22 \\ -27 \end{bmatrix} r$$

$$y = \begin{bmatrix} 1 & 1 \end{bmatrix} \begin{bmatrix} x_1 \\ x_2 \end{bmatrix}$$

(b) $T(s)=\dfrac{3s^2-2}{(s+1)(s+4)(s+10)}$

$$ans.\quad \begin{bmatrix}\dot{x}_1\\ \dot{x}_2\\ \dot{x}_3\end{bmatrix}=\begin{bmatrix}-1 & 0 & 0\\ 0 & -4 & 0\\ 0 & 0 & -10\end{bmatrix}\begin{bmatrix}x_1\\ x_2\\ x_3\end{bmatrix}+\begin{bmatrix}\frac{1}{27}\\ -\frac{23}{9}\\ \frac{149}{27}\end{bmatrix}r$$

$$y=\begin{bmatrix}1 & 1 & 1\end{bmatrix}\begin{bmatrix}x_1\\ x_2\\ x_3\end{bmatrix}$$

(c) $T(s)=\dfrac{4}{s^3+3s^2+2s}$

$$ans.\quad \begin{bmatrix}\dot{x}_1\\ \dot{x}_2\\ \dot{x}_3\end{bmatrix}=\begin{bmatrix}0 & 0 & 0\\ 0 & -1 & 0\\ 0 & 0 & -2\end{bmatrix}\begin{bmatrix}x_1\\ x_2\\ x_3\end{bmatrix}+\begin{bmatrix}1\\ 1\\ 1\end{bmatrix}r$$

$$y=\begin{bmatrix}2 & -4 & 2\end{bmatrix}\begin{bmatrix}x_1\\ x_2\\ x_3\end{bmatrix}$$

D7-7. The following systems have real characteristic roots. Find alternative diagonal state equations, and then draw a simulation diagram for the new equations.

$$ans.\ (a)\quad \begin{bmatrix}\dot{x}'_1\\ \dot{x}'_2\end{bmatrix}=\begin{bmatrix}-3 & 0\\ 0 & -4\end{bmatrix}\begin{bmatrix}x'_1\\ x'_2\end{bmatrix}+\begin{bmatrix}8\\ 4\end{bmatrix}r$$

$$y=\begin{bmatrix}1 & 1\end{bmatrix}\begin{bmatrix}x'_1\\ x'_2\end{bmatrix}$$

$$(b)\quad \begin{bmatrix}\dot{x}'_1\\ \dot{x}'_2\end{bmatrix}=\begin{bmatrix}-2 & 0\\ 0 & -4\end{bmatrix}\begin{bmatrix}x'_1\\ x'_2\end{bmatrix}+\begin{bmatrix}1\\ 1\end{bmatrix}r$$

$$y=\begin{bmatrix}6 & -9\end{bmatrix}\begin{bmatrix}x'_1\\ x'_2\end{bmatrix}$$

$$(c)\quad \begin{bmatrix}\dot{x}'_1\\ \dot{x}'_2\\ \dot{x}'_3\end{bmatrix}=\begin{bmatrix}0 & 0 & 0\\ 0 & -1 & 0\\ 0 & 0 & -2\end{bmatrix}\begin{bmatrix}x'_1\\ x'_2\\ x'_3\end{bmatrix}+\begin{bmatrix}1\\ 1\\ 1\end{bmatrix}r$$

$$y=\begin{bmatrix}\frac{3}{2} & -3 & -\frac{5}{2}\\ 0 & -1 & -1\end{bmatrix}\begin{bmatrix}x'_1\\ x'_2\\ x'_3\end{bmatrix}$$

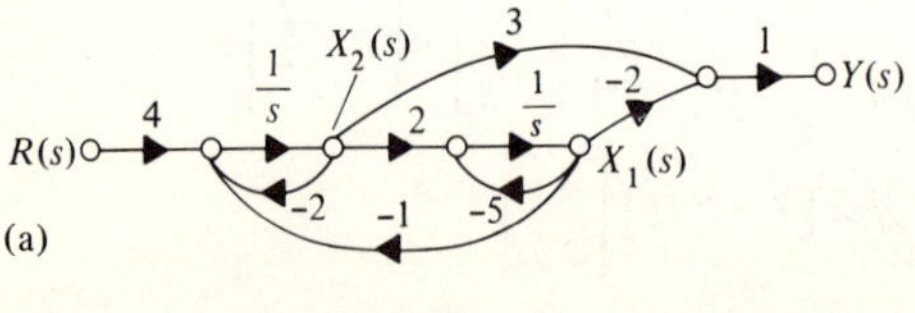

(a)

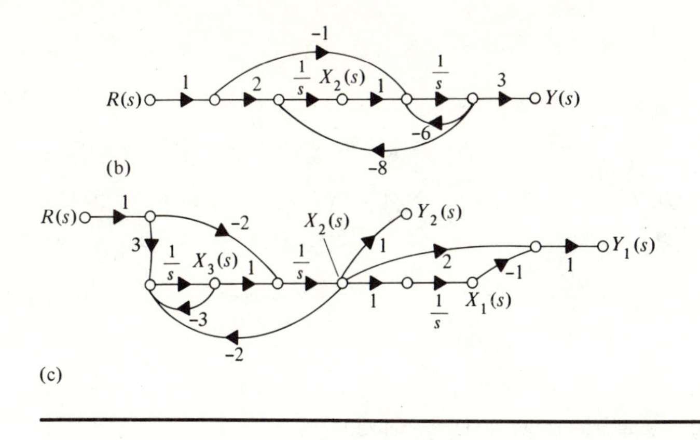

7.4.3 Complex Conjugate Characteristic Roots

In general, diagonalized state equations for systems with complex characteristic roots involve state equations with complex coefficients. For example, the single-input, single-output system with transfer function

$$T(s)=\frac{6s^2+26s+8}{(s+2)(s^2+2s+10)}=\frac{-2}{s+2}+\frac{4+j}{s+1+j3}+\frac{4-j}{s+1-j3}$$

may be represented in terms of state variables as in the simulation diagram of Fig. 7-10a. The gains associated with the complex characteristic roots are generally complex numbers. The state equations, in terms of the indicated state variables, are given by

$$\begin{aligned}
sX_1(s)&=-2X_1(s)+R(s)\\
sX_2(s)&=(-1+j3)X_2(s)+R(s)\\
sX_3(s)&=(-1-j3)X_3(s)+R(s)\\
Y(s)&=-2X_1(s)+(4+j)X_2(s)+(4-j)X_3(s)
\end{aligned}$$

or

$$\begin{bmatrix}\dot{x}_1\\ \dot{x}_2\\ \dot{x}_3\end{bmatrix}=\begin{bmatrix}-2 & 0 & 0\\ 0 & -1+j3 & 0\\ 0 & 0 & -1-j3\end{bmatrix}\begin{bmatrix}x_1\\ x_2\\ x_3\end{bmatrix}+\begin{bmatrix}1\\ 1\\ 1\end{bmatrix}r$$

$$y=[-2 \quad 4+j \quad 4-j]\begin{bmatrix}x_1\\ x_2\\ x_3\end{bmatrix}$$

Although the individual physical components of this representation cannot be assembled, involving complex numbers as they do, the mathematical relation-

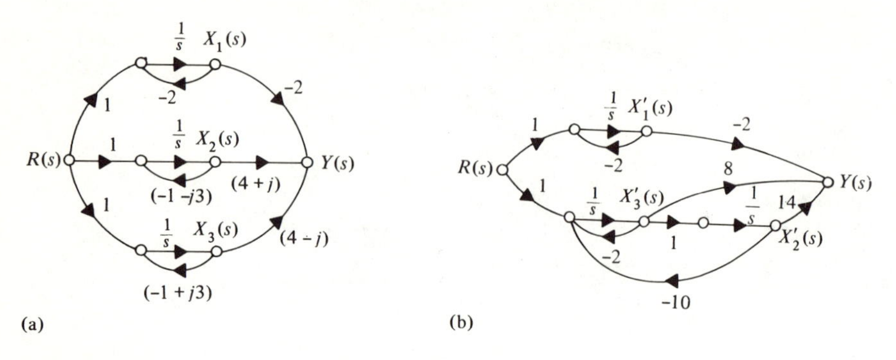

FIGURE 7-10
A system with complex characteristic roots. (a) Diagonalized system. (b) Alternative form for the diagonalized system, where the complex conjugate root terms have been combined and placed in phase-variable form.

ships are valid. To build such a system, or to represent it in a convenient form which does not involve complex numbers, the two complex conjugate component parts may be combined just as one commonly combines the corresponding conjugate partial fraction terms:

$$\frac{4+j}{s+1+j3}+\frac{4-j}{s+1-j3}=\frac{8s+14}{s^2+2s+10}$$

This portion of the system may be represented in phase-variable form, giving the real-number simulation diagram of Fig. 7-10b. The state equations for this alternative arrangement are given by

$$\begin{aligned}
sX_1'(s)&=-2X_1'(s)+R(s)\\
sX_2'(s)&=X_3'(s)\\
sX_3'(s)&=-10X_2'(s)-2X_3'(s)+R(s)\\
Y(s)&=-2X_1'(s)+14X_2'(s)+8X_3'(s)
\end{aligned}$$

or

$$\begin{bmatrix}\dot{x}_1'\\ \dot{x}_2'\\ \dot{x}_3'\end{bmatrix}=\begin{bmatrix}-2 & 0 & 0\\ 0 & 0 & 1\\ 0 & -10 & -2\end{bmatrix}\begin{bmatrix}x_1\\ x_2\\ x_3\end{bmatrix}+\begin{bmatrix}1\\ 0\\ 1\end{bmatrix}r$$

$$y=\begin{bmatrix}-2 & 14 & 8\end{bmatrix}\begin{bmatrix}x_1\\ x_2\\ x_3\end{bmatrix}$$

It is thus possible to represent systems with one or more pairs of complex conjugate characteristic roots with diagonalized state equations involving complex numbers or in block diagonal form involving real numbers. The state equations

$$\begin{bmatrix} \dot{x}_1 \\ \dot{x}_2 \\ \dot{x}_3 \\ \dot{x}_4 \\ \dot{x}_5 \\ \dot{x}_6 \end{bmatrix} = \begin{bmatrix} 3 & 0 & 0 & 0 & 0 & 0 \\ 0 & -4 & 0 & 0 & 0 & 0 \\ 0 & 0 & 0 & 1 & 0 & 0 \\ 0 & 0 & -17 & -2 & 0 & 0 \\ 0 & 0 & 0 & 0 & 0 & 1 \\ 0 & 0 & 0 & 0 & -10 & 3 \end{bmatrix} \begin{bmatrix} x_1 \\ x_2 \\ x_3 \\ x_4 \\ x_5 \\ x_6 \end{bmatrix} + \begin{bmatrix} 1 \\ 1 \\ 0 \\ 1 \\ 0 \\ 1 \end{bmatrix} r$$

$$y = \begin{bmatrix} 6 & -8 & 1 & -5 & 0 & 7 \end{bmatrix} \begin{bmatrix} x_1 \\ x_2 \\ x_3 \\ x_4 \\ x_5 \\ x_6 \end{bmatrix}$$

for example, which are in block diagonal form, represent a system with transfer function

$$T(s) = \frac{6}{s-3} + \frac{-8}{s+4} + \frac{-5s+1}{s^2+2s+17} + \frac{7s}{s^2-3s+10}$$

DRILL PROBLEM

D7-8. The following transfer functions for single-input, single-output systems involve complex characteristic roots. Find diagonal state equations for these systems. Then find an alternative block diagonal representation which does not involve complex numbers.

(a) $T(s) = \dfrac{10}{s^3+2s^2+5s}$

ans.
$$\begin{bmatrix} \dot{x}_1 \\ \dot{x}_2 \\ \dot{x}_3 \end{bmatrix} = \begin{bmatrix} 0 & 0 & 0 \\ 0 & 0 & 1 \\ 0 & -5 & -2 \end{bmatrix} \begin{bmatrix} x_1 \\ x_2 \\ x_3 \end{bmatrix} + \begin{bmatrix} 2 \\ 0 \\ 1 \end{bmatrix} r$$

$$y = \begin{bmatrix} 1 & -4 & -2 \end{bmatrix} \begin{bmatrix} x_1 \\ x_2 \\ x_3 \end{bmatrix}$$

(b) $T(s)=\dfrac{3s^2-1}{(s^2+4)(s^2+4s+5)}$

$$ans.\quad \begin{bmatrix}\dot{x}_1\\ \dot{x}_2\\ \dot{x}_3\\ \dot{x}_4\end{bmatrix}=\begin{bmatrix}0&1&0&0\\ -4&0&0&0\\ 0&0&0&1\\ 0&0&-5&-4\end{bmatrix}\begin{bmatrix}x_1\\ x_2\\ x_3\\ x_4\end{bmatrix}+\begin{bmatrix}0\\ 1\\ 0\\ 1\end{bmatrix}r$$

$$y=\begin{bmatrix}-\frac{1}{5} & \frac{4}{5} & 0 & -\frac{4}{5}\end{bmatrix}\begin{bmatrix}x_1\\ x_2\\ x_3\\ x_4\end{bmatrix}$$

(c) $T(s)=\dfrac{s^2-4s+10}{(s+2)(s^2+6s+13)}$

$$ans.\quad \begin{bmatrix}\dot{x}_1\\ \dot{x}_2\\ \dot{x}_3\end{bmatrix}=\begin{bmatrix}-2&0&0\\ 0&0&1\\ 0&-13&-6\end{bmatrix}\begin{bmatrix}x_1\\ x_2\\ x_3\end{bmatrix}+\begin{bmatrix}1\\ 0\\ 1\end{bmatrix}r$$

$$y=\begin{bmatrix}\frac{22}{5} & -\frac{118}{5} & -\frac{17}{5}\end{bmatrix}\begin{bmatrix}x_1\\ x_2\\ x_3\end{bmatrix}$$

7.4.4 Repeated Characteristic Roots

The state equations for a system with repeated characteristic roots cannot be diagonalized. A block diagonal form, termed a *Jordan canonical form,* is commonly used when a simple representation is desired. For example, the single-input, single-output system with transfer function

$$T(s)=\frac{10s^2+51s+56}{(s+4)(s+2)^2}=\frac{3}{s+4}+\frac{-6}{s+2}+\frac{7}{(s+2)^2}$$

may be represented as in Fig. 7-11a. A simplification results when the $1/(s+2)$ transmittance is used in common by two paths, as shown in Fig. 7-11b. A corresponding state variable representation is given in Fig. 7-11c.

The state equations are given by

$$\begin{aligned}
sX_1(s)&=-4X_1(s)+R(s)\\
sX_2(s)&=-2X_2(s)+X_3(s)\\
sX_3(s)&=-2X_3(s)+R(s)\\
Y(s)&=3X_1(s)+7X_2(s)-6X_3(s)
\end{aligned}$$

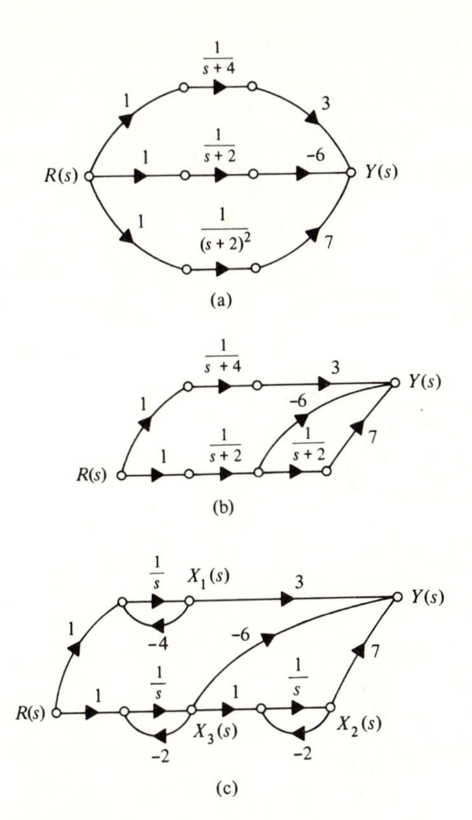

FIGURE 7-11
State equations for a system with repeated characteristic roots. (a) Diagram showing each partial fraction term. (b) Diagram with common signal path through a repeated transmittance. (c) Diagram showing state variables.

or

$$\begin{bmatrix} \dot{x}_1 \\ \dot{x}_2 \\ \dot{x}_3 \end{bmatrix} = \begin{bmatrix} -4 & 0 & 0 \\ 0 & -2 & 1 \\ 0 & 0 & -2 \end{bmatrix} \begin{bmatrix} x_1 \\ x_2 \\ x_3 \end{bmatrix} + \begin{bmatrix} 1 \\ 0 \\ 1 \end{bmatrix} r$$

$$y = \begin{bmatrix} 3 & 7 & -6 \end{bmatrix} \begin{bmatrix} x_1 \\ x_2 \\ x_3 \end{bmatrix}$$

For three repetitions of a characteristic root, the corresponding transfer function partial fraction terms are

$$\frac{k_1}{s+a} + \frac{k_2}{(s+a)^2} + \frac{k_3}{(s+a)^3}$$

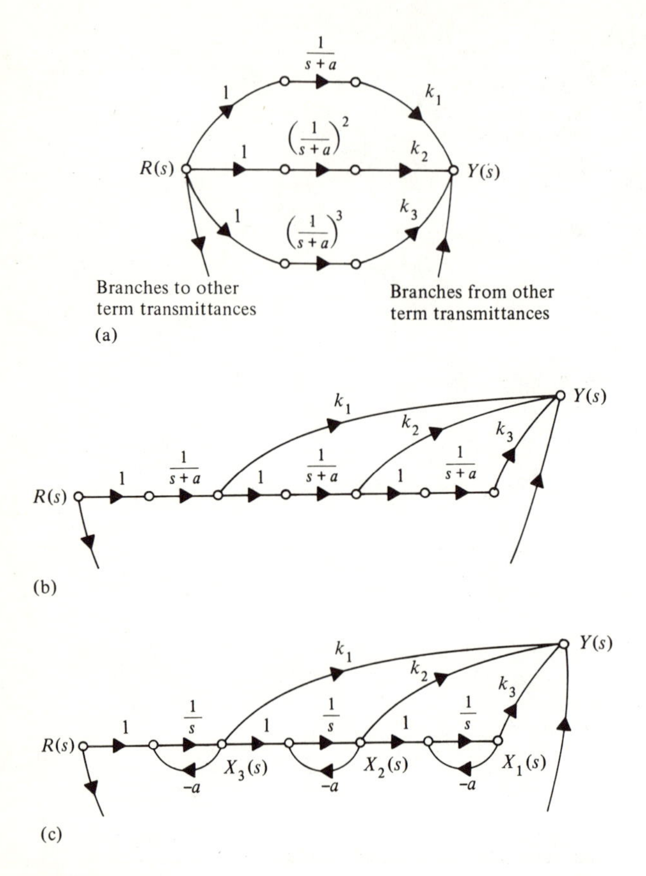

FIGURE 7-12
State variables for three repeated roots. (a) Diagram showing each partial fraction term. (b) Diagram using common signal paths. (c) Diagram showing state variables.

and the state variables may be defined as in Fig. 7-12. The resulting Jordan block has the following structure:

$$\begin{bmatrix} \dot{x}_1 \\ \dot{x}_2 \\ \dot{x}_3 \\ \\ \vdots \end{bmatrix} = \begin{bmatrix} -a & 1 & 0 & 0 & 0 & \cdots \\ 0 & -a & 1 & 0 & 0 & \cdots \\ 0 & 0 & -a & 0 & 0 & \cdots \\ 0 & 0 & 0 & & & \\ \vdots & \vdots & \vdots & \vdots & \vdots & \end{bmatrix} \begin{bmatrix} x_1 \\ x_2 \\ x_3 \\ \\ \vdots \end{bmatrix} + \begin{bmatrix} 0 \\ 0 \\ 1 \\ \\ \vdots \end{bmatrix} r$$

$$y = [k_3 \quad k_2 \quad k_1 \quad \cdots] \begin{bmatrix} x_1 \\ x_2 \\ x_3 \\ \vdots \end{bmatrix}$$

The state variable equations

$$\begin{bmatrix} \dot{x}_1 \\ \dot{x}_2 \\ \dot{x}_3 \\ \dot{x}_4 \\ \dot{x}_5 \\ \dot{x}_6 \end{bmatrix} = \begin{bmatrix} -2 & 1 & 0 & 0 & 0 & 0 \\ 0 & -2 & 0 & 0 & 0 & 0 \\ 0 & 0 & -3 & 0 & 0 & 0 \\ 0 & 0 & 0 & 4 & 1 & 0 \\ 0 & 0 & 0 & 0 & 4 & 1 \\ 0 & 0 & 0 & 0 & 0 & 4 \end{bmatrix} \begin{bmatrix} x_1 \\ x_2 \\ x_3 \\ x_4 \\ x_5 \\ x_6 \end{bmatrix} + \begin{bmatrix} 0 \\ 1 \\ 1 \\ 0 \\ 0 \\ 1 \end{bmatrix} r$$

$$y = [4 \quad -5 \quad 6 \quad 7 \quad -8 \quad 9] \begin{bmatrix} x_1 \\ x_2 \\ x_3 \\ x_4 \\ x_5 \\ x_6 \end{bmatrix}$$

for example, are in Jordan canonical form. They represent a system with transfer function

$$T(s) = \frac{-5}{s+2} + \frac{4}{(s+2)^2} + \frac{6}{s+3} + \frac{9}{s-4} + \frac{-8}{(s-4)^2} + \frac{7}{(s-4)^3}$$

DRILL PROBLEM

D7-9. The following systems have repeated characteristic roots. Find an alternate set of state equations in Jordan canonical form.

(a) $$\begin{bmatrix} \dot{x}_1 \\ \dot{x}_2 \end{bmatrix} = \begin{bmatrix} 2 & 9 \\ -1 & -4 \end{bmatrix} \begin{bmatrix} x_1 \\ x_2 \end{bmatrix} + \begin{bmatrix} 4 \\ -3 \end{bmatrix} r$$

$$y = [2 \quad -6] \begin{bmatrix} x_1 \\ x_2 \end{bmatrix}$$

ans. $$\begin{bmatrix} \dot{x}'_1 \\ \dot{x}'_2 \end{bmatrix} = \begin{bmatrix} -1 & 1 \\ 0 & -1 \end{bmatrix} \begin{bmatrix} x'_1 \\ x'_2 \end{bmatrix} + \begin{bmatrix} 0 \\ 1 \end{bmatrix} r$$

$$y = [-60 \quad 26] \begin{bmatrix} x'_1 \\ x'_2 \end{bmatrix}$$

(b) $$\begin{bmatrix} \dot{x}_1 \\ \dot{x}_2 \\ \dot{x}_3 \end{bmatrix} = \begin{bmatrix} -6 & 1 & 0 \\ -9 & 0 & 1 \\ 0 & 0 & 0 \end{bmatrix} \begin{bmatrix} x_1 \\ x_2 \\ x_3 \end{bmatrix} + \begin{bmatrix} 1 \\ 2 \\ -1 \end{bmatrix} r$$

$$y=\begin{bmatrix}3 & -2 & 1\end{bmatrix}\begin{bmatrix}x_1\\x_2\\x_3\end{bmatrix}$$

$$ans.\quad \begin{bmatrix}\dot{x}'_1\\\dot{x}'_2\\\dot{x}'_3\end{bmatrix}=\begin{bmatrix}-3 & 1 & 0\\0 & -3 & 1\\0 & 0 & -3\end{bmatrix}\begin{bmatrix}x'_1\\x'_2\\x'_3\end{bmatrix}+\begin{bmatrix}0\\0\\1\end{bmatrix}r$$

$$y=\begin{bmatrix}2 & -2 & 0\end{bmatrix}\begin{bmatrix}x'_1\\x'_2\\x'_3\end{bmatrix}$$

7.5 CONTROLLABILITY AND OBSERVABILITY

7.5.1 Uncontrollable and Unobservable Modes

When a system is placed in diagonal form, its state equations are decoupled from one another. The natural response terms in each of the individual equation solutions are called *modes*. The natural component of each system output consists of a linear combination of the system modes. For example, the diagonal system

$$\begin{bmatrix}\dot{x}_1\\\dot{x}_2\\\dot{x}_3\end{bmatrix}=\begin{bmatrix}8 & 0 & 0\\0 & -4 & 0\\0 & 0 & -10\end{bmatrix}\begin{bmatrix}x_1\\x_2\\x_3\end{bmatrix}+\begin{bmatrix}1\\-2\\3\end{bmatrix}r$$

$$y=\begin{bmatrix}4 & -5 & -6\end{bmatrix}\begin{bmatrix}x_1\\x_2\\x_3\end{bmatrix}$$

is described by the decoupled differential equations

$$\dot{x}_1=8x_1+r$$
$$\dot{x}_2=-4x_2-2r$$
$$\dot{x}_3=-10x_3+3r$$

The state variables have natural response components of the following form:

$$x_{1\ \text{natural}}(t)=K_1e^{8t}$$
$$x_{2\ \text{natural}}(t)=K_2e^{-4t}$$
$$x_{3\ \text{natural}}(t)=K_3e^{-10t}$$

If, for a *diagonalized* system, there is no path from any input to one of the decoupled equations, as in Fig. 7-13a, the corresponding mode is termed

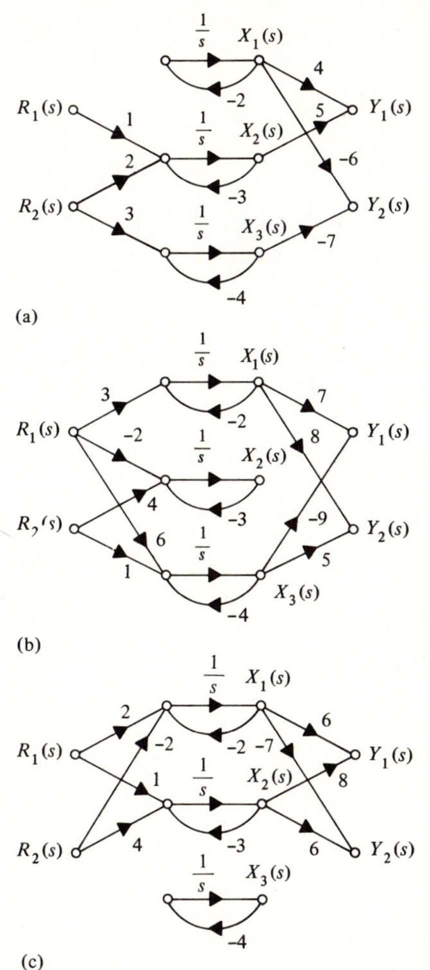

FIGURE 7-13
Systems with uncontrollable and unobservable modes. (a) Systems where the e^{-2t} mode is uncontrollable. (b) System where the e^{-3t} mode is unobservable. (c) System where the e^{-4t} mode is uncontrollable and unobservable.

uncontrollable. That portion of the system is not affected by any input. If it has no uncontrollable modes, a system is said to be completely controllable.

For a *diagonalized* system, if one of the state variables of the decoupled equations is not added to any of the system outputs, as in Fig. 7-13b, the corresponding mode is termed *unobservable.* The response of that first-order differential equation is not visible from any output. A system with no unobservable modes is said to be completely observable. Modes may be both uncontrollable and unobservable, as in the example of Fig. 7-13c.

The system with diagonalized state equations

$$\begin{bmatrix} \dot{x}_1 \\ \dot{x}_2 \\ \dot{x}_3 \end{bmatrix} = \begin{bmatrix} 2 & 0 & 0 \\ 0 & -3 & 0 \\ 0 & 0 & -4 \end{bmatrix} \begin{bmatrix} x_1 \\ x_2 \\ x_3 \end{bmatrix} + \begin{bmatrix} 1 & -2 \\ 0 & 0 \\ 0 & 0 \end{bmatrix} \begin{bmatrix} r_1 \\ r_2 \end{bmatrix}$$

is not completely controllable. For this system, both the e^{-3t} and the e^{-4t} modes are not coupled to any input. The diagonal system

$$\begin{bmatrix} \dot{x}_1 \\ \dot{x}_2 \\ \dot{x}_3 \end{bmatrix} = \begin{bmatrix} 3+j & 0 & 0 \\ 0 & 3-j & 0 \\ 0 & 0 & 6 \end{bmatrix} \begin{bmatrix} x_1 \\ x_2 \\ x_3 \end{bmatrix} + \begin{bmatrix} -1 & 2j \\ 1 & -2j \\ 0 & 4 \end{bmatrix} \begin{bmatrix} r_1 \\ r_2 \end{bmatrix}$$

is completely controllable since the rows of its input coupling matrix are each nonzero.

The system with diagonalized state equations

$$\begin{bmatrix} \dot{x}_1 \\ \dot{x}_2 \\ \dot{x}_3 \end{bmatrix} = \begin{bmatrix} -2 & 0 & 0 \\ 0 & 1 & 0 \\ 0 & 0 & 4 \end{bmatrix} \begin{bmatrix} x_1 \\ x_2 \\ x_3 \end{bmatrix} + \begin{bmatrix} 1 \\ 0 \\ 0 \end{bmatrix} r$$

$$\begin{bmatrix} y_1 \\ y_2 \end{bmatrix} = \begin{bmatrix} 1 & 0 & 3 \\ 0 & -2 & 3 \end{bmatrix} \begin{bmatrix} x_1 \\ x_2 \\ x_3 \end{bmatrix}$$

is completely observable since there are no columns of zeros in its output coupling matrix. The diagonalized system

$$\begin{bmatrix} \dot{x}_1 \\ \dot{x}_2 \end{bmatrix} = \begin{bmatrix} -3 & 0 \\ 0 & -10 \end{bmatrix} \begin{bmatrix} x_1 \\ x_2 \end{bmatrix} + \begin{bmatrix} 1 & 2 \\ 0 & 4 \end{bmatrix} \begin{bmatrix} r_1 \\ r_2 \end{bmatrix}$$

$$y = \begin{bmatrix} 1 & 0 \end{bmatrix} \begin{bmatrix} x_1 \\ x_2 \end{bmatrix}$$

is not completely observable since the e^{-10t} mode does not couple to the system output.

The system with diagonalized state equations

$$\begin{bmatrix} \dot{x}_1 \\ \dot{x}_2 \\ \dot{x}_3 \end{bmatrix} = \begin{bmatrix} -3 & 0 & 0 \\ 0 & 2 & 0 \\ 0 & 0 & 4 \end{bmatrix} \begin{bmatrix} x_1 \\ x_2 \\ x_3 \end{bmatrix} + \begin{bmatrix} -1 \\ 0 \\ 0 \end{bmatrix} r$$

$$y = \begin{bmatrix} 0 & 6 & 0 \end{bmatrix} \begin{bmatrix} x_1 \\ x_2 \\ x_3 \end{bmatrix}$$

is neither completely controllable nor completely observable. The e^{-3t} mode is controllable but not observable; the e^{2t} mode is observable but not controllable; while the e^{4t} mode is both uncontrollable and unobservable.

For a system which is not diagonal, the presence or absence of complete controllability or complete observability is not obvious. A row of zeros in the input coupling matrix does *not* indicate lack of complete controllability; when diagonalized, such a system may or may not have a row of zeros in the transformed input coupling matrix. Similarly, a column of zeros in the output coupling matrix of a nondiagonal system does not necessarily indicate absence of complete observability.

For systems with repeated characteristic roots, which cannot be diagonalized, controllability and observability may be determined by examining the system matrices in Jordan form.

7.5.2 The Controllability Matrix

Fortunately, there is a much simpler method of determining system controllability than diagonalization. It can be shown that an nth-order system, with or without repeated characteristic roots,

$$\dot{\mathbf{x}} = \mathbf{A}\mathbf{x} + \mathbf{B}\mathbf{r}$$

is completely controllable if and only if its controllability matrix

$$\mathbf{M}_c = [\mathbf{B} \mid \mathbf{A}\mathbf{B} \mid \cdots \mid \mathbf{A}^{n-1}\mathbf{B}]$$

is of full rank. The controllability matrix consists of the columns of **B** followed by the columns of **AB**, and so on.

For the system

$$\begin{bmatrix} \dot{x}_1 \\ \dot{x}_2 \\ \dot{x}_3 \end{bmatrix} = \begin{bmatrix} -2 & 1 & 2 \\ 4 & 0 & 3 \\ 1 & -1 & 0 \end{bmatrix} \begin{bmatrix} x_1 \\ x_2 \\ x_3 \end{bmatrix} + \begin{bmatrix} 0 & 4 \\ -5 & 0 \\ 0 & 0 \end{bmatrix} \begin{bmatrix} r_1 \\ r_2 \end{bmatrix}$$

A is 3×3 so

$$\mathbf{M}_c = [\mathbf{B} \mid \mathbf{A}\mathbf{B} \mid \mathbf{A}^2\mathbf{B}]$$

Using

$$\mathbf{A}\mathbf{B} = \begin{bmatrix} -2 & 1 & 2 \\ 4 & 0 & 3 \\ 1 & -1 & 0 \end{bmatrix} \begin{bmatrix} 0 & 4 \\ -5 & 0 \\ 0 & 0 \end{bmatrix} = \begin{bmatrix} -5 & -8 \\ 0 & 16 \\ 5 & 4 \end{bmatrix}$$

$$\mathbf{A}^2\mathbf{B} = \mathbf{A}(\mathbf{AB}) = \begin{bmatrix} -2 & 1 & 2 \\ 4 & 0 & 3 \\ 1 & -1 & 0 \end{bmatrix} \begin{bmatrix} -5 & -8 \\ 0 & 16 \\ 5 & 4 \end{bmatrix} = \begin{bmatrix} 20 & 40 \\ -5 & -20 \\ -5 & -24 \end{bmatrix}$$

then

$$\mathbf{M}_c = \begin{bmatrix} 0 & 4 & -5 & -8 & 20 & 40 \\ -5 & 0 & 0 & 16 & -5 & -20 \\ 0 & 0 & 5 & 4 & -5 & -24 \end{bmatrix}$$

To be of full rank, the controllability matrix must have three linearly independent columns, which it does, since

$$\begin{vmatrix} 0 & 4 & -5 \\ -5 & 0 & 0 \\ 0 & 0 & 5 \end{vmatrix} \neq 0$$

The system

$$\begin{bmatrix} \dot{x}_1 \\ \dot{x}_2 \end{bmatrix} = \begin{bmatrix} 2 & 3 \\ 6 & -1 \end{bmatrix} \begin{bmatrix} x_1 \\ x_2 \end{bmatrix} + \begin{bmatrix} 1 \\ -2 \end{bmatrix} r$$

has controllability matrix

$$\mathbf{M}_c = [\mathbf{B} \,\vdots\, \mathbf{AB}]$$

where

$$\mathbf{AB} = \begin{bmatrix} 2 & 3 \\ 6 & -1 \end{bmatrix} \begin{bmatrix} 1 \\ -2 \end{bmatrix} = \begin{bmatrix} -4 \\ 8 \end{bmatrix}$$

Thus

$$\mathbf{M}_c = \begin{bmatrix} 1 & -4 \\ -2 & 8 \end{bmatrix}$$

which is not of rank 2 since

$$\begin{vmatrix} 1 & -4 \\ -2 & 8 \end{vmatrix} = 0$$

This system is not completely controllable.

The rank test with the controllability matrix does not indicate which mode is uncontrollable, but it is far simpler to apply the test than to diagonalize the state equations.

7.5.3 The Observability Matrix

To determine whether or not a nondiagonalized nth-order system is completely observable, its observability matrix

$$\mathbf{M}_o = \begin{bmatrix} \mathbf{C} \\ \hdashline \mathbf{CA} \\ \hdashline \vdots \\ \hdashline \mathbf{CA}^{n-1} \end{bmatrix}$$

may be formed. The system is completely observable if and only if the observability matrix is of full rank, that is, if $\mathbf{M}_o$ has n linearly independent rows.

The system

$$\begin{bmatrix} \dot{x}_1 \\ \dot{x}_2 \\ \dot{x}_3 \end{bmatrix} = \begin{bmatrix} 2 & 1 & 0 \\ -3 & 0 & 1 \\ 4 & 0 & 0 \end{bmatrix} \begin{bmatrix} x_1 \\ x_2 \\ x_3 \end{bmatrix} = \begin{bmatrix} 1 \\ 1 \\ 1 \end{bmatrix} r$$

$$y = \begin{bmatrix} 0 & 0 & 1 \end{bmatrix} \begin{bmatrix} x_1 \\ x_2 \\ x_3 \end{bmatrix}$$

for example, is completely observable:

$$\mathbf{CA} = \begin{bmatrix} 0 & 0 & 1 \end{bmatrix} \begin{bmatrix} 2 & 1 & 0 \\ -3 & 0 & 1 \\ 4 & 0 & 0 \end{bmatrix} = \begin{bmatrix} 4 & 0 & 0 \end{bmatrix}$$

$$\mathbf{CA}^2 = (\mathbf{CA})\mathbf{A} = \begin{bmatrix} 4 & 0 & 0 \end{bmatrix} \begin{bmatrix} 2 & 1 & 0 \\ -3 & 0 & 1 \\ 4 & 0 & 0 \end{bmatrix}$$

$$= \begin{bmatrix} 8 & 4 & 0 \end{bmatrix}$$

$$\mathbf{M}_o = \begin{bmatrix} 0 & 0 & 1 \\ 4 & 0 & 0 \\ 8 & 4 & 0 \end{bmatrix}$$

As another example, the system

$$\begin{bmatrix} \dot{x}_1 \\ \dot{x}_2 \end{bmatrix} = \begin{bmatrix} 1 & 0 \\ 1 & 1 \end{bmatrix} \begin{bmatrix} x_1 \\ x_2 \end{bmatrix} + \begin{bmatrix} 1 \\ 1 \end{bmatrix} r$$

$$\begin{bmatrix} y_1 \\ y_2 \end{bmatrix} = \begin{bmatrix} 1 & -1 \\ -2 & 2 \end{bmatrix} \begin{bmatrix} x_1 \\ x_2 \end{bmatrix}$$

is not completely observable:

$$\mathbf{M}_o = \begin{bmatrix} 1 & -1 \\ -2 & 2 \\ 1 & -1 \\ -1 & 1 \end{bmatrix}$$

The observability matrix does not have two linearly independent rows.

It is a very special case when a system is not completely observable or not completely controllable. The parameters describing the system must be "just right" for such situations to occur. Nevertheless, the special cases do occur frequently enough in practice to warrant careful consideration.

DRILL PROBLEM

D7-10. Use controllability and observability matrices to determine whether the following systems are completely controllable and whether these systems are completely observable.

ans. (a) not completely controllable and not completely observable
(b) completely controllable and completely observable
(c) completely controllable but not completely observable

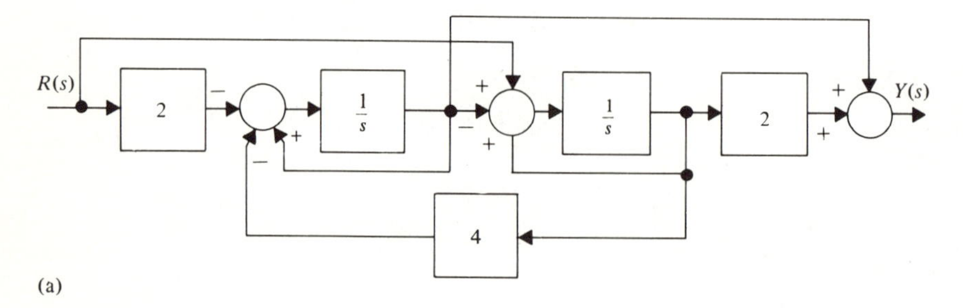

(a)

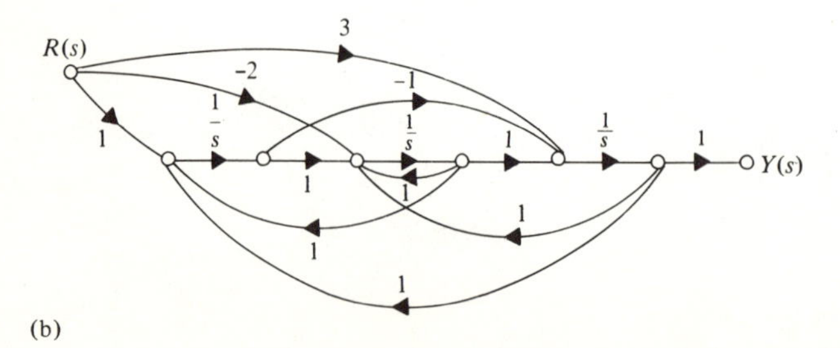

(b)

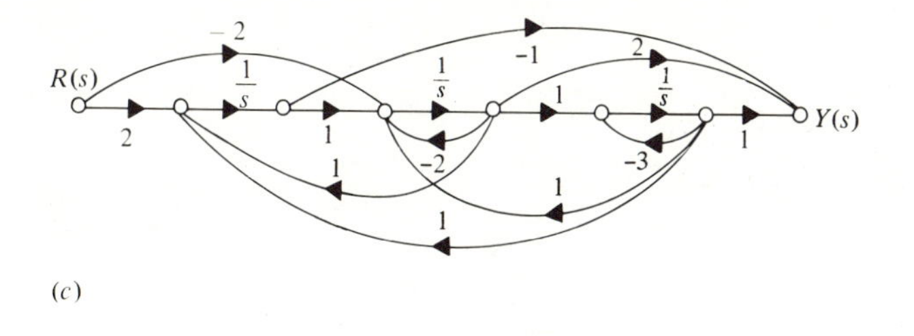

(c)

7.6 TIME RESPONSE FROM STATE EQUATIONS

7.6.1 Laplace Transform Solution

One method of calculating the state of a system as a function of time is to Laplace-transform the equations, solve for the transform of the signals of interest, then invert the transforms. The system outputs, being linear combinations of the state signals, are easily found from the state.

For example, consider the system

$$\begin{bmatrix} \dot{x}_1 \\ \dot{x}_2 \end{bmatrix} = \begin{bmatrix} -6 & 1 \\ -5 & 0 \end{bmatrix} \begin{bmatrix} x_1 \\ x_2 \end{bmatrix} + \begin{bmatrix} 0 \\ 1 \end{bmatrix} r$$

$$y = [1 \quad -2] \begin{bmatrix} x_1 \\ x_2 \end{bmatrix}$$

with initial state

$$\begin{bmatrix} x_1(0^-) \\ x_2(0^-) \end{bmatrix} = \begin{bmatrix} -3 \\ 1 \end{bmatrix}$$

and input

$$r(t) = 7u(t)$$

where $u(t)$ is the unit step function.

The Laplace-transformed state equations are as follows:

$$\begin{cases} sX_1(s) + 3 = -6X_1(s) + X_2(s) \\ sX_2(s) - 1 = -5X_1(s) + \dfrac{7}{s} \end{cases}$$

$$\begin{cases} (s+6)X_1(s) - X_2(s) = -3 \\ 5X_1(s) + sX_2(s) = 1 + \dfrac{7}{s} = \dfrac{s+7}{s} \end{cases}$$

$$X_1(s)=\frac{\begin{vmatrix} -3 & -1 \\ \dfrac{s+7}{s} & s \end{vmatrix}}{\begin{vmatrix} s+6 & -1 \\ 5 & s \end{vmatrix}}=\frac{-3s+(s+7)/s}{s^2+6s+5}$$

$$=\frac{-3s^2+s+7}{s(s+1)(s+5)}=\frac{\frac{7}{5}}{s}+\frac{-\frac{3}{4}}{s+1}+\frac{-\frac{73}{20}}{s+5}$$

$$x_1(t)=\tfrac{7}{5}-\tfrac{3}{4}e^{-t}-\tfrac{73}{20}e^{-5t} \qquad t\geqslant 0$$

Similarly,

$$X_2(s)=\frac{\begin{vmatrix} s+6 & -3 \\ 5 & \dfrac{s+7}{s} \end{vmatrix}}{s^2+6s+5}=\frac{(s^2+13s+42)/s+15}{s^2+6s+5}$$

$$=\frac{s^2+28s+42}{s(s+1)(s+5)}=\frac{\frac{42}{5}}{s}+\frac{-\frac{15}{4}}{s+1}+\frac{-\frac{73}{20}}{s+5}$$

$$x_2(t)=\tfrac{42}{5}-\tfrac{15}{4}e^{-t}-\tfrac{73}{20}e^{-5t} \qquad t\geqslant 0$$

The system output is then

$$y(t)=x_1-2x_2=-\tfrac{77}{5}+(+\tfrac{27}{4})e^{-t}+\tfrac{73}{20}e^{-5t}$$

7.6.2 Time-Domain Response of First-Order Systems

In many situations, it is advantageous to have an expression for the solution of a set of state equations as functions of time rather than in terms of Laplace transforms. For a first-order state variable system,

$$\frac{dx}{dt}=ax+br$$

$$sX(s)-x(0^-)=aX(s)+bR(s)$$

$$X(s)=\frac{x(0^-)}{s-a}+bR(s)\frac{1}{s-a}$$

$$x(t)=\mathscr{L}^{-1}\left\{\frac{x(0^-)}{s-a}+bR(s)\frac{1}{s-a}\right\}$$

$$=e^{-at}x(0^-)+\text{convolution}\left[br(t),e^{-at}\right]$$

$$=e^{-at}x(0^-)+\int_{0^-}^{t}e^{-a(t-\tau)}br(\tau)\,d\tau$$

the inverse transform of a product of Laplace transforms being the convolution of the corresponding time functions.

An alternate derivation of this result is as follows. Multiplying the first-order equation

$$\frac{dx}{dt}=ax+br$$

by the integrating factor e^{-at} gives

$$e^{-at}\frac{dx}{dt}-axe^{-at}=e^{-at}br(t)$$

The left side of this equation is seen to be the derivative of a product:

$$\frac{d}{dt}[e^{-at}x(t)]=e^{-at}br(t)$$

Integrating both sides, there results

$$e^{-at}x(t)=\int e^{-at}br(t)\,dt+C$$

where C is an arbitrary constant of integration.

If the integration is begun at time $t=0^-$,

$$e^{-at}x(t)=\int_{0^-}^{t} e^{-a\tau}br(\tau)\,d\tau + C \qquad t\geqslant 0$$

where the variable of integration has been written as τ to avoid confusion with the integral's upper limit t. Substituting $t=0$, the constant of integration is seen to be

$$C=x(0^-)$$

giving

$$\begin{aligned}x(t)&=e^{at}\int_{0^-}^{t} e^{-a\tau}br(\tau)\,d\tau + e^{at}x(0^-)\\ &=e^{at}x(0^-)+\int_{0^-}^{t} e^{a(t-\tau)}br(\tau)\,d\tau \qquad t\geqslant 0\end{aligned}$$

The integral is the convolution of the function e^{at} and the input $r(t)$.

As a numerical example, consider the first-order system

$$\dot{x} = -2x + 3r$$
$$y = 4x$$

The general solution for $x(t)$ is

$$x(t) = e^{-2t}x(0^-) + \int_{0^-}^{t} 3e^{-2(t-\tau)} r(\tau)\, d\tau$$

If

$$x(0^-) = 10 \quad \text{and} \quad r = 5$$

then

$$\begin{aligned} x(t) &= 10e^{-2t} + \int_{0^-}^{t} 15e^{-2(t-\tau)}\, d\tau \\ &= 10e^{-2t} + 15e^{-2t} \left. \frac{e^{2\tau}}{2} \right|_{0^-}^{t} \\ &= 10e^{-2t} + 15e^{-2t} \frac{e^{2t} - 1}{2} \\ &= \tfrac{5}{2}e^{-2t} + \tfrac{15}{2} \qquad t \geqslant 0 \end{aligned}$$

and the system output is

$$y(t) = 4x(t) = 10e^{-2t} + 30 \qquad t \geqslant 0$$

7.6.3 Time-Domain Response of Higher-Order Systems

In general, a state variable system

$$\dot{\mathbf{x}} = \mathbf{A}x + \mathbf{B}\mathbf{r}$$

has state response given by

$$s\mathbf{X}(s) - \mathbf{x}(0^-) = \mathbf{A}\mathbf{X}(s) + \mathbf{B}\mathbf{R}(s)$$
$$[s\mathbf{I} - \mathbf{A}]\mathbf{X}(s) = \mathbf{x}(0^-) + \mathbf{B}\mathbf{R}(s)$$
$$\mathbf{X}(s) = [s\mathbf{I} - \mathbf{A}]^{-1}\mathbf{x}(0^-) + [s\mathbf{I} - \mathbf{A}]^{-1}\mathbf{B}\mathbf{R}(s)$$

Denoting the *state transition matrix* by

$$\mathbf{\Phi}(t) = \mathscr{L}^{-1}\{[s\mathbf{I} - \mathbf{A}]^{-1}\}$$

then

$$\mathbf{x}(t)=\mathbf{\Phi}(t)\mathbf{x}(0^-)+\text{convolution}\,[\mathbf{Br}(t),\,\mathbf{\Phi}(t)]$$
$$=\mathbf{\Phi}(t)\mathbf{x}(0^-)+\int_{0^-}^{t}\mathbf{\Phi}(t-\tau)\mathbf{Br}(\tau)\,d\tau$$

For example, for the system

$$\begin{bmatrix}\dot{x}_1\\ \dot{x}_2\end{bmatrix}=\begin{bmatrix}-3 & 1\\ -2 & 0\end{bmatrix}\begin{bmatrix}x_1\\ x_2\end{bmatrix}+\begin{bmatrix}2\\ -1\end{bmatrix}r$$

the state transition matrix is given by

$$\mathbf{\Phi}(t)=\mathscr{L}^{-1}\{[s\mathbf{I}-\mathbf{A}]^{-1}\}$$
$$=\mathscr{L}^{-1}\left\{\begin{bmatrix}s+3 & -1\\ 2 & s\end{bmatrix}^{-1}\right\}$$
$$=\mathscr{L}^{-1}\begin{bmatrix}\dfrac{s}{s^2+3s+2} & \dfrac{1}{s^2+3s+2}\\[2ex] \dfrac{-2}{s^2+3s+2} & \dfrac{s+3}{s^2+3s+2}\end{bmatrix}$$
$$=\mathscr{L}^{-1}\begin{bmatrix}\dfrac{-1}{s+1}+\dfrac{2}{s+2} & \dfrac{1}{s+1}+\dfrac{-1}{s+2}\\[2ex] \dfrac{-2}{s+1}+\dfrac{2}{s+2} & \dfrac{2}{s+1}+\dfrac{-1}{s+2}\end{bmatrix}$$
$$=\begin{bmatrix}-e^{-t}+2e^{-2t} & e^{-t}-e^{-2t}\\ -2e^{-t}+2e^{-2t} & 2e^{-t}-e^{-2t}\end{bmatrix}$$

The system state is, in terms of initial conditions and the inputs,

$$\begin{bmatrix}x_1(t)\\ x_2(t)\end{bmatrix}=\begin{bmatrix}(-e^{-t}+2e^{-2t}) & (e^{-t}-e^{-2t})\\ (-2e^{-t}+2e^{-2t}) & (2e^{-t}-e^{-2t})\end{bmatrix}\begin{bmatrix}x_1(0^-)\\ x_2(0^-)\end{bmatrix}$$
$$+\int_{0^-}^{t}\begin{bmatrix}-e^{-(t-\tau)}+2e^{-2(t-\tau)} & e^{-(t-\tau)}-e^{-2(t-\tau)}\\ -2e^{-(t-\tau)}+2e^{-2(t-\tau)} & 2e^{-(t-\tau)}-e^{-2(t-\tau)}\end{bmatrix}\begin{bmatrix}2\\ -1\end{bmatrix}r(\tau)\,d\tau$$

DRILL PROBLEMS

D7-11. Use Laplace transform methods to find the outputs of the following systems for $t\geqslant 0$ with the given inputs and initial conditions:

(a) $\dot{x}=-2x+r(t)$
$y=10x$
$x(0^-)=3$
$r(t)=4e^{5t}$

ans. $\frac{170}{7}e^{-2t}+\frac{40}{7}e^{5t}$

(b) $$\begin{bmatrix}\dot{x}_1\\ \dot{x}_2\end{bmatrix}=\begin{bmatrix}0 & 1\\ -12 & -7\end{bmatrix}\begin{bmatrix}x_1\\ x_2\end{bmatrix}+\begin{bmatrix}1\\ 1\end{bmatrix}r$$

$$y=\begin{bmatrix}1 & -1\end{bmatrix}\begin{bmatrix}x_1\\ x_2\end{bmatrix}$$

$$\begin{bmatrix}x_1(0^-)\\ x_2(0^-)\end{bmatrix}=\begin{bmatrix}10\\ 0\end{bmatrix}$$

$r(t)=u(t)$, where $u(t)$ is the unit step function.

ans. $Y(s)=(10s^2+190s+20)/s(s+3)(s+4)$
$y(t)=\frac{5}{3}+\frac{460}{3}e^{-3t}-145e^{-4t} \quad t\geqslant 0$

(c) $$\begin{bmatrix}\dot{x}_1\\ \dot{x}_2\\ \dot{x}_3\end{bmatrix}=\begin{bmatrix}-5 & 1 & 0\\ -6 & 0 & 1\\ 0 & 0 & 0\end{bmatrix}\begin{bmatrix}x_1\\ x_2\\ x_3\end{bmatrix}+\begin{bmatrix}0\\ 0\\ 1\end{bmatrix}r$$

$$y=\begin{bmatrix}1 & 0 & 0\end{bmatrix}\begin{bmatrix}x_1\\ x_2\\ x_3\end{bmatrix}$$

$\mathbf{x}(0)=\mathbf{0}$
$r(t)=\delta(t)$, where $\delta(t)$ is the unit inpulse.

ans. $\frac{1}{6}+\frac{1}{3}e^{-3t}-\frac{1}{2}e^{-2t} \quad t\geqslant 0$

D7-12. Calculate state transition matrices for systems with the following state coupling matrices **A**, using
$\mathbf{\Phi}(t)=\mathscr{L}^{-1}\{[s\mathbf{I}-\mathbf{A}]^{-1}\}$:

(a) $$\begin{bmatrix}-9 & 1\\ -14 & 0\end{bmatrix}$$

ans. $$\begin{bmatrix}-\frac{2}{5}e^{-2t}+\frac{7}{5}e^{-7t} & \frac{1}{5}e^{-2t}-\frac{1}{5}e^{-7t}\\ -\frac{14}{5}e^{-2t}+\frac{14}{5}e^{-7t} & \frac{7}{5}e^{-2t}-\frac{2}{5}e^{-7t}\end{bmatrix}$$

(b) $$\begin{bmatrix}1 & -1\\ 2 & -4\end{bmatrix}$$

ans. $$\begin{bmatrix}1.11e^{0.56t}-0.11e^{-3.56t} & -0.24e^{0.56t}+0.24e^{-3.56t}\\ 0.48e^{0.56t}-0.48e^{-3.56t} & -0.11e^{0.56t}+1.11e^{-3.56t}\end{bmatrix}$$

7.7 SYSTEM RESPONSE COMPUTATION

One advantage of placing system equations in a state variable form is that it is well suited to digital computer calculations. Computers are not particularly efficient at equation manipulation, Laplace transformation, and the like, but they excel at such repetitive tasks as matrix addition and multiplication. The capability of simulating a system, that is, investigating and testing its performance by modeling, is important to the designer, particularly for the common situation in which the plant is expensive and the design has to be correct when it is first installed.

A simple method of response calculation is to approximate the time derivatives by

$$\frac{dx_i}{dt} = \frac{x_i(t+\Delta t) - x_i(t)}{\Delta t}$$

where Δt is a small time increment. The state equations become, approximately,

$$\frac{\mathbf{x}(t+\Delta t) - \mathbf{x}(t)}{\Delta t} \cong \mathbf{A}\mathbf{x}(t) + \mathbf{B}\mathbf{r}(t)$$

or

$$\mathbf{x}(t+\Delta t) \cong (\mathbf{I} + \mathbf{A}\,\Delta t)\mathbf{x}(t) + (\mathbf{B}\,\Delta t)\mathbf{r}(t)$$

For sufficiently small time increments Δt, one can start with the initial state $\mathbf{x}(0)$ and calculate $\mathbf{x}(\Delta t)$ as follows:

$$\mathbf{x}(\Delta t) \cong (\mathbf{I} + \mathbf{A}\,\Delta t)\mathbf{x}(0) + (\mathbf{B}\,\Delta t)\mathbf{r}(0)$$

then $\mathbf{x}(2\Delta t)$ may be calculated from $\mathbf{x}(\Delta t)$,

$$\mathbf{x}(2\Delta t) \cong (\mathbf{I} + \mathbf{A}\,\Delta t)\mathbf{x}(\Delta t) + (\mathbf{B}\,\Delta t)\mathbf{r}(\Delta t)$$

and so on, obtaining approximate solutions for the state,

$$\mathbf{x}\{(k+1)\Delta t\} \cong (\mathbf{I} + \mathbf{A}\,\Delta t)\mathbf{x}(k\,\Delta t) + (\mathbf{B}\,\Delta t)\mathbf{r}(k\,\Delta t)$$

For example, the response of the first-order system

$$\dot{x} = -2x + r$$
$$y = x$$

with

$$x(0^-)=10$$
$$r(t)=3 \sin t$$

is approximated by

$$x\{(k+1)\Delta t\} \cong (1-2\Delta t)x(k\,\Delta t)+3\Delta t \sin(k\,\Delta t)$$

with

$$x(0 \cdot \Delta t)=10$$

Representative computer-generated plots of $x(t)$ are given in Fig. 7-14 for various choices of Δt. For a sufficiently small time increment Δt, the approximate response is very nearly the actual system response.

Another example system is the following:

$$\begin{bmatrix} \dot{x}_1 \\ \dot{x}_2 \end{bmatrix} = \begin{bmatrix} -2 & 1 \\ -3 & 0 \end{bmatrix} \begin{bmatrix} x_1 \\ x_2 \end{bmatrix} + \begin{bmatrix} 2 \\ -1 \end{bmatrix} r$$

$$y = \begin{bmatrix} 1 & -\frac{1}{2} \end{bmatrix} \begin{bmatrix} x_1 \\ x_2 \end{bmatrix}$$

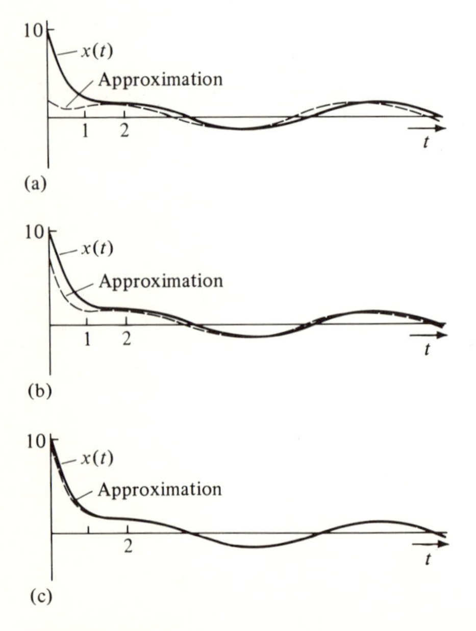

FIGURE 7-14
Computer-generated response plots for a first-order system.
(a) Step size $\Delta t=0.4$. (b) Step size $\Delta t=0.2$. (c) Step size $\Delta t=0.05$.

with

$$\begin{bmatrix} x_1(0^-) \\ x_2(0^-) \end{bmatrix} = \begin{bmatrix} -4 \\ 5 \end{bmatrix}$$
$$r(t) = \cos 0.25t$$

is approximated by

$$\begin{bmatrix} x_1\{(k+1)\Delta t\} \\ x_2\{(k+1)\Delta t\} \end{bmatrix} = \begin{bmatrix} 1-2\Delta t & \Delta t \\ -3\Delta t & 1 \end{bmatrix} \begin{bmatrix} x_1(k\,\Delta t) \\ x_2(k\,\Delta t) \end{bmatrix} + \begin{bmatrix} 2 \\ -1 \end{bmatrix} \Delta t \cos(0.25k\,\Delta t)$$
$$y\{(k+1)\Delta t\} = \begin{bmatrix} 1 & -\frac{1}{2} \end{bmatrix} \begin{bmatrix} x_1\{(k+1)\Delta t\} \\ x_2\{(k+1)\Delta t\} \end{bmatrix}$$

or

$$\begin{cases} x_1\{(k+1)\Delta t\} = (1-2\Delta t)x_1(k\,\Delta t) + \Delta t\; x_2(k\,\Delta t) + 2\Delta t \cos(0.25k\,\Delta t) \\ x_2\{(k+1)\Delta t\} = -3\Delta t\; x_1(k\,\Delta t) + x_2(k\,\Delta t) - \Delta t \cos(0.25k\,\Delta t) \\ \;\; y\{(k+1)\Delta t\} = x_1\{(k+1)\Delta t\} - \frac{1}{2}x_2\{(k+1)\Delta t\} \end{cases}$$

with

$$\begin{cases} x_1(0\cdot\Delta t) = -4 \\ x_2(0\cdot\Delta t) = 5 \end{cases}$$

Computer-generated response plots for this system are given in Fig. 7-15.

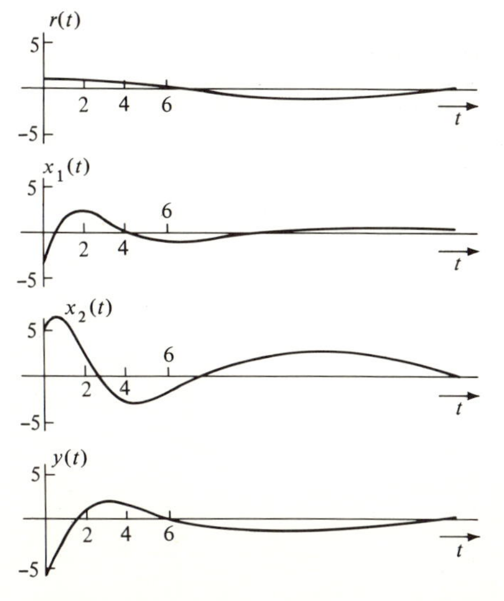

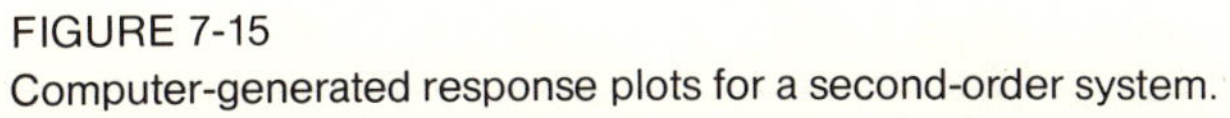
FIGURE 7-15
Computer-generated response plots for a second-order system.

TABLE 7-1
Fortran program to compute response of the example system.

```
 100 FORMAT (29H SECOND ORDER SYSTEM RESPONSE)
a110 WRITE (4,100)
 120 FORMAT (5H TIME, 6X, 5HINPUT, 5X, 8HSTATE X1, 2X,
    18HSTATE X2, 2X, 6HOUTPUT)
a130 WRITE (4, 120)
 140 FORMAT (5F10.6)
 150 D=0.05
 160 T=0.0
 170 X1=-4.0
 180 X2=5.0
b190 R=ACOS (0.25*T)
 200 Y=X1-0.5*X2
a210 WRITE (4,140) T, R, X1, X2, Y
 220 Z1=(1.0-2.0*D)*X1 + D*X2 + 2.0*D*R
 230 Z2=-3.0*D*X1 + X2-D*R
 240 X1=Z1
 250 X2=Z2
 260 T=T+D
 270 GO TO 190
 280 STOP
 290 END
```

[a]Printing device is here taken to be device #4.
[b]The library routine for floating-point computation is here taken to be the function ACOS( ).

TABLE 7-2
Basic program to compute response of the example system.

```
100 PRINT "SECOND ORDER SYSTEM RESPONSE"
110 PRINT "TIME", "INPUT", "STATE X1", "STATE X2", "OUTPUT"
120 D=0.05
130 T=0
140 X1=-4
150 X2=5
160 R=COS (0.25*T)
170 Y=X1-0.5*X2
180 PRINT T, R, X1, X2, Y
190 Z1=(1-2*D) *X1 + D*X2 + 2*D*R
200 Z2=-3*D*X1 + X2-D*R
210 X1=Z1
220 X2=Z2
230 T=T+D
240 GO TO 160
250 STOP
260 END
```

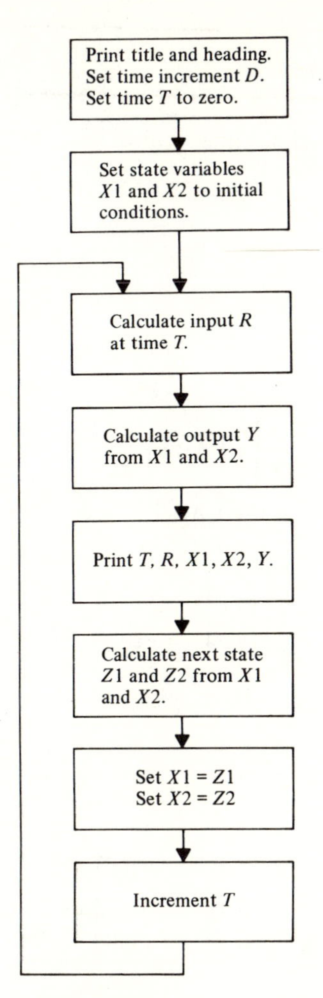

FIGURE 7-16
Flowchart for response calculation programming.

For these plots, a time increment of $\Delta t = 0.05$ was used. A flowchart for the calculation and printing of response values for this system is given in Fig. 7-16. Response calculation programs in Fortran and in Basic are given in Tables 7-1 and 7-2.

Improved accuracy and reduced computation time may result from using more involved approximations—for example, matrix power series, predictor correctors, or Runge-Kutta methods.

DRILL PROBLEMS

D7-13. For the following systems, develop discrete-time approximation equations using the indicated time steps Δt.

(a) $$\begin{bmatrix} \dot{x}_1 \\ \dot{x}_2 \end{bmatrix} = \begin{bmatrix} 3 & 1 \\ 2 & -1 \end{bmatrix} \begin{bmatrix} x_1 \\ x_2 \end{bmatrix} + \begin{bmatrix} 1 \\ 4 \end{bmatrix} r$$

$$y = \begin{bmatrix} -3 & -1 \end{bmatrix} \begin{bmatrix} x_1 \\ x_2 \end{bmatrix}$$

$$\Delta t = 0.2$$

ans. $$\mathbf{x}[(k+1)\Delta t] = \begin{bmatrix} 1.6 & 0.2 \\ 0.4 & 0.8 \end{bmatrix} \mathbf{x}(k\,\Delta t) + \begin{bmatrix} 0.2 \\ 0.8 \end{bmatrix} r(k\,\Delta t)$$

$$y = \begin{bmatrix} -3 & -1 \end{bmatrix} \mathbf{x}(k\,\Delta t)$$

(b) $$\begin{bmatrix} \dot{x}_1 \\ \dot{x}_2 \\ \dot{x}_3 \end{bmatrix} = \begin{bmatrix} 1 & 2 & 3 \\ 7 & -2 & -3 \\ 6 & 0 & 4 \end{bmatrix} \begin{bmatrix} x_1 \\ x_2 \\ x_3 \end{bmatrix} + \begin{bmatrix} 1 & -2 \\ -1 & 3 \\ 0 & 4 \end{bmatrix} \begin{bmatrix} r_1 \\ r_2 \end{bmatrix}$$

$$y = \begin{bmatrix} 5 & -2 & 1 \end{bmatrix} \begin{bmatrix} x_1 \\ x_2 \\ x_3 \end{bmatrix}$$

$$\Delta t = 0.01$$

ans. $$\mathbf{x}[(k+1)\Delta t] = \begin{bmatrix} 1.01 & 0.02 & 0.03 \\ 0.07 & 0.98 & -0.03 \\ 0.06 & 0 & 1.04 \end{bmatrix} \mathbf{x}(k\,\Delta t)$$

$$+ \begin{bmatrix} 0.01 & -0.02 \\ -0.01 & 0.03 \\ 0 & 0.04 \end{bmatrix} \mathbf{r}(k\,\Delta t)$$

$$y = \begin{bmatrix} 5 & -2 & 1 \end{bmatrix} \mathbf{x}(k\,\Delta t)$$

D7-14. For the set of state equations

$$\begin{bmatrix} \dot{x}_1 \\ \dot{x}_2 \end{bmatrix} = \begin{bmatrix} -2 & 1 \\ -3 & 0 \end{bmatrix} \begin{bmatrix} x_1 \\ x_2 \end{bmatrix} + \begin{bmatrix} 1 \\ 4 \end{bmatrix} r$$

$$y = \begin{bmatrix} 1 & -1 \end{bmatrix} \begin{bmatrix} x_1 \\ x_2 \end{bmatrix}$$

a discrete-time approximation is

$$\begin{bmatrix} x_1\{(k+1)\Delta t\} \\ x_2\{(k+1)\Delta t\} \end{bmatrix} = \begin{bmatrix} (1-2\Delta t) & \Delta t \\ -3\Delta t & 1 \end{bmatrix} \begin{bmatrix} x_1(k\,\Delta t) \\ x_2(k\,\Delta t) \end{bmatrix} + \begin{bmatrix} \Delta t \\ 4\Delta t \end{bmatrix} r(k\,\Delta t)$$

$$y(k\,\Delta t) = [1 \quad -1] \begin{bmatrix} x_1(k\,\Delta t) \\ x_2(k\,\Delta t) \end{bmatrix}$$

If

$$\begin{bmatrix} x_1(0^-) \\ x_2(0^-) \end{bmatrix} = \begin{bmatrix} 10 \\ 0 \end{bmatrix} \quad \text{and} \quad r(t) = 2u(t)$$

where $u(t)$ is the unit step function, calculate approximate values for $\mathbf{x}(\Delta t)$, $\mathbf{x}(2\Delta t)$, and $\mathbf{x}(3\Delta t)$ for the following:

(a) $\Delta t = 0.2$

$$ans. \begin{bmatrix} 6.4 \\ -4.4 \end{bmatrix}, \begin{bmatrix} 3.36 \\ -6.64 \end{bmatrix}, \begin{bmatrix} 1.09 \\ -7.06 \end{bmatrix}$$

(b) $\Delta t = 0.1$

$$ans. \begin{bmatrix} 8.2 \\ -2.2 \end{bmatrix}, \begin{bmatrix} 6.54 \\ -3.86 \end{bmatrix}, \begin{bmatrix} 6.37 \\ -5.02 \end{bmatrix}$$

(c) $\Delta t = 0.02$

$$ans. \begin{bmatrix} 9.84 \\ -0.44 \end{bmatrix}, \begin{bmatrix} 9.67 \\ -0.87 \end{bmatrix}, \begin{bmatrix} 9.503 \\ -1.29 \end{bmatrix}$$

7.8 STATE VARIABLE DESIGN

7.8.1 System Description

The first step in using state variable methods for control system design is to express the system equations in state variable form. Systematic steps for converting a system block diagram to a set of state equations is illustrated with an example system in Fig. 7-17. The block diagram of Fig. 7-17a is converted to an equivalent signal flow graph in Fig. 7-17b. In Fig. 7-17c, simulation diagrams replace the individual transmittances. In each of these, the phase-variable form is used because it is simple and easily related to the coefficients of the transmittances. Care must be taken to preserve the signal relationships. For instance, the unit transmittance in the lower feedback loop is necessary; if it were removed, the $X_4(s)$ signal would couple through the -10 transmittance into $Y(s)$.

State equations for the system are then written by labeling each integrator

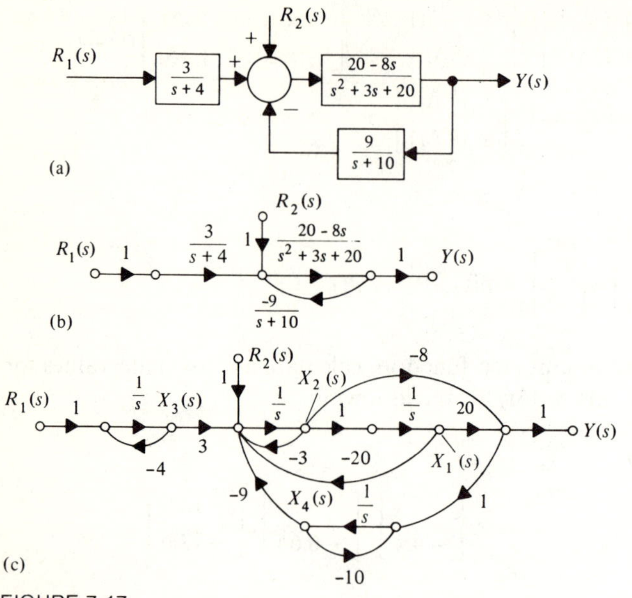

FIGURE 7-17
Constructing a simulation diagram. (a) Block diagram of the system. (b) Equivalent signal flow graph. (c) Signal flow graph expanded into a simulation diagram.

output as a state variable and equating the sum of the input signals to each integrator to the derivative of the appropriate state variable:

$$\dot{x}_1 = x_2$$
$$\dot{x}_2 = -20x_1 - 3x_2 + 3x_3 - 9x_4 + r_2$$
$$\dot{x}_3 = -4x_3 + r_1$$
$$\dot{x}_4 = -10x_4 + y = 20x_1 - 8x_2 - 10x_4$$
$$y = 20x_1 - 8x_2$$

Occasionally, it is necessary to substitute, in terms of the inputs and state variables, for an intermediate signal such as y in the $\dot{x}_4$ equation above. In vector-matrix notation, these equations are as follows:

$$\begin{bmatrix} \dot{x}_1 \\ \dot{x}_2 \\ \dot{x}_3 \\ \dot{x}_4 \end{bmatrix} = \begin{bmatrix} 0 & 1 & 0 & 0 \\ -20 & -3 & 3 & -9 \\ 0 & 0 & -4 & 0 \\ 20 & -8 & 0 & -10 \end{bmatrix} \begin{bmatrix} x_1 \\ x_2 \\ x_3 \\ x_4 \end{bmatrix} + \begin{bmatrix} 0 & 0 \\ 0 & 1 \\ 1 & 0 \\ 0 & 0 \end{bmatrix} \begin{bmatrix} r_1 \\ r_2 \end{bmatrix}$$

$$y = \begin{bmatrix} 20 & -8 & 0 & 0 \end{bmatrix} \begin{bmatrix} x_1 \\ x_2 \\ x_3 \\ x_4 \end{bmatrix}$$

As another example, consider the system with the signal flow graph of Fig. 7-18a. The transmittance $G(s)$ does not have numerator degree less than the denominator degree, so an individual simulation diagram for it cannot be drawn. However, dividing denominator into numerator for one step gives

$$G(s)=2+\frac{-1}{s+4}$$

which is represented in the equivalent signal flow graph of Fig. 7-18b.

For variety, the transmittance $G_2(s)$ is expanded in the diagonal form corresponding to

$$G_2(s)=\frac{3}{s+2}+\frac{-3}{s+3}$$

in the simulation diagram of Fig. 7-18c for which

$$\begin{aligned}\dot{x}_1&=-2x_1+y'\\ \dot{x}_2&=-3x_2+y'\\ \dot{x}_3&=-4x_3+r\\ y&=3x_1-3x_2\end{aligned}$$

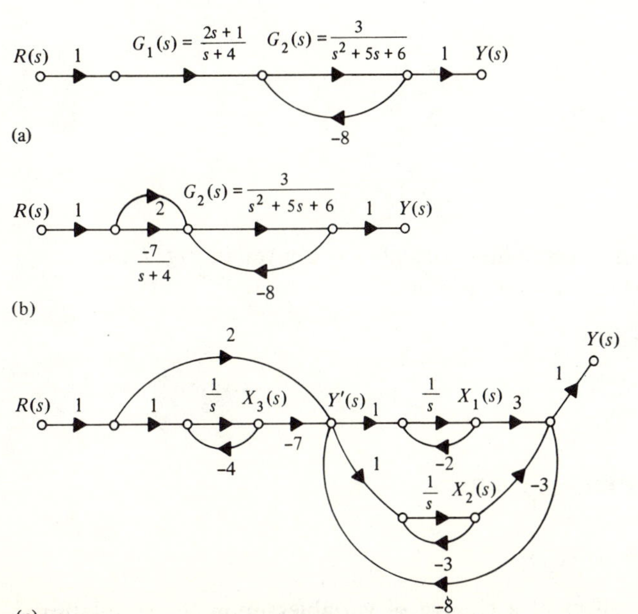

FIGURE 7-18
Another example of simulation diagram construction. (a) Signal flow graph of the system. (b) Improper rational transmittance expanded. (c) Simulation diagram.

Substituting for the intermediate signal

$$y' = -2r - 7x_3 - 8y = 2r - 7x_3 - 24x_1 + 24x_2$$

the following state equations result:

$$\begin{bmatrix} \dot{x}_1 \\ \dot{x}_2 \\ \dot{x}_3 \end{bmatrix} = \begin{bmatrix} -26 & 24 & -7 \\ -24 & 21 & -7 \\ 0 & 0 & -4 \end{bmatrix} \begin{bmatrix} x_1 \\ x_2 \\ x_3 \end{bmatrix} + \begin{bmatrix} 2 \\ 2 \\ 1 \end{bmatrix} r$$

$$y = \begin{bmatrix} 3 & -3 & 0 \end{bmatrix} \begin{bmatrix} x_1 \\ x_2 \\ x_3 \end{bmatrix}$$

7.8.2 State Feedback

A compact diagram of the state variable description of a system is given in Fig. 7-19a. The bold arrows represent signal vectors rather than individual signals, and the blocks show matrix operations. In this diagram,

$$s\mathbf{X}(s) = \mathbf{AX}(s) + \mathbf{BR}(s)$$
$$\mathbf{Y}(s) = \mathbf{CX}(s)$$

In the time domain,

$$\frac{d\mathbf{x}}{dt} = \mathbf{Ax} + \mathbf{Br}$$

$$\mathbf{y} = \mathbf{Cx}$$

An alternative diagram in terms of functions of time is given in Fig. 7-19b.

When a change of state variables is made,

$$\mathbf{x}' = \mathbf{Px} \qquad \mathbf{x} = \mathbf{P}^{-1}\mathbf{x}'$$

the state equations become

$$\frac{d\mathbf{x}'}{dt} = (\mathbf{PAP}^{-1})\mathbf{x}' + (\mathbf{PB})\mathbf{r} = \mathbf{A}'\mathbf{x}' + \mathbf{B}'\mathbf{r}$$

$$\mathbf{y} = (\mathbf{CP}^{-1})\mathbf{x}' = \mathbf{C}'\mathbf{x}'$$

Figure 7-19c shows the effect of a change of variables upon the simulation diagram.

For feedback system control an important design strategy is to sense the state variable signals and feed them back to the input through appropriate

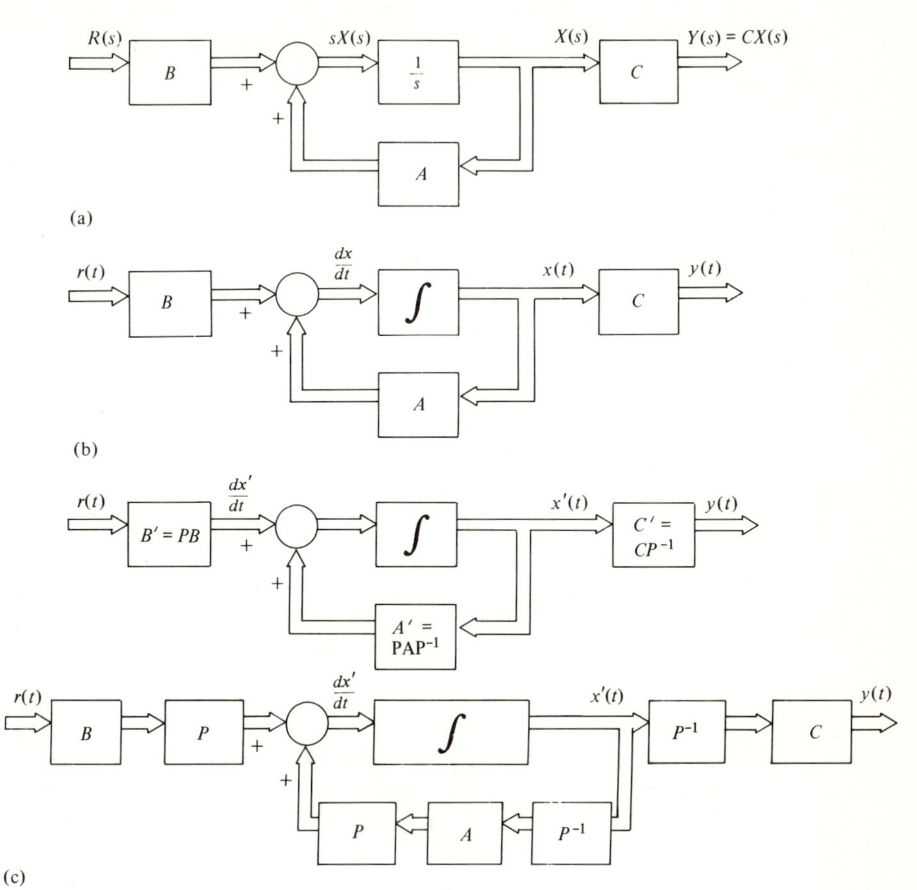

FIGURE 7-19
Vector-matrix simulation diagrams. (a) Diagram in terms of Laplace transforms. (b) Diagram in terms of functions of time. (c) Diagram with change of state variables.

gains. Provided that the system is completely controllable, state feedback may be used to place the system's characteristic roots, the poles of its transfer functions, at any desired locations. State feedback is illustrated in Fig. 7-20a. The open-loop system input is driven with a linear combination of the state signals and a reference input in a tracking configuration:

$$\mathbf{r} = \mathbf{Kx} + \mathbf{u}$$

Figures 7-20b and c show the simulation diagram rearranged so that it is evident that the state coupling matrix of the feedback system is

$$\mathscr{A} = \mathbf{A} + \mathbf{BK}$$

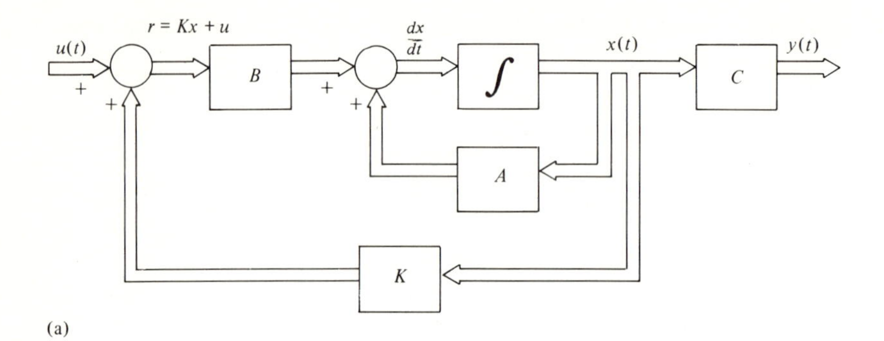

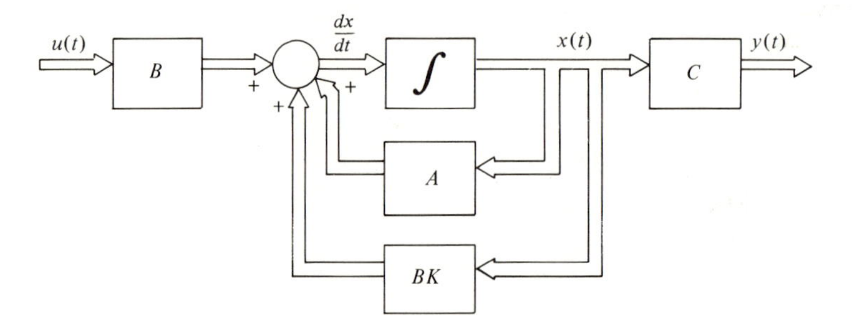

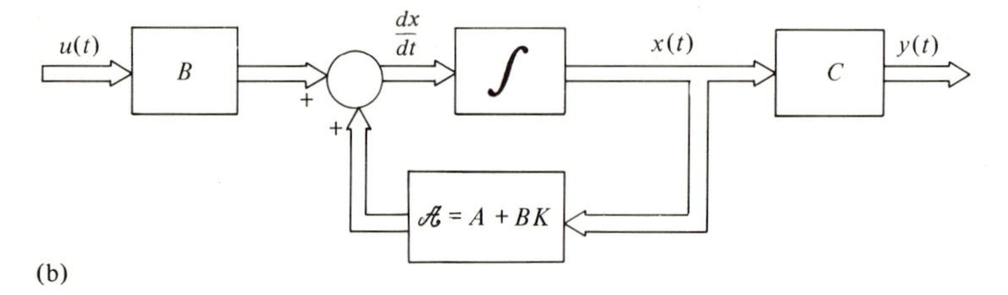

FIGURE 7-20
State feedback. (a) Feedback of the state in a tracking configuration, $\mathbf{r}=\mathbf{kx}+\mathbf{u}$. (b) Simulation diagram rearranged to show feedback system state coupling matrix, $\mathscr{A}=\mathbf{A}+\mathbf{BK}$.

As a numerical example of placing the system poles as desired with state feedback, consider the following single-input, single-output system which is described in phase-variable form:

$$\begin{bmatrix} \dot{x}_1 \\ \dot{x}_2 \\ \dot{x}_3 \end{bmatrix} = \begin{bmatrix} 0 & 1 & 0 \\ 0 & 0 & 1 \\ -5 & -7 & -3 \end{bmatrix} \begin{bmatrix} x_1 \\ x_2 \\ x_3 \end{bmatrix} + \begin{bmatrix} 0 \\ 0 \\ 1 \end{bmatrix} r(t)$$

$$y = [-2 \quad 4 \quad 3] \begin{bmatrix} x_1 \\ x_2 \\ x_3 \end{bmatrix}$$

This system is shown in the simulation diagram of Fig. 7-21a and, using Mason's gain rule, has transfer function

$$T(s)=\frac{3/s+4/s^2+-2/s^3}{1+3/s+7/s^2+5/s^3}$$
$$=\frac{3s^2+4s-2}{s^3+3s^2+7s+5}$$

Since the characteristic equation factors as

$$s^3+3s^2+7s+5=(s+1+j2)(s+1-j2)(s+1)$$

its poles are at $s=-1-j2$, $-1+j2$, and -1.

With state feedback,

$$r(t)=k_1x_1+k_2x_2+k_3x_3+u(t)$$

the state equations are of the form

$$\begin{bmatrix}\dot{x}_1\\ \dot{x}_2\\ \dot{x}_3\end{bmatrix}=\begin{bmatrix}0 & 1 & 0\\ 0 & 0 & 1\\ k_1-5 & k_2-7 & k_3-3\end{bmatrix}\begin{bmatrix}x_1\\ x_2\\ x_3\end{bmatrix}+\begin{bmatrix}0\\ 0\\ 1\end{bmatrix}u(t)$$

$$y=[-2 \quad 4 \quad 3]\begin{bmatrix}x_1\\ x_2\\ x_3\end{bmatrix}$$

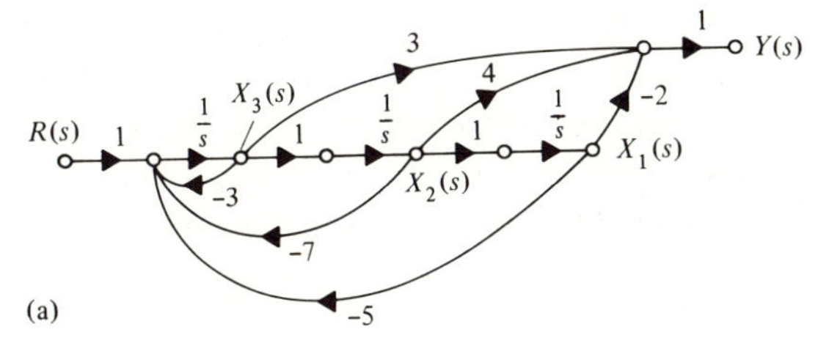

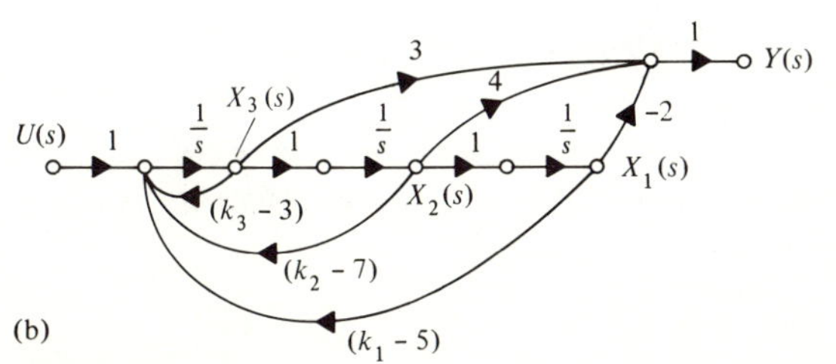

FIGURE 7-21
State feedback example. (a) Open-loop system. (b) System with state feedback.

as diagramed in Fig. 7-21b. The feedback system has transfer function, in terms of the feedback gain constants k,

$$T(s)=\frac{3/s+4/s^2-2/s^3}{1+(3-k_3)/s+(7-k_2)/s^2+(5-k_1)/s^3}$$

$$=\frac{3s^2+4s-2}{s^3+(3-k_3)s^2+(7-k_2)s+(5-k_1)}$$

The coefficients of the characteristic equation may be chosen at will, by appropriately selecting k_1, k_2, and k_3. If, for instance, it is desired that the system poles be located at $s=-4$, -4, and -5, the characteristic polynomial should then be

$$\begin{aligned}(s+4)(s+4)(s+5)&=s^3+13s^2+56s+80\\&=s^3+(3-k_3)s^2+(7-k_2)s+(5-k_1)\end{aligned}$$

which will be the case for

$$\begin{aligned}k_1&=-75\\k_2&=-49\\k_3&=-10\end{aligned}$$

It is true in general that appropriate state feedback will place the poles of any completely controllable system arbitrarily. For a single-input system, the phase-variable form of the state equations is especially convenient because the required values of the feedback gain constants may be determined by inspection, as was the case with the above example. In a multiple-input system, additional design freedom exists. For any completely observable system, when all of the state variables cannot be sensed directly, and fed back for control, they may be estimated (at the expense of increased system order) and the estimates fed back to achieve pole placement. State estimators for this purpose are termed *observers*.

7.8.3 Steady State Response to Power of Time Inputs

The steady state response to step or constant inputs may, as always, be found using system transfer functions and the final-value theorem. A generally easier procedure in terms of state equations

$$\begin{aligned}\dot{\mathbf{x}}&=\mathbf{Ax}+\mathbf{Br}\\\mathbf{y}&=\mathbf{Cx}\end{aligned}$$

is the following: Assuming the system is stable, all steady state signals due to constant inputs will be constant, and

$$\dot{\mathbf{x}}=\mathbf{0}$$

giving

$$\begin{aligned}\mathbf{0}&=\mathbf{Ax}+\mathbf{Br}\\ \mathbf{x}&=-\mathbf{A}^{-1}\mathbf{Br}\\ \mathbf{y}&=-\mathbf{CA}^{-1}\mathbf{Br}\end{aligned}$$

The existence of $\mathbf{A}^{-1}$ is guaranteed for a stable system since

$$|\mathbf{A}|=0$$

if and only if $s=0$ is a characteristic root of **A**, that is, if the system has a pole at $s=0$.

Consider the system

$$\begin{bmatrix}\dot{x}_1\\ \dot{x}_2\\ \dot{x}_3\end{bmatrix}=\begin{bmatrix}-2 & 1 & 0\\ -3 & 0 & 1\\ -4 & 0 & 0\end{bmatrix}\begin{bmatrix}x_1\\ x_2\\ x_3\end{bmatrix}+\begin{bmatrix}-1\\ 5\\ 0\end{bmatrix}r(t)$$

$$\begin{bmatrix}y_1\\ y_2\end{bmatrix}=\begin{bmatrix}1 & 0 & -1\\ 2 & -2 & 0\end{bmatrix}\begin{bmatrix}x_3\\ x_2\\ x_3\end{bmatrix}$$

which is stable. For a unit step input $r(t)$, the steady state behavior of the system is governed by

$$\begin{bmatrix}0\\ 0\\ 0\end{bmatrix}=\begin{bmatrix}-2 & 1 & 0\\ -3 & 0 & 1\\ -4 & 0 & 0\end{bmatrix}\begin{bmatrix}x_1\\ x_2\\ x_3\end{bmatrix}+\begin{bmatrix}-1\\ 5\\ 0\end{bmatrix}r(t)$$

$$\begin{aligned}\begin{bmatrix}x_1\\ x_2\\ x_3\end{bmatrix}&=-\begin{bmatrix}-2 & 1 & 0\\ -3 & 0 & 1\\ -4 & 0 & 0\end{bmatrix}^{-1}\begin{bmatrix}-1\\ 5\\ 0\end{bmatrix}\\ &=-\begin{bmatrix}0 & 0 & -\frac{1}{4}\\ 1 & 0 & -\frac{1}{2}\\ 0 & 1 & -\frac{3}{4}\end{bmatrix}\begin{bmatrix}-1\\ 5\\ 0\end{bmatrix}=\begin{bmatrix}0\\ 1\\ -5\end{bmatrix}\end{aligned}$$

The steady state values of the outputs are

$$\begin{bmatrix}y_1\\ y_2\end{bmatrix}=\begin{bmatrix}1 & 0 & -1\\ 2 & -2 & 0\end{bmatrix}\begin{bmatrix}0\\ 1\\ -5\end{bmatrix}=\begin{bmatrix}5\\ -2\end{bmatrix}$$

This method may be applied to ramp and higher power of t inputs if desired. For a unit ramp input to the example system,

$$r(t)=t$$

the system state will consist of constant and ramp components,

$$x_1(t)=\alpha_1 t+\beta_1$$
$$x_2(t)=\alpha_2 t+\beta_2$$
$$x_3(t)=\alpha_3 t+\beta_3$$

where the α and β terms are constants. Then

$$\begin{bmatrix}\dot{x}_1\\ \dot{x}_2\\ \dot{x}_3\end{bmatrix}=\begin{bmatrix}\alpha_1\\ \alpha_2\\ \alpha_3\end{bmatrix}=\begin{bmatrix}-2 & 1 & 0\\ -3 & 0 & 1\\ -4 & 0 & 0\end{bmatrix}\begin{bmatrix}\alpha_1 t+\beta_1\\ \alpha_2 t+\beta_2\\ \alpha_3 t+\beta_3\end{bmatrix}+\begin{bmatrix}-1\\ 5\\ 0\end{bmatrix}t$$

Equating coefficients of the powers of t, the α terms are governed by

$$\begin{bmatrix}0\\ 0\\ 0\end{bmatrix}=\begin{bmatrix}-2 & 1 & 0\\ -3 & 0 & 1\\ -4 & 0 & 0\end{bmatrix}\begin{bmatrix}\alpha_1\\ \alpha_2\\ \alpha_3\end{bmatrix}+\begin{bmatrix}-1\\ 5\\ 0\end{bmatrix}$$

$$\begin{bmatrix}\alpha_1\\ \alpha_2\\ \alpha_3\end{bmatrix}=-\begin{bmatrix}-2 & 1 & 0\\ -3 & 0 & 1\\ -4 & 0 & 0\end{bmatrix}^{-1}\begin{bmatrix}-1\\ 5\\ 0\end{bmatrix}=\begin{bmatrix}0\\ 1\\ -5\end{bmatrix}$$

which are always identical to the constant steady state response of the state to a unit step input. The β terms satisfy

$$\begin{bmatrix}\alpha_1\\ \alpha_2\\ \alpha_3\end{bmatrix}=\begin{bmatrix}-2 & 1 & 0\\ -3 & 0 & 1\\ -4 & 0 & 0\end{bmatrix}\begin{bmatrix}\beta_1\\ \beta_2\\ \beta_3\end{bmatrix}$$

$$\begin{bmatrix}\beta_1\\ \beta_2\\ \beta_3\end{bmatrix}=\begin{bmatrix}-2 & 1 & 0\\ -3 & 0 & 1\\ -4 & 0 & 0\end{bmatrix}^{-1}\begin{bmatrix}\alpha_1\\ \alpha_2\\ \alpha_3\end{bmatrix}=\begin{bmatrix}0 & 0 & -\frac{1}{4}\\ 1 & 0 & -\frac{1}{2}\\ 0 & 1 & -\frac{3}{4}\end{bmatrix}\begin{bmatrix}0\\ 1\\ -5\end{bmatrix}$$

$$=\begin{bmatrix}\frac{5}{4}\\ \frac{5}{2}\\ \frac{19}{4}\end{bmatrix}$$

and the steady state outputs are

$$\begin{bmatrix}y_1\\ y_2\end{bmatrix}=\begin{bmatrix}1 & 0 & -1\\ 2 & -2 & 0\end{bmatrix}\begin{bmatrix}\frac{5}{4}\\ t+\frac{5}{2}\\ -5t+\frac{19}{4}\end{bmatrix}=\begin{bmatrix}(5t-\frac{7}{2})\\ (-2t-\frac{5}{2})\end{bmatrix}$$

DRILL PROBLEMS

D7-15. Find state variable equations for systems with the following signal flow graphs.

ans. (a) One possibility is the following:

$$\begin{bmatrix} \dot{x}_1 \\ \dot{x}_2 \\ \dot{x}_3 \\ \dot{x}_4 \end{bmatrix} = \begin{bmatrix} 0 & 1 & 0 & 0 \\ -6 & 0 & 1 & 0 \\ 0 & 0 & 0 & 4 \\ 0 & -10 & 0 & -3 \end{bmatrix} \begin{bmatrix} x_1 \\ x_2 \\ x_3 \\ x_4 \end{bmatrix} + \begin{bmatrix} 0 \\ 0 \\ 0 \\ 1 \end{bmatrix} r$$

$$y = [0 \quad 5 \quad 0 \quad 0] \begin{bmatrix} x_1 \\ x_2 \\ x_3 \\ x_4 \end{bmatrix}$$

(b) One possibility is the following:

$$\begin{bmatrix} \dot{x}_1 \\ \dot{x}_2 \\ \dot{x}_3 \\ \dot{x}_4 \end{bmatrix} = \begin{bmatrix} -2 & 0 & 0 & 0 \\ 3 & -1 & 0 & 0 \\ 0 & 0 & 0 & 1 \\ 0 & 0 & -4 & 0 \end{bmatrix} \begin{bmatrix} x_1 \\ x_2 \\ x_3 \\ x_4 \end{bmatrix} = \begin{bmatrix} 1 & -6 \\ 0 & 0 \\ 0 & 0 \\ 0 & 10 \end{bmatrix} \begin{bmatrix} r_1 \\ r_2 \end{bmatrix}$$

$$\begin{bmatrix} y_1 \\ y_2 \end{bmatrix} = \begin{bmatrix} 3 & 0 & 0 & 0 \\ 0 & 1 & 1 & 0 \end{bmatrix} \begin{bmatrix} x_1 \\ x_2 \\ x_3 \\ x_4 \end{bmatrix}$$

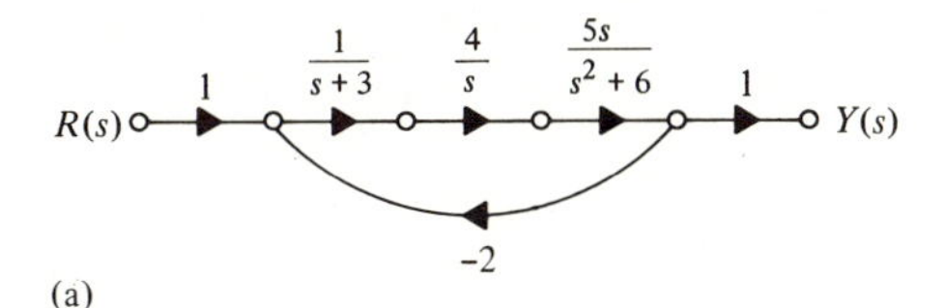

(a)

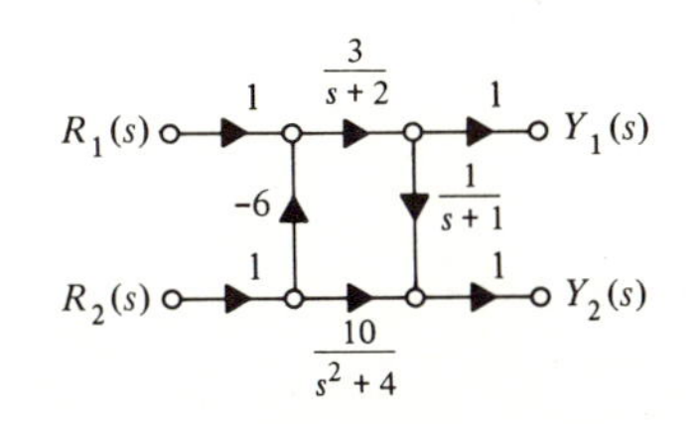

(b)

D7-16. For the state feedback systems described by the following equations, choose the feedback gain constants k_i to place the closed-loop system poles at the indicated locations:

(a) $$\begin{bmatrix}\dot{x}_1\\ \dot{x}_2\\ \dot{x}_3\end{bmatrix}=\begin{bmatrix}0 & 1 & 0\\ 0 & 0 & 1\\ -3 & -6 & -7\end{bmatrix}\begin{bmatrix}x_1\\ x_2\\ x_3\end{bmatrix}+\begin{bmatrix}0\\ 0\\ 1\end{bmatrix}r$$

$$r=[k_1 \quad k_2 \quad k_3]\begin{bmatrix}x_1\\ x_2\\ x_3\end{bmatrix}+u$$

$$y=[2 \quad 0 \quad -1]\begin{bmatrix}x_1\\ x_2\\ x_3\end{bmatrix}$$

Closed-loop poles at $s=-3$, -4, and -5

ans. $k_1=-57$, $k_2=-41$, $k_3=-5$

(b) $$\begin{bmatrix}\dot{x}_1\\ \dot{x}_2\\ \dot{x}_3\end{bmatrix}=\begin{bmatrix}-2 & 1 & 0\\ 4 & 0 & 1\\ 0 & 0 & 0\end{bmatrix}\begin{bmatrix}x_1\\ x_2\\ x_3\end{bmatrix}+\begin{bmatrix}0\\ 0\\ 1\end{bmatrix}r$$

$$r=[k_1 \quad k_2 \quad k_3]\begin{bmatrix}x_1\\ x_2\\ x_3\end{bmatrix}$$

$$y=\begin{bmatrix}2 & 0 & -1\\ 1 & 1 & 0\end{bmatrix}\begin{bmatrix}x_1\\ x_2\\ x_3\end{bmatrix}$$

Closed-loop poles at $s=-3\pm j3$, -3

ans. $k_1=-30$, $k_2=-26$, $k_3=-7$

D7-17. For a unit step input, find the steady state output, if it exists, of each of the following systems:

(a) $$\begin{bmatrix}\dot{x}_1\\ \dot{x}_2\end{bmatrix}=\begin{bmatrix}-2 & 2\\ -3 & 0\end{bmatrix}\begin{bmatrix}x_1\\ x_2\end{bmatrix}+\begin{bmatrix}1\\ 4\end{bmatrix}r$$

$$y=[-1 \quad 7]\begin{bmatrix}x_1\\ x_2\end{bmatrix}$$

ans. $\frac{27}{6}$

(b $$\begin{bmatrix}\dot{x}_1\\ \dot{x}_2\\ \dot{x}_3\end{bmatrix}=\begin{bmatrix}-1 & 1 & 0\\ 2 & 2 & 0\\ 3 & -1 & 4\end{bmatrix}\begin{bmatrix}x_1\\ x_2\\ x_3\end{bmatrix}+\begin{bmatrix}1\\ 0\\ -1\end{bmatrix}r$$

$$y=[3 \quad -2 \quad 1]\begin{bmatrix}x_1\\ x_2\\ x_3\end{bmatrix}$$

ans. system unstable

(c) $$\begin{bmatrix} \dot{x}_1 \\ \dot{x}_2 \\ \dot{x}_3 \end{bmatrix} = \begin{bmatrix} 0 & 1 & 0 \\ 0 & 0 & 1 \\ 4 & -4 & -3 \end{bmatrix} \begin{bmatrix} x_1 \\ x_2 \\ x_3 \end{bmatrix} + \begin{bmatrix} 1 \\ 1 \\ 1 \end{bmatrix} r$$

$$y = \begin{bmatrix} 2 & -1 & 0 \\ 0 & 0 & 1 \end{bmatrix} \begin{bmatrix} x_1 \\ x_2 \\ x_3 \end{bmatrix}$$

$$ans. \begin{bmatrix} -3 \\ -1 \end{bmatrix}$$

7.9 A MAGNETIC LEVITATION SYSTEM

Beginning in 1969, West Germany sought to develop a high-speed electric train system to span central Europe. Using state space analysis and aircraft technology, a train has now been designed, built, and tested for operation at speeds as high as 400 km/hr (248 mi/hr). The train is suspended in midair by magnetic fields. This type of suspension is called magnetic levitation or MAGLEV.

Figure 7-22 shows the cross section of a MAGLEV vehicle. The track is a T-shaped concrete guideway. Once underway, the train does not touch the guideway, resulting in greatly reduced friction and reduced guideway construction costs. Electromagnets are distributed along the guideway and along the length of the train in matched pairs. The magnetic attraction of the vertically paired magnets balances the force of gravity and levitates the vehicle above the guideway. The horizontally paired magnets stabilize the vehicle against sideways forces. Forward propulsion is produced by linear induction motor action

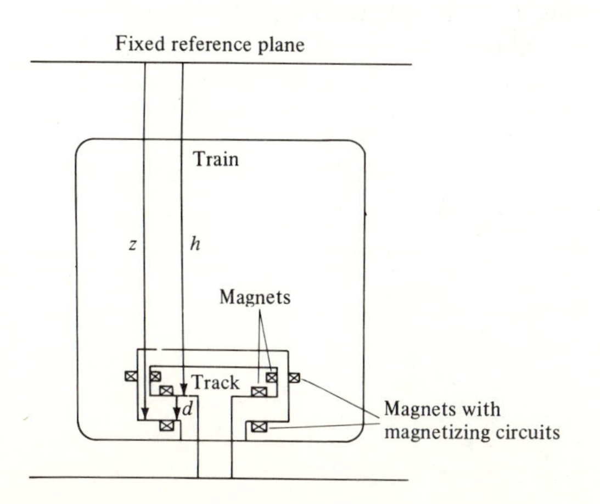

FIGURE 7-22
Cross-section of a MAGLEV train.

between train and guideway. Only the vertical motion and control of the suspended vehicle will be considered here.

The equations characterizing the train's vertical motion are now developed. It is desired to control the gap distance d within a close tolerance in normal operation of the train. The gap distance d between the track and the train magnets is

$$d = z - h$$

Then

$$\dot{d} = \dot{z} - \dot{h}$$
$$\ddot{d} = \ddot{z} - \ddot{h}$$

where the dots denote time derivatives. The magnet produces a force that is dependent upon residual magnetism and upon the current passing through the magnetizing circuit. For small changes in the magnetizing current i and the gap distance d, that force is approximately

$$f_1 = -Gi + Hd$$

where G and H are positive constants. That force acts to accelerate the mass M of the train in a vertical direction, so

$$f_1 = M\ddot{z} = -Gi + Hd$$

For increased current, the distance z diminishes, reducing d as the vehicle is attracted to the guideway.

A network model for the magnetizing circuit is given in Fig. 7-23. This circuit represents a generator driving a coil wrapped around the magnet on the vehicle. The voltage induced in the coil by the vehicle motion is represented by the term $(LH/G)\dot{d}$, for which it is assumed that the magnetic flux loss is negligible. For that circuit

$$Ri + L\dot{i} - \frac{LH}{G}\dot{d} = v$$

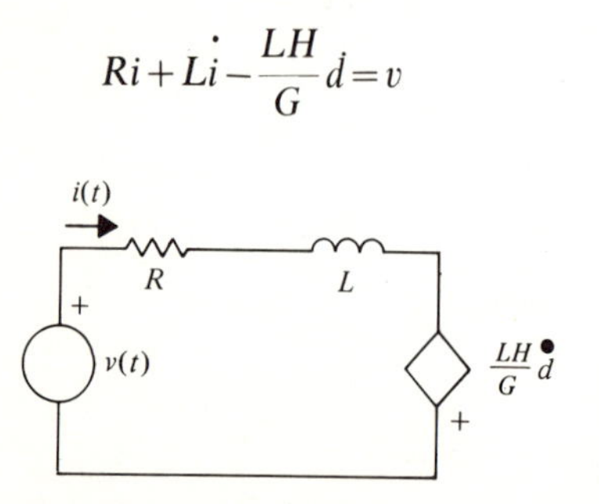

FIGURE 7-23
Magnetizing circuit model.

The three state variables

$$x_1 = d$$
$$x_2 = \dot{d}$$
$$x_3 = i$$

are convenient, and in terms of them the vertical motion state equations are

$$\begin{bmatrix} \dot{x}_1 \\ \dot{x}_2 \\ \dot{x}_3 \end{bmatrix} = \begin{bmatrix} 0 & 1 & 0 \\ \frac{H}{M} & 0 & -\frac{G}{M} \\ 0 & \frac{H}{G} & -\frac{R}{L} \end{bmatrix} \begin{bmatrix} x_1 \\ x_2 \\ x_3 \end{bmatrix} + \begin{bmatrix} 0 & 0 \\ 0 & -1 \\ \frac{1}{L} & 0 \end{bmatrix} \begin{bmatrix} v \\ f \end{bmatrix}$$

where

$$f = \ddot{h}$$

If the gap distance d is considered to be the system output, then the state variable output equation is

$$d = x_1$$

The voltage v is considered to be a control input, while guideway irregularities $f = \ddot{h}$ constitute a disturbance. Figure 7-24 shows block diagram and signal flow graph representations of the state equations.

The characteristic equation for the system, the roots of which are the transfer function poles, is given by

$$|s\mathbf{I} - \mathbf{A}| = \begin{vmatrix} s & -1 & 0 \\ -\frac{H}{M} & s & \frac{G}{M} \\ 0 & -\frac{H}{G} & s+\frac{R}{L} \end{vmatrix} = 0$$

$$= s \begin{vmatrix} s & \frac{G}{M} \\ -\frac{H}{G} & s+\frac{R}{L} \end{vmatrix} + \begin{vmatrix} -\frac{H}{M} & \frac{G}{M} \\ 0 & s+\frac{R}{L} \end{vmatrix}$$

$$= s\left(s^2 + \frac{R}{L}s + \frac{H}{M}\right) - \frac{H}{M}\left(s + \frac{R}{L}\right)$$

$$= s^3 + \frac{R}{L}s^2 - \frac{HR}{ML} = 0$$

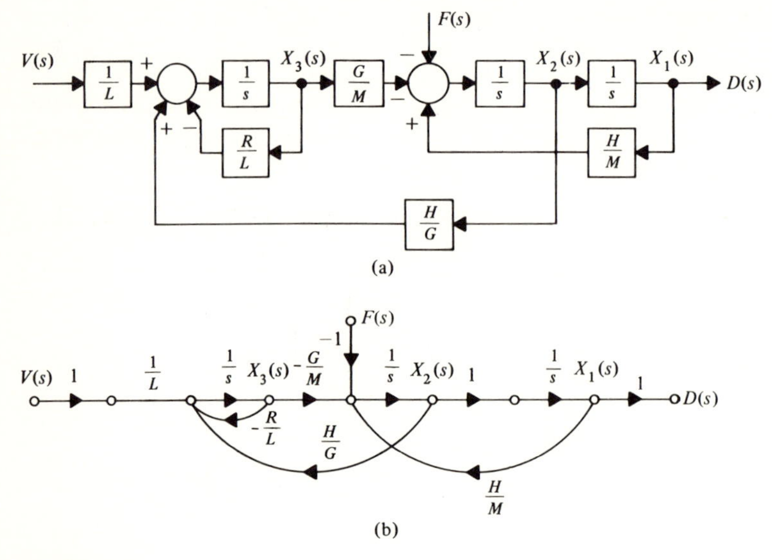

FIGURE 7-24
Diagrams of the state equations. (a) Block diagram. (b) Signal flow graph.

The system is thus unstable since its characteristic polynomial has coefficients with differing algebraic signs. Also, the coefficient of s in the characteristic equation is zero. The system instability is quite understandable when one considers the action of the magnets. If the gap distance d should increase slightly, the magnetic attraction decreases, tending to further increase the gap, and so on.

To control the system, the magnetizing circuit voltage is chosen to be a linear combination of the state signals plus a tracking input $r(t)$:

$$v = k_1 x_1 + k_2 x_2 + k_3 x_3 + r(t)$$

The feedback signals are produced from sensors that monitor the state variables, namely gap distance d, gap velocity $\dot{d}$, and magnetizing current i. The resulting feedback system is described by

$$\begin{bmatrix} \dot{x}_1 \\ \dot{x}_2 \\ \dot{x}_3 \end{bmatrix} = \begin{bmatrix} 0 & 1 & 0 \\ \dfrac{H}{M} & 0 & -\dfrac{G}{M} \\ \dfrac{k_1}{L} & \dfrac{H}{G} + \dfrac{k_2}{L} & -\dfrac{R}{L} + \dfrac{k_3}{L} \end{bmatrix} \begin{bmatrix} x_1 \\ x_2 \\ x_3 \end{bmatrix} + \begin{bmatrix} 0 & 0 \\ 0 & -1 \\ \dfrac{1}{L} & 0 \end{bmatrix} \begin{bmatrix} r \\ f \end{bmatrix}$$

$$d = x_1$$

Appropriate choice of the feedback gain constants k_1, k_2, and k_3, that is, the feedback gain matrix,

$$K=[k_1 \quad k_2 \quad k_3]$$

will place the system poles at any desired locations.

Consider now some specific values for the system parameters. These values are not those of the actual train design, but are chosen to illustrate the concepts of state space design:

$$M=100$$
$$G=5$$
$$H=1$$
$$L=10$$
$$R=3$$

For these, the feedback system equations are

$$\begin{bmatrix} \dot{x}_1 \\ \dot{x}_2 \\ \dot{x}_3 \end{bmatrix} = \begin{bmatrix} 0 & 1 & 0 \\ 0.01 & 0 & -0.05 \\ 0.1k_1 & 0.2+0.1k_2 & -0.3+0.1k_3 \end{bmatrix} \begin{bmatrix} x_1 \\ x_2 \\ x_3 \end{bmatrix} + \begin{bmatrix} 0 & 0 \\ 0 & -1 \\ 0.1 & 0 \end{bmatrix} \begin{bmatrix} r \\ f \end{bmatrix}$$
$$d=x_1$$

The characteristic equation for the feedback system is given by

$$\begin{vmatrix} s & -1 & 0 \\ -0.01 & s & 0.05 \\ -0.1k_1 & -0.2-0.1k_2 & s+0.3-0.1k_3 \end{vmatrix}$$
$$=s\begin{vmatrix} s & 0.05 \\ -0.2-0.1k_2 & s+0.3-0.1k_3 \end{vmatrix} + \begin{vmatrix} -0.01 & 0.05 \\ -0.1k_1 & s+0.3-0.1k_3 \end{vmatrix}$$
$$=s(s^2+0.3s-0.1k_3s+0.01+0.005k_2)-0.01s-0.003+0.001k_3+0.005k_1$$
$$=s^3+(0.3-0.1k_3)s^2+(0.005k_2)s+(-0.003+0.005k_1+0.001k_3)$$
$$=0$$

The feedback gains k_1, k_2, and k_3 may be chosen to give any desired coefficients of the characteristic equation of the feedback system. For example, if it is desired to have the system poles at $s=-1+j2$, $-1-j2$, and -3, the characteristic polynomial would be

$$(s+1-j2)(s+1+j2)(s+3)=s^3+5s^2+11s+15$$

which is achieved with

$$0.005k_2=11 \qquad k_2=2200$$
$$0.3-0.1k_3=5 \qquad k_3=-47$$
$$-0.003+0.005k_1+0.001(-47)=15 \qquad k_1=3010$$

For this choice of feedback gains, the feedback system model is

$$\begin{bmatrix}\dot{x}_1\\ \dot{x}_2\\ \dot{x}_3\end{bmatrix}=\begin{bmatrix}0 & 1 & 0\\ 0.01 & 0 & -0.05\\ 301 & 220.2 & -5\end{bmatrix}\begin{bmatrix}x_1\\ x_2\\ x_3\end{bmatrix}+\begin{bmatrix}0 & 0\\ 0 & -1\\ 0.1 & 0\end{bmatrix}\begin{bmatrix}r\\ f\end{bmatrix}$$

$$d=[1\quad 0\quad 0]\begin{bmatrix}x_1\\ x_2\\ x_3\end{bmatrix}$$

The steady state output d due to a unit step disturbance input f is given by

$$\begin{bmatrix}0\\ 0\\ 0\end{bmatrix}=\begin{bmatrix}0 & 1 & 0\\ 0.01 & 0 & -0.05\\ 301 & 220.2 & -5\end{bmatrix}\begin{bmatrix}x_1\\ x_2\\ x_3\end{bmatrix}+\begin{bmatrix}0 & 0\\ 0 & -1\\ 0.1 & 0\end{bmatrix}\begin{bmatrix}0\\ 1\end{bmatrix}$$

$$d=[1\quad 0\quad 0]\begin{bmatrix}x_1\\ x_2\\ x_3\end{bmatrix}$$

$$=-[1\quad 0\quad 0]\begin{bmatrix}0 & 1 & 0\\ 0.01 & 0 & -0.05\\ 301 & 220.2 & -5\end{bmatrix}^{-1}\begin{bmatrix}0 & 0\\ 0 & -1\\ 0.1 & 0\end{bmatrix}\begin{bmatrix}0\\ 1\end{bmatrix}$$

$$=-[1\quad 0\quad 0]\begin{bmatrix}-0.734 & -0.333 & 0.0033\\ 1 & 0 & 0\\ -0.1468 & -20.07 & 0.00067\end{bmatrix}\begin{bmatrix}0\\ -1\\ 0\end{bmatrix}$$

$$=-[1\quad 0\quad 0]\begin{bmatrix}0.333\\ 0\\ 20.07\end{bmatrix}=-0.333$$

Further study would be needed to determine if this amount of disturbance rejection from track irregularities is sufficient. The negative algebraic sign above simply means that a positive step in $f=\ddot{h}$ results in a steady state decrease in the gap disturbance. Other types of disturbances than constant ones should also be considered in the design.

The reference input r would normally be a constant which sets the nominal gap distance. The steady state gap distance d due to a constant reference input r is given by

$$\begin{bmatrix}0\\ 0\\ 0\end{bmatrix}=\begin{bmatrix}0 & 1 & 0\\ 0.01 & 0 & -0.05\\ 301 & 220.2 & -5\end{bmatrix}\begin{bmatrix}x_1\\ x_2\\ x_3\end{bmatrix}+\begin{bmatrix}0 & 0\\ 0 & -1\\ 0.1 & 0\end{bmatrix}\begin{bmatrix}r\\ 0\end{bmatrix}$$

$$d=\begin{bmatrix}1 & 0 & 0\end{bmatrix}\begin{bmatrix}x_1\\x_2\\x_3\end{bmatrix}$$

$$=-\begin{bmatrix}1 & 0 & 0\end{bmatrix}\begin{bmatrix}0 & 1 & 0\\0.01 & 0 & -0.05\\301 & 220.2 & -5\end{bmatrix}^{-1}\begin{bmatrix}0 & 0\\0 & -1\\0.1 & 0\end{bmatrix}\begin{bmatrix}r\\0\end{bmatrix}$$

$$=-\begin{bmatrix}1 & 0 & 0\end{bmatrix}\begin{bmatrix}-0.734 & -0.333 & 0.0033\\1 & 0 & 0\\-0.1468 & -20.07 & 0.00067\end{bmatrix}\begin{bmatrix}0\\0\\0.1r\end{bmatrix}$$

$$=-0.00033r$$

For a nominal gap distance $d=0.05$, a reference input

$$r=-151.5$$

would be applied.

Figure 7-25a shows calculated system response where the train accelerates from a standstill and traverses an irregular guideway with a downgrade followed by an upgrade. In the West German system, the nominal airgap distance is 14 mm (about $\frac{1}{2}$ in). Improvement in disturbance rejection is obtained by modeling the track irregularities by differential equations that are included as state

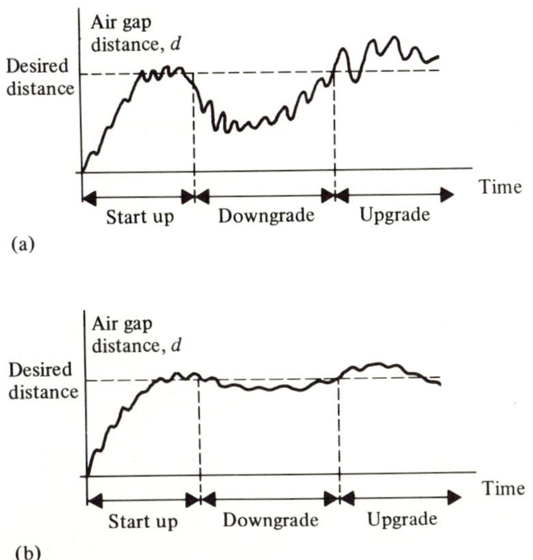

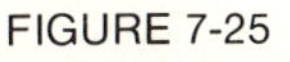

FIGURE 7-25
MAGLEV system response. (a) Response of the system with state feedback. (b) Improved response with disturbance modeling and feedback.

equations, giving a total of five state variables and equations. The additional two state variables, vertical track velocity and acceleration, are sensed and fed back also, improving performance at the expense of increased system complexity. Figure 7-25b indicates the improvement attained.

7.10 SUMMARY

Simulation diagrams were first developed for synthesizing transfer functions with integrators. The phase-variable canonical form was shown to be especially convenient for single- and multiple-output system transfer function synthesis, and the dual phase-variable form is convenient for multiple-input system transfer functions. Simulation diagrams are not only useful in transfer function synthesis; they also give a standard, systematic, and compact description of a system.

The relationships between signals in a simulation diagram were shown to be a set of coupled first-order differential state equations and linear algebraic output equations, relating the system outputs to the state variables. These state variable equations are compactly expressed using matrix notation:

$$\dot{\mathbf{x}} = \mathbf{A}\mathbf{x} + \mathbf{B}\mathbf{r}$$
$$\mathbf{y} = \mathbf{C}\mathbf{x}$$

System transfer functions were calculated systematically from the state variable equations using matrix algebra:

$$\mathbf{T}(s) = \mathbf{C}[s\mathbf{I} - \mathbf{A}]^{-1}\mathbf{B}$$

It was seen that all transfer functions of a system share a common characteristic polynomial,

$$|s\mathbf{I} - \mathbf{A}|$$

A nonsingular change of state variables gives a new representation for a system, but leaves the system's input-output relations, its transfer functions, unchanged. Hence a system characterized by a set of transfer functions may be represented in countless different ways, each differing in the choice of state variables.

A very special set of state variables for a system are those for which the state equations, each of first order, are decoupled from one another. A system so represented is said to be in *normal* or *diagonal* form:

$$\mathbf{A} = \begin{bmatrix} s_1 & 0 & 0 & \cdots & 0 \\ 0 & s_2 & 0 & \cdots & 0 \\ \vdots & & & & \\ 0 & 0 & 0 & & s_n \end{bmatrix}$$

Determining the change of variables which places a system in diagonal form is the *characteristic value problem* of matrix algebra. An alternative transformation method involves expansion of a system transfer function into partial fractions. Systems with repeated characteristic roots cannot be diagonalized; however, they may be placed in a related *Jordan* form, where the repeated root terms involve a distinctive nonzero "block" along the diagonal of the state coupling matrix.

The modes of a system are the exponential terms in the natural component of its response. It is possible that one or more system modes are not affected by any system input. These modes are termed uncontrollable. A system with all modes controllable is said to be completely controllable. Similarly, if a system mode is not coupled to any output of the system, that mode is unobservable. A system with all modes observable is termed completely observable. For a system represented in diagonal form, controllability and observability are apparent from inspection of the input and output coupling matrices. For systems in other than diagonal form, simple rank tests of the controllability and observability matrices,

$$M_c = [\mathbf{B} \mid \mathbf{AB} \mid \cdots \mid \mathbf{A}^{n-1}\mathbf{B}]$$

$$M_o = \begin{bmatrix} \mathbf{C} \\ \hline \mathbf{CA} \\ \hline \vdots \\ \hline \mathbf{CA}^{n-1} \end{bmatrix}$$

may be used.

The response of a first-order state variable system

$$\dot{x} = ax + br$$

is

$$x(t) = e^{at}x(0^-) + \int_{0^-}^{t} e^{a(t-\tau)}br(\tau)\,d\tau \qquad t \geqslant 0$$

For an nth-order system,

$$\mathbf{x}(t) = \mathbf{\Phi}(t)\mathbf{x}(0^-) + \int_{0^-}^{t} \mathbf{\Phi}(t-\tau)\mathbf{Br}(\tau)\,d\tau$$

where $\mathbf{\Phi}(t)$ is the $n \times n$ state transition matrix

$$\mathbf{\Phi}(t) = \mathscr{L}^{-1}\{(s\mathbf{I} - \mathbf{A})^{-1}\}$$

Numerical response calculation may be performed by approximating time derivatives by differences, yielding

$$\mathbf{x}\{(k+1)\Delta t\} \cong (\mathbf{I} + \mathbf{A}\,\Delta t)\mathbf{x}(k\,\Delta t) + (\mathbf{B}\,\Delta t)\mathbf{r}(k\,\Delta t)$$

which may be evaluated repeatedly, starting with $\mathbf{x}(0)$ to find $\mathbf{x}(\Delta t)$, using $\mathbf{x}(\Delta t)$ to find $\mathbf{x}(2\Delta t)$, and so on.

The following are basic state variable design considerations:

State equations. To develop a set of state equations for a system from a block diagram or signal flow graph, individual transmittances are expanded into simulation subdiagrams and state variables are assigned to each integrator output signal.

Pole placement. For any completely controllable system state feedback will place the closed-loop poles at any desired locations on the complex plane. With single-input feedback, the phase-variable form of the state equations is especially convenient for this purpose.

Steady state response. For any stable system, the steady state response to a step input is found directly from the state equations using

$$\dot{\mathbf{x}} = \mathbf{0} = \mathbf{A}\mathbf{x} + \mathbf{B}\mathbf{r}$$

Similar methods apply to higher-power-of-time inputs.

REFERENCES

Simulation Diagrams

Jackson, A. S. *Analog Computation.* New York: McGraw-Hill, 1960.

Korn, G. A., and Korn, T. M. *Electronic Analog Computers.* New York: McGraw-Hill, 1952.

State Variables

Brockett, R. W. "Poles, Zeros and Feedback: State Space Interpretation." *IEEE Trans. Auto. Contr.*, April 1965.

De Russo, P. M.; Roy, R. J.; and Close, C. M. *State Variables for Engineers.* New York: Wiley, 1965.

Gupta, S. C. *Transform and State Variable Methods in Linear Systems.* New York: Wiley, 1966.

Horowitz, I. C., and Shaked, U. "Superiority of Transfer Function over State Variable Methods in Linear Time-Invariant Feedback System Design." *IEEE Trans. Auto. Contr.*, February 1975.

Kalman, R. E. "Mathematical Description of Linear Dynamical Systems." *SIAM J. Contr. ser. A*, 1 (1963).

Ogata, K. *State Space Analysis of Control Systems.* Englewood Cliffs, N.J.: Prentice-Hall, 1967.

Timothy, L. K. and Bona, B. E. *State Space Analysis: An Introduction.* New York: McGraw-Hill, 1968.

Zadeh, L. A., and Desoer, C. A. *Linear System Theory: The State Space Approach.* New York: McGraw-Hill, 1963.

Matrix Algebra and the Characteristic Value Problem

Bellman, R. *Introduction to Matrix Analysis.* New York: McGraw-Hill, 1960.
Guillemin, E. A. *The Mathematics of Circuit Analysis.* New York: Wiley, 1949.
Pipes, L. A. *Matrix Methods for Engineering.* Englewood Cliffs, N.J.: Prentice-Hall, 1963.
Wylie, C. R. *Advanced Engineering Mathematics*, 4th ed. New York: McGraw-Hill, 1975.

Controllability and Observability

Gilbert, E. G. "Controllability and Observability in Multivariable Control Systems." *J. Soc. Ind. Appl. Math.*, ser. A, 1963, pp. 128–51.
Kalman, R. E. "Canonical Structure of Linear Dynamical Systems." *Proc. Nat. Acad. Sci.*, April 1962, pp. 596–600.
Stubberud, A. R. "A Controllability Criterion for a Class of Linear Systems." *IEEE Trans. Appl. Ind.* 68 (1964): 411–13.

Computational Methods

Faddeeva, D. K., and Faddeeva, V. N. *Computational Methods of Linear Algebra.* San Francisco: Freeman, 1963.
James, M. L.; Smith, G. M.; and Wolford, J. C. *Applied Numerical Methods for Digital Computations*, 2nd ed. New York, Harper & Row, 1977.
Melsa, J. L., and Jones, S. K. *Computer Programs for Computational Assistance in the Study of Linear Control Theory*, 2nd ed. New York: McGraw-Hill, 1973.

State Feedback and Observers

Anderson, B. D. O., and Moore, J. B. *Linear Optimal Control.* Englewood Cliffs, N.J.: Prentice-Hall, 1971.
Davison, E. J. "On Pole Assignment in Multivariable Linear Systems." *IEEE Trans. Auto. Contr.* AC-13 (December 1968): 747–48.
Luenberger, D. G. "Observers for Multivariable Systems." *IEEE Trans. Auto. Contr.* AC-11 (April 1966): 190–97.
———, "An Introduction to Observers." *IEEE Trans. Auto. Contr.* AC-16 (December 1971): 596–602.
Wonham, W. M. "On Pole Assignment in Multi-Input Controllable Linear Systems." *IEEE Trans. Auto. Contr.* AC-12 (December 1967): 660–65.

Magnetic Levitation of Trains

Brock, K. H.; Gottzein, E.; Pfefferl, J.; and Schneider, E. "Control Aspects of a Tracked Magnetic Levitation High Speed Test Vehicle." *Automatica*, vol. 13, no. 3, pp. 205–223 (1977).
Glatzel, K.; Khurdok, G.; and Rogg, D. "The Development of the Magnetically Suspended Transportation System in the Federal Republic of Germany." *IEEE Trans. Vehic. Technol.* February 1980, pp. 3–17.
Glatzel, K., and Schulz, H. "Transportation: The Promise of MAGLEV." *IEEE Spec.*, March 1980, pp. 63–66.
Gottzein, E.; Meisinger, R.; and Miller, L. "The Magnetic Wheel in the Suspension of High Speed Ground Transportation Vehicles." *IEEE Trans. Vehic. Technol.*, February 1980, pp. 17–22.

PROBLEMS

1. Draw phase-variable form simulation diagrams for systems with the following transfer functions. Then write the state variable equations in matrix form.

(a) $T(s)=\dfrac{-3s+4}{s^2+10}$

(b) $T(s)=\dfrac{11s}{s^3+4s^2+3s+2}$

ans.
$$\begin{bmatrix}\dot{x}_1\\ \dot{x}_2\\ \dot{x}_3\end{bmatrix}=\begin{bmatrix}0 & 1 & 0\\ 0 & 0 & 1\\ -2 & -3 & -4\end{bmatrix}\begin{bmatrix}x_1\\ x_2\\ x_3\end{bmatrix}+\begin{bmatrix}0\\ 0\\ 1\end{bmatrix}r$$

$$y=\begin{bmatrix}0 & 11 & 0\end{bmatrix}\begin{bmatrix}x_1\\ x_2\\ x_3\end{bmatrix}$$

(c) $T(s)=\dfrac{5s^3-2s^2+s}{2s^4+4s^3+8s^2+s+1}$

(d) Two outputs:

$$T_{11}(s)=\frac{-s^2+4}{s^3+3s^2+s+2}$$

$$T_{21}(s)=\frac{s^2+s+5}{s^3+3s^2+s+2}$$

ans.
$$\begin{bmatrix}\dot{x}_1\\ \dot{x}_2\\ \dot{x}_3\end{bmatrix}=\begin{bmatrix}0 & 1 & 0\\ 0 & 0 & 1\\ -2 & -1 & -3\end{bmatrix}\begin{bmatrix}x_1\\ x_2\\ x_3\end{bmatrix}+\begin{bmatrix}0\\ 0\\ 1\end{bmatrix}r$$

$$\begin{bmatrix}y_1\\ y_2\end{bmatrix}=\begin{bmatrix}4 & 0 & 1\\ 5 & 1 & 1\end{bmatrix}\begin{bmatrix}x_1\\ x_2\\ x_3\end{bmatrix}$$

2. Draw dual phase-variable form simulation diagrams for systems with the following transfer functions. Then write the state variable equations in matrix form.

(a) $T(s)=\dfrac{-3s+4}{s^2+10}$

(b) $T(s)=\dfrac{2s+6}{3s^3+2s^2+8s+10}$

ans.
$$\begin{bmatrix}\dot{x}_1\\ \dot{x}_2\\ \dot{x}_3\end{bmatrix}=\begin{bmatrix}-\frac{2}{3} & 1 & 0\\ -\frac{8}{3} & 0 & 1\\ -\frac{10}{3} & 0 & 0\end{bmatrix}\begin{bmatrix}x_1\\ x_2\\ x_3\end{bmatrix}+\begin{bmatrix}0\\ \frac{2}{3}\\ 2\end{bmatrix}r$$

$$y=\begin{bmatrix}1 & 0 & 0\end{bmatrix}\begin{bmatrix}x_1\\ x_2\\ x_3\end{bmatrix}$$

(c) $T(s)=\frac{-7s^3+8s^2-9s+10}{s^4+3s^3+2s^2+5s+6}$

(d) Two inputs:

$$T_{11}(s)=\frac{3s^2+2}{s^3+3s^2+s+2}$$

$$T_{12}(s)=\frac{s-10}{s^3+3s^2+s+2}$$

ans. $$\begin{bmatrix}\dot{x}_1\\ \dot{x}_2\\ \dot{x}_3\end{bmatrix}=\begin{bmatrix}-3 & 1 & 0\\ -1 & 0 & 1\\ -2 & 0 & 0\end{bmatrix}\begin{bmatrix}x_1\\ x_2\\ x_3\end{bmatrix}+\begin{bmatrix}3 & 0\\ 2 & 0\\ 0 & -10\end{bmatrix}\begin{bmatrix}r_1\\ r_2\end{bmatrix}$$

$$y=[1\quad 0\quad 0]\begin{bmatrix}x_1\\ x_2\\ x_3\end{bmatrix}$$

3. Draw simulation diagrams to represent the following systems:

(a) $$\begin{bmatrix}\dot{x}_1\\ \dot{x}_2\\ \dot{x}_3\end{bmatrix}=\begin{bmatrix}1 & 3 & -1\\ 2 & 0 & 4\\ -2 & 1 & 5\end{bmatrix}\begin{bmatrix}x_1\\ x_2\\ x_3\end{bmatrix}+\begin{bmatrix}1\\ 0\\ 0\end{bmatrix}r$$

$$y=[1\quad -2\quad 0]\begin{bmatrix}x_1\\ x_2\\ x_3\end{bmatrix}$$

(b) $$\begin{bmatrix}\dot{x}_1\\ \dot{x}_2\\ \dot{x}_3\end{bmatrix}=\begin{bmatrix}0 & 1 & 0\\ -1 & -1 & 1\\ 2 & 0 & 3\end{bmatrix}\begin{bmatrix}x_1\\ x_2\\ x_3\end{bmatrix}+\begin{bmatrix}1\\ 4\\ -5\end{bmatrix}r$$

$$y=[0\quad -2\quad 6]\begin{bmatrix}x_1\\ x_2\\ x_3\end{bmatrix}$$

(c) $$\begin{bmatrix}\dot{x}_1\\ \dot{x}_2\end{bmatrix}=\begin{bmatrix}2 & 1\\ -3 & 0\end{bmatrix}\begin{bmatrix}x_1\\ x_2\end{bmatrix}+\begin{bmatrix}1 & 0 & 3\\ 2 & 1 & 1\end{bmatrix}\begin{bmatrix}r_1\\ r_2\\ r_3\end{bmatrix}$$

$$y=[1\quad 0]\begin{bmatrix}x_1\\ x_2\end{bmatrix}$$

(d) $$\begin{bmatrix}\dot{x}_1\\ \dot{x}_2\\ \dot{x}_3\end{bmatrix}=\begin{bmatrix}0 & 0 & 1\\ -1 & 2 & 1\\ -3 & 1 & 0\end{bmatrix}\begin{bmatrix}x_1\\ x_2\\ x_3\end{bmatrix}+\begin{bmatrix}1 & 0\\ 0 & 0\\ -1 & 2\end{bmatrix}\begin{bmatrix}r_1\\ r_2\end{bmatrix}$$

$$\begin{bmatrix}y_1\\ y_2\\ y_3\end{bmatrix}=\begin{bmatrix}1 & 0 & 0\\ 0 & 1 & 0\\ 1 & 1 & 1\end{bmatrix}\begin{bmatrix}x_1\\ x_2\\ x_3\end{bmatrix}$$

4. Find state equations in matrix form for the systems described by the simulation diagrams of Fig. P7-1.

ans. (b)
$$\begin{bmatrix}\dot{x}_1\\ \dot{x}_2\\ \dot{x}_3\end{bmatrix}=\begin{bmatrix}-1 & 3 & 0\\ 0 & -4 & 2\\ 0 & 0 & -3\end{bmatrix}\begin{bmatrix}x_1\\ x_2\\ x_3\end{bmatrix}+\begin{bmatrix}0 & 7\\ -2 & 0\\ 0 & 1\end{bmatrix}\begin{bmatrix}r_1\\ r_2\end{bmatrix}$$

$$y=[5 \quad 2 \quad 0]\begin{bmatrix}x_1\\ x_2\\ x_3\end{bmatrix}$$

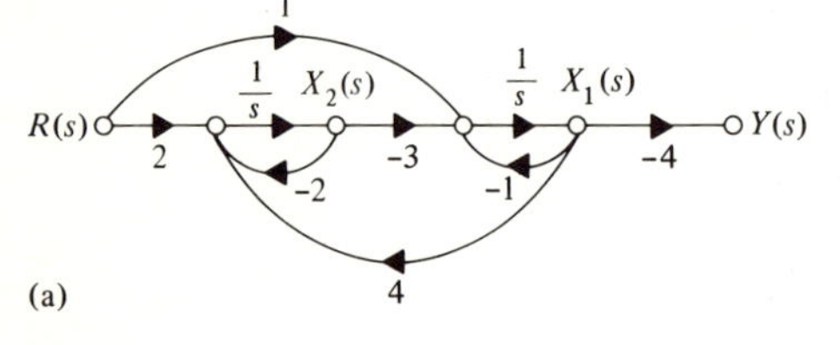

(a)

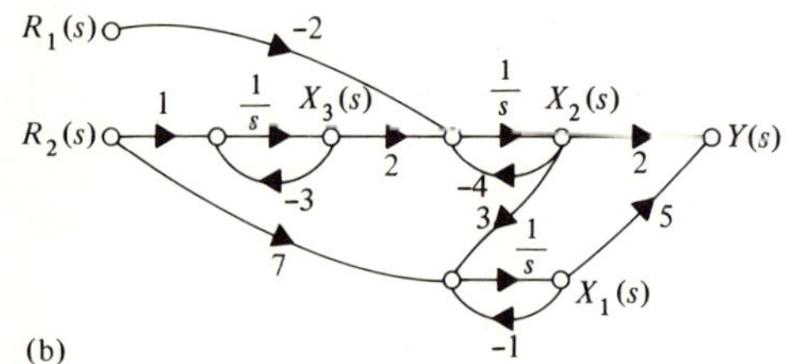

(b)

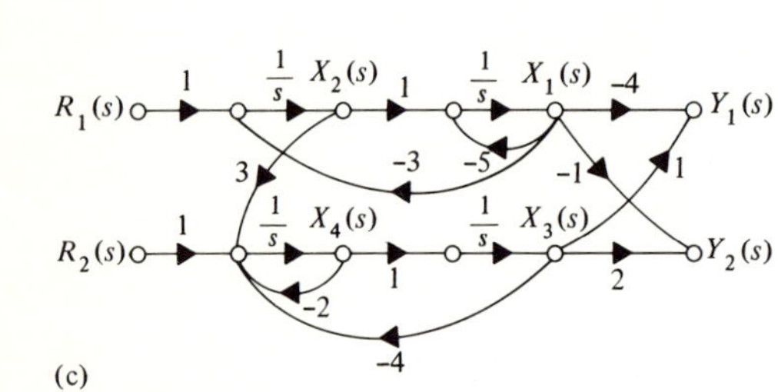

(c)

FIGURE P7-1

5. For the following systems, find the transfer function matrices:

(a)
$$\begin{bmatrix}\dot{x}_1\\ \dot{x}_2\end{bmatrix}=\begin{bmatrix}-2 & 1\\ -3 & 0\end{bmatrix}\begin{bmatrix}x_1\\ x_2\end{bmatrix}+\begin{bmatrix}2\\ -1\end{bmatrix}r(t)$$

$$\begin{bmatrix}y_1\\ y_2\end{bmatrix}=\begin{bmatrix}1 & -1\\ 0 & 1\end{bmatrix}\begin{bmatrix}x_1\\ x_2\end{bmatrix}$$

(b)
$$\begin{bmatrix}\dot{x}_1\\ \dot{x}_2\end{bmatrix}=\begin{bmatrix}0 & 1\\ -3 & 2\end{bmatrix}\begin{bmatrix}x_1\\ x_2\end{bmatrix}+\begin{bmatrix}0 & 1\\ 4 & -6\end{bmatrix}\begin{bmatrix}r_1\\ r_2\end{bmatrix}$$

$$y=[-1 \quad 1]\begin{bmatrix}x_1\\ x_2\end{bmatrix}$$

ans.
$$\begin{bmatrix}\dfrac{4(s-1)}{s^2-2s+3} & \dfrac{5-7s}{s^2-2s+3}\end{bmatrix}$$

(c)
$$\begin{bmatrix}\dot{x}_1\\ \dot{x}_2\\ \dot{x}_3\end{bmatrix}=\begin{bmatrix}0 & 1 & 0\\ 0 & 0 & 1\\ -2 & -3 & -4\end{bmatrix}\begin{bmatrix}x_1\\ x_2\\ x_3\end{bmatrix}+\begin{bmatrix}1 & 0\\ 0 & 1\\ 6 & -8\end{bmatrix}\begin{bmatrix}r_1\\ r_2\end{bmatrix}$$
$$\begin{bmatrix}y_1\\ y_2\end{bmatrix}=\begin{bmatrix}1 & 0 & 0\\ 0 & 1 & -2\end{bmatrix}\begin{bmatrix}x_1\\ x_2\\ x_3\end{bmatrix}$$

6. Find the characteristic equations of the following systems. Then for each, determine if they are stable.

(a)
$$\begin{bmatrix}\dot{x}_1\\ \dot{x}_2\end{bmatrix}=\begin{bmatrix}1 & 2\\ -3 & 5\end{bmatrix}\begin{bmatrix}x_1\\ x_2\end{bmatrix}+\begin{bmatrix}1 & 1\\ 2 & -3\end{bmatrix}\begin{bmatrix}r_1\\ r_2\end{bmatrix}$$
$$\begin{bmatrix}y_1\\ y_2\end{bmatrix}=\begin{bmatrix}1 & 2\\ 0 & 4\end{bmatrix}\begin{bmatrix}x_1\\ x_2\end{bmatrix}$$

(b)
$$\begin{bmatrix}\dot{x}_1\\ \dot{x}_2\\ \dot{x}_3\end{bmatrix}=\begin{bmatrix}0 & -2 & 3\\ 0 & 4 & -1\\ 0 & 1 & 7\end{bmatrix}\begin{bmatrix}x_1\\ x_2\\ x_3\end{bmatrix}\begin{bmatrix}7\\ -8\\ 9\end{bmatrix}r$$
$$y=[4\quad 5\quad 6]\begin{bmatrix}x_1\\ x_2\\ x_3\end{bmatrix}$$

ans. s^3-11s^2+29s, unstable

(c)
$$\begin{bmatrix}\dot{x}_1\\ \dot{x}_2\\ \dot{x}_3\end{bmatrix}=\begin{bmatrix}1 & -2 & 3\\ 4 & 0 & 6\\ -1 & 2 & 1\end{bmatrix}\begin{bmatrix}x_1\\ x_2\\ x_3\end{bmatrix}+\begin{bmatrix}2 & -1\\ 0 & 0\\ 3 & 6\end{bmatrix}\begin{bmatrix}r_1\\ r_2\end{bmatrix}$$
$$y=[1\quad 0\quad -1]\begin{bmatrix}x_1\\ x_2\\ x_3\end{bmatrix}$$

7. Although the following transfer functions do not share a common denominator polynomial, they may be made to have a common denominator by multiplying their numerators and denominators by appropriate factors. Find simulation diagram and a matrix state variable representation for a single-input, two-output system with the following two transfer functions:

$$T_{11}(s)=\frac{3s+1}{(s+2)(s+3)}$$

$$T_{21}(s)=\frac{16s}{(s+1)(s+3)}$$

The best solutions will involve only three integrators.

8. Use the partial fraction method to find diagonal state equations for single-input, single-output systems with the following transfer functions:

(a) $T(s)=\dfrac{-3s+6}{s^2+8s+12}$

(b) $T(s)=\dfrac{6s^2+3s-10}{(s+4)(s+3)(s+5)}$

ans.
$$\begin{bmatrix}\dot{x}'_1\\ \dot{x}'_2\\ \dot{x}'_3\end{bmatrix}=\begin{bmatrix}-3 & 0 & 0\\ 0 & -4 & 0\\ 0 & 0 & -5\end{bmatrix}\begin{bmatrix}x'_1\\ x'_2\\ x'_3\end{bmatrix}+\begin{bmatrix}1\\ 1\\ 1\end{bmatrix}r$$

$$y=\begin{bmatrix}\frac{35}{2} & -74 & \frac{125}{2}\end{bmatrix}\begin{bmatrix}x'_1\\ x'_2\\ x'_3\end{bmatrix}$$

(c) $T(s)=\dfrac{10}{s^3+9s^2+20s}$

9. The following transfer functions for single-input, single-output systems involve complex characteristic roots. Find diagonal state equations for these systems. Then find an alternative block diagonal representation which does not involve complex numbers.

(a) $T(s)=\dfrac{10s}{s^2+2s+10}$

(b) $T(s)=\dfrac{2s^2+3s-4}{(s+2)(s+3+j)(s+3-j)}$

ans.
$$\begin{bmatrix}\dot{x}'_1\\ \dot{x}'_2\\ \dot{x}'_3\end{bmatrix}=\begin{bmatrix}-2 & 0 & 0\\ 0 & -3-j & 0\\ 0 & 0 & -3+j\end{bmatrix}\begin{bmatrix}x'_1\\ x'_2\\ x'_3\end{bmatrix}+\begin{bmatrix}1\\ 1\\ 1\end{bmatrix}r$$

$$y=\begin{bmatrix}-1 & \dfrac{52-24j}{40} & \dfrac{52+24j}{40}\end{bmatrix}\begin{bmatrix}x'_1\\ x'_2\\ x'_3\end{bmatrix}$$

$$\begin{bmatrix}\dot{x}''_1\\ \dot{x}''_2\\ \dot{x}''_3\end{bmatrix}=\begin{bmatrix}-2 & 0 & 0\\ 0 & 0 & 1\\ 0 & -10 & -6\end{bmatrix}\begin{bmatrix}x''_1\\ x''_2\\ x''_3\end{bmatrix}+\begin{bmatrix}1\\ 0\\ 1\end{bmatrix}r$$

$$y=\begin{bmatrix}-1 & 3 & 3\end{bmatrix}\begin{bmatrix}x''_1\\ x''_2\\ x''_3\end{bmatrix}$$

(c) $T(s)=\dfrac{9}{(s+2)(s^2+4s+13)}$

10. The following transfer functions for single-input, single-output systems involve repeated characteristic roots. Find block diagonal Jordan canonical form state equations for these systems.

(a) $T(s)=\dfrac{3s-1}{s^2+4s+4}$

(b) $T(s)=\dfrac{s^3-4s^2+s-2}{(s+2)(s+3)^3}$

ans.
$$\begin{bmatrix}\dot{x}'_1\\ \dot{x}'_2\\ \dot{x}'_3\\ \dot{x}'_4\end{bmatrix}=\begin{bmatrix}-2 & 0 & 0 & 0\\ 0 & -3 & 1 & 0\\ 0 & 0 & -3 & 1\\ 0 & 0 & 0 & -3\end{bmatrix}\begin{bmatrix}x'_1\\ x'_2\\ x'_3\\ x'_4\end{bmatrix}+\begin{bmatrix}1\\ 0\\ 0\\ 1\end{bmatrix}r$$

$$y=[-28\quad 68\quad 16\quad 58]\begin{bmatrix}x'_1\\ x'_2\\ x'_3\\ x'_4\end{bmatrix}$$

(c) $T(s)=\dfrac{7s^3}{(s+2)^2(s+6)^2}$

11. The following systems have real characteristic roots. Find alternative diagonal state equations.

(a)
$$\begin{bmatrix}\dot{x}_1\\ \dot{x}_2\end{bmatrix}=\begin{bmatrix}-9 & 1\\ -20 & 0\end{bmatrix}\begin{bmatrix}x_1\\ x_2\end{bmatrix}+\begin{bmatrix}1\\ 4\end{bmatrix}r$$

$$y=[2\quad -3]\begin{bmatrix}x_1\\ x_2\end{bmatrix}$$

(b)
$$\begin{bmatrix}\dot{x}_1\\ \dot{x}_2\end{bmatrix}=\begin{bmatrix}0 & 1\\ -6 & -5\end{bmatrix}\begin{bmatrix}x_1\\ x_2\end{bmatrix}+\begin{bmatrix}1 & 0\\ 0 & 1\end{bmatrix}\begin{bmatrix}r_1\\ r_2\end{bmatrix}$$

$$y=[0\quad 1]\begin{bmatrix}x_1\\ x_2\end{bmatrix}$$

ans.
$$\begin{bmatrix}\dot{x}'_1\\ \dot{x}'_2\end{bmatrix}=\begin{bmatrix}-2 & 0\\ 0 & -3\end{bmatrix}\begin{bmatrix}x'_1\\ x'_2\end{bmatrix}+\begin{bmatrix}3 & 1\\ -2 & -1\end{bmatrix}r$$

$$y=[-2\quad -3]\begin{bmatrix}x'_1\\ x'_2\end{bmatrix}$$

(c)
$$\begin{bmatrix}\dot{x}_1\\ \dot{x}_2\end{bmatrix}=\begin{bmatrix}-5 & 1\\ -4 & 0\end{bmatrix}\begin{bmatrix}x_1\\ x_2\end{bmatrix}+\begin{bmatrix}1 & 0\\ 1 & -1\end{bmatrix}\begin{bmatrix}r_1\\ r_2\end{bmatrix}$$

$$\begin{bmatrix}y_1\\ y_2\end{bmatrix}=\begin{bmatrix}1 & 1\\ -1 & 0\end{bmatrix}\begin{bmatrix}x_1\\ x_2\end{bmatrix}$$

(d)
$$\begin{bmatrix}\dot{x}_1\\ \dot{x}_2\\ \dot{x}_3\end{bmatrix}=\begin{bmatrix}-3 & 1 & 0\\ -2 & 0 & 1\\ 0 & 0 & 0\end{bmatrix}\begin{bmatrix}x_1\\ x_2\\ x_3\end{bmatrix}+\begin{bmatrix}1\\ 1\\ 1\end{bmatrix}r$$

$$\begin{bmatrix}y_1\\ y_2\end{bmatrix}=\begin{bmatrix}1 & 1 & 0\\ 0 & 1 & 1\end{bmatrix}\begin{bmatrix}x_1\\ x_2\\ x_3\end{bmatrix}$$

ans.
$$\begin{bmatrix}\dot{x}'_1\\ \dot{x}'_2\\ \dot{x}'_3\end{bmatrix}=\begin{bmatrix}0 & 0 & 0\\ 0 & -1 & 0\\ 0 & 0 & -2\end{bmatrix}\begin{bmatrix}x'_1\\ x'_2\\ x'_3\end{bmatrix}+\begin{bmatrix}3\\ -3\\ 1\end{bmatrix}r$$

$$\begin{bmatrix}y_1\\ y_2\end{bmatrix}=\begin{bmatrix}1 & 0 & -1\\ 0 & 0 & 2\end{bmatrix}\begin{bmatrix}x'_1\\ x'_2\\ x'_3\end{bmatrix}$$

An F-16 aircraft performs a nine-g climbing turn. State variable control system design methods are essential in the aerospace industry, where exacting performance of extremely complex systems is critical. (Photo courtesy of General Dynamics Corporation.)

12. The following system has a set of complex conjugate characteristic roots. Find an alternative diagonal set of state equations. Then find another alternative set of state equations where the complex root terms are placed in real number block diagonal form.

$$\begin{bmatrix} \dot{x}_1 \\ \dot{x}_2 \\ \dot{x}_3 \end{bmatrix} = \begin{bmatrix} 0 & 1 & 0 \\ 0 & 0 & 1 \\ 0 & -17 & -2 \end{bmatrix} \begin{bmatrix} x_1 \\ x_2 \\ x_3 \end{bmatrix} + \begin{bmatrix} 2 \\ -1 \\ 0 \end{bmatrix} r$$

$$y = \begin{bmatrix} 1 & 1 & 1 \end{bmatrix} \begin{bmatrix} x_1 \\ x_2 \\ x_3 \end{bmatrix}$$

13. The following system has a repeated characteristic root. Find an alternate set of state equations in Jordan form:

$$\begin{bmatrix} \dot{x}_1 \\ \dot{x}_2 \\ \dot{x}_3 \end{bmatrix} = \begin{bmatrix} 0 & 1 & 0 \\ 0 & 0 & 1 \\ -9 & -15 & -7 \end{bmatrix} \begin{bmatrix} x_1 \\ x_2 \\ x_3 \end{bmatrix} + \begin{bmatrix} 0 \\ 2 \\ 3 \end{bmatrix} r$$

$$y = \begin{bmatrix} 1 & 1 & 0 \\ 0 & 1 & 1 \end{bmatrix} \begin{bmatrix} x_1 \\ x_2 \\ x_3 \end{bmatrix}$$

14. Find diagonal state equations for systems with the following transfer function matrices:

(a)
$$\mathbf{T}(s) = \begin{bmatrix} \dfrac{-6s}{s^2+5s+4} & \dfrac{8}{s^2+5s+4} \end{bmatrix}$$

(b) $\mathbf{T}(s)=\begin{bmatrix} \dfrac{s^2-2}{s^3+3s^2+2s} & \dfrac{4s-5}{s^3+3s^2+2s} & \dfrac{s^2+3s-1}{s^3+3s^2+2s} \end{bmatrix}$

ans. One possibility is the following:

$$\begin{bmatrix} \dot{x}_1 \\ \dot{x}_2 \\ \dot{x}_3 \end{bmatrix}=\begin{bmatrix} 0 & 0 & 0 \\ 0 & -1 & 0 \\ 0 & 0 & -2 \end{bmatrix}\begin{bmatrix} x_1 \\ x_2 \\ x_3 \end{bmatrix}+\begin{bmatrix} 1 & \frac{5}{2} & \frac{1}{2} \\ 1 & 9 & 3 \\ 1 & \frac{13}{2} & -\frac{3}{2} \end{bmatrix}\begin{bmatrix} r_1 \\ r_2 \\ r_3 \end{bmatrix}$$

$$y=[-1 \quad 1 \quad 1]\begin{bmatrix} x_1 \\ x_2 \\ x_3 \end{bmatrix}$$

(c) $\mathbf{T}(s)=\begin{bmatrix} \dfrac{4s-2}{s^2+4} \\ \dfrac{-s+3}{s^2+4} \end{bmatrix}$

(d) $\mathbf{T}(s)=\dfrac{\begin{bmatrix} 2s \\ -3s^2-1 \\ 4 \end{bmatrix}}{s^3+3s^2+2s}$

ans. One possibility is the following:

$$\begin{bmatrix} \dot{x}_1 \\ \dot{x}_2 \\ \dot{x}_3 \end{bmatrix}=\begin{bmatrix} 0 & 0 & 0 \\ 0 & -1 & 0 \\ 0 & 0 & -2 \end{bmatrix}\begin{bmatrix} x_1 \\ x_2 \\ x_3 \end{bmatrix}+\begin{bmatrix} 1 \\ 1 \\ 1 \end{bmatrix} r$$

$$\begin{bmatrix} y_1 \\ y_2 \\ y_3 \end{bmatrix}=\begin{bmatrix} 0 & 2 & -2 \\ -\frac{1}{2} & 4 & -\frac{13}{2} \\ 2 & -4 & 2 \end{bmatrix}\begin{bmatrix} x_1 \\ x_2 \\ x_3 \end{bmatrix}$$

15. Find a simulation diagram and a matrix state variable representation for a two-input, two-output system with the following transfer function matrix:

$$\mathbf{T}(s)=\begin{bmatrix} \dfrac{6s}{s^2+3s+2} & \dfrac{3s-2}{s^2+3s+2} \\ \dfrac{-4}{s^2+3s+2} & \dfrac{s+1}{s^2+3s+2} \end{bmatrix}$$

16. A transfer function with equal numerator and denominator polynomial degrees may be expanded as a constant plus a proper remainder, as in the following example:

$$T(s)=\frac{3s^2+2s-4}{s^2+3s+2}=3+\frac{-7s-10}{s^2+3s+2}$$

It may be realized by adding a term to the system output which is proportional to the system input. The resulting state variable equations have the form

$$\dot{\mathbf{x}}=\mathbf{A}\mathbf{x}+\mathbf{B}\mathbf{r}$$
$$\mathbf{y}=\mathbf{C}\mathbf{x}+\mathbf{D}\mathbf{r}$$

Find matrices **A**, **B**, **C**, and **D** for a system with the above transfer function. For such a second-order single-input, single-output system, the state variable equations will be of the form

$$\begin{bmatrix} \dot{x}_1 \\ \dot{x}_2 \end{bmatrix} = \begin{bmatrix} a_{11} & a_{12} \\ a_{21} & a_{22} \end{bmatrix} \begin{bmatrix} x_1 \\ x_2 \end{bmatrix} + \begin{bmatrix} b_1 \\ b_2 \end{bmatrix} r$$

$$y = \begin{bmatrix} c_1 & c_2 \end{bmatrix} \begin{bmatrix} x_1 \\ x_2 \end{bmatrix} + dr$$

17. Use controllability and observability matrices to determine whether the following systems are completely controllable and whether these systems are completely observable:

(a) $$\begin{bmatrix} \dot{x}_1 \\ \dot{x}_2 \end{bmatrix} = \begin{bmatrix} 2 & -4 \\ 0 & 1 \end{bmatrix} \begin{bmatrix} x_1 \\ x_2 \end{bmatrix} + \begin{bmatrix} 1 \\ 0 \end{bmatrix} r$$

$$y = \begin{bmatrix} 1 & 1 \end{bmatrix} \begin{bmatrix} x_1 \\ x_2 \end{bmatrix}$$

(b) $$\begin{bmatrix} \dot{x}_1 \\ \dot{x}_2 \\ \dot{x}_3 \end{bmatrix} = \begin{bmatrix} 3 & 0 & -5 \\ -2 & 1 & 5 \\ 0 & 0 & -2 \end{bmatrix} \begin{bmatrix} x_1 \\ x_2 \\ x_3 \end{bmatrix} + \begin{bmatrix} 1 & 0 \\ 2 & 0 \\ 0 & -1 \end{bmatrix} \begin{bmatrix} r_1 \\ r_2 \end{bmatrix}$$

$$\begin{bmatrix} y_1 \\ y_2 \end{bmatrix} = \begin{bmatrix} 4 & 1 & -3 \\ 3 & 2 & -1 \end{bmatrix} \begin{bmatrix} x_1 \\ x_2 \\ x_3 \end{bmatrix}$$

ans. completely controllable but not completely observable

(c) $$\begin{bmatrix} \dot{x}_1 \\ \dot{x}_2 \\ \dot{x}_3 \end{bmatrix} = \begin{bmatrix} 1 & 0 & -2 \\ 3 & -3 & 0 \\ 0 & 0 & 1 \end{bmatrix} \begin{bmatrix} x_1 \\ x_2 \\ x_3 \end{bmatrix} + \begin{bmatrix} 1 & -1 \\ 2 & 0 \\ 0 & 0 \end{bmatrix} \begin{bmatrix} r_1 \\ r_2 \end{bmatrix}$$

$$\begin{bmatrix} y_1 \\ y_2 \end{bmatrix} = \begin{bmatrix} 0 & 4 & 1 \\ 0 & -2 & 3 \end{bmatrix} \begin{bmatrix} x_1 \\ x_2 \\ x_3 \end{bmatrix}$$

18. Write state equations for systems, each with modes e^{2t}, e^{-3t}, $e^{(-4+j)t}$, and $e^{(-4-j)t}$ which have the following properties:
   (a) The mode e^{2t} is uncontrollable.
   (b) The mode e^{-3t} is unobservable.
   (c) The mode e^{2t} is both uncontrollable and unobservable.
   (d) The modes $e^{(-4+j)t}$ and $e^{(-4-j)t}$ are uncontrollable.
   (e) The mode e^{-3t} is uncontrollable and the mode e^{2t} is unobservable.
19. The system

$$\begin{bmatrix} \dot{x}_1 \\ \dot{x}_2 \end{bmatrix} = \begin{bmatrix} -1 & 1 \\ 2 & 0 \end{bmatrix} \begin{bmatrix} x_1 \\ x_2 \end{bmatrix} + \begin{bmatrix} 0 \\ 1 \end{bmatrix} r$$

$$y = \begin{bmatrix} 1 & 1 \end{bmatrix} \begin{bmatrix} x_1 \\ x_2 \end{bmatrix}$$

is unstable. Can the instability be detected from input-output measurements? Determine whether or not the system is completely observable. Then calculate the system transfer function. A common factor in the numerator and the denominator should cancel.

Repeat if instead the output equation is

$$y=[-2 \quad 1]\begin{bmatrix} x_1 \\ x_2 \end{bmatrix}$$

20. Find a third-order system, in phase-variable form, which is not completely controllable.
21. Show that an nth-order system with n outputs is completely observable if its $n \times n$ output coupling matrix is nonsingular.
22. Solve

$$\dot{x} = -3x + r(t)$$

with

$$x(0^-)=4$$
$$r(t)=5e^{3t}$$

using the time-domain method involving convolution.

23. Use Laplace transform methods to find the state response of the following systems for $t \geqslant 0$ with the given inputs and initial conditions. Also find the system output.

(a) $\dot{x} = -3x + 4r(t)$

$y=6x$

$x(0^-)=2$

$r(t)=5u(t)$, where $u(t)$ is the unit step function

(b) $$\begin{bmatrix} \dot{x}_1 \\ \dot{x}_2 \end{bmatrix} = \begin{bmatrix} -4 & 1 \\ -3 & 0 \end{bmatrix} \begin{bmatrix} x_1 \\ x_2 \end{bmatrix} + \begin{bmatrix} 1 \\ 0 \end{bmatrix} r$$

$$y=[1 \quad 0]\begin{bmatrix} x_1 \\ x_2 \end{bmatrix}$$

$$\begin{bmatrix} x_1(0^-) \\ x_2(0^-) \end{bmatrix} = \begin{bmatrix} 0 \\ 0 \end{bmatrix}$$

$r(t)=\delta(t)$, the unit inpulse function

ans. $\begin{bmatrix} (-\frac{1}{2}e^{-t}+\frac{3}{2}e^{-3t}) \\ (-\frac{3}{2}e^{-t}+\frac{3}{2}e^{-3t}) \end{bmatrix}$, $y(t) = -\frac{1}{2}e^{-t}+\frac{3}{2}e^{-3t}$

(c) $$\begin{bmatrix} \dot{x}_1 \\ \dot{x}_2 \end{bmatrix} = \begin{bmatrix} 0 & 1 \\ -8 & -6 \end{bmatrix} \begin{bmatrix} x_1 \\ x_2 \end{bmatrix} + \begin{bmatrix} 0 & 1 \\ 1 & -1 \end{bmatrix} \begin{bmatrix} r_1 \\ r_2 \end{bmatrix}$$

$$y=[1 \quad 1]\begin{bmatrix} x_1 \\ x_2 \end{bmatrix}$$

$$\begin{bmatrix} x_1(0^-) \\ x_2(0^-) \end{bmatrix} = \begin{bmatrix} 2 \\ -3 \end{bmatrix}$$

$$\begin{bmatrix} r_1(t) \\ r_2(t) \end{bmatrix} = \begin{bmatrix} e^t \\ 5 \end{bmatrix}$$

(d) $$\begin{bmatrix} \dot{x}_1 \\ \dot{x}_2 \\ \dot{x}_3 \end{bmatrix} = \begin{bmatrix} -3 & 1 & 0 \\ -2 & 0 & 1 \\ 0 & 0 & 0 \end{bmatrix} \begin{bmatrix} x_1 \\ x_2 \\ x_3 \end{bmatrix} + \begin{bmatrix} 0 \\ 1 \\ 0 \end{bmatrix} r$$

$$\begin{bmatrix} y_1 \\ y_2 \end{bmatrix} = \begin{bmatrix} 1 & 0 & 1 \\ 0 & 0 & 1 \end{bmatrix} \begin{bmatrix} x_1 \\ x_2 \\ x_3 \end{bmatrix}$$

$$\begin{bmatrix} x_1(0^-) \\ x_2(0^-) \\ x_3(0^-) \end{bmatrix} = \begin{bmatrix} 1 \\ 0 \\ 0 \end{bmatrix}$$

$$r(t) = 2u(t)$$

24. The state transition matrix for a certain system is

$$\mathbf{\Phi}(t) = \begin{bmatrix} \frac{1}{2}e^{-t} + \frac{1}{2}e^{-2t} & e^{-t} - e^{-2t} \\ 3e^{-t} - 3e^{-2t} & -e^{-t} + 2e^{-2t} \end{bmatrix}$$

Find the state $\mathbf{x}(t)$ for $t \geqslant 0$ if all system inputs are zero and

$$\mathbf{x}(0^-) = \begin{bmatrix} x_1(0^-) \\ x_2(0^-) \end{bmatrix} = \begin{bmatrix} -4 \\ 3 \end{bmatrix}$$

25. Calculate state transition matrices for systems with the following state coupling matrices $\mathbf{A}$, using

$$\mathbf{\Phi}(t) = \mathscr{L}^{-1}\{[s\mathbf{I} - \mathbf{A}]^{-1}\}$$

(a) $$\begin{bmatrix} 0 & 1 \\ -10 & -7 \end{bmatrix}$$

(b) $$\begin{bmatrix} -1 & 1 \\ 6 & -2 \end{bmatrix}$$

ans. $$\begin{bmatrix} \frac{3}{5}e^{t} + \frac{2}{5}e^{-4t} & \frac{1}{5}e^{t} - \frac{1}{5}e^{-4t} \\ \frac{6}{5}e^{t} - \frac{6}{5}e^{-4t} & \frac{2}{5}e^{t} + \frac{3}{5}e^{-4t} \end{bmatrix}$$

(c) $$\begin{bmatrix} -6 & 1 & 0 \\ -8 & 0 & 1 \\ 0 & 0 & 0 \end{bmatrix}$$

26. Show that the state transition matrix for a diagonalized system is diagonal, with the system modes along the diagonal.
27. Find state variable equations for each of the systems of Fig. P7-2. Then find the transfer function(s) from the original drawing and compare with the transfer function(s) of the state variable model.
28. For the state feedback systems described by the following equations, choose the feedback gain constants k_i to place the closed-loop system poles at the indicated locations. Then, for the feedback system, find the steady state outputs due to a unit step input $u(t)$.

(a)
$$\begin{bmatrix} \dot{x}_1 \\ \dot{x}_2 \\ \dot{x}_3 \\ \dot{x}_4 \end{bmatrix} = \begin{bmatrix} 0 & 1 & 0 & 0 \\ 0 & 0 & 1 & 0 \\ 0 & 0 & 0 & 1 \\ 0 & 0 & -4 & -5 \end{bmatrix} \begin{bmatrix} x_1 \\ x_2 \\ x_3 \\ x_4 \end{bmatrix} + \begin{bmatrix} 0 \\ 0 \\ 0 \\ 1 \end{bmatrix} r$$

$$r = [k_1 \quad k_2 \quad k_3 \quad k_4] \begin{bmatrix} x_1 \\ x_2 \\ x_3 \\ x_4 \end{bmatrix} + u$$

$$y = \begin{bmatrix} 2 & -1 & 0 & 3 \\ 1 & 0 & 1 & -2 \end{bmatrix} \begin{bmatrix} x_1 \\ x_2 \\ x_3 \\ x_4 \end{bmatrix}$$

Closed-loop poles at $s = -5 \pm j3,\ -4 \pm j4$

(b)
$$\begin{bmatrix} \dot{x}_1 \\ \dot{x}_2 \\ \dot{x}_3 \end{bmatrix} = \begin{bmatrix} 0 & 0 & 1 \\ 3 & -2 & -1 \\ 0 & 1 & 0 \end{bmatrix} \begin{bmatrix} x_1 \\ x_2 \\ x_3 \end{bmatrix} + \begin{bmatrix} 0 & -1 \\ 1 & 0 \\ 0 & 1 \end{bmatrix} \begin{bmatrix} r \\ u \end{bmatrix}$$

$$r = [k_1 \quad k_2 \quad k_3] \begin{bmatrix} x_1 \\ x_2 \\ x_3 \end{bmatrix}$$

$$y = [1 \quad 0 \quad 0] \begin{bmatrix} x_1 \\ x_2 \\ x_3 \end{bmatrix}$$

Closed-loop poles at $s = -4$ and $-4 \pm j2$

ans. $k_1 = 77,\ k_2 = -10,\ k_3 = 53;\ -\frac{4}{5}$

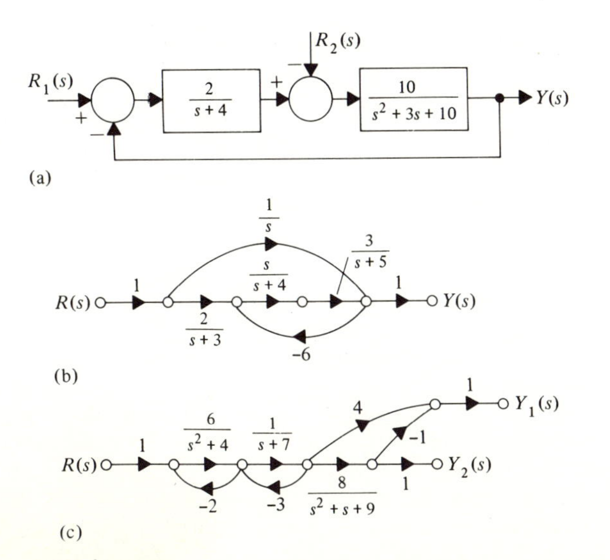

FIGURE P7-2

(c) $$\begin{bmatrix} \dot{x}_1 \\ \dot{x}_2 \end{bmatrix} = \begin{bmatrix} 0 & 1 \\ -1 & 4 \end{bmatrix} \begin{bmatrix} x_1 \\ x_2 \end{bmatrix} + \begin{bmatrix} 1 \\ 1 \end{bmatrix} r$$

$$r = [k_1 \quad k_2] \begin{bmatrix} x_1 \\ x_2 \end{bmatrix} + u$$

$$y = [1 \quad 1] \begin{bmatrix} x_1 \\ x_2 \end{bmatrix}$$

Both closed-loop poles at $s = -3$

29. For the MAGLEV system (Sec. 7.9), choose instead values of the feedback gain constants k_1, k_2, and k_3 to place all three of the overall system poles at $s = -5$. For this system, find the steady state response d to a unit step disturbance f and the value of constant reference input r to give a nominal gap distance $d = 0.05$.

30. For the open-loop MAGLEV system (Sec. 7.9), suppose the vertical track elevation varies sinusoidally with time as the train is in motion, according to

$$\dot{h}(t) = 0.2 \sin \frac{\pi t}{10}$$

Find the second-order differential equation satisfied by $\dot{h}(t)$, then augment the original state equations with two more equations and two more state variables

$$x_4 = \dot{h}(t)$$
$$x_5 = \ddot{h}(t)$$

in place of the disturbance input $f = \ddot{h}$. With additional sensors for the signals x_4 and x_5 and feedback of the form

$$v = -k_1 x_1 - k_2 x_2 - k_3 x_3 - k_4 x_4 - k_5 x_6 + r$$

find the state equations for the feedback system in terms of the k constants.

eight

DIGITAL CONTROL

8.1 PREVIEW

The terms *continuous-time* and *analog* are identical in meaning when applied to signals and systems. Analog signals are functions of a continuous time variable, and analog systems are systems described in terms of such signals. Similarly, *discrete-time* and *digital* have the same meaning; they refer to signals which are defined only for specified instants of time. The use of the words *analog* and *digital* in this sense dates from an era when large-scale analog computers were common.

Past years have seen an exponential growth in the capability and application of digital computers, and there is every indication that a high rate of growth will continue far into the future. In the field of control systems, digital computers were first applied in military and space applications where the high costs were justified by new capabilities. As computer costs dropped, their use for control began in large-scale industry such as chemical processing, heavy manufacturing, and telecommunications plants which could afford the large investment. Later, minicomputers, faster, more powerful, and much less expensive than their predecessors, began to revolutionize industry everywhere. No longer was a huge central computing installation necessary; general- and special-purpose computers could be economically distributed and tailored to specific tasks. Now, with the widespread availability and low cost of microcomputers, every process is a candidate for digital control, and sophisticated control systems that were impractically expensive only a few years ago are feasible.

The possibilities for the future are fabulous! It is no longer farfetched to imagine a digital-computer-based control system that monitors plant behavior to determine a mathematical model of the plant, then through repeated simulation or other calculation determines an optimum control strategy and proceeds to effect control according to programmed objectives.

In view of the ongoing explosion in digital technology, it is tempting to consider scrapping analog concepts altogether, until one realizes that the macroscopic physical world is overwhelmingly analog.

The popularity of digital control components is likely to accelerate even further, and that is good because it means that designers will have new tools that will result in greatly improved performance at low cost. Control system design will, however, continue to require abilities with both the controller and the controlled. While highly specialized work predominantly in the digital domain (or the analog domain) will probably exist, the major long-term need will be for a blend of both and an understanding of the analog-digital interface.

We feel that the best preparation for these exciting eventualities involves fundamental study of both analog and digital signals, systems, and control. It is fitting that the analog concepts precede the digital because the transition from analog to digital is much easier than the reverse or than a parallel development. In this final chapter, basic digital concepts are presented in a manner which forms an introduction to a complete course of study of the subject.

The description of digital systems and control is given here in two stages. The first consists of a summary view of basic hardware and software considerations. Data acquisition is studied in Sec. 8.2, with emphasis upon analog-digital conversion processes. Fundamental concepts relating to digital computer processing are introduced in Sec. 8.3, with particular attention given to microprocessor-based control systems and to the outlook for the future.

In the second stage, a careful analytical foundation is developed and design methods are introduced. Z-transformation is defined in terms of sample sequences and basic transforms and properties are shown in Sec. 8.4. The relations between Z-transform and Laplace transform are developed in connection with the description of sampling in Sec. 8.5. Further insight is given in Sec. 8.6, which concerns the reconstruction of analog signals from samples.

Discrete-time signal processing is described in terms of difference equations in Sec. 8.7 and in state variable format in Sec. 8.8. The state variable description parallels the continuous-time development of Chap. 7.

The chapter concludes with two approaches to design. In Sec. 8.9, an existing or model analog system is digitized. The approach is conventional, with the step-invariant approximation used for conversion of the analog compensator to an approximating digital one. In Sec. 8.10, a digital controller of an analog satellite-tracking system is designed directly.

8.2 ANALOG-DIGITAL CONVERSION

An analog signal such as a voltage can be expressed as a binary number, suitable for computer processing, by assigning weights to each bit position. Table 8-1 gives a 4-bit coding of an analog signal which may range between 0 and 10 V. Each binary increment represents $2^{-4}=\frac{1}{16}$ the maximum representable voltage of 10 V. The information in the table is shown in graphical form in Fig. 8-1.

Each binary number represents a *range* of analog voltage; hence there is a

TABLE 8-1
Representing a non-negative analog voltage with a binary number

Analog Voltage	*Binary Representation*
0 to 0.625	0000
0.625 to 1.25	0001
1.25 to 1.875	0010
1.875 to 2.5	0011
2.5 to 3.125	0100
3.125 to 3.75	0101
3.75 to 4.375	0110
4.375 to 5.0	0111
5.0 to 5.625	1000
5.625 to 6.25	1001
6.25 to 6.875	1010
6.875 to 7.5	1011
7.5 to 8.125	1100
8.125 to 8.75	1101
8.75 to 9.375	1110
9.375 to 10.0	1111

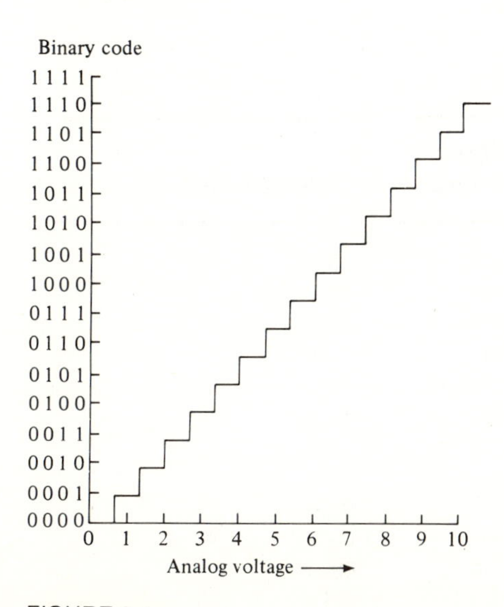

FIGURE 8-1
Binary coding of an analog voltage.

quantization error associated with the conversion. For a 4-bit conversion, the maximum quantization error is $2^{-4} = 6.25\%$. Table 8-2 shows quantization error percentages for various numbers of bits in the digital representation. The quantization error in 16-bit conversion, for example, corresponds to a signal-to-noise ratio (SNR) of

$$\text{SNR (in dB)} = 20 \log_{10} 2^{16} = 96.3 \text{ dB}$$

TABLE 8-2
Quantization error for analog-digital conversion

Number of Bits	*Maximum Percent Error*
1	50
2	25
4	6.25
6	1.56
8	0.391
10	0.0977
12	0.0244
14	0.0061
16	0.0015

TABLE 8-3
Bipolar analog voltage representations

Analog Voltage	*Sign and Magnitude*	*Offset Binary*	*Two's Complement*
−5.0 to −4.375	1111	0000	1001
−4.375 to −3.75	1110	0001	1010
−3.75 to −3.125	1101	0010	1011
−3.125 to −2.5	1100	0011	1100
−2.5 to −1.875	1011	0100	1101
−1.875 to −1.25	1010	0101	1110
−1.25 to −0.625	1001	0110	1111
−0.625 to 0	1000	0111	1000
0 to 0.625	0000	1000	0000
0.625 to 1.25	0001	1001	0001
1.25 to 1.875	0010	1010	0010
1.875 to 2.5	0011	1011	0011
2.5 to 3.125	0100	1100	0100
3.125 to 3.75	0101	1101	0101
3.75 to 4.375	0110	1110	0110
4.375 to +5.0	0111	1111	0111

By comparison, typical signal-to-noise ratios in quality audio recording and reproduction are 60–70 dB, which may be accurately portrayed by only 12-bit coding.

For bipolar signals, the three types of binary codes shown in Table 8-3 are the most commonly used. In the sign and magnitude arrangement, the most significant bit of the binary code represents the algebraic sign of the signal, with a zero meaning a positive number. The remaining bits are the binary representation of the signal's magnitude. The offset binary code is equivalent to adding a fixed constant (or bias) to the signal to be converted so that the sum is always nonnegative. In two's complement coding, negative signals are represented as the two's complement of their magnitude, in the same manner as negative numbers are commonly manipulated in digital computers. In applications involving digital displays, binary-coded-decimal (BCD) coding may be used, where the signal is represented as *decimal* digits and then each decimal digit is individually converted to a 4-bit binary equivalent.

8.2.1 D/A Conversion

Conversion from a binary code to an analog voltage is relatively simple in principle. An operational amplifier may be used to sum voltages with the proper binary weights, as indicated in Fig. 8-2 for a 4-bit, 0–10-V range converter. In general, the operational amplifier output voltage in Fig. 8-2a is

$$v_{\text{out}} = -\frac{R_f}{R_1}v_1 - \frac{R_f}{R_2}v_2 - \frac{R_f}{R_3}v_3 - \frac{R_f}{R_4}v_4$$

In the weighted sum arrangement of Fig. 8-2b, the electronic switches (typically JFET or MOS devices), controlled by the individual bits of the digital representation, add or do not add the voltage weight to v_{out} according to whether the bit is a binary one or a zero.

For conversion involving more than a few bits, the weighted-sum digital-to-analog (D/A) converter is difficult to build because of the wide range of highly accurate resistors required. The ladder arrangement of Fig. 8-2c is widely used because it requires only matched resistors of values R and $2R$.

8.2.2 A/D Conversion

Most analog-to-digital (A/D) converters use a D/A converter as a component part. A schematic symbol for a D/A converter is given in Fig. 8-3a. A set of binary digital input voltages, one for each bit of the code, forms the converter input. The single quantized analog output signal is produced. The alternate symbol in which the collection of binary inputs is shown as a wide arrow is

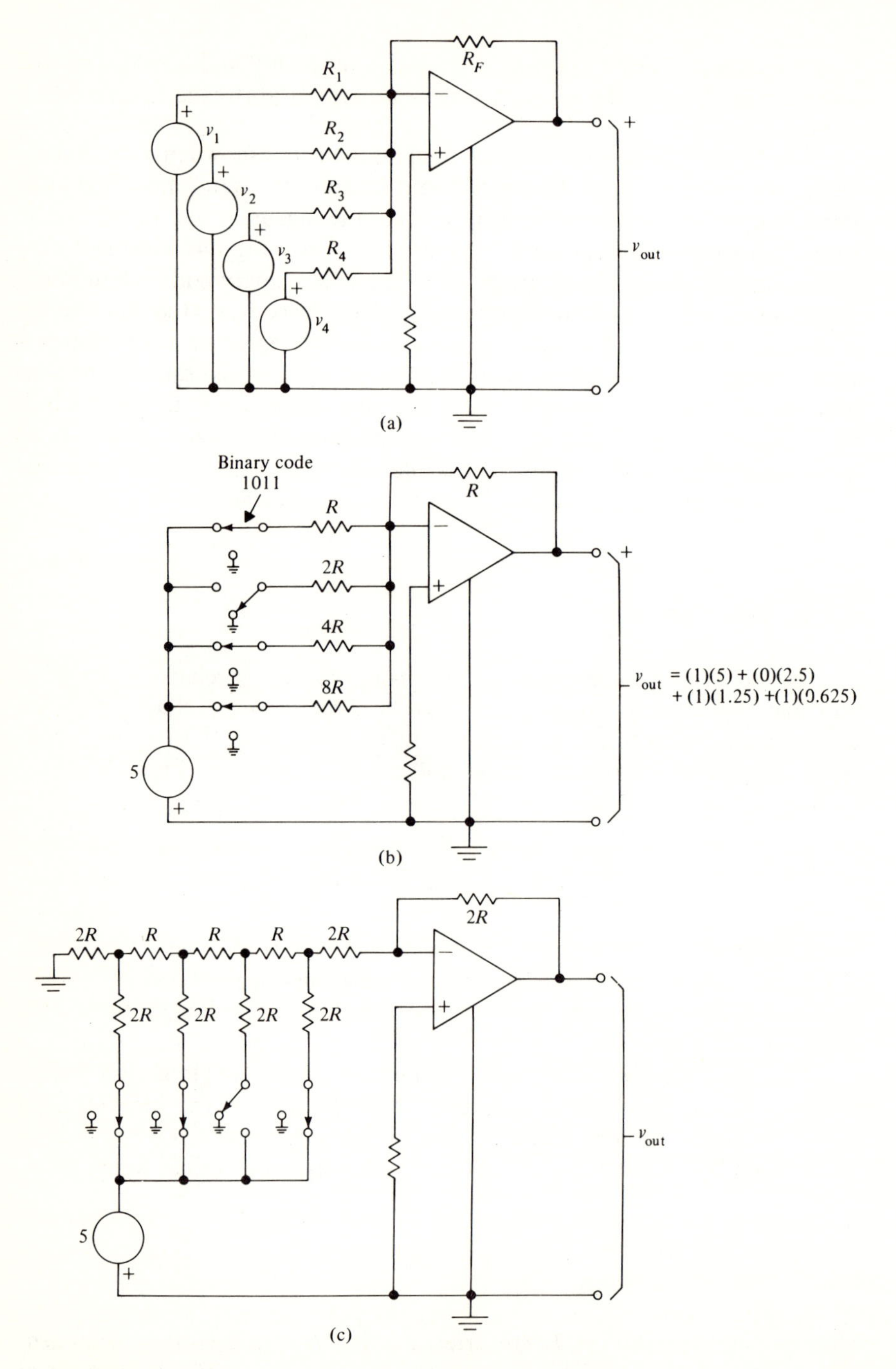

FIGURE 8-2
Digital-to-analog (D/A) conversion. (a) Operational amplifier summing circuit with four inputs. (b) A four-bit weighted sum D/A converter. (c) A four-bit ladder D/A converter.

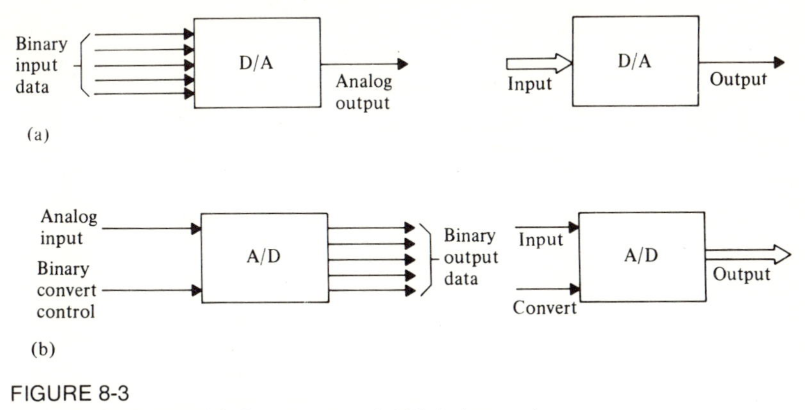

FIGURE 8-3
Symbols for D/A and A/D converters. (a) Digital-to-analog conversion. (b) Analog-to-digital conversion.

convenient when an unspecified or large number of bits are involved. When the binary input to the converter changes, the converter output changes, although a short interval of time, termed the *settling time* of the converter, is required before the conversion is accurate. An A/D converter symbol is given in Fig. 8-3b. An analog input signal is converted to a binary representation in terms of a set of binary digital output voltages, one for each bit of the code. A digital input signal, the convert control, determines when the conversion is to begin.

In the tracking counter A/D converter, Fig. 8-4a, a progression of binary codes is generated by a digital counter and converted to an analog signal by a D/A converter. When the converter voltage passes the input voltage, the direction of count is reversed so that the binary code tends to track the input. Variations on this arrangement are used in practice, where the conversion time is greatly reduced by stepping the binary code in large increments at first, then refining the code by small steps. This "dual slope" conversion is popular for the A/D conversion involved in digital voltmeters.

The successive approximation converter, Fig. 8-4b, performs a short sequence of tests to determine the binary coding of an analog input signal. For an input signal in the range 0–10 V, it would first generate the binary code for 5 V and determine if the input signal is greater than 5 V. If it is greater, the code for 7.5 V would be generated, compared, and so on.

A series of voltage comparators as in Fig. 8-4c is termed a *flash converter* because it is capable of operating at very high speed. The input signal is simultaneously compared with each of the possible coded voltage levels. For applications requiring moderate conversion speed, the flash converter is seldom used because of its complexity. An 8-bit converter of this type, for example, requires 256 comparators. For 16 bits, $2^{16} = 65{,}536$ comparators are required.

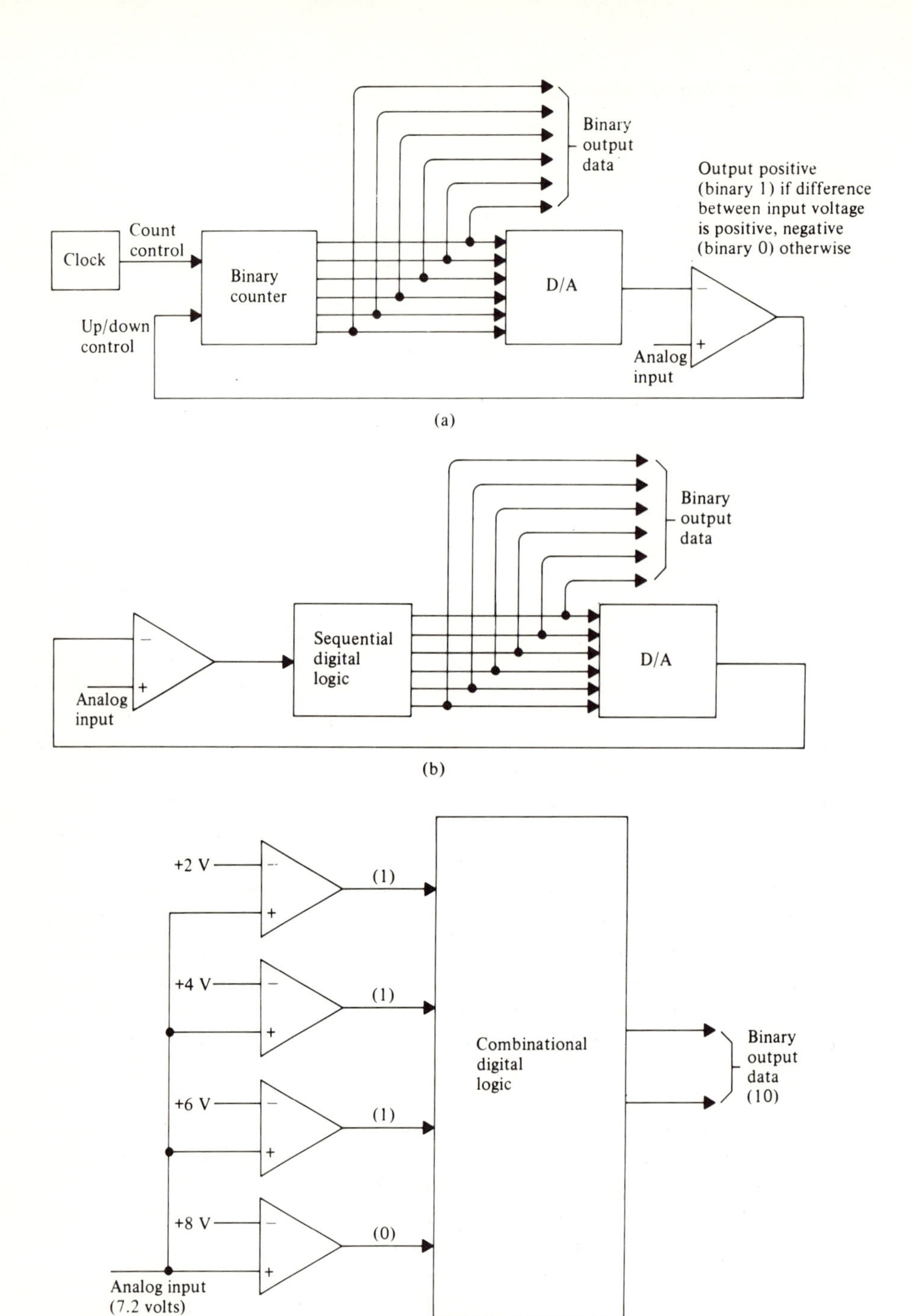

FIGURE 8-4
Common types of A/D converters. (a) Tracking counter.
(b) Successive approximation. (c) Flash conversion.

8.2.3 Sample-and-Hold

A/D and D/A converters are generally used to repetitively perform conversions. For analog-to-digital conversion it is often desirable to "freeze" the analog signal while the conversion is taking place. A sample-and-hold device, with symbol given in Fig. 8-5a, may be used to hold an analog signal steady while conversion proceeds, as in Fig. 8-5b.

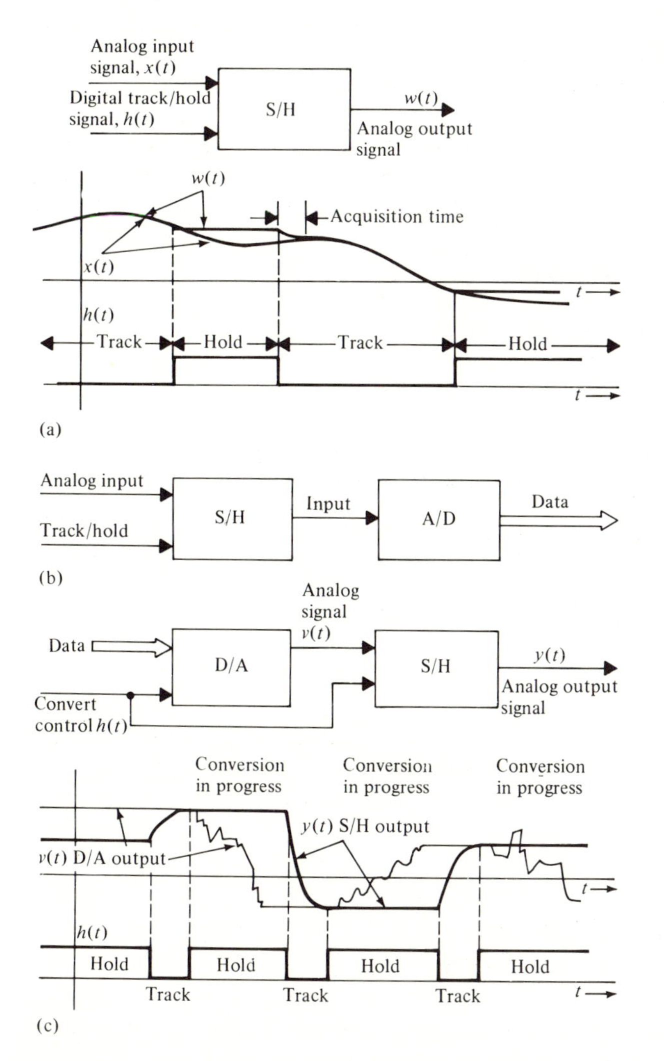

FIGURE 8-5
Sample-and-hold used in analog/digital conversion.
(a) Sample-and-hold symbol and representative signals. (b) S/H used to "freeze" an analog signal while D/A conversion takes place. (c) S/H used to hold the analog output of a D/A converter while a new conversion takes place.

In converting from digital to analog, a D/A converter output may fluctuate wildly while conversion is taking place. A sample-and-hold device is conveniently used to hold the previously converted signal while a new conversion takes place, as in Fig. 8-5c. The result is an output signal which changes in nearly a stepwise fashion each time a conversion occurs.

DRILL PROBLEMS

D8-1. A 12-bit D/A converter has minimum output voltage -10 and maximum output voltage $+10$. After the binary code 010110101001 is applied, what is the output voltage if the converter is of the following type:

(a) Sign and magnitude
(b) Offset binary
(c) Two's complement

ans. (a) 7.08; (b) -2.92; (c) 7.08

D8-2. What is the maximum percentage error if a binary number is truncated to 10 bits? What if the number is rounded?

ans. 0.098%; 0.049%

D8-3. The sinusoidal signal

$$f(t) = 10 \sin t$$

is tracked for $t < 1$, held for $1 \leq t < 2$, tracked for $2 \leq t < 5$, then held thereafter to form the signal $g(t)$. Sketch both $f(t)$ and $g(t)$.

8.3 COMPUTER PROCESSING

8.3.1 Computer History and Trends

A brief digital computer chronology is given in Table 8-4. Although their basic concepts are credited largely to Charles Babbage (c. 1830), today's computers became practical only after the invention and development of the transistor. Rapid technological advances in solid state physics have been the primary driving force behind the rapid evolution of digital computers since 1960. The emphasis here is upon stored program general-purpose mini- and microcomputers rather than special-purpose digital logic (which will accomplish the

TABLE 8-4
A brief history of digital computers

Date	*Development*
B.C.	Abacus is in use. It becomes widespread in Europe and Asia.
c. 1650	Pascal builds a mechanical desk calculator for addition.
c. 1670	Leibniz builds a calculator also capable of subtraction, multiplication, division, and root extraction.
c. 1800	Jacquard perfects automatic looms which weave designs programmed by punched cards.
c. 1830	Babbage develops modern computer principles, including memory, program control, and branching capabilities.
1890	Hollerith uses a punched card system for the U.S. census. Hollerith's company later becomes IBM.
1940	Aiken builds an electromechanical programmed computer, Mark I, used for ballistics calculations by the U.S. Army. It is capable of several additions per second. Similar work is done by Stibitz at Bell Telephone Laboratories, who notices similarity between telephone switching and computation.
1946	Eckert and Mauchly build the first vacuum tube digital computer, ENIAC, at the University of Pennsylvania. It is capable of several thousand additions per second.
1948	Von Neumann directs construction of the IAS stored program computer at Princeton. Memory involves charge storage on cathode-ray tube targets and a rotating magnetic drum. Addition is performed in approximately 65 μsec. EDSAC at Cambridge University is completed first, becoming the first stored program computer. IBM also builds a stored program computer, the SSEC.
1950	Sperry Rand Corporation builds the first commercial data-processing computer, using semiconductor diodes and vacuum tubes, the UNIVAC I.
1952	IBM begins marketing the 701 digital computer commercially.
1960	The "second generation" of computers is introduced to the market. These use solid state components in place of vacuum tubes.
1964	The "third generation" of computers begins. Integrated circuit hardware predominates as in the IBM System 360.
1965	Digital Equipment Corporation markets the PDP-8 at about $50,000. The minicomputer industry begins.
1973	Intel markets the first microcomputer system using the 8080 microprocessor chip.
1975	Ten different microprocessors are on the market, including the Fairchild F8, the Intel 8080A, the Motorola 6800, and the Signetics 2650. Others soon enter the field.
1977	Motorola, Texas Instruments, Intel, and Zilog each begin to market competing 16-bit microprocessors; 32-bit microprocessors are announced.
1978	Rockwell International markets the AIM-65 microprocessor system, priced at about $500. System includes alphanumeric display, small printer, cassette tape interface, and input-output ports. ROM-based software includes system monitor, assembler, and BASIC compiler.
1982	Very-large-scale integration (VLSI) technology is capable of integrating a million devices on a tiny semiconductor chip, approximately the number of devices in a very large computer processing unit and roughly the complexity level of primitive organisms.

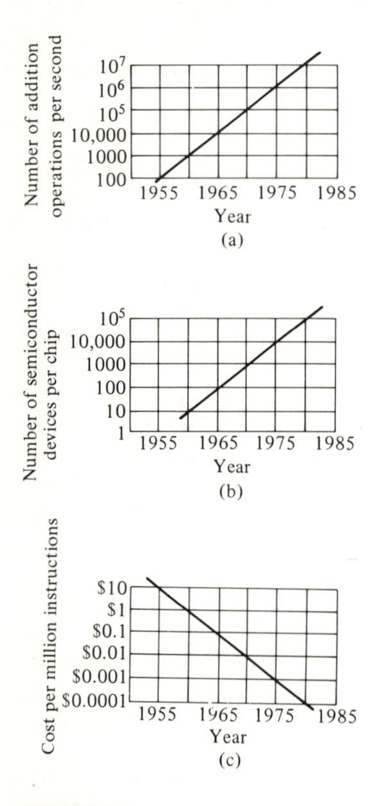

FIGURE 8-6
Digital computer trends. (a) Approximate speed increase of high-speed computers. (b) Approximate semiconductor device density increase. (c) Approximate decrease in computation cost.

same purpose) because it appears that the vast majority of future control applications will be with low-cost, mass-produced hardware.

The charts given in Fig. 8-6 show three dramatic trends in computer evolution. Their speed of operation has increased by about an order of magnitude every five years since 1955. The density of their electronic circuits has increased by roughly the same factor of 10 every five years since the early use of solid state components in 1960. With this compactness has come a large decrease in power requirements. The total costs of digital computation have similarly plummeted.

8.3.2 Concepts and Terminology

Today's digital computer is an electronic device with input and output capability, control, arithmetic, and memory circuits. A program, stored in memory, controls the machine's processing. The numbers stored in memory which constitute the program represent *instructions*, with each possible instruction coded as a unique number. The collection of all usable instructions for a particular computer is termed its *instruction set*.

Some of the instructions allow branching; hence, the computation may proceed in different ways, determined by the outcome of previous results. Other instructions are capable of changing memory contents, including (if allowed) the contents of the program itself.

Entry of a computer program into memory may be done directly, by storing strings of binary ones and zeros in the desired memory locations, or with various degrees of aid from intermediate computer programs. These levels of aid are classified as follows:

Machine Language. *The collection of binary ones and zeros stored in memory which constitute the computer program that controls the machine. For example, the two 9-bit binary codes below, stored in successive memory locations, might be an instruction to add the contents of decimal memory address 0200 to the accumulator register:*

```
Ø 1 Ø 1 Ø Ø 1 1
1 1 Ø Ø 1 Ø Ø Ø
```

The binary code 01010011 is the operation code part of the instruction which specifies the operation to be performed. The following code is the instruction's operand, the address 0200 in binary.

A variation on this arrangement which gives a more compact notation is to write each group of 4 binary digits in the binary machine language as a single hexadecimal (base 16) digit. The above instruction would be written in hex as

```
53
C8
```

The 16 hexadecimal digits are 0 through 9 and A through F. It is a relatively simple matter to arrange a computing system so that hexadecimal (or other variations such as octal) characters entered from a keyboard or other input device cause the corresponding binary entries to be placed sequentially in memory.

Mnemonic Machine Language. *A collection of alphabetic and numerical symbols used to represent machine language in pleasing form for humans. For example, the above instruction to add the contents of memory location 0200 to the accumulator register might be written in mnemonics as*

```
ADD  C8
```

A program which converts mnemonic language to machine language is termed a translator. With a translator, a mnemonic program may be entered and automatically converted to the corresponding binary machine code and placed in memory.

Assembly Language. *In addition to mnemonics, symbols may be used to represent addresses and data. The example instruction could be given in assembly language in a form similar to the following:*

```
X=$Ø2ØØ
ADD X
```

The variable X *is, somewhere in the program, assigned decimal value 0200. The $ sign indicates that the value is given in decimal, not hex or binary. The* ADD X *instruction would be interpreted as the* ADD *instruction using memory location* X=0200. *Assemblers may also incorporate other features such as complicated addressing modes and simplified subroutine handling.*

Generally, assemblers are machine language computer programs which convert relatively simple alphanumeric entries into another machine language program.

Higher-Level Languages. *Basic, Fortran, Cobol, Pascal, and APL are examples of higher-level languages. These involve sophisticated instructions which are defined independently from any particular machine language. A compiler for such a language is a program which generates assembly or machine language code for a specific processor's instruction set, often producing long strings of machine instructions for a single higher-level instruction.*

Operating Systems. *The operation of most digital computers is overseen by a supervisory program called an operating system. The operating system handles initialization of the computer at power-up and the loading of programs and data into memory.*

The electronic devices in a digital computer nowadays are largely integrated circuits, called *chips*, that perform binary logic functions. Numbers, in the form of groups of binary voltage signals, termed *words*, on wires are manipulated and combined to form resulting new binary words. To keep them in proper synchronization, changes in the binary signals are controlled by an electronic oscillator, or *clock*.

Each instruction of the computer is typically composed of a series of *microoperations*, each executed in one or more separate clock cycles. To transfer a binary number from one register to another, for example, the data transfer direction would first be set, then the transmitting register would place the data on wires joining the two registers. Finally, the receiving register would latch (accept) the data, replacing its old content with the new. In larger computers, primitive instructions such as register transfer are termed *microinstructions* and are further joined to form the instructions accessible to the programmer.

8.3.3 Computer Signal Processing

When a digital computer is combined with A/D and D/A capability, as in Fig. 8-7, the result is a signal-processing device with extraordinary potential.

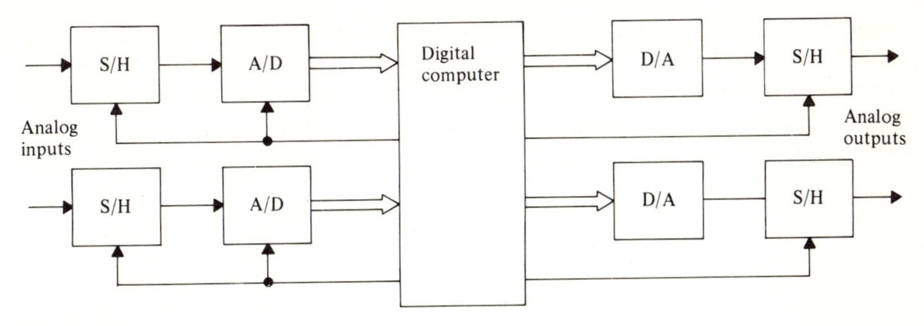

FIGURE 8-7
A simple digital controller.

Analog signals are input, analog signals are output, and in between may be placed powerful computational ability. Operations such as square-rooting, correlation, function generation, and spectral analysis which are a nightmare in analog hardware are simply and routinely done digitally. Furthermore, if a general-purpose programmable digital computer is used, changes in objectives and improvements in design may require only changes in the stored program (the software), not changes to the equipment itself (the hardware).

Digital computation is subject to numerical *roundoff* or *truncation errors* because numbers are represented by a finite number of bits. For example, the product of two 16-bit numbers involves potentially 32 bits. To represent such a product as another 16-bit number, the least significant 16 bits of the product must be eliminated. Ordinarily, finite arithmetic precision is not of great concern when the number of bits used to represent a number far exceeds the required numerical precision. However, the power of a computer to perform huge numbers of calculations in a short time means that it is possible for tiny errors to quickly accumulate to large proportions. One must also be aware that certain kinds of calculations, such as forming differences between large but nearly identical numbers, are especially susceptible to error.

8.3.4 The Microprocessor Revolution

In 1969, Datapoint Corporation contracted with Intel, a semiconductor device manufacturer, to produce a computerlike component for their data terminal logic. The resulting design, on a single semiconductor chip, worked well but executed instructions considerably more slowly than was needed for the application. Datapoint declined to buy, instead using existing logic components for their product. Intel was left with an extraordinary logic device whose development had been paid for. They chose to market it, as the Intel 8008, which was soon followed by the Intel 8080 and a flood of microprocessors from other manufacturers.

Microprocessors differ from general-purpose computers and minicomputers in that they are very inexpensive, with relatively simple organization, and are intended to become components in a larger system. With such powerful computational capability available at an extremely low cost, it is now feasible to digitally control even relatively small systems such as those found in appliances and automobiles. With a microprocessor system in place, additional control and other functions may also be provided at little additional cost.

A typical microprocessor system, suitable for medium-scale control

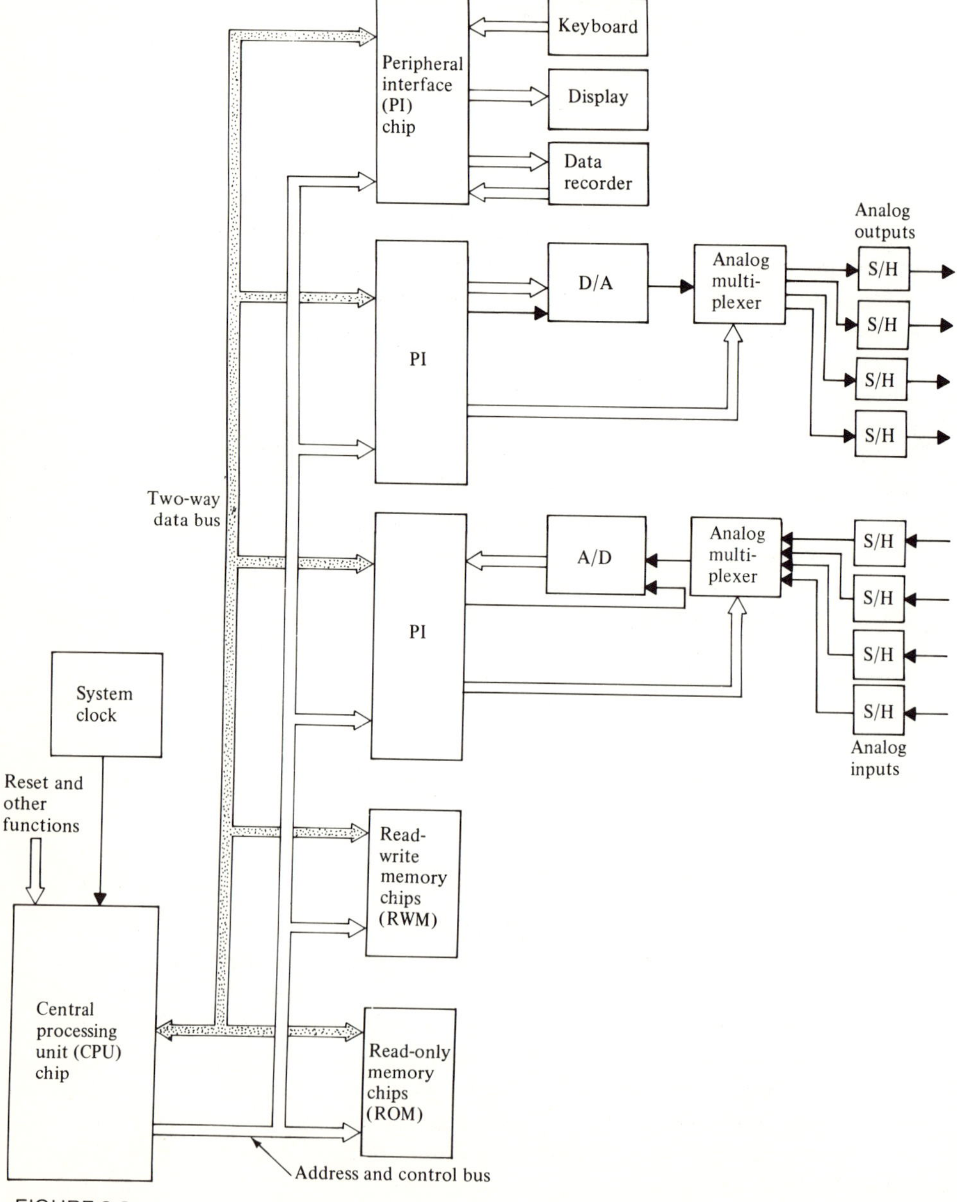

FIGURE 8-8
A microprocessor system.

applications, is diagramed in Fig. 8-8. The heart of the system is the microprocessor central processing unit (CPU) chip which is driven by an oscillator or clock. It provides a set of digital signals, on a set of wires termed the address bus, to select which peripheral device if any is to receive or provide binary data signals on the data bus. The read-write memory (RWM) is accessed in this manner, as is permanent read-only memory (ROM). Usually, the program is stored in ROM so that it cannot be inadvertently altered, while data and the results of calculations are stored in RWM. Many variations on this arrangement are possible, including those where some memory is provided on the CPU chip and where the data and address busses share the same wires.

Digital interfaces to the system are made via peripheral interface (PI) chips, which store and manipulate data, then transfer it to and from the CPU via the data bus at the proper times. In this example system, a keyboard, display, and tape recorder connect to the system through one PI, while D/A and A/D converters are connected through other PIs. The analog-digital conversion arrangements shown are typical of many systems where a single converter is periodically switched (or *multiplexed*) between several signals, under digital control.

8.4 DISCRETE-TIME SIGNALS

8.4.1 Representing Sequences

Periodic samples of a continuous-time signal, as are generated by an A/D converter, form a sequence of numbers termed a *discrete-time signal.* Figure 8-9 shows a continuous-time signal $f(t)$ and the corresponding sequence of samples

$$f(t=0),\ f(t=T),\ f(t=2T),\ f(t=3T), \ldots$$

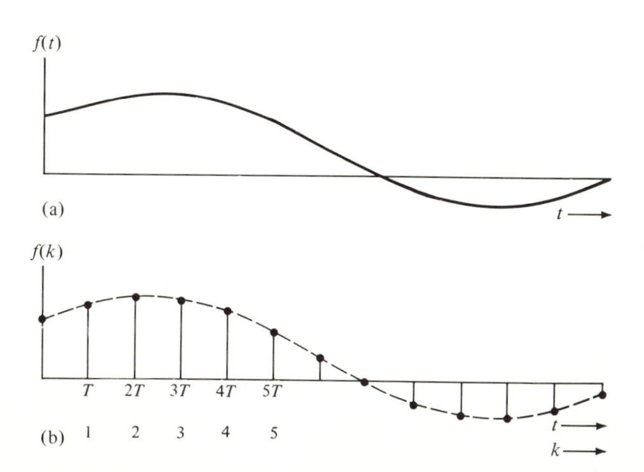

FIGURE 8-9
Sampling a continuous-time signal. (a) A continuous-time signal. (b) Samples of the continuous-time signal.

Although it results in an ambiguity that is only resolved by context, it is common practice to denote the sequence by $f(k)$, where k is the sample number.

Some important sequences are shown in Fig. 8-10. All of these basic sequences consist of zero samples prior to $k=0$. The unit pulse sequence, $\delta(k)$ in Fig. 8-10a, has unit sample value for $k=0$ and all other values zero. The unit step sequence $u(k)$, Fig. 8-10b, has samples which are all unity for $k=0$ and

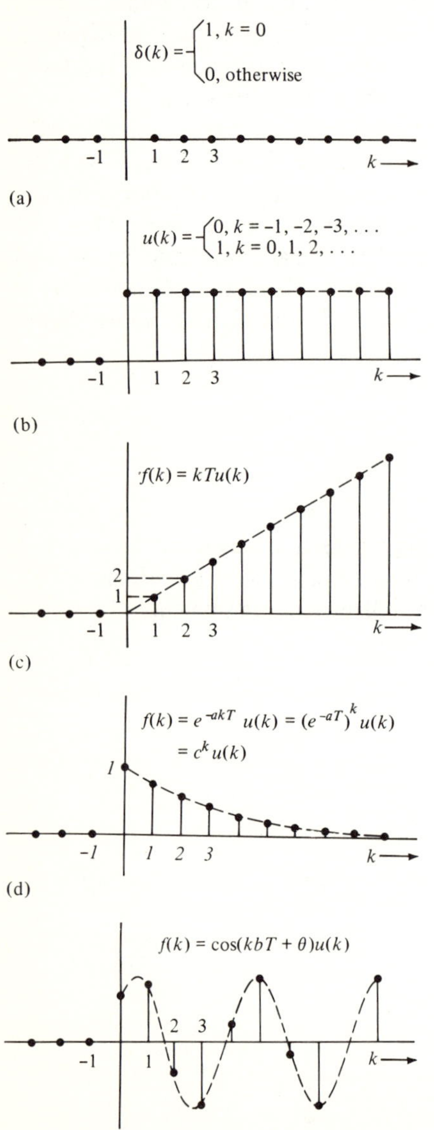

FIGURE 8-10

Some basic sequences. (a) Unit sample. (b) Unit step. (c) Ramp. (d) Exponential (or geometric). (e) Sinusoidal.

thereafter. The ramp, Fig. 8-10c, consists of samples of the continuous-time unit ramp function.

The sampled exponential, Fig. 8-10d, has samples that are progressive powers of the number

$$c = e^{-aT}$$

and so forms a geometric sequence:

$$f(0) = c^0 = 1$$
$$f(1) = c$$
$$f(2) = c^2$$
$$f(n) = c^n$$

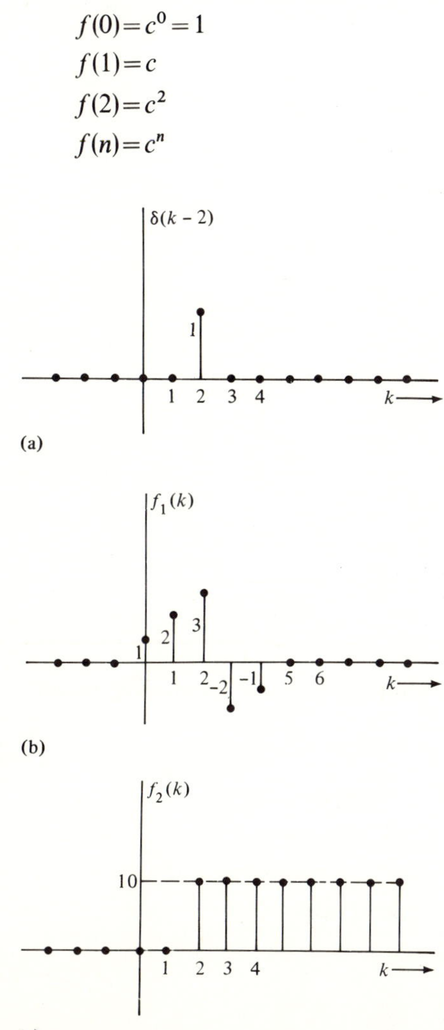

FIGURE 8-11
Shifting and summing sequences. (a) A shifted unit pulse sequence. (b) A finite sum of shifted pulses. (c) Shifted step sequence or a step sequence with samples at $k = 0$ and $k = 1$ cancelled by pulses.

Geometric sequences (or sampled exponential functions) have fundamental importance to discrete-time systems in the same way that exponential functions are basic to continuous-time system. A sampled sinusoidal function as in Fig. 8-10e is termed a *sinusoidal sequence.*

More complicated sequences may often be represented as shifts and sums of the basic sequences. For example, $\delta(k-2)$ is the unit pulse sequence shifted two samples to the right, as in Fig. 8-11a. The sequence of Fig. 8-11b is thus

$$f_1(k)=\delta(k)+2\delta(k-1)+3\delta(k-2)-2\delta(k-3)-\delta(k-4)$$

The sequence

$$f_2(k)=\begin{cases}10 & k=2, 3, 4, \ldots \\ 0 & \text{otherwise}\end{cases}$$

drawn in Fig. 8-11c, is

$$f_2(k)=10u(k)-10\delta(k)-10\delta(k-1)$$

Alternatively,

$$f_2(k)=10u(k-2)$$

DRILL PROBLEM

D8-4. Sketch the function $f(t)$ and the samples $f(k)$ for a sampling period $T=0.5$ sec.

(a) $f(t)=e^{-0.5t}u(t)$

ans. $f(k)=e^{-0.25k}u(k)$

(b) $f(t)=(\sin \pi t)u(t)$

ans. $f(k)=(\sin k\pi/2)u(k)$

(c) $f(t)=(\cos 2\pi t)u(t)$

ans. $f(k)=(-1)^k u(k)$

(d) $f(t)=(\sin 2\pi t)u(t)$

ans. $f(k)=0$

8.4.2 z-Transformation and Properties

The z-transform of a sequence $f(k)$ is defined as the infinite series

$$Z[f(k)] = F(z) = \sum_{k=0}^{\infty} f(k) z^{-k}$$

It plays much the same role in the description of discrete-time signals as the Laplace transform does with continuous-time signals.

Table 8-5 lists basic z-transform pairs, together with the corresponding

TABLE 8-5
Some Laplace and z-transform pairs

$f(t)$	$F(s)$	$f(k)$	$F(z)$
		$\delta(k)$, unit pulse	1
$u(t)$, unit step	$\frac{1}{s}$	$u(k)$, unit step	$\frac{z}{z-1}$
$tu(t)$	$\frac{1}{s^2}$	$kTu(k)$	$\frac{Tz}{(z-1)^2}$
$e^{-at}u(t)$	$\frac{1}{s+a}$	$(e^{-aT})^k u(k) = c^k u(k)$ where $c = e^{-aT}$	$\frac{z}{z-e^{-aT}} = \frac{z}{z-c}$
$te^{-at}u(t)$	$\frac{1}{(s+a)^2}$	$kT(e^{-aT})^k u(k) = kTc^k u(k)$	$\frac{Tze^{-aT}}{(z-e^{-aT})^2} = \frac{Tcz}{(z-c)^2}$
$(\sin bt)u(t)$	$\frac{b}{s^2+b^2}$	$(\sin kbT)u(k)$	$\frac{z \sin bT}{z^2 - 2z \cos bT + 1}$
$(\cos bt)u(t)$	$\frac{s}{s^2+b^2}$	$(\cos kbT)u(k)$	$\frac{z(z-\cos bT)}{z^2 - 2z \cos bT + 1}$
$e^{-at}(\sin bt)u(t)$	$\frac{b}{(s+a)^2+b^2}$	$(e^{-aT})^k(\sin kbT)u(k) = c^k(\sin kbT)u(k)$	$\frac{z(e^{-aT} \sin bT)}{(z-e^{(-a+jb)T})(z-e^{(-a-jb)T})} = \frac{zc \sin bT}{z^2-(2c \cos bT)z+c^2}$
$e^{-at}(\cos bt)u(t)$	$\frac{s+a}{(s+a)^2+b^2}$	$(e^{-aT})^k(\cos kbT)u(k) = c^k(\cos kbT)u(k)$	$\frac{z(z-e^{-aT} \cos bT)}{(z-e^{(-a+jb)T})(z-e^{(-a-jb)T})} = \frac{z(z-c \cos bT)}{z^2-(2c \cos bT)z+c^2}$

continuous-time function which gives the sequence when sampled with period T. Using the z-transform definition, the transform of the unit pulse is

$$Z[\delta(k)]=\sum_{k=0}^{\infty}\delta(k)z^{-k}=z^{-0}=1$$

It should be noted that the unit pulse sequence $\delta(k)$, while analogous to the unit impulse $\delta(t)$, does not consist of samples of $\delta(t)$, which is infinite for $k=t=0$.

The z-transform of the unit step sequence is as follows:

$$Z[u(k)]=\sum_{k=0}^{\infty}1\cdot z^{-k}$$

using the fact that for a geometric series,

$$\sum_{k=0}^{\infty}x^k=\frac{1}{1-x} \qquad |x|<1$$

then

$$Z[u(k)]=\sum_{k=0}^{\infty}z^{-k}=\sum_{k=0}^{\infty}\left(\frac{1}{z}\right)^k$$

$$=\frac{1}{1-1/z}=\frac{z}{z-1} \qquad \left|\frac{1}{z}\right|<1$$

Conditions for z-transform convergence are satisfied by all but the most pathological sequences and so will not be emphasized here.

A sampled exponential function, whether decaying or expanding, is of the form

$$f(k)=e^{-kaT}=(e^{-aT})^k$$

Its z-transform is given by

$$Z[e^{-kaT}]=\sum_{k=0}^{\infty}e^{-kaT}z^{-k}$$

$$=\sum_{k=0}^{\infty}\left(\frac{1}{ze^{aT}}\right)^k=\frac{1}{1-1/ze^{aT}}$$

$$=\frac{z}{z-e^{-aT}}$$

Samples of an exponential function form a geometric series since by defining

$$c=e^{-aT}$$

$$f(k)=e^{-kaT}=(e^{-aT})^k=c^k$$

In terms of c, the z-transform is

$$Z[c^k]=\frac{z}{z-c}$$

The transforms of sampled sinusoids given in Table 8-5 follow easily from expanding the sinusoidal function into Euler components and applying the result for sampled exponentials. For the sampled sine,

$$\begin{aligned}
Z[\sin kbT] &= \sum_{k=0}^{\infty}\left(\frac{e^{jkbT}-e^{-jkbT}}{2j}\right)z^{-k} \\
&= \frac{1}{2j}\sum_{k=0}^{\infty} e^{jkbT}z^{-k}-\frac{1}{2j}\sum_{k=0}^{\infty} e^{-jkbT}z^{-k} \\
&= \frac{z/2j}{z-e^{jbT}}-\frac{z/2j}{z-e^{-jbT}} \\
&= \frac{(z/2j)(z-e^{-jbT}-z+e^{jbT})}{z^2-(e^{jbT}+e^{-jbT})z+1} \\
&= \frac{z\left(\dfrac{e^{jbT}-e^{-jbT}}{2j}\right)}{z^2-2\left(\dfrac{e^{jbT}+e^{-jbT}}{2}\right)z+1} \\
&= \frac{z\sin bT}{z^2-2z\cos bT+1}
\end{aligned}$$

Basic z-transform properties are listed in Table 8-6. The transform of a sequence scaled by a multiplicative constant is that constant times the original z-transform. The z-transform of a sample-by-sample sum of sequences is the sum of their individual z-transforms. A sequence weighted by the step number k has z-transform

$$\begin{aligned}
Z[kf(k)] &= \sum_{k=0}^{\infty} kf(k)z^{-k}=\sum_{k=0}^{\infty} f(k)(kz^{-k}) \\
&= \sum_{k=0}^{\infty} f(k)\frac{d}{dz}(-z^{-k})=-\frac{d}{dz}\sum_{k=0}^{\infty} f(k)z^{-k} \\
&= -\frac{d}{dz}F(z)
\end{aligned}$$

A sequence weighted by successive powers of a constant c has z-transform as follows:

$$\begin{aligned}
Z[c^kf(k)] &= \sum_{k=0}^{\infty} c^kf(k)z^{-k}=\sum_{k=0}^{\infty} f(k)\left(\frac{z}{c}\right)^{k} \\
&= F\left(\frac{z}{c}\right)
\end{aligned}$$

TABLE 8-6
Some z-transform properties

$$Z[cf(k)]=cF(z),\ c \text{ a constant}$$
$$Z[f(k)+g(k)]=F(z)+G(z)$$
$$Z[kf(k)]=-\frac{dF(z)}{dz}$$
$$Z[c^k f(k)]=F\left(\frac{z}{c}\right),\quad c \text{ a constant}$$
$$Z[f(k-1)]=f(-1)+z^{-1}F(z)$$
$$Z[f(k-2)]=f(-2)+z^{-1}f(-1)+z^{-2}F(z)$$
$$Z[f(k-n)]=f(-n)+z^{-1}f(1-n)+z^{-2}f(2-n)t+\cdots+z^{-n+1}f(-1)+z^{-n}F(z)$$
$$Z[f(k+1)]=zF(z)-zf(0)$$
$$Z[f(k+2)]=z^2F(z)-z^2f(0)-zf(1)$$
$$Z[f(k+n)]=z^nF(z)-z^nf(0)-z^{n-1}f(1)-\cdots-z^2f(n-2)-zf(n-1)$$

$$f(0)=\lim_{z\to\infty} F(z)$$

If $\lim_{k\to\infty} f(k)$ exists and is finite,

$$\lim_{k\to\infty} f(k)=\lim_{z\to 1}\left[\frac{z-1}{z}F(z)\right]$$

Figure 8-12a shows an example of a sequence which is shifted one step to the right. Its z-transform is given by

$$Z[f(k-1)]=\sum_{k=0}^{\infty} f(k-1)z^{-k}=\sum_{k=-1}^{\infty} f(k)z^{-(k+1)}$$
$$=f(-1)+z^{-1}\sum_{k=0}^{\infty} f(k)z^{-k}$$
$$=f(-1)+z^{-1}F(z)$$

Then

$$Z[f(k-2)]=f(-2)+z^{-1}Z[f(k-1)]$$
$$=f(-2)+z^{-1}[f(-1)+z^{-1}F(z)]$$
$$=f(-2)+z^{-1}f(-1)+z^{-2}F(z)$$

and, similarly,

$$Z[f(k-n)]=f(-n)+z^{-1}f(1-n)+\cdots+z^{-n+1}f(-1)+z^{-n}F(z)$$

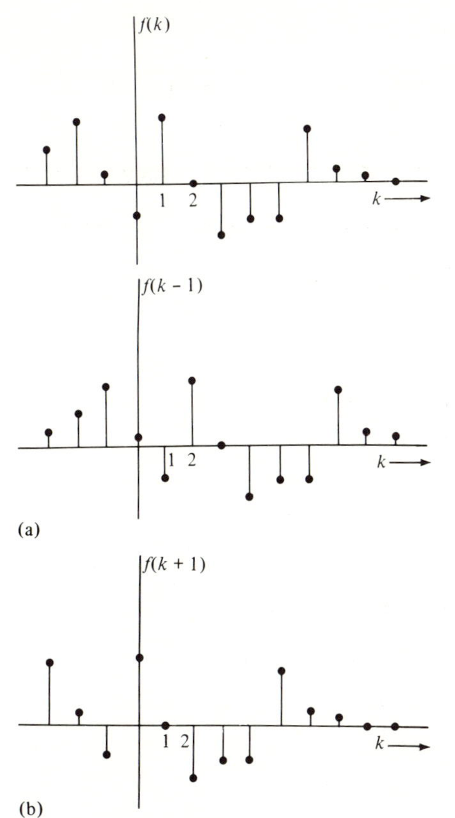

FIGURE 8-12
Right- and left-shifted sequences. (a) A sequence and the same sequence shifted right one step. (b) The sequence shifted left one step.

For a left shift of the sequence, Fig. 8-12b,

$$
\begin{aligned}
Z[f(k+1)] &= \sum_{k=0}^{\infty} f(k+1)z^{-k} = \sum_{k=1}^{\infty} f(k)z^{-(k-1)} \\
&= z\sum_{k=1}^{\infty} f(k)z^{k} = z\sum_{k=0}^{\infty} f(k)z^{k} - zf(0) \\
&= zF(z) - zf(0)
\end{aligned}
$$

Similarly,

$$
\begin{aligned}
Z[f(k+2)] &= z[zF(z) - zf(0)] - zf(1) \\
&= z^2F(z) - z^2f(0) - zf(1)
\end{aligned}
$$

and

$$Z[f(k+n)] = z^n F(z) - z^n f(0) - z^{n-1} f(1) - \cdots - z^2 f(n-2) - z f(n-1)$$

DRILL PROBLEM

D8-5. Find the z-transforms of the following sequences:

(a) $f(k) = [(-0.5)^k - 4(0.2)^k]u(k)$

$$ans. \frac{z}{z+0.5} - \frac{4z}{z-0.2}$$

(b) $f(k) = \begin{cases} (-1)^k & k = 3, 4, 5, \ldots \\ 0 & \text{otherwise} \end{cases}$

$$ans. \frac{-z^{-2}}{1+z}$$

(c) $f(1) = 2$, $f(4) = -3$, $f(7) = 8$, and all other samples are zero

$$ans. 2z^{-1} - 3z^{-4} + 8z^{-7}$$

(d) $f(k) = u(k) \sin 3k - 2u(k-4) \sin 3(k-4)$

$$ans. \frac{(1-2z^{-4})z \sin 3}{z^2 - 2z \cos 3 + 1}$$

8.4.3 Inverse z-Transform

The sequence of samples represented by a rational z-transform may be obtained, if desired, by long division. Consider the z-transform

$$F(z) = \frac{4z}{z^2 - z + 0.5}$$

For example:

$$\begin{array}{r|l} & 4z^{-1} + 4z^{-2} + 2z^{-3} + \cdots \\ \hline z^2 - z + 0.5 & 4z \\ & \underline{4z - 4 + 2z^{-1}} \\ & 4 - 2z^{-1} \\ & \underline{4 - 4z^{-1} + 2z^{-2}} \\ & 2z^{-1} - 2z^{-2} \end{array}$$

Since

$$F(z)=\frac{4z}{z^2-z+0.5}=0z^0+4z^{-1}+4z^{-2}+2z^{-3}+\cdots$$

$$f(k)=4\delta(k-1)+4\delta(k-2)+2\delta(k-3)+\cdots$$

and

$$\begin{aligned} f(0)&=0\\ f(1)&=4\\ f(2)&=4\\ f(3)&=2\\ &\vdots \end{aligned}$$

Repeated steps of long division do not give a closed-form expression for the sequence represented by a z-transform, although in principle as many terms in the sequence as desired may be found in this manner.

To find a formula for the sequence of samples, partial fraction expansion may be used. Rather than expanding a z-transform $F(z)$ directly in partial fractions, $F(z)/z$ is expanded so that terms of the form

$$\frac{z}{z-e^{-aT}}$$

result. For example, for the z-transform

$$F(z)=\frac{-2z^2+2z}{z^2+4z+3}$$

$$\frac{F(z)}{z}=\frac{-2z+2}{(z+1)(z+3)}=\frac{2}{z+1}+\frac{-4}{z+3}$$

giving

$$F(z)=\frac{2z}{z+1}+\frac{-4z}{z+3}$$

$$f(k)=2(-1)^k-4(-3)^k \qquad k=0, 1, 2, 3, \ldots$$

Another example is the following:

$$F(z)=\frac{z^3-3}{z(z-0.25)(z-0.5)}$$

$$\frac{F(z)}{z}=\frac{z^3-3}{z^2(z-0.25)(z-0.5)}$$

$$=\frac{-144}{z}+\frac{-24}{z^2}+\frac{191}{z-0.25}+\frac{-46}{z-0.5}$$

$$F(z) = -144z^0 - 24z^{-1} + \frac{-191z}{z-0.25} + \frac{-46z}{z-0.5}$$

$$f(k) = -144\delta(k) - 24\delta(k-1) - 191(0.25)^k + 46(0.5)^k$$

A set of complex conjugate root terms should be manipulated into the form of the last two entries of Table 8-5:

$$F(z) = \frac{4z^2 - 3z}{z^2 + 2z + 2}$$
$$= \frac{K_1 zc \sin bT}{z^2 - 2c \cos bT + c^2} + \frac{K_2 z(z - c \cos bT)}{z^2 - 2c \cos bT + c^2}$$

Equating,

$$c^2 = 2 \qquad c = \sqrt{2}$$
$$2c \cos bT = 2\sqrt{2} \cos bT = -2 \qquad bT = 3\pi/4$$
$$\sin bT = \sin(3\pi/4) = 1/\sqrt{2}$$

so

$$F(z) = \frac{4z^2 - 3z}{z^2 + 2z + 2} = \frac{K_1 z + K_2 z(z+1)}{z^2 + 2z + 2}$$

giving

$$K_2 = 4$$
$$K_1 + K_2 = -3 \qquad K_1 = -7$$
$$f(k) = 4(\sqrt{2})^k \cos\frac{3\pi k}{4} - 7(\sqrt{2})^k \sin\frac{3\pi k}{4} \qquad k = 0, 1, 2, \ldots$$

DRILL PROBLEMS

D8-6. Use long division to show that the inverse z-transform of

$$F(z) = \frac{10z}{(z-1)^2}$$

is

$$f(k) = 10k$$

D8-7. Find the inverse z-transforms for $k \geqslant 0$:

(a) $F(z)=\dfrac{1}{z+0.3}$

ans. $\frac{10}{3}\delta(k)-\frac{10}{3}(-0.3)^k$

(b) $F(z)=\dfrac{-6z^2+z}{z^2+5z+6}$

ans. $-19(-3)^k+13(-2)^k$

(c) $F(z)+\dfrac{4z^2-3z+2}{z^2+4z+4}$

ans. $\frac{1}{2}\delta(k)+\frac{7}{2}(-2)^k+6k(-2)^k$

(d) $F(z)=\dfrac{3z^2-z}{z^2+2z+10}$

ans. $3(\sqrt{10})^k \cos 1.89k-1.33(\sqrt{10})^k \sin 1.89k$

8.5 SAMPLING

When an analog signal $f(t)$ is sampled to form the sequence $f(k)$, there is a direct relationship between the Laplace transform $F(s)$ of the analog signal and the z-transform $F(z)$ of the sequence. If a rational Laplace transform is expanded into a sum of terms of the type given in Table 8-5, the z-transform of the sample sequence is obtained by simply summing the corresponding z-transform terms from the table.

For example, for a continuous-time signal with Laplace transform

$$F(s)=\frac{4s^2+13s+18}{s^3+5s^2+6s}$$

$$=\frac{3}{s}+\frac{-4}{s+2}+\frac{5}{s+3}$$

$$f(t)=(3-4e^{-2t}+5e^{-3t})u(t)$$

For a sampling interval $T=0.2$,

$$f(k)=(3-4e^{-0.4k}+5e^{-0.6k})u(k)$$

$$[3(1)^k-4(e^{-0.4})^k+5(e^{-0.6})^k]u(k)$$

$$F(z)=3\left(\frac{z}{z-1}\right)-4\left(\frac{z}{z-e^{-0.4}}\right)+5\left(\frac{z}{z-e^{-0.6}}\right)$$

When the Laplace transform involves delay operations in multiples of the sampling interval T, the remainder of the transform is expanded into partial fraction terms, as in the following example, for which $T=0.1$:

$$F(s)=\frac{e^{-0.1s}+2}{s(s+3)}=(e^{-0.1s}+2)\left(\frac{\frac{1}{3}}{s}+\frac{-\frac{1}{3}}{s+3}\right)$$

Denoting

$$G(s)=\frac{\frac{1}{3}}{s}+\frac{-\frac{1}{3}}{s+3}$$

$$F(s)=(e^{-0.1s}+2)G(s)$$

$$f(t)=g(t-0.1)u(t-0.1)+2g(t)u(t)$$

then

$$f(k)=g(k-1)u(k-1)+2g(k)u(k)$$

$$F(z)=(z^{-1}+2)G(z)$$

$$=(z^{-1}+2)\left(\frac{1}{3}\frac{z}{z-1}-\frac{1}{3}\frac{z}{z-e^{-0.3}}\right)$$

$$=\frac{0.173z^2+0.086z}{z(z-1)(z-0.74)}$$

Finding the z-transform of the corresponding sequence thus involves separating the time delay operations from the rational part of the Laplace transform, then substituting z^{-1} for each unit of time delay in the delay operation and substituting from the entries of Table 8-5 for each term in the partial fraction expansion of the remainder. Another example is the following, for which the sampling period is $T=0.05$:

$$F(s)=\frac{10}{s^2+4}+\frac{6e^{-0.2s}}{s^2+3s}$$

$$=\frac{10}{4}\left(\frac{4}{s^2+4}\right)+e^{-0.2s}\left(\frac{2}{s}+\frac{-2}{s+3}\right)$$

$$F(z)=\frac{10}{4}\left(\frac{z\sin 0.1}{z^2-2z\cos 0.1+1}\right)+z^{-4}\left(\frac{2z}{z-1}-\frac{2z}{z-e^{-0.15}}\right)$$

$$=\frac{10}{4}\left(\frac{0.0998z}{z^2-1.99z+1}\right)+\frac{1}{z^4}\left(\frac{2z}{z-1}-\frac{2z}{z-0.86}\right)$$

DRILL PROBLEM

D8-8. For each of the analog signals with the given Laplace transform $F(s)$, find the z-transform $F(z)$ of the corresponding sample sequence with the given sampling interval T:

(a) $F(s)=\dfrac{-4s+1}{s^2+7s+12}$ $\quad T=0.2$

ans. $(-4z^2+3.48)/(z-0.5488)(z-0.45)$

(b) $F(s)=\dfrac{100}{s^2+2s+10}$ $\quad T=0.5$

ans. $20.2z/(z^2-0.086z+0.368)$

(c) $F(s)=\dfrac{e^{-0.1s}-3e^{-0.3s}+2}{s^2}\left(\dfrac{10}{s+1}\right)$ $\quad T=0.1$

ans. $\dfrac{(2z^3+z^2-3)(0.05z+0.0045)}{z^2(z-1)^2(z-0.905)}$

(d) $F(s)=\dfrac{se^{-s}+4}{s^2+9}$ $\quad T=0.2$

ans. $(z-0.825)/z^4(z^2-1.65z+1)$
$+0.753z/(z^2-1.65z+1)$

8.6 RECONSTRUCTION OF SIGNALS FROM SAMPLES

8.6.1 Representing Sampled Signals with Impulses

Description of A/D conversion involves discrete-time representation of continuous-time signals, the process termed *sampling*. Describing D/A conversion requires, conversely, continuous-time representation of discrete-time signals. This process is termed *reconstruction*.

Although very often the analog signal reconstructed from digital samples is a sampled-and-held waveform, that is not always the case. A more fundamental continuous-time signal related to a sequence of samples is a train of impulses, timed periodically at intervals T, with strengths equal to the corresponding samples. Figure 8-13 shows an original analog signal $f(t)$, the corresponding sample sequence $f(k)$, and the impulse train $f^*(t)$ that is a useful representation of the samples in the analog domain. The objective of reconstruction is to recover from the samples $f(k)$ the analog signal $f(t)$ or a sufficiently close approximation to it. Of course, the signal $f(t)$ may not actually exist anywhere in the system. Typically the samples $f(k)$ are computed from combinations of samples of other signals, present and delayed. From the samples

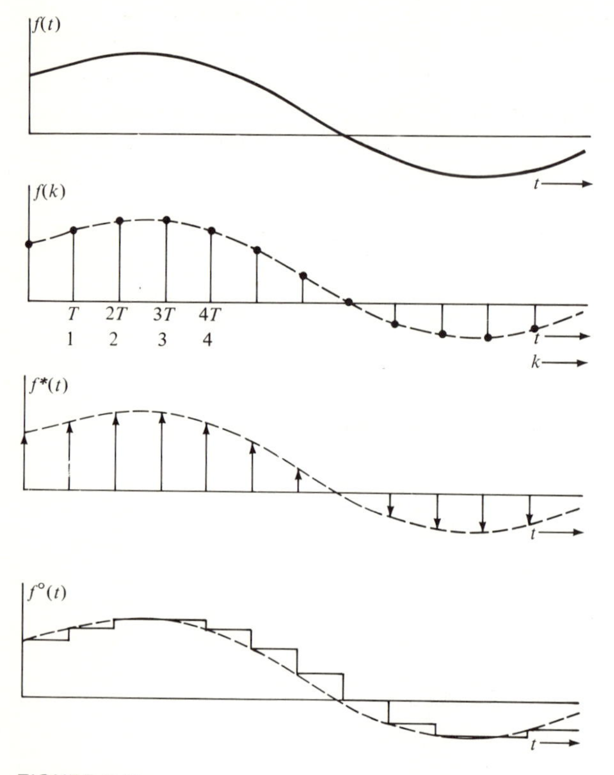

FIGURE 8-13
Signal reconstruction.

$f(k)$, it is desired to generate an analog signal which *could have been* sampled to obtain $f(k)$.

The sampled-and-held waveform $f^0(t)$ in Fig. 8-14 is a reconstructed signal with samples $f(k)$. It may be derived from the impulse train by passing the impulses through an appropriate transmittance, termed a *zero-order hold*. The impulse train is related to the samples by

$$f^*(t) = f(0)\delta(t) + f(1)\delta(t-T) + f(2)\delta(t-2T) + \cdots$$
$$= \sum_{k=0}^{\infty} f(k)\delta(t-kT)$$

To obtain the sample-and-hold waveform from $f^*(t)$ requires a linear, time-invariant analog system with the impulse response given in Fig. 8-14a. A unit impulse input to this transmittance causes a unit rectangular pulse output of duration T as shown. A delayed impulse with different amplitude produces a delayed pulse with that amplitude, as in Fig. 8-14b, and an impulse train, Fig. 8-14c, results in the desired sampled-and-held reconstruction.

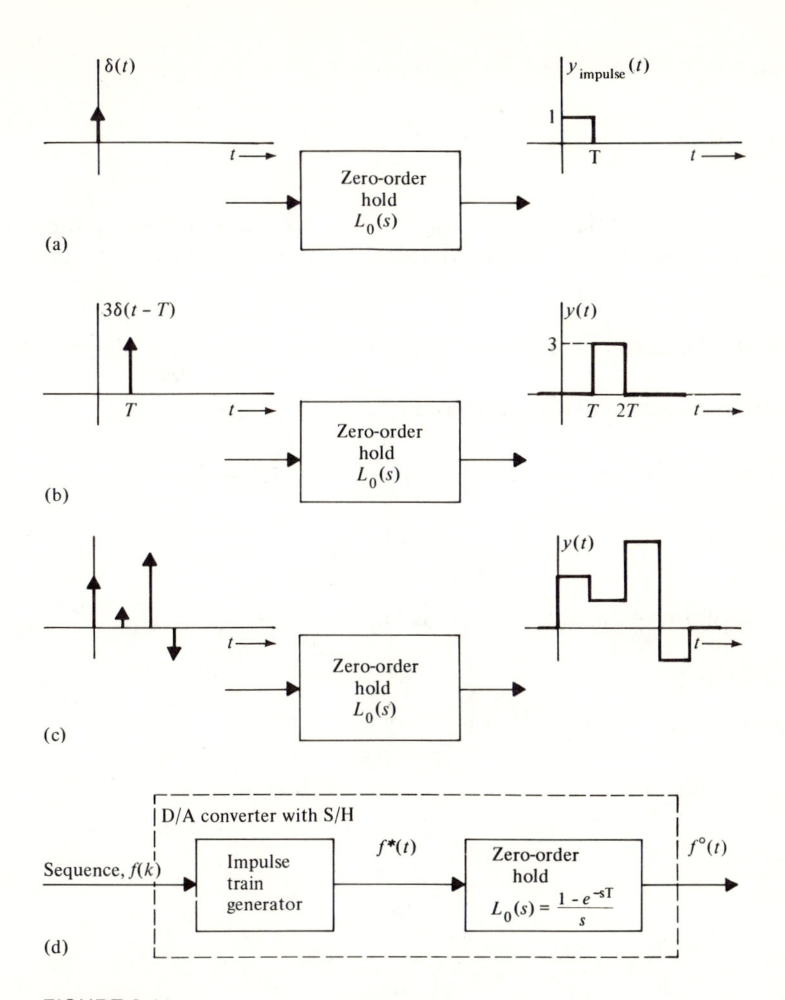

FIGURE 8-14
Response of the zero-order hold transmittance. (a) Impulse response. (b) Response to a scaled and delayed impulse. (c) Response to an impulse train. (d) Model of a D/A converter with S/H output.

The required impulse response of the zero order hold is

$$y_{\text{impulse}}(t) = u(t) - u(t - T)$$

and its Laplace transform is

$$Y_{\text{impulse}}(s) = \frac{1 - e^{-sT}}{s}$$

Since the Laplace transform of the unit impulse is unity, the zero-order hold transmittance, which is the ratio of the transforms, is

$$L_0(s) = Y_{\text{impulse}}(s) = \frac{1 - e^{-sT}}{s}$$

A D/A converter with sample-and-hold output may thus be modeled by the idealized impulse train generator followed by $L_0(s)$, as in Fig. 8-14d.

8.6.2 Relation Between the z-Transform and the Laplace Transform

The impulse train associated with a sequence of samples $f(k)$ is

$$\begin{aligned} f^*(t) &= f(0)\delta(t) + f(1)\delta(t-T) + f(2)\delta(t-2T) + f(3)\delta(t-3T) + \cdots \\ &= \sum_{k=0}^{\infty} f(k)\delta(t-kT) \end{aligned}$$

The Laplace transform of the impulse train is

$$\begin{aligned} F^*(s) &= \mathscr{L}[f^*(t)] \\ &= f(0) + f(1)e^{-sT} + f(2)e^{-2sT} + f(3)e^{-3sT} + \cdots \\ &= \sum_{k=0}^{\infty} f(k)(e^{sT})^{-k} \end{aligned}$$

Letting

$$z = e^{sT}$$

there results

$$F^*(s)\Big|_{e^{sT}=z} = \sum_{k=0}^{\infty} f(k)z^{-k} = Z[f(k)] = F(z)$$

One interpretation of the z-transformation is that it is the Laplace transform of the impulse train with e^{sT} replaced by z.

A sequence $f(k)$ with a rational z-transform $F(z)$ has a corresponding impulse train $f^*(t)$ with Laplace transform that may be obtained simply by substitution:

$$F^*(s) = F(z)\big|_{z=e^{sT}}$$

For example, the sequence with z-transform

$$F(z) = \frac{-4z^3 + 5z^2 - 6z}{z^3 + 2z^2 - z + 3}$$

has the following related impulse train transform when the sampling period $T=0.1$:

$$F^*(s)=\frac{-4e^{0.3s}+5e^{0.2s}-6e^{0.1s}}{e^{0.3s}+2e^{0.2s}-e^{0.1s}+3}$$

8.6.3 The Sampling Theorem

In applications such as communications, it is especially important to establish conditions for which a signal $g(t)$ is completely specified by (and thus recoverable from) its samples. Communication signals are typically band-limited, or nearly so, meaning that they contain no frequencies higher than a certain band limit frequency f_B. The frequency content of a signal $g(t)$ is given by its Fourier transform

$$G(\omega)=\int_{-\infty}^{\infty} g(t)e^{-j\omega t}\,dt$$

a calculation similar to the Laplace transformation with $s=j\omega$ but extending over all time, not just from $t=0$ and thereafter. A signal band-limited beyond frequency f_B is one for which

$$G(\omega)=0 \qquad |\omega|>2\pi f_B$$

A statement of the sampling theorem is as follows:

A signal $g(t)$ that is band-limited above (hertz) frequency f_B can be recovered from an infinite sequence of its periodic samples $g(k)$ if and only if the sampling interval T is less than $1/2f_B$.

That is, a band-limited signal must be sampled at a rate over twice that of its highest component frequency in order for the samples to be unique. The rate $2f_B$, where f_B is the highest frequency in a band-limited signal, is termed the *Nyquist rate* for that signal.

For a single sinusoidal signal of radian frequency b,

$$g(t)=A\cos(bt+\theta)$$

the sample sequence is, in terms of T,

$$g(k)=A\cos(kbT+\theta)$$

If the sampling interval is less than $1/2f_B$,

$$T<\frac{1}{2f_B}=\frac{2\pi}{2b}$$

$$bT<\pi$$

then the samples are unique, there being at least two per cycle of $g(t)$. If this condition is not met, then any higher-frequency sinusoid

$$h(t) = A\cos(b't + \theta)$$

for which

$$b'T = bT + n2\pi \qquad n = 1, 2, 3, \ldots$$

could be present, as it produces precisely the same sample sequence:

$$\begin{aligned} h(k) &= A\cos[k(bT + n2\pi) + \theta] \\ &= A\cos(kbT + kn2\pi + \theta) \\ &= A\cos(kbT + \theta) = g(k) \end{aligned}$$

The effects of any of these higher frequencies, being indistinguishable from those below the presumed band limit, are termed *aliasing distortion.*

A constructive statement of how to recover a band-limited signal from its samples is as follows:

To recover a suitably band-limited signal $g(t)$ from its samples $g(k)$, form the impulse train $g^*(t)$ and pass it through a low-pass filter that passes, unchanged, all frequencies in $g(t)$ below its band limit frequency f_B and removes all frequencies above $1/2T$.

This arrangement is shown in Fig. 8-15a. The required low-pass filter has the frequency response shown in Fig. 8-15b if the analog signal is to be reconstructed without delay. In practice, a phase shift proportional to frequency, representing a time delay in the reconstruction, is approximated. The frequency response of the zero-order hold is given in Fig. 8-15c and is seen to approximate a low-pass characteristic with time delay.

The sampling theorem does not apply directly to most control system design problems for the following reasons:

1. Many of the signals used in control system analysis, such as those involving step changes in amplitude and slope, are not band-limited.
2. For a band-limited signal, perfect reconstruction requires an infinite number of samples. Another way of stating this fact is to note that the low-pass filter required for reconstruction is a physical impossibility. It can be approximated only if a delay is introduced into the signal processing. Better approximation requires longer delays.
3. In control, good reconstruction of signals from their samples is only occasionally of primary interest compared to such concerns as stability, relative stability, and steady state errors.

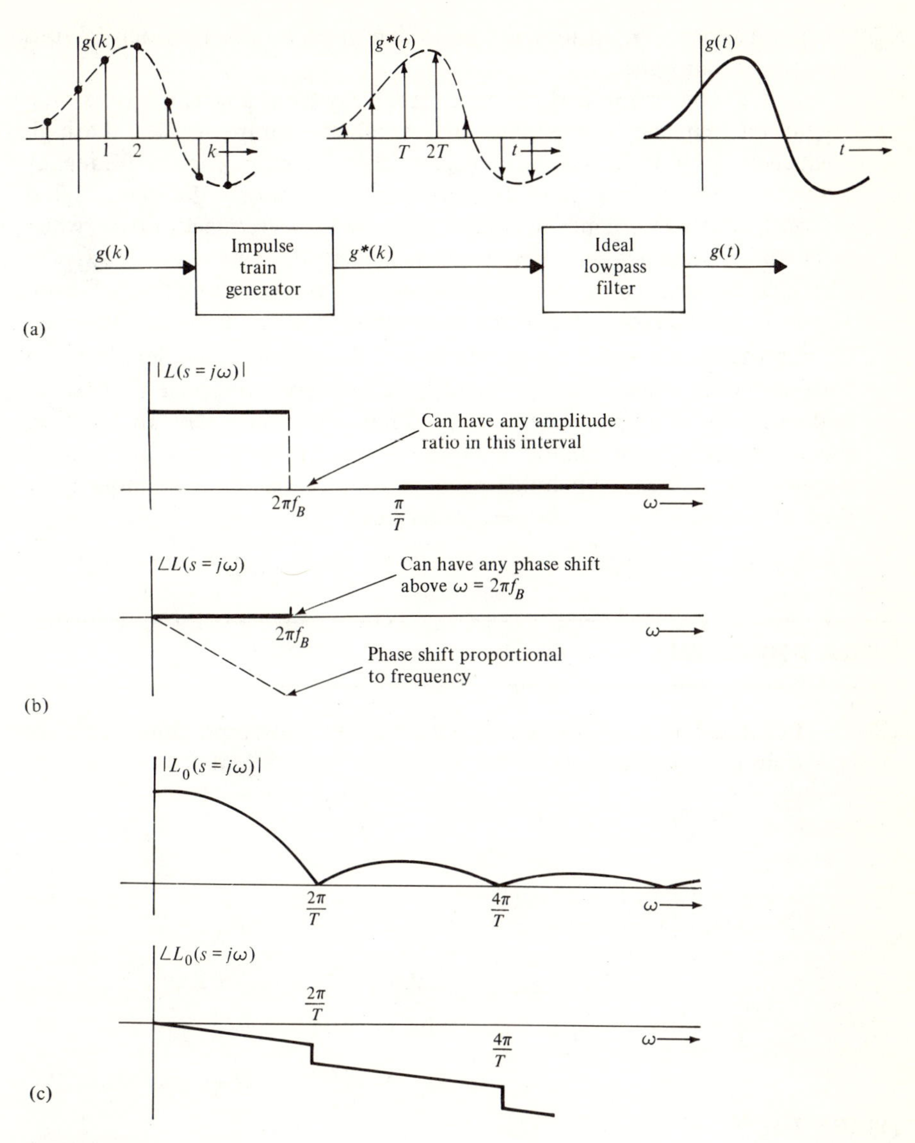

FIGURE 8-15
Reconstruction from impulses. (a) Reconstruction of a band-limited signal. (b) Ideal lowpass filter frequency response. (c) Frequency response of the zero-order hold.

Nevertheless, the sampling theorem is useful to digital control because it shows important properties of the analog-digital interface such as the following:

1. Samples of a signal uniquely determine that signal only under special circumstances, the sampling theorem stating one such situation. In particular, large transients and high-frequency oscillations in an analog signal

such as a system output may not be evident from relatively widely spaced samples of that signal.

2. When A/D conversion is done on a signal, say from a sensor, containing significant frequency components above half the sampling rate, the high frequencies produce errors equivalent to the presence of lower-frequency sinusoidal components. It is thus possible for a poorly designed digital feedback system to attempt to correct presumed low-frequency errors when in fact high-frequency sensor noise is the culprit. For this reason, low-pass filters, termed *antialiasing filters*, are commonly placed before the A/D converters to greatly reduce high-frequency sensor noise in many applications.
3. To improve the smoothing of reconstructed signals, the equivalent impulse train-to-output transmittance should have frequency response that better approximates an appropriate low-pass filter with time delay. In practice, this is achieved with analog low-pass filters and/or higher-order hold circuits. The higher-order holds produce outputs based upon more than the single sample used by the zero-order hold.

DRILL PROBLEMS

D8-9. For the following sequences $f(k)$, find the corresponding impulse train Laplace transforms $F^*(s)$ for a sampling interval T:

(a) $f(k)=[1-3(-1)^k+4(\frac{1}{2})^k]u(k) \qquad T=1$

ans. $(2e^{3s}+5e^{2s}-3e^s)/[e^{2s}-1][e^s-\frac{1}{2}]$

(b) $f(k)=\left[(-1)^k-\sin\dfrac{\pi k}{2}\right]u(k) \qquad T=0.2$

ans. $(e^{0.6s}-e^{0.4s})/(e^{0.2s}+1)(e^{0.4s}+1)$

(c) $f(k)=(\frac{1}{2})^k(10\sin 4k-8\cos 4k)u(k) \qquad T=0.1$

ans. $(-6.38e^{0.1s}-8e^{0.2s})/(e^{0.2s}+0.65e^{0.1s}+0.25)$

D8-10. For the continuous-time function

$$f(t)=10+3\cos \pi t-7\sin 6t$$

determine which of the following functions have the same sample sequence as $f(k)$ for a sampling interval of $T=0.2$:

(a) $g_1(t)=10\cos 10\pi t+3\cos 11\pi t-7\sin 6t$
(b) $g_2(t)=10+\sin 5\pi t+3\cos \pi t-7\sin[(6+10\pi)t]$
(c) $g_3(t)=10\cos 20\pi t-3\cos 6\pi t+7\sin[(6+5\pi)t]$
(d) $g_4(t)=5+6\cos 10\pi t-2\cos 20\pi t+\cos 30\pi t$
$+6\cos 11\pi t-3\cos \pi t-7\sin 6t$

(e) $$g_5(t) = 10\sqrt{2}\sin\frac{170\pi t}{8} + 3\cos 51\pi t - 8\sin 40\pi t - 7\sin[(6+30\pi)t]$$

Then find five other functions which, when sampled at this rate, have the same sample sequence.

8.7 DISCRETE-TIME SYSTEMS

Computer processing of input signal samples to produce output signal samples may be described by difference equations, analogous to the differential equations that characterize continuous-time systems. In this introductory treatment, only linear, step-invariant (or constant-coefficient) difference equations are considered.

8.7.1 Difference Equations and Response

Discrete-time systems are described by difference equations, of the form

$$y(k+n) + a_{n-1}y(k+n-1) + a_{n-2}y(k+n-2) + \cdots + a_1 y(k+1) + a_0 y(k) = b_m r(k+m) + b_{m-1}r(k+m-1) + \cdots + b_1 r(k+1) + b_0 r(k)$$

where $y(k)$ is the output sequence and $r(k)$ is the input sequence. Solving this nth-order difference equation for $y(k+n)$ gives

$$y(k+n) = -a_{n-1}y(k+n-1) - \cdots - a_1 y(k+1) - a_0 y(k) + b_m r(k+m) + \cdots + b_0 r(k)$$

In other words the $(k+n)$th sample of the output is a linear combination of the previous output samples through $y(k)$ and of the input samples from step $(k+m)$ through step k.

For example,

$$y(k+2) = 3y(k+1) - 2y(k) + 2r(k+1) - r(k)$$

is a discrete-time system described by a second-order difference equation. Given the input sequence $r(k)$ and two initial values of the sequence $y(k)$, the entire output sequence can be calculated recursively. If $r(k) = u(k)$, the unit step sequence, and

$$y(0) = 4$$
$$y(1) = -1$$

then

$$\begin{aligned} y(2) &= 3y(1) - 2y(0) + 2u(1) - u(0) \\ &= 12 - 2 + 2 - 1 = 11 \\ y(3) &= 3y(2) - 2y(1) + 2u(2) - u(1) \\ &= 33 - 8 + 2 - 1 = 26 \\ y(4) &= 3y(3) - 2y(2) + 2u(3) - u(2) \\ &= 78 - 22 + 2 - 1 = 57 \\ &\vdots \end{aligned}$$

and so forth.

A closed-form expression for the response of a discrete-time system may be obtained by z-transform methods. For the discrete-time system

$$y(k+1) = -0.5y(k) + 3r(k)$$

for example, with

$$\begin{aligned} y(0) &= 4 \\ r(k) &= u(k) \end{aligned}$$

the unit step sequence, z-transforming using the sequence shift relation of Table 8-6 gives

$$zY(z) - y(0) = -0.5Y(z) + 3R(z)$$

$$(z+0.5)Y(z) = 4 + \frac{3z}{z-1}$$

$$Y(z) = \frac{7z-4}{(z+0.5)(z-1)}$$

$$\frac{Y(z)}{z} = \frac{7z-4}{z(z+0.5)(z-1)} = \frac{8}{z} + \frac{-10}{z+0.5} + \frac{2}{z-1}$$

$$Y(z) = 8 + \frac{-10z}{z+0.5} + \frac{2z}{z-1}$$

$$y(k) = 8\delta(k) - 10(-0.5)^k + 2u(k) \qquad k = 0, 1, 2, \ldots$$

8.7.2 z-Transfer Functions

The z-transfer function of a discrete-time system is the ratio of the z-transform of the output to the z-transform of the input when all initial conditions are zero:

$$D(z) = \left. \frac{Y(z)}{R(z)} \right|_{\substack{\text{initial} \\ \text{conditions}=0}}$$

For the discrete-time system

$$y(k+3) = -0.3y(k+2) + y(k+1) - 0.5y(k) + 4r(k+3) - r(k+1) - 0.6r(k)$$

z-transforming with zero initial conditions gives

$$(z^3 + 0.3z^2 - z + 0.5)Y(z) = (4z^3 - z - 0.6)R(z)$$

$$D(z) = \left.\frac{Y(z)}{R(z)}\right|_{\substack{\text{initial}\\ \text{conditions}=0}} = \frac{4z^3 - z - 0.6}{z^3 + 0.3z^2 - z + 0.5}$$

A discrete-time system with z-transfer function

$$D(z) = \frac{3z+1}{z^2 - z + 2}$$

is described by the difference equation

$$y(k+2) - y(k+1) + 2y(k) = 3r(k+1) + r(k)$$

If the initial conditions are zero and

$$R(z) = \frac{5}{z+4}$$

the z-transform of the output is

$$Y(z) = D(z)R(z) = \left(\frac{3z+1}{z^2 - z + 2}\right)\left(\frac{5}{z+4}\right)$$

Nonzero initial conditions and multiple inputs and outputs may be accommodated in a manner analogous to the transfer function development for continuous-time systems given in Chap. 1.

8.7.3 Block Diagrams and Signal Flow Graphs

The block diagram and signal flow graph manipulations used for continuous-time system components also apply to discrete-time systems. For example, Fig. 8-16a shows the description of a discrete-time system by a block diagram. Reduction of the block diagram to determine the overall system z-transfer function is shown in Fig. 8-16b. A discrete-time system signal flow graph is given in Fig. 8-16c. Using Mason's gain rule, the system z-transfer function is

$$D(z) = \frac{3/z(z+1)}{1 + 12/(z+1) - (5z+2)/(z - \frac{1}{2})}$$

$$= \frac{3(z-0.5)}{z^3 + 7.5z^2 - 13.5z - 2}$$

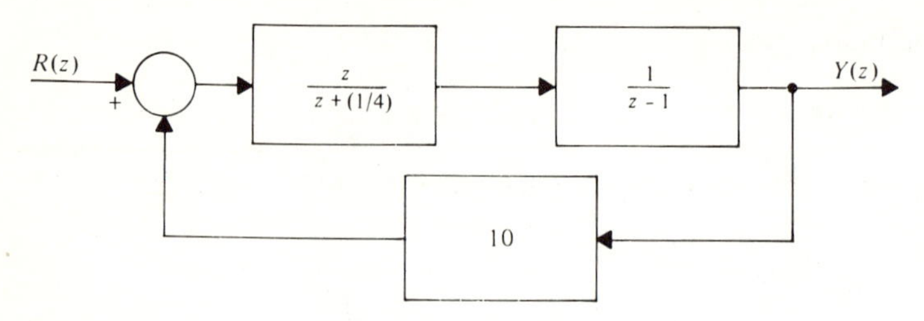

(a)

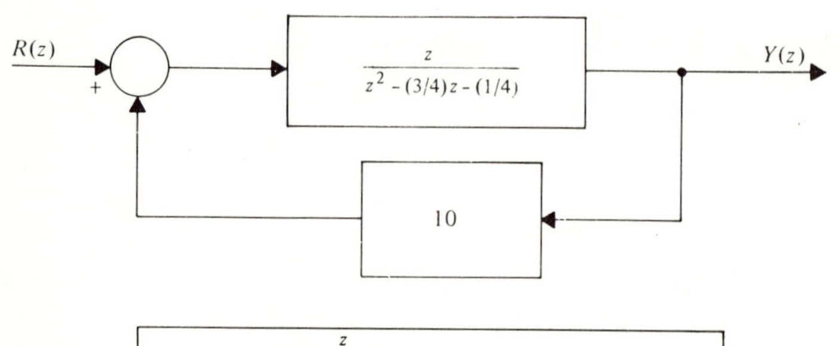

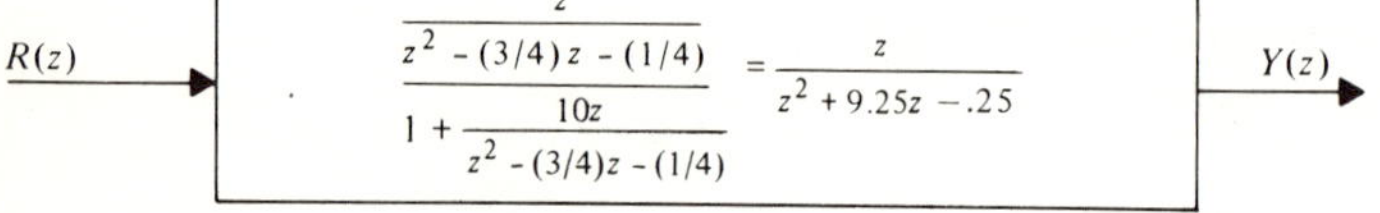

(b)

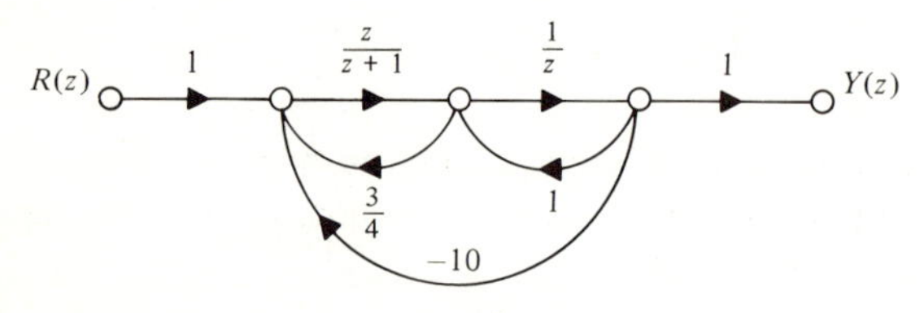

(c)

FIGURE 8-16
Discrete-time systems represented by block diagrams and signal flow graphs. (a) Block diagram of a discrete-time system. (b) Reduction of the block diagram. (c) Signal flow graph for a discrete-time system.

8.7.4 Stability and the Bilinear Transformation

Table 8-7 shows the characters of sequences corresponding to various complex plane locations of the denominator roots of a z-transformed sequence $F(z)$. Denominator roots within the unit circle on the complex plane give rise to sequences which decay with k, while roots outside the unit circle represent response terms that grow in magnitude with k.

A discrete-time system is said to be stable if and only if its unit pulse

TABLE 8-7
Sequences corresponding to various z-transform denominator polynomial root locations

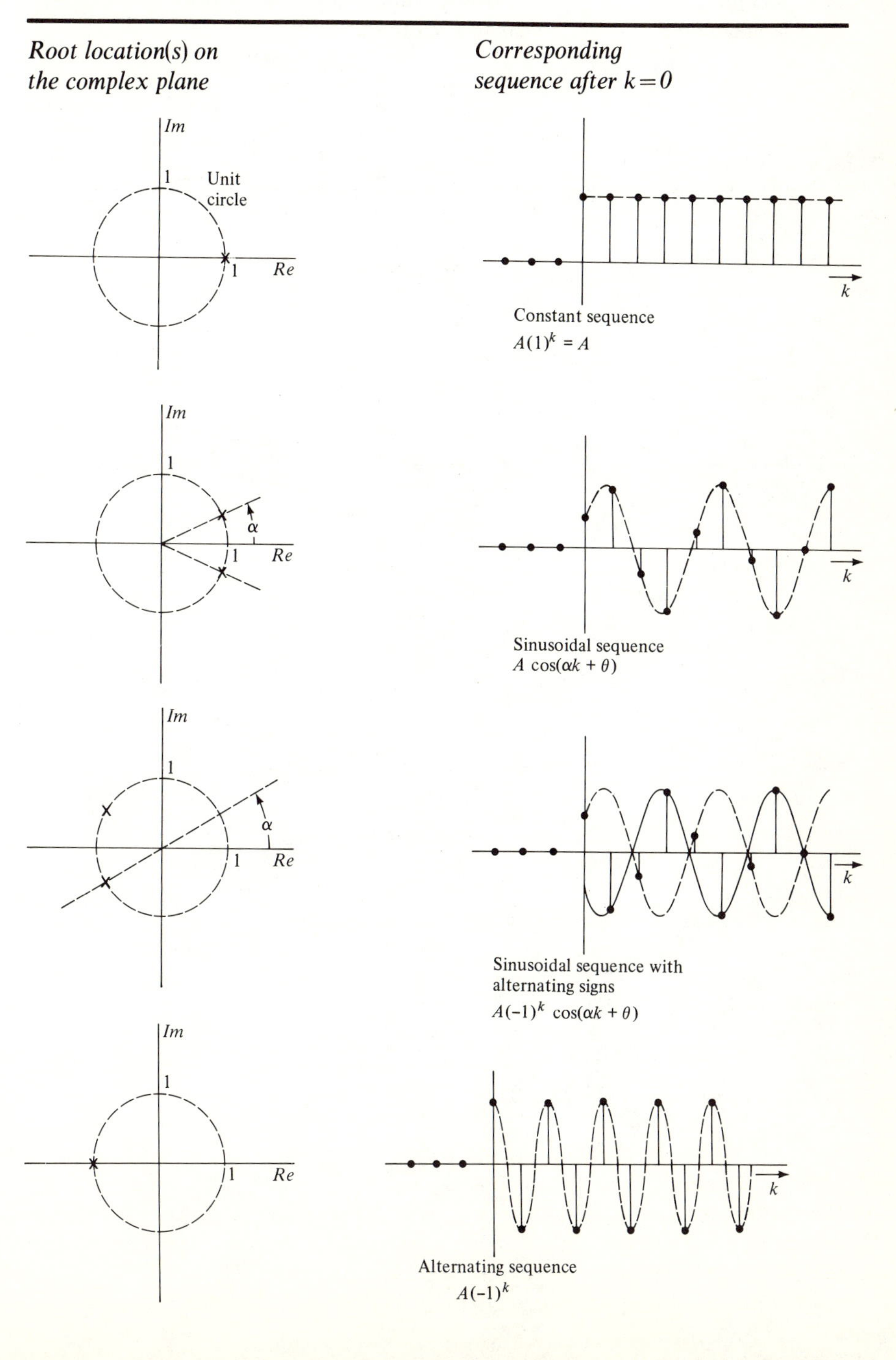

TABLE 8-7 (Continued).

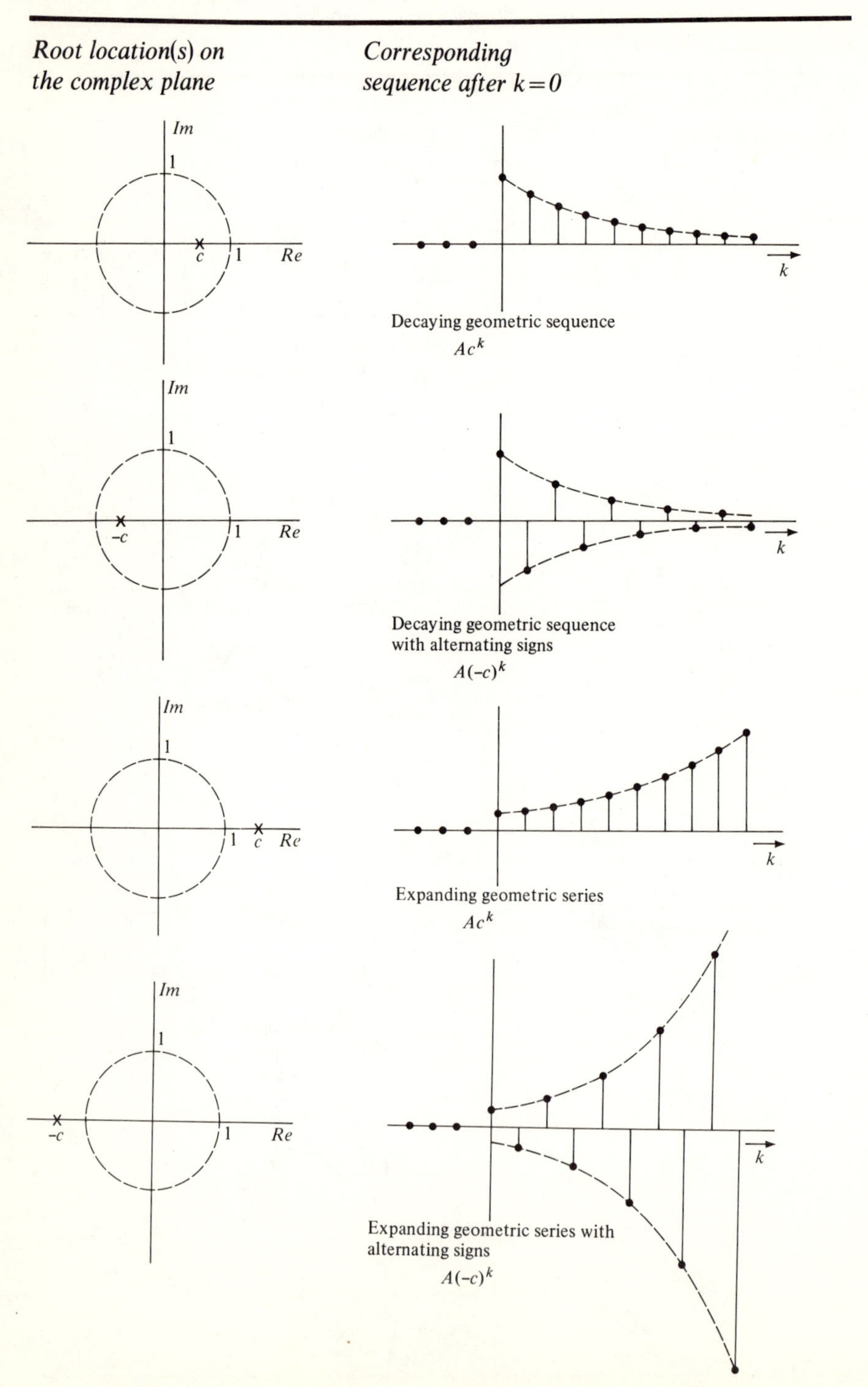

TABLE 8-7 (Continued).

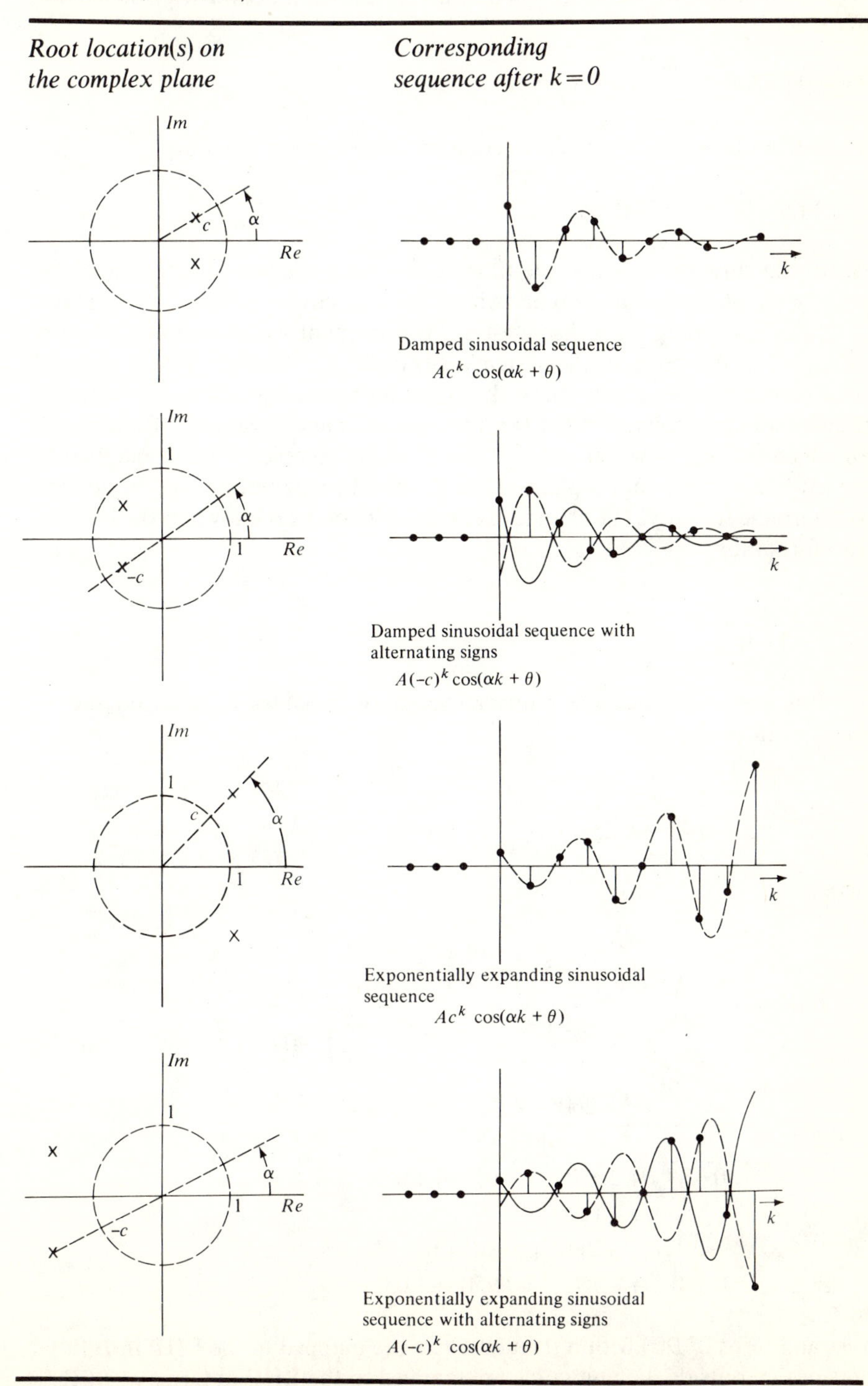

response decays with k. If a system with z-transfer function $D(z)$ has a unit pulse input

$$r(k)=\delta(k) \qquad R(z)=1$$

the system output has z-transform equal to the z-transfer function

$$Y(z)=D(z)\cdot 1=D(z)$$

Hence the stability of a discrete-time system hinges upon whether all of the poles of its z-transfer function are within the unit circle on the complex plane.

Stability testing for a discrete-time system involves determining whether or not all of the poles of the system's z-transfer function are within the unit circle on the complex plane. One stability-testing method which avoids factoring the denominator polynomial of $D(z)$ involves a change of variables from z to W for which the region within the unit circle on the complex plane is mapped to the left half of the complex plane. Then Routh-Hurwitz testing may be applied to determine stability. The change of variables involved is known as the bilinear transformation:

$$z=\frac{1+W}{1-W}$$

For example, when the bilinear change of variables is made on the z-transfer function

$$D(z)=\frac{8z^3-3z^2+z}{z^3+0.4z^2-0.25z-0.1}$$

there results

$$D(W)=\frac{8\left(\frac{1+W}{1-W}\right)^3-3\left(\frac{1+W}{1-W}\right)^2+\left(\frac{1+W}{1-W}\right)}{\left(\frac{1+W}{1-W}\right)^3+0.4\left(\frac{1+W}{1-W}\right)^2-0.25\left(\frac{1+W}{1-W}\right)-0.1}$$

$$=\frac{\dfrac{12W^3+26W^2+20W+6}{(1-W)^3}}{\dfrac{0.45W^3+2.55W^2+3.95W+1.05}{(1-W)^3}}$$

$$=\frac{12W^3+28W^2+22W+6}{0.45W^3+2.55W^2+3.95W+1.05}$$

Poles and zeros of $D(z)$ within the unit circle are mapped to the LHP in $D(W)$; roots of $D(z)$ outside the unit circle are mapped to the RHP in terms of $D(W)$;

and roots of $D(z)$ located precisely on the unit circle are mapped to the imaginary axis in $D(W)$.

A Routh-Hurwitz test of the poles of $D(W)$ is as follows:

$$\begin{array}{l|ll} W^3 & 0.45 & 3.95 \\ W^2 & 2.55 & 1.05 \\ W^1 & 3.76 & \\ W^0 & 1.05 & \end{array}$$

There are no left column sign changes in the array, so all poles of $D(W)$ are in the LHP. All poles of $D(z)$ are then within the unit circle on the complex plane. The system represented by $D(z)$ is thus stable.

The bilinear transformation, because it converts a digital problem to a related analog one, is also very useful for applying root locus and frequency response methods to digital systems.

8.7.5 Computer Software

Programming a digital computer with A/D and D/A capability as a discrete-time system is straightforward. For example, a system with z-transfer function

$$D(z)=\frac{2z^2+5}{z^2+3z+2}$$

TABLE 8-8
Fortran program to realize a discrete-time system

```
   100 FORMAT (F16.8)
   110 Y0=0.
   120 Y1=0.
   130 R1=0.
   140 R0=0.
a  150 READ (1, 100)R2
   160 Y2=3.*Y1-2.*Y0+2.*R2+5.*R0
   170 Y0=Y1
   180 Y1=Y2
   190 R0=R1
   200 R1=R2
b  210 WRITE (2,100)Y2
   220 GO TO 150
   230 STOP
   240 END
```

[a]The A/D converter is taken to be device number 1 with F16.8 format. It is here assumed that the processor waits at step 15Ø at each looping until a new sample is ready, just as it would wait for a character to be input from a keyboard device.

[b]The D/A converter is taken to be device number 2 with F16.8 format. It is here assumed that the D/A device contains a buffer that stores each new output sample for conversion at the sample time.

is described by the difference equation

$$y(k+2) = -3y(k+1) - 2y(k) + 2r(k+2) + 5r(k)$$

A Fortran program for this system is outlined in the flow diagram of Fig. 8-17 and listed in Table 8-8. The variables Y2, Y1, and YØ are used for $y(k+2)$, $y(k+1)$, and $y(k)$ respectively, while R2, R1, and RØ represent $r(k+2)$, $r(k+1)$, and $r(k)$. The initial conditions YØ, Y1, RØ, and R1 are first set to zero. Then an input value R2 is read from the A/D converter. Y2 is calculated and the

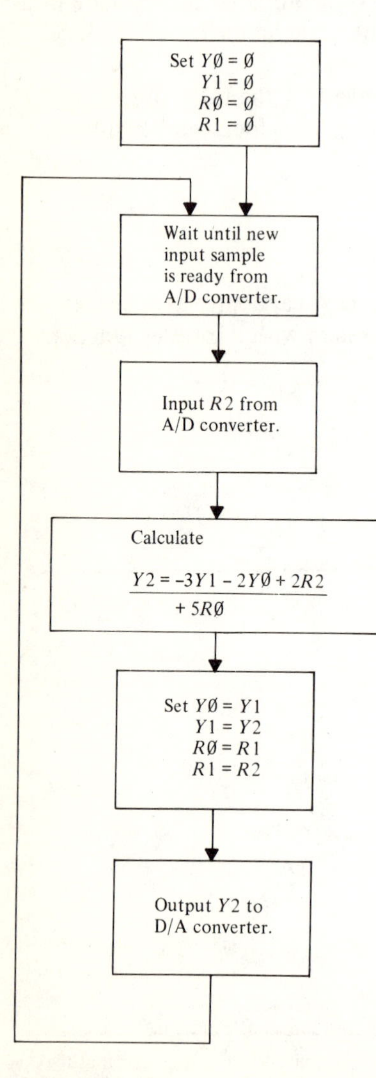

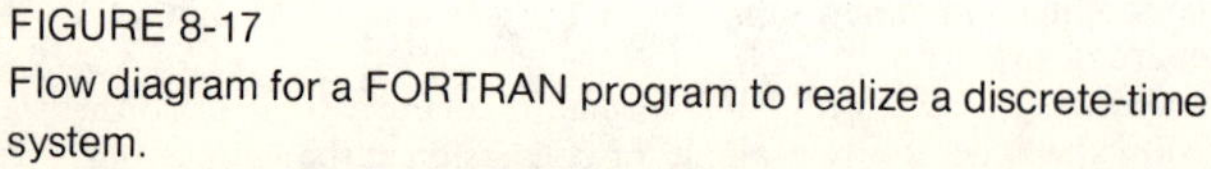

FIGURE 8-17
Flow diagram for a FORTRAN program to realize a discrete-time system.

values of YØ, Y1, RØ, and R1 are updated for the next calculation cycle. Y2 is then output to the D/A converter. Assuming that there is sufficient time between samples to perform the calculations, the program waits for a new input sample, computes the next Y2 sample, and so forth.

There are, of course, many other functions the computer could perform. It could limit the output signal, check that the input signal samples are "reasonable" (that is, they are within some predetermined bounds), and trigger alarms in the event that a malfunction is detected.

DRILL PROBLEMS

D8-11. Find the z-transfer functions of the following discrete-time systems:

(a) $y(k+3)+3y(k+2)-2y(k+1)+y(k)=r(k+2)+r(k+1)-4r(k)$

$$\text{ans. } D(z)=\frac{z^2+z-4}{z^3+3z^2-2z+1}$$

(b) $y(k+4)=0.2r(k+4)-0.3r(k+3)+0.1r(k+2)+0.7r(k+1)-0.5r(k)$

$$\text{ans. } D(z)=\frac{0.2z^4-0.3z^3+0.1z^2+0.7z-0.5}{z^4}$$

(c) $y(k+3)=0.5y(k+2)-y(k+1)+0.125y(k)+10r(k+3)$

$$\text{ans. } D(z)=\frac{10z^3}{z^3-0.5z^2+z-0.125}$$

D8-12. For

$$y(k+2)-5y(k+1)-6y(k)=2r(k)$$

recursively find $y(2)$, $y(3)$, $y(4)$, and $y(5)$ if

$$r(k)=(-1)^k$$
$$y(0)=-4$$
$$y(1)=\ \ 7$$

ans. 13, 105, 605, 3653

D8-13. For discrete-time systems with the following z-transfer functions and input sequences, find the output sequences for $k=0$ and thereafter if the initial conditions are zero:

(a) $D(z)=\dfrac{4}{z+3}$

$r(k)=5\delta(k)$

$$\text{ans. } \tfrac{20}{3}\delta(k)-\tfrac{20}{3}(-3)^k$$

(b) $D(z) = \dfrac{z}{z - \frac{1}{10}}$

$r(k) = u(k)$

ans. $-\frac{1}{9}\left(\frac{1}{10}\right)^k + \frac{10}{9}$

(c) $D(z) = \dfrac{-8}{z + \frac{1}{3}}$

$r(k) = \left(\frac{1}{4}\right)^k$

ans. $-\frac{96}{7}\left(\frac{1}{4}\right)^k + \frac{96}{7}\left(-\frac{1}{3}\right)^k$

(d) $D(z) = \dfrac{z}{2z - 1}$

$r(k) = \delta(k - 3)$

ans. $\frac{1}{2}\left(\frac{1}{2}\right)^{k-3} u(k-3)$

D8-14. Determine whether or not each of the following discrete-time systems is stable:

(a) $D(z) = \dfrac{-3z^2 + 1}{4z^2 + 2z - 1}$

ans. denominator roots 0.31 and -0.81; stable

(b) $D(z) = \dfrac{z^3}{z^3 + 0.3z^2 - 0.25z - 0.075}$

ans. $D(W) = \dfrac{W^3 + 3W^2 + 3W + 1}{0.525W^3 + 2.725W^2 + 3.775W + 0.975}$; stable

(c) $D(z) = \dfrac{5(z - 0.2)}{z^3 - 2.8z^2 + 1.75z - 0.3}$

ans. $D(W) = \dfrac{1.026(W + 0.667)(1 - W)^2}{W^3 + 0.538W^2 - 0.111W - 0.060}$; unstable

---

8.8 STATE VARIABLE DESCRIPTIONS OF DISCRETE-TIME SYSTEMS

8.8.1 Simulation Diagrams and Equations

Simulation diagrams for discrete-time systems involve as a basic element blocks or branches having transmittance $1/z$. A simulation diagram, in phase-variable canonical form, for a system with z-transfer function

$$D(z)=\frac{3z^2-2z+8}{z^3+0.5z^2-0.25z+0.75}$$
$$=\frac{3/z-2/z^2+8/z^3}{1+0.5/z-0.25/z^2+0.75/z^3}$$

is given in Fig. 8-18a. In terms of the indicated state variables, the z-transformed equations describing the system are as follows:

$$zX_1(z)=X_2(z)$$
$$zX_2(z)=X_3(z)$$
$$zX_3(z)=-0.75X_1(z)+0.25X_2(z)-0.5X_3(z)+R(z)$$
$$Y(z)=8X_1(z)-2X_2(z)+3X_3(z)$$

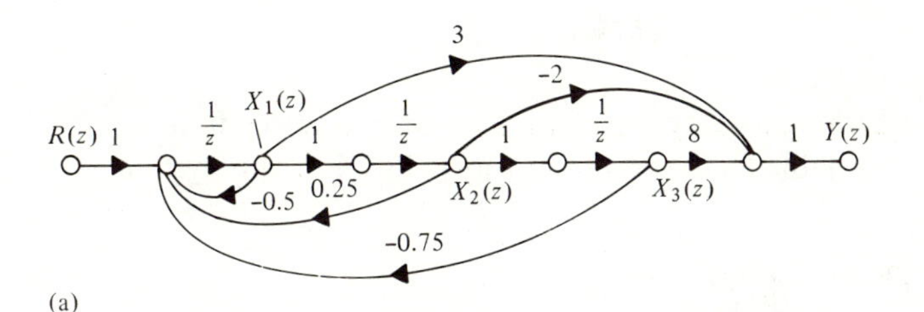

(a)

(b)

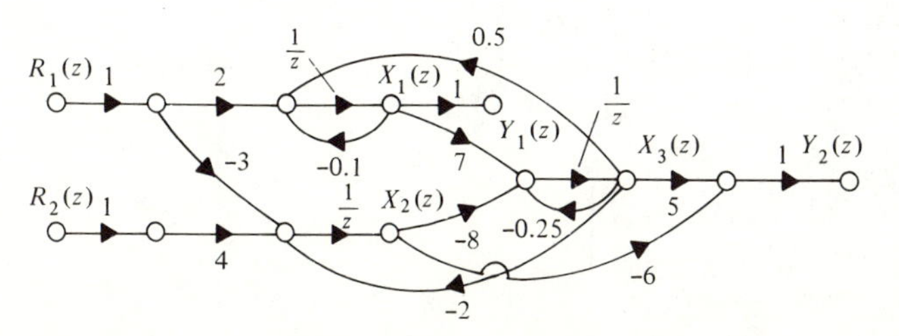

(c)

FIGURE 8-18
Simulation diagrams for discrete-time systems. (a) A system in phase-variable form. (b) A system in dual phase-variable form. (c) A multiple-input, multiple-output system.

In terms of the step, k, these equations are

$$\begin{aligned} x_1(k+1) &= x_2(k) \\ x_2(k+1) &= x_3(k) \\ x_3(k+1) &= -0.75x_1(k)+0.25x_2(k)-0.5x_3(k)+r(k) \\ y(k) &= 8x_1(k)-2x_2(k)+3x_3(k) \end{aligned}$$

or

$$\begin{bmatrix} x_1(k+1) \\ x_2(k+1) \\ x_3(k+1) \end{bmatrix} = \begin{bmatrix} 0 & 1 & 0 \\ 0 & 0 & 1 \\ -0.75 & 0.25 & -0.5 \end{bmatrix} \begin{bmatrix} x_1(k) \\ x_2(k) \\ x_3(k) \end{bmatrix} + \begin{bmatrix} 0 \\ 0 \\ 1 \end{bmatrix} r(k)$$

$$y(k) = [8 \quad -2 \quad 3] \begin{bmatrix} x_1(k) \\ x_2(k) \\ x_3(k) \end{bmatrix}$$

A simulation diagram for a system with the same transfer function but using the dual phase-variable form is given in Fig. 8-18b. For this system,

$$X_1(z) = \frac{1}{z}[-0.5X_1(z)+X_2(z)+3R(z)]$$

$$X_2(z) = \frac{1}{z}[0.25X_1(z)+X_3(z)-2R(z)]$$

$$X_3(z) = \frac{1}{z}[-0.75X_1(z)+8R(z)]$$

$$Y(z) = X_1(z)$$

or

$$\begin{bmatrix} x_1(k+1) \\ x_2(k+1) \\ x_3(k+1) \end{bmatrix} = \begin{bmatrix} -0.5 & 1 & 0 \\ 0.25 & 0 & 1 \\ -0.75 & 0 & 0 \end{bmatrix} \begin{bmatrix} x_1(k) \\ x_2(k) \\ x_3(k) \end{bmatrix} + \begin{bmatrix} 3 \\ -2 \\ 8 \end{bmatrix} r(k)$$

$$y(k) = [1 \quad 0 \quad 0] \begin{bmatrix} x_1(k) \\ x_2(k) \\ x_3(k) \end{bmatrix}$$

A simulation diagram for a multiple-input, multiple-output system is given in Fig. 8-18c. It consists of constant transmittances and unit delay transmittances $1/z$. The output of each delay is a state variable, and each delay input consists of a linear combination of the inputs and the state variables. For the example system,

$$\begin{cases} zX_1(z) = -0.1X_1(z) + 0.5X_3(z) + 2R_1(z) \\ zX_2(z) = -2X_3(z) - 3R_1(z) + 4R_2(z) \\ zX_3(z) = 7X_1(z) - 8X_2(z) - 0.25X_3(z) \\ Y_1(z) = X_1(z) \\ Y_2(z) = -6X_2(z) + 5X_3(z) \end{cases}$$

$$\begin{bmatrix} x_1(k+1) \\ x_2(k+1) \\ x_3(k+1) \end{bmatrix} = \begin{bmatrix} -0.1 & 0 & 0.5 \\ 0 & 0 & -2 \\ 7 & -8 & -0.25 \end{bmatrix} \begin{bmatrix} x_1(k) \\ x_2(k) \\ x_3(k) \end{bmatrix} + \begin{bmatrix} 2 & 0 \\ -3 & 4 \\ 0 & 0 \end{bmatrix} + \begin{bmatrix} r_1(k) \\ r_2(k) \end{bmatrix}$$

$$\begin{bmatrix} y_1(k) \\ y_2(k) \end{bmatrix} = \begin{bmatrix} 1 & 0 & 0 \\ 0 & -6 & 5 \end{bmatrix} \begin{bmatrix} x_1(k) \\ x_2(k) \\ x_3(k) \end{bmatrix}$$

In general, the state equations for a discrete-time system are of the form

$$\begin{aligned} \mathbf{x}(k+1) &= \mathbf{F}x(k) + \mathbf{G}r(k) \\ \mathbf{y}(k) &= \mathbf{H}\mathbf{x}(k) \end{aligned} \qquad (8\text{-}1)$$

When written out, these are

$$\begin{bmatrix} x_1(k+1) \\ x_2(k+1) \\ x_3(k+1) \\ \vdots \\ x_n(k+1) \end{bmatrix} = \begin{bmatrix} f_{11} & f_{12} & f_{13} \cdots & f_{1n} \\ f_{21} & f_{22} & f_{23} \cdots & f_{2n} \\ f_{31} & f_{32} & f_{33} & f_{3n} \\ \vdots & & & \\ f_{n1} & f_{n2} & f_{n3} \cdots & f_{nn} \end{bmatrix} \begin{bmatrix} x_1(k) \\ x_2(k) \\ x_3(k) \\ \vdots \\ x_n(k) \end{bmatrix} + \begin{bmatrix} g_{11} & g_{12} \cdots g_{1i} \\ g_{21} & g_{22} \cdots g_{2i} \\ g_{31} & g_{32} \cdots g_{3i} \\ \vdots & \\ g_{n1} & g_{n2} \cdots g_{ni} \end{bmatrix} \begin{bmatrix} r_1(k) \\ r_2(k) \\ \vdots \\ r_i(k) \end{bmatrix}$$

$$\begin{bmatrix} y_1(k) \\ y_2(k) \\ \vdots \\ y_m(k) \end{bmatrix} = \begin{bmatrix} h_{11} & h_{12} \cdots h_{1n} \\ h_{21} & h_{22} \cdots h_{2n} \\ \vdots & \\ h_{m1} & h_{m2} \cdots h_{mn} \end{bmatrix} \begin{bmatrix} x_1(k) \\ x_2(k) \\ \vdots \\ x_n(k) \end{bmatrix}$$

8.8.2 Response and Stability

The response of a discrete-time system may be calculated recursively, starting with an initial state and repeatedly using the state equations (8-1). From $\mathbf{x}(0)$ and $\mathbf{r}(0)$, $\mathbf{x}(1)$ may be calculated:

$$\mathbf{x}(1) = \mathbf{F}\mathbf{x}(0) + \mathbf{G}\mathbf{r}(0)$$

Then, using $\mathbf{x}(1)$ and $\mathbf{r}(1)$, $\mathbf{x}(2)$ is calculated:

$$\begin{aligned} \mathbf{x}(2) &= \mathbf{F}\mathbf{x}(1) + \mathbf{G}\mathbf{r}(1) \\ &= \mathbf{F}^2\mathbf{x}(0) + \mathbf{F}\mathbf{G}\mathbf{r}(0) + \mathbf{G}\mathbf{r}(1) \end{aligned}$$

Continuing,

$$\begin{aligned}\mathbf{x}(3) &= \mathbf{F}\mathbf{x}(2) + \mathbf{G}\mathbf{r}(2)\\ &= \mathbf{F}^3\mathbf{x}(0) + \mathbf{F}^2\mathbf{G}\mathbf{r}(0) + \mathbf{F}\mathbf{G}\mathbf{r}(1) + \mathbf{G}\mathbf{r}(2)\\ &\vdots\\ \mathbf{x}(k) &= \mathbf{F}^k\mathbf{x}(0) + \mathbf{F}^{k-1}\mathbf{G}\mathbf{r}(0) + \mathbf{F}^{k-2}\mathbf{G}\mathbf{r}(1)\\ &\quad + \cdots + \mathbf{F}\mathbf{G}\mathbf{r}(k-2) + \mathbf{G}\mathbf{r}(k-1)\\ &= \mathbf{F}^k\mathbf{x}(0) + \sum_{n=1}^{k} \mathbf{F}^{k-n}\mathbf{G}\mathbf{r}(n-1)\end{aligned}$$

As a numerical example, the system

$$\begin{bmatrix} x_1(k+1) \\ x_2(k+2) \end{bmatrix} = \begin{bmatrix} 0 & -2 \\ -1 & 3 \end{bmatrix}\begin{bmatrix} x_1(k) \\ x_2(k) \end{bmatrix} + \begin{bmatrix} 2 \\ 1 \end{bmatrix} r(k)$$

$$y(k) = [3 \quad -2]\begin{bmatrix} x_1(k) \\ x_2(k) \end{bmatrix} \tag{8-2}$$

with

$$\begin{bmatrix} x_1(0) \\ x_2(0) \end{bmatrix} = \begin{bmatrix} 5 \\ -7 \end{bmatrix}$$

and

$$r(k) = \delta(k)$$

has response as follows:

$$y(0) = [3 - 2]\begin{bmatrix} 5 \\ -7 \end{bmatrix} = 29$$

$$\begin{bmatrix} x_1(1) \\ x_2(1) \end{bmatrix} = \begin{bmatrix} 0 & -2 \\ -1 & 3 \end{bmatrix}\begin{bmatrix} 5 \\ -7 \end{bmatrix} + \begin{bmatrix} 2 \\ 1 \end{bmatrix} = \begin{bmatrix} 16 \\ -25 \end{bmatrix}$$

$$y(1) = [3 \quad -2]\begin{bmatrix} 16 \\ -25 \end{bmatrix} = 98$$

$$\begin{bmatrix} x_1(2) \\ x_2(2) \end{bmatrix} = \begin{bmatrix} 0 & -2 \\ -1 & 3 \end{bmatrix}\begin{bmatrix} 16 \\ -25 \end{bmatrix} + \begin{bmatrix} 0 \\ 0 \end{bmatrix} = \begin{bmatrix} 50 \\ -91 \end{bmatrix}$$

$$y(2) = [3 \quad -2]\begin{bmatrix} 50 \\ -91 \end{bmatrix} = 332$$

$$\begin{bmatrix} x_1(3) \\ x_2(3) \end{bmatrix} = \begin{bmatrix} 0 & -2 \\ -1 & 3 \end{bmatrix}\begin{bmatrix} 50 \\ -91 \end{bmatrix} + \begin{bmatrix} 0 \\ 0 \end{bmatrix} = \begin{bmatrix} 182 \\ -323 \end{bmatrix}$$

$$y(3) = [3 \quad -2]\begin{bmatrix} 182 \\ -323 \end{bmatrix} = 1192$$

$$\vdots$$

A discrete-time system's z-transfer function is found by z-transforming the state equations (8-1) with zero initial conditions and solving for the ratio of output to input transforms:

$$\begin{cases} z\mathbf{X}(z)=\mathbf{F}\mathbf{X}(z)+\mathbf{G}\mathbf{R}(z) \\ \mathbf{Y}(z)=\mathbf{H}\mathbf{X}(z) \end{cases}$$

$$(z\mathbf{I}-\mathbf{F})\mathbf{X}(z)=\mathbf{G}\mathbf{R}(z)$$

$$\mathbf{X}(z)=(z\mathbf{I}-\mathbf{F})^{-1}\mathbf{G}\mathbf{R}(z)$$

$$\mathbf{Y}(z)=\mathbf{H}\mathbf{X}(z)=\mathbf{H}(z\mathbf{I}-\mathbf{F})^{-1}\mathbf{G}\mathbf{R}(z)$$

The z-transfer function matrix of the system is thus

$$\mathbf{D}(z)=\mathbf{H}(z\mathbf{I}-\mathbf{F})^{-1}\mathbf{G}=\frac{\mathbf{H}\ \text{adj}\ (z\mathbf{I}-F)\mathbf{G}}{|z\mathbf{I}-\mathbf{F}|}$$

and the system is stable if and only if all of the roots of the characteristic polynomial

$$|z\mathbf{I}-\mathbf{F}|=0$$

are within the unit circle on the complex plane.

For the single-input, single-output example system (8-2) the transfer function is

$$D(z)=[3\quad -2]\begin{bmatrix} z & 2 \\ 1 & z-3 \end{bmatrix}^{-1}\begin{bmatrix} 2 \\ 1 \end{bmatrix}$$

$$=\frac{[3\quad -2]\begin{bmatrix} z-3 & -2 \\ -1 & z \end{bmatrix}\begin{bmatrix} 2 \\ 1 \end{bmatrix}}{z^2-3z-2}$$

$$=\frac{[3\quad -2]\begin{bmatrix} (2z-8) \\ (z-2) \end{bmatrix}}{z^2-3z-2}$$

$$=\frac{4z-20}{z^2-3z-2}$$

This system is unstable, since the roots of the characteristic equation are

$$z^2-3z-2=(z-1)(z-2)=0$$

$$z_1,\ z_2=1,\ 2$$

which do not lie within the unit circle on the complex plane.

8.8.3 Controllability and Observability

For discrete-time systems with nonrepeated characteristic roots, an appropriate change of state variables

$$\mathbf{x}' = \mathbf{P}x \qquad \mathbf{x} = \mathbf{P}^{-1}\mathbf{x}'$$

determined as for continuous-time systems with the methods of Sec. 7.4, will decouple the state equations:

$$\mathbf{x}'(k+1) = \mathbf{P}\mathbf{F}\mathbf{P}^{-1}\mathbf{x}'(k) + \mathbf{P}\mathbf{G}\mathbf{r}(k) = \mathbf{F}'\mathbf{x}'(k) + \mathbf{G}'\mathbf{r}(k)$$
$$\mathbf{y}(k) = \mathbf{H}\mathbf{P}^{-1}\mathbf{x}'(k) = \mathbf{H}'\mathbf{x}'(k)$$

where

$$\mathbf{F}' = \begin{bmatrix} z_1 & 0 & 0 & \cdots & 0 \\ 0 & z_2 & 0 & \cdots & 0 \\ \vdots & & & & \\ 0 & 0 & 0 & \cdots & z_n \end{bmatrix}$$

In terms of the new state variables $\mathbf{x}'$, the state coupling matrix $\mathbf{F}'$ is diagonal, with the system's characteristic roots along the diagonal.

If, when the state equations for a discrete-time system are diagonalized, any row of the new input coupling matrix

$$\mathbf{G}' = \mathbf{P}\mathbf{G}$$

is zero, the corresponding discrete-time system mode is uncontrollable. If any column of

$$\mathbf{H}' = \mathbf{H}\mathbf{P}^{-1}$$

is zero, the corresponding system mode is unobservable. The same rank tests for complete controllability and complete observability that apply to continuous-time systems apply here because controllability and observability are algebraic properties of the system matrices.

An nth-order discrete-time system $(8-1)$ is completely controllable if and only if its controllability matrix

$$\mathbf{M}_c = [\mathbf{G} \mid \mathbf{F}\mathbf{G} \mid \mathbf{F}^2\mathbf{G} \mid \cdots \mid \mathbf{F}^{n-1}\mathbf{G}]$$

is of full rank. The system is completely observable if and only if the observability matrix

$$\mathbf{M}_o = \begin{bmatrix} \mathbf{H} \\ \hdashline \mathbf{HF} \\ \hdashline \mathbf{HF}^2 \\ \hdashline \vdots \\ \hdashline \mathbf{HF}^{n-1} \end{bmatrix}$$

is of full rank.

DRILL PROBLEMS

D8-15. Draw discrete-time simulation diagrams in phase-variable canonical form for systems with the following z-transfer function:

(a) $D(z) = \dfrac{4z}{z^2 + z + 0.5}$

(b) $D(z) = \dfrac{10z^3 - 4z^2 + 5z}{z^3 + 0.5z^2 - 0.2z + 0.3}$

D8-16. Find state equations in matrix form for the systems with the following discrete-time simulation diagrams.

ans. (a) $$\begin{bmatrix} x_1(k+1) \\ x_2(k+1) \end{bmatrix} = \begin{bmatrix} \frac{1}{6} & 5 \\ -\frac{1}{10} & -1 \end{bmatrix} \begin{bmatrix} x_1(k) \\ x_2(k) \end{bmatrix} + \begin{bmatrix} 4 \\ 3 \end{bmatrix} r(k)$$

$$y(k) = [2 \quad 0] \begin{bmatrix} x_1(k) \\ x_2(k) \end{bmatrix}$$

(b) $$\begin{bmatrix} x_1(k+1) \\ x_2(k+1) \\ x_3(k+1) \end{bmatrix} = \begin{bmatrix} \frac{1}{5} & 0 & 0 \\ 0 & -\frac{1}{6} & 0 \\ 0 & 0 & -\frac{1}{7} \end{bmatrix} \begin{bmatrix} x_1(k) \\ x_2(k) \\ x_3(k) \end{bmatrix} + \begin{bmatrix} 2 \\ -3 \\ 4 \end{bmatrix} r(k)$$

$$y(k) = [1 \quad 1 \quad 1] \begin{bmatrix} x_1(k) \\ x_2(k) \\ x_3(k) \end{bmatrix}$$

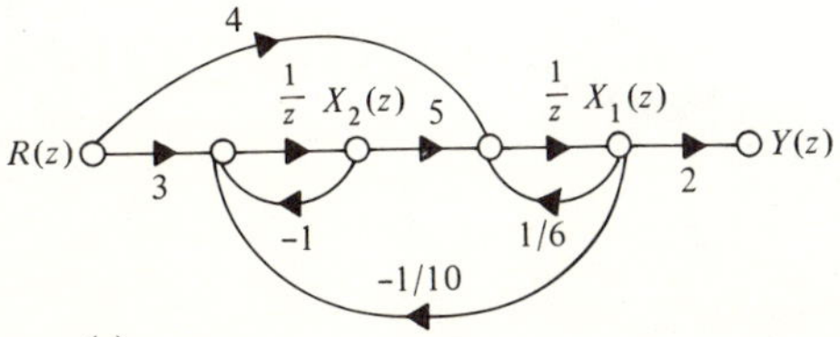

(a)

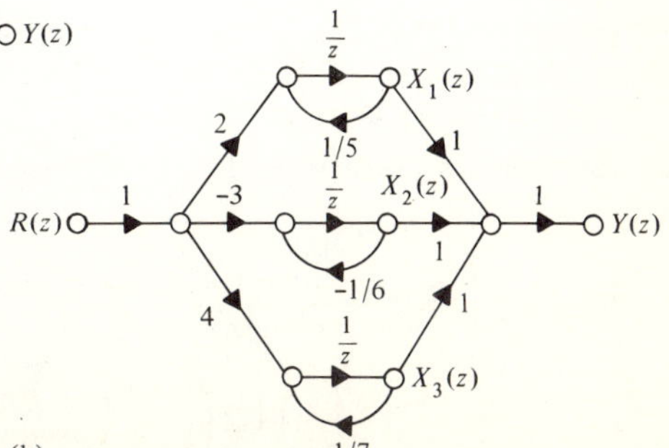

(b)

D8-17. For the system

$$\begin{bmatrix} x_1(k+1) \\ x_2(k+1) \end{bmatrix} = \begin{bmatrix} -2 & 1 \\ 1 & -3 \end{bmatrix} \begin{bmatrix} x_1(k) \\ x_2(k) \end{bmatrix} + \begin{bmatrix} 1 \\ 2 \end{bmatrix} r(k)$$

$$y(k) = [-1 \quad 1] \begin{bmatrix} x_1(k) \\ x_2(k) \end{bmatrix}$$

with

$$\begin{bmatrix} x_1(0) \\ x_2(0) \end{bmatrix} = \begin{bmatrix} 2 \\ 0 \end{bmatrix}$$

and $r(k) = u(k)$, find $y(0)$, $y(1)$, $y(2)$, and $y(3)$.

ans. $-2, 7, -24, 86$

8.9 DIGITIZING CONTROL SYSTEMS

8.9.1 Step-Invariant Approximation

An important technique in the design of digital control systems is to require that the unit step sequence response of a digital transmittance be samples of the continuous-time unit step response of a model analog transmittance. Often in practice, the analog transmittance is a working component of the system and it is desired to replace the analog component with a digital one that performs similarly. In Fig. 8-19a, an analog transmittance $G(s)$ and its unit step response $f_{\text{step}}(t)$ are indicated. The step-invariant digital approximation to $G(s)$, Fig. 8-19b, has a unit step response sequence $f_{\text{step}}(k)$ that consists of samples of the analog step response $f_{\text{step}}(t)$.

The conversion from $F_{\text{step}}(s)$ to $F_{\text{step}}(z)$, where the sequence consists of samples of the time function, is the sampling conversion of Sec. 8.5. For example, the analog transmittance

$$G(s) = \frac{-s+2}{s^2+3s+2}$$

has unit step response given by

$$F_{\text{step}}(s) = \frac{1}{s}\, G(s) = \frac{-s+2}{s(s^2+3s+2)}$$

$$= \frac{1}{s} - 3\left(\frac{1}{s+1}\right) + 2\left(\frac{1}{s+2}\right)$$

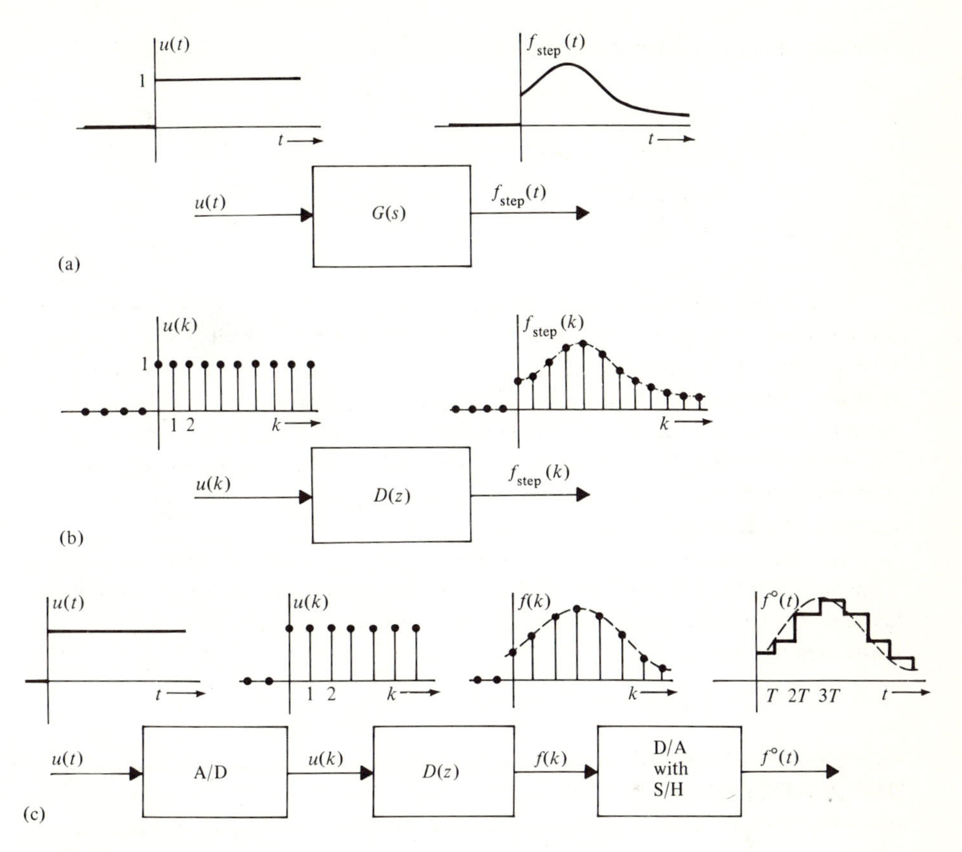

FIGURE 8-19
Step-invariant approximation. (a) Model analog transmittance and unit step response. (b) Digital system derived from $G(s)$ using the step-invariant approximation. (c) Digitization of the analog transmittance.

Samples of this step response at a sampling interval $T=0.1$ are given by

$$F_{\text{step}}(z)=\frac{z}{z-1}-3\left(\frac{z}{z-e^{-0.1}}\right)+2\left(\frac{z}{z-e^{-0.2}}\right)$$

$$=\frac{z}{z-1}+\frac{-3z}{z-0.905}+\frac{2z}{z-0.82}$$

so the step-invariant digital system with $T=0.1$ is to have unit step sequence response

$$F_{\text{step}}(z)=U(z)D(z)=\frac{z}{z-1}\,D(z)$$

The step-invariant z-transmittance for $T=0.1$ is thus

$$D(z)=1+\frac{-3(z-1)}{z-0.905}+\frac{2(z-1)}{z-0.82}$$

$$=\frac{-0.07z+0.092}{(z-0.905)(z-0.82)}$$

The digital system with analog input and analog output in Fig. 8-19c then approximates the performance of $G(s)$. Generally, the smaller the sampling interval, the better the approximation. Further improvement in the approximation may be obtained if desired by further smoothing of the output waveform rather than simply holding it between samples. Step-invariance design has the properties that the digital transmittance is stable if the analog model is stable and that the resulting z-transmittance is of the same order as the original continuous-time transmittance.

Step invariance is but one of several useful approximations of an analog transmittance by a digital one. Other commonly used approximations include impulse invariance, ramp invariance, matched z-transformation, and bilinear transformation.

DRILL PROBLEM

D8-18. Find the step-invariant approximations, for the given sampling interval T, to the following continuous-time transmittances:

(a) $G(s)=\dfrac{1}{s}$ $T=0.5$

ans. $0.5/(z-1)$

(b) $G(s)=\dfrac{2}{s+4}$ $T=0.1$

ans. $0.165/(z-0.67)$

(c) $G(s)=\dfrac{2}{s+4}$ $T=0.03$

ans. $0.057/(z-0.887)$

(d) $G(s)=\dfrac{e^{-s}}{s+3}$ $T=0.1$

ans. $0.086/z^{10}(z-0.74)$

8.9.2 z-Transfer Functions of Systems with Analog Measurements

Often in digital control system analysis and design, the situation of Fig. 8-20a occurs, where it is desired to find the z-transfer function of a system or subsystem with discrete-time input and output but intervening analog components. For this basic situation, the Laplace transform of the impulse train is

$$F^*(s) = F(z)|_{z=e^{sT}}$$

and the analog output $y(t)$ is given by

$$Y(s) = F^*(s)D(s) = [F(z)|_{z=e^{sT}}]D(s)$$

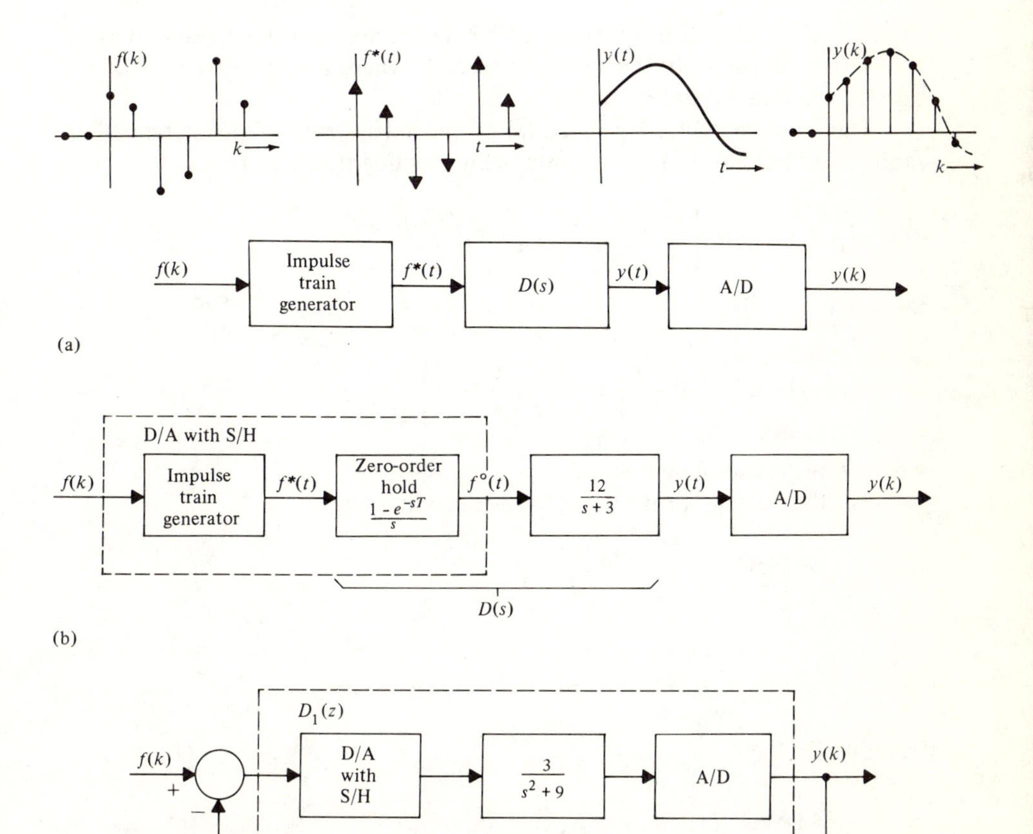

FIGURE 8-20
Digital subsystems with analog components. (a) Basic arrangement. (b) An example subsystem. (c) An example system with digital feedback.

To obtain the z-transform $Y(z)$ of the output sequence, the time shift terms are separated from the rational part of the transform and the usual z-transform term substitutions made. The e^{sT} terms involved in $F^*(s)$ are part of the time shift portion of $Y(s)$, so in forming $Y(z)$, the e^{sT} terms in $F^*(s)$ are simply converted back to $e^{sT}=z$, giving

$$Y(z)=F(z)D(z)$$

The z-transmittance of the basic subsystem is thus

$$\frac{Y(z)}{F(z)}=D(z)$$

That is, the analog transmittance $D(s)$ is converted to the corresponding z-transmittance just as in the sampling process where a signal given by $D(s)$ is sample, yielding a z-transform $D(z)$.

The example digital-input, digital-output subsystem of Fig. 8-20b involves sample-and-hold and so has the intervening analog transmittance

$$D(s)=(1-e^{-sT})\left[\frac{12}{s(s+3)}\right]=(1-e^{-sT})\left(\frac{4}{s}-\frac{4}{s+3}\right)$$

for which

$$D(z)=(1-z^{-1})\left[4\left(\frac{z}{z-1}\right)-4\left(\frac{z}{z-e^{-3T}}\right)\right]$$

where T is the sampling interval.

For the system of Fig. 8-20c, the z-transmittance

$$D_1(z)=\frac{Y(z)}{E(z)}$$

is given by

$$D_1(s)=\frac{1-e^{-sT}}{s}\left(\frac{3}{s^2+9}\right)$$
$$=(1-e^{-sT})\left[\frac{\frac{1}{3}}{s}+\frac{-\frac{1}{3}s}{s^2+9}\right]$$

$$D_1(z)=(1-z^{-1})\left\{\frac{1}{3}\left(\frac{z}{z-1}\right)-\frac{1}{3}\left[\frac{z(z-\cos 3T)}{z^2-2z\cos 3T+1}\right]\right\}$$
$$=\frac{1}{3}\left(\frac{1-z}{z}\right)\frac{(1-\cos 3T)(z^2+z)}{(z-1)(z^2-2z\cos 3T+1)}$$
$$=\frac{\frac{1}{3}(1-\cos 3T)(z+1)}{z^2-2z\cos 3T+1}$$

The overall feedback system z-transfer function is

$$D(z)=\frac{Y(z)}{F(z)}=\frac{D_1(z)}{1+D_1(z)}$$

$$=\frac{\frac{1}{3}(1-\cos 3T)(z+1)}{z^2+[\frac{1}{3}-\frac{7}{3}\cos 3T]z+[\frac{4}{3}-\frac{1}{3}\cos 3T]}$$

DRILL PROBLEM

D8-19. Find the z-transfer functions of the following systems in terms of the sampling interval T.

ans. (a) $0.6(e^{-5T}-1)/(z-e^{-5T})$

(b) $\dfrac{2(z-1)(e^{-T}-e^{-2T})}{(z-e^{-T})(z-e^{-2T})}$

(c) $(2e^{-4T}-2)/(-z+3e^{-4T}-2)$

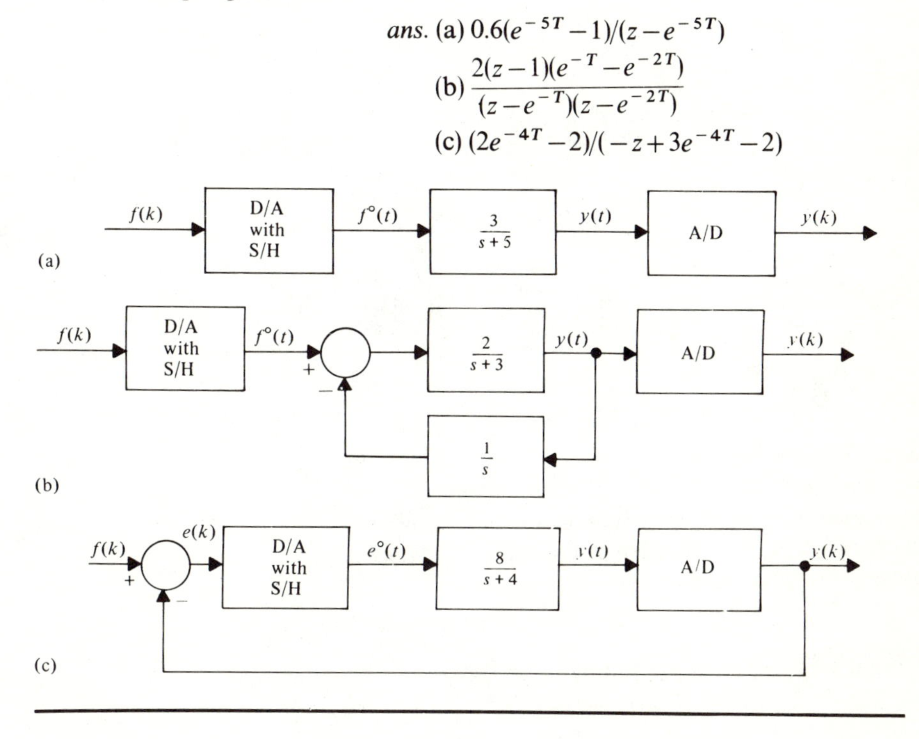

8.9.3 A Design Example

Figure 8-21a shows a unity feedback continuous-time system consisting of a controller with compensator $G_1(s)$ and the plant $G_2(s)$. The system transfer function is

$$T(s)=\frac{G_1(s)G_2(s)}{1+G_1(s)G_2(s)}=\frac{1}{s^2+s+1}$$

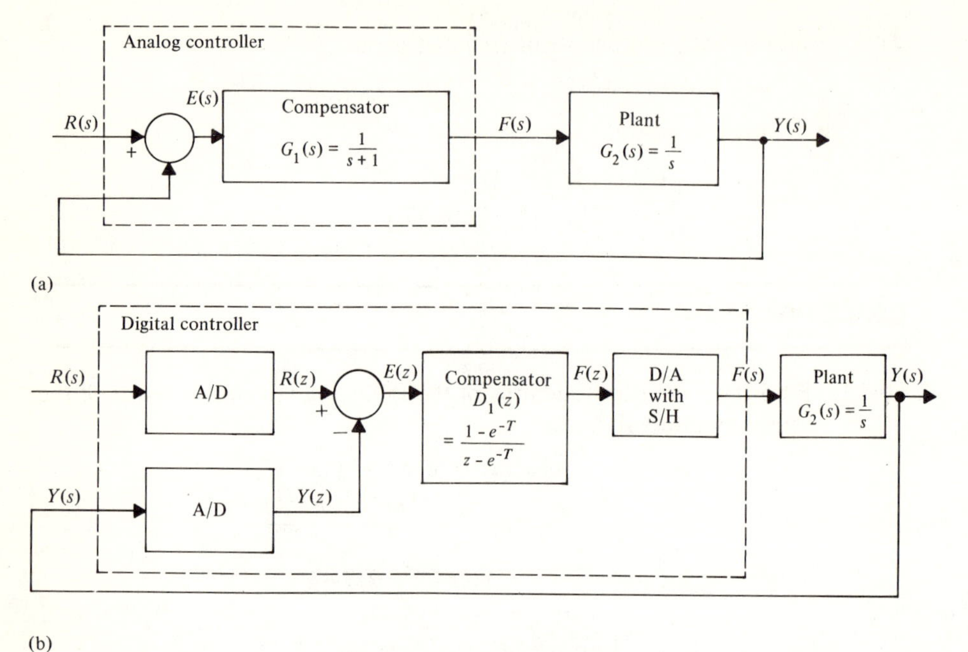

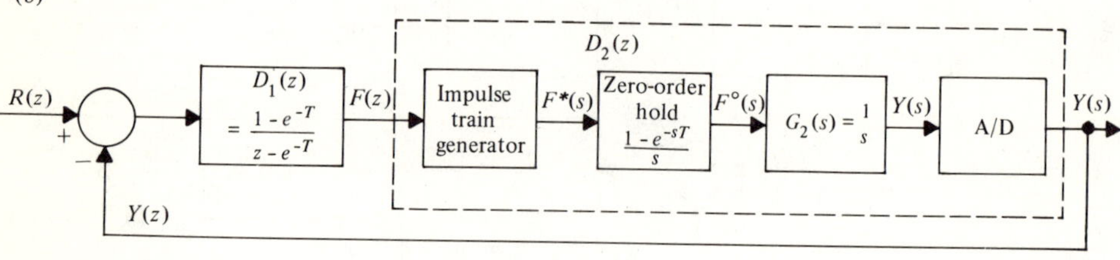

(c)

FIGURE 8-21
Converting analog to digital control. (a) A continuous-time feedback system. (b) Conversion to digital control. (c) Relations between the digital signals.

Its unit step response is given by

$$Y_{\text{step}}(s) = \frac{1}{s}\,T(s) = \frac{1}{s(s^2+s+1)}$$

$$= \frac{1}{s} + \frac{0.58e^{-j150^\circ}}{s+0.5+j0.866} + \frac{0.58e^{j150^\circ}}{s+0.5-j0.866}$$

$$Y_{\text{step}}(t) = [1 + 1.16e^{-0.5t}\cos(0.866t + 150^\circ)]u(t)$$

To convert the analog controller to a digital one, analog-to-digital converters are used to sample the input and the output signals, a digital-to-analog converter drives the plant, and the compensator transmittance is replaced by a

z-transmittance as in Fig. 8-21b. The unit step response of the original analog is given by

$$F_{\text{step}}(s)=\frac{1}{s}\left(\frac{1}{s+1}\right)=\frac{1}{s}-\frac{1}{s+1}$$

$$f_{\text{step}}(t)=(1-e^{-t})u(t)$$

Requiring that the unit step response of the discrete-time compensator consist of samples of the analog unit step response gives, in terms of the sampling interval T,

$$f_{\text{step}}(k)=(1-e^{-kT})u(k)=u(k)-(e^{-T})^k u(k)$$

$$F_{\text{step}}(z)=\frac{z}{z-1}-\frac{z}{z-e^{-T}}$$

$$=\frac{z}{z-1}\left(1-\frac{z-1}{z-e^{-T}}\right)=U(z)D_1(z)$$

Hence the step-invariant design for $D_1(z)$ is

$$D_1(z)=1-\frac{z+1}{z-e^{-T}}=\frac{1-e^{-T}}{z-e^{-T}}$$

The relationships between the digital signals in the system with digital control are shown in Fig. 8-21c. The z-transmittance relating the sequence $y(k)$ to $f(k)$ is given by

$$D_2(s)=\frac{1-e^{-sT}}{s}\left(\frac{1}{s}\right)=(1-e^{-sT})\frac{1}{s^2}$$

$$D_2(z)=(1-z^{-1})\frac{Tz}{(z-1)^2}=\frac{T}{z-1}$$

where T is the sampling interval. The feedback system thus has overall z-transfer function

$$D(z)=\frac{Y(z)}{R(z)}=\frac{D_1(z)D_2(z)}{1+D_1(z)D_2(z)}$$

$$=\frac{\dfrac{1-e^{-T}}{z-e^{-T}}\left(\dfrac{T}{z-1}\right)}{1+\dfrac{1-e^{-T}}{z-e^{-T}}\left(\dfrac{T}{z-1}\right)}$$

$$=\frac{T(1-e^{-T})}{z^2-(1+e^{-T})z+T(1-e^{-T})+e^{-T}}$$

Response samples of the step response of the digitized system for various sampling rates are plotted in Fig. 8-22, where they are compared with the continuous-time step response of the original analog system. For a one-half second sampling interval, $T=0.5$,

$$Y(z)=U(z)D(z)=\frac{0.197z}{z^3-2.607z+2.41z-0.804}$$
$$=0.197z^{-2}+0.51z^{-3}+0.86z^{-4}+\cdots$$

For $T=0.2$,

$$Y(z)=U(z)D(z)=\frac{0.036z}{z^3-2.819z^2+2.674z-1.655}$$
$$=0.036z^{-2}+0.102z^{-3}+0.191z^{-4}+\cdots$$

For $T=0.05$,

$$Y(z)=U(z)D(z)=\frac{0.0024z}{z^3-2.95z^2+1.95z+0.954}$$
$$=0.0024z^{-2}+0.00708z^{-3}+0.014z^{-4}+\cdots$$

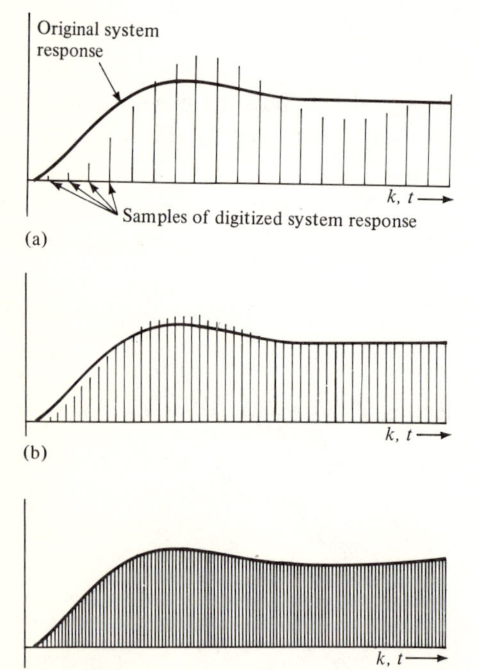

FIGURE 8-22
Response of the digitized system. (a) $T=0.5$. (b) $T=0.2$. (c) $T=0.05$.

The higher the sampling rate, the more closely the digital system approximates the behavior of the original analog system.

Since $G_1(s)$ is an integrator driven by the piecewise constant sample-and-hold (S/H) waveform $f^0(t)$ and since the continuous-time output $y(t)$ passes through the sample points $y(k)$, $y(t)$ is a piecewise linear signal through the sample points.

The final value of the unit step response is, from the formula in Table 8-6,

$$\lim_{k\to\infty} y(k) = \lim_{z\to 1}\left[\frac{z-1}{z}Y(z)\right]$$
$$= \lim_{z\to 1}\left[\frac{z-1}{z}\frac{z}{z-1}D(z)\right]$$
$$= \lim_{z\to 1}\left[\frac{T(1-e^{-T})}{z^2-(1+e^{-T})z+T(1-e^{-T})+e^{-T}}\right]$$
$$= \frac{T(1-e^{-T})}{T(1-e^{-T})} = 1$$

8.10 DIRECT DIGITAL DESIGN

8.10.1 Steady State Response

If a discrete-time signal $f(k)$ reaches a finite, constant steady state value, it is given by

$$\lim_{k\to 0} f(k) = \lim_{z\to 1}\frac{z-1}{z}F(z)$$

which is the discrete-time version of the final value theorem. If $F(z)/z$ were expanded into partial fractions,

$$\frac{F(z)}{z} = \frac{K_1}{z-1} + \cdots$$

the residue of the $(z-1)$ pole represents the constant term in the sequence $f(k)$, which may be found by multiplying $F(z)/z$ by $(z-1)$ and evaluating at $z=1$.

For example, the sequence with z-transform

$$F(z) = \frac{3z^2+4}{z^2-\frac{1}{2}z-\frac{1}{2}}$$
$$= \frac{3z^2+4}{(z-1)(z+\frac{1}{2})} = \frac{K_1}{z-1} + \frac{K_2}{z+\frac{1}{2}}$$

has steady state value given by

$$\lim_{k\to 0} f(k) = \lim_{z\to 1} \frac{z-1}{z} F(z) = \lim_{z\to 1} \frac{3z^2+4}{z^2+\frac{1}{2}z}$$
$$= \tfrac{14}{3}$$

The sequence with z-transform

$$F(z) = \frac{z^3-4z+1}{z^3+z^2-2z} = \frac{K_1}{z-1} + \frac{K_2}{z} + \frac{K_3}{z+2}$$

does not reach a finite steady state because the $K_3/(z+2)$ term in its partial fraction expansion represents an expanding term in the sequence $f(k)$. The limit exists and is finite, however:

$$\lim_{z\to 1} \frac{z-1}{z} F(z) = \lim_{z\to 1} \frac{z^3-4z+1}{z^3+2z^2} = -\frac{2}{3}$$

For a system with stable z-transfer function

$$D(z) = \frac{4z^2+8z-1}{(3z+2)(4z-1)}$$

the error between input and output, when the input is a unit step sequence, is

$$E(z) = R(z) - Y(z) = R(z)[1 - D(z)]$$
$$= \frac{z}{z-1}\left(1 - \frac{4z^2+8z-1}{12z^2+5z-2}\right)$$

which has a final value

$$\lim_{k\to 0} e(k) = \lim_{z\to 1} \frac{z-1}{z} E(z)$$
$$= \lim_{z\to 1} \frac{8z^2-3z-1}{12z^2+5z-2}$$
$$= \tfrac{4}{15}$$

DRILL PROBLEM

D8-20. Each of the following systems is stable. Find the steady state error between input and output for each when the input is a unit step sequence.

ans. (a) 0.2
(b) 0
(c) −0.25

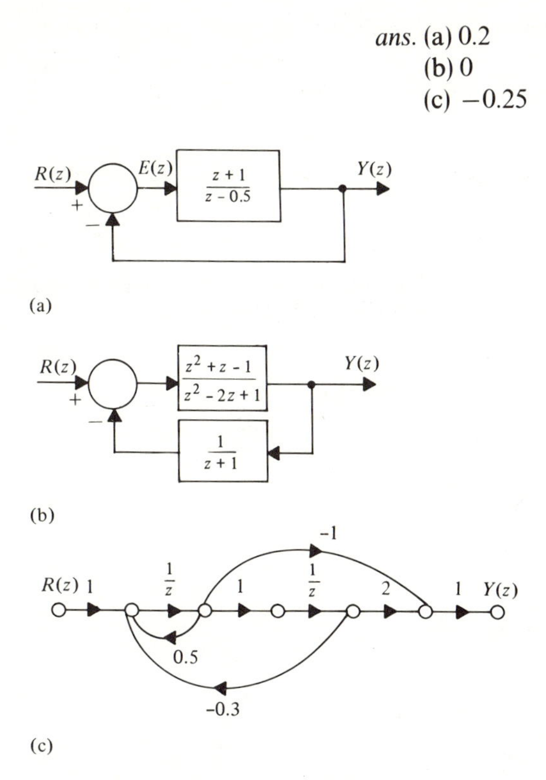

8.10.2 Deadbeat Systems

A digital system with all of its poles at $z=0$ is termed *deadbeat*. Such systems have a remarkable property: Their pulse response is zero after n steps, where n is the order of the system. For example, the system with z-transfer function

$$D(z)=\frac{z^3+3z^2-2z+4}{z^3}$$

is deadbeat. Its unit pulse response is

$$\begin{aligned} Y_{\text{pulse}}(z)&=1\cdot D(z)\\ &=1+3z^{-1}-2z^{-2}+4z^{-3}\\ y_{\text{pulse}}(k)&=\delta(k)+3\delta(k-1)-2\delta(k-2)+4\delta(k-3) \end{aligned}$$

Deadbeat systems are also commonly termed *finite-duration impulse response* (FIR) systems.

As pulse response is representative of the natural component of a system's response in general, deadbeat systems have a natural response that goes to zero after n steps. There is no counterpart in analog systems; analog natural response may only decay asymptotically to zero. In many practical situations, digital systems are designed to be deadbeat if possible.

DRILL PROBLEM

D8-21. Find the response $y(k)$ of each of the following systems if the input is a unit step sequence and the initial conditions are zero.

ans. (a) $3u(k-1)+u(k-3)$
(b) $\frac{5}{6}u(k-1)-\frac{1}{6}u(k-2)$
(c) $u(k-2)$

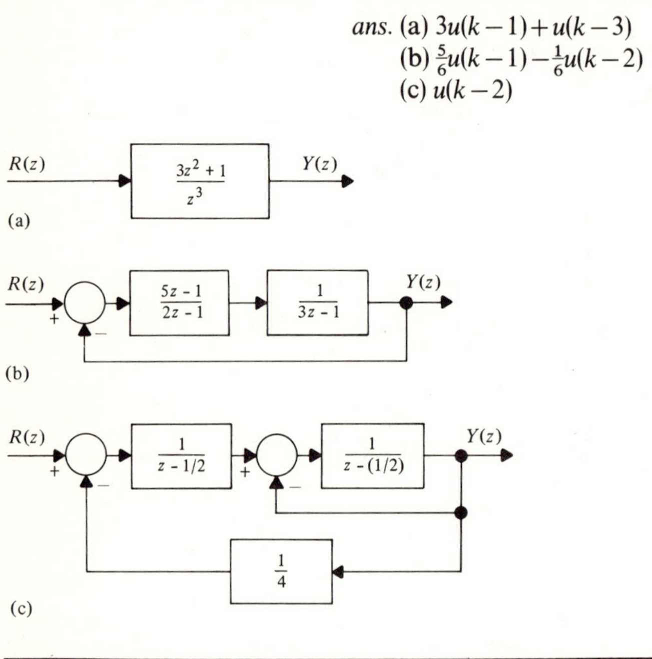

8.10.3 A Design Example

Figure 8-23a shows a simplified model of a satellite-tracking control system. Computer-generated digital commands, based on orbital calculations, are applied to the system consisting of a digital controller driving the analog positioning subsystem. The diagram is rearranged in Fig. 8-23b to show the relations between digital signals. The sampling interval is $T=0.1$. It is desired to design the controller transmittance $D_1(z)$ so that, if possible,

1. the overall digital system $D(z)=Y(z)/R(z)$ is deadbeat; and
2. the error between output and input, $E(z)$, to a step input $R(z)$ is zero in steady state.

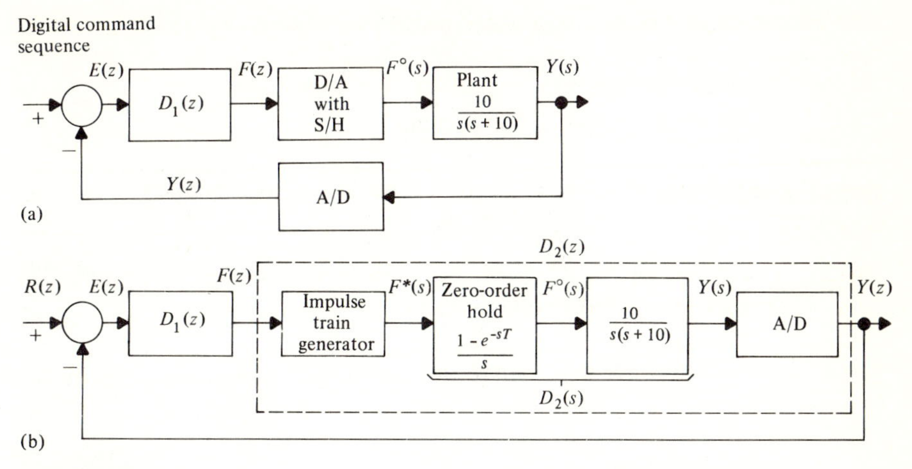

FIGURE 8-23
Satellite tracking system. (a) System block diagram. (b) Block diagram rearranged to show digital relations.

In addition, it is important that the analog position output $y(t)$ be well behaved between samples, particularly as a constant steady state is approached.

The analog transmittance $D_2(s)$ in Fig. 8-23b is

$$D_2(s)=\frac{10(1-e^{-0.1s})}{s^2(s+10)}=(1-e^{-0.1s})\left(\frac{-\frac{1}{10}}{s}+\frac{1}{s^2}+\frac{\frac{1}{10}}{s+10}\right)$$

hence the corresponding z-transmittance is

$$\begin{aligned}D_2(z)&=(1-z^{-1})\left[\frac{-\frac{1}{10}z}{z-1}+\frac{0.1z}{(z-1)^2}+\frac{\frac{1}{10}z}{z-0.368}\right]\\&=\frac{(z-1)(0.0368z^2+0.0264z)}{z(z-1)^2(z-0.368)}\\&=\frac{0.0368z+0.0264}{(z-1)(z-0.368)}\end{aligned}$$

In terms of the controller transmittance $D_1(z)$, the overall system transfer function is

$$\begin{aligned}D(z)&=\frac{D_1(z)D_2(z)}{1+D_1(z)D_2(z)}\\&=\frac{D_1(z)\dfrac{0.0368z+0.0264}{(z-1)(z-0.368)}}{1+D_1(z)\dfrac{0.0368z+0.0264}{(z-1)(z-0.368)}}\end{aligned}$$

Let $D(z)$ have n poles, all of which are at $z=0$. Then

$$1+D_1(z)\frac{0.0368z+0.0264}{(z-1)(z-0.368)}=\frac{z^n}{\text{polynomial}}$$

where "polynomial" denotes an as-yet-unspecified polynomial in z. Solving for $D_1(z)$,

$$D_1(z)=\frac{(z-1)(z-0.368)}{0.0368z+0.0264}\left(\frac{z^n}{\text{polynomial}}-1\right)$$

Any such z-transmittance $D_1(z)$ will result in an overall system that is deadbeat:

$$D(z)=\frac{\dfrac{z^n}{\text{polynomial}}-1}{1+\left(\dfrac{z^n}{\text{polynomial}}-1\right)}$$
$$=\frac{z^n-\text{polynomial}}{z^n}$$

The error between input and output is

$$E(z)=R(z)-Y(z)=R(z)[1-D(z)]$$
$$=R(z)\left(1-\frac{z^n-\text{polynomial}}{z^n}\right)$$
$$=R(z)\left(\frac{\text{polynomial}}{z^n}\right)$$

For a step input

$$R(z)=\frac{z}{z-1}$$

the error signal has z-transform

$$E(z)=\frac{1}{z-1}\left(\frac{\text{polynomial}}{z^n}\right)$$

and steady state value given by

$$\lim_{k\to\infty} e(k)=\lim_{z\to 1}\left(\frac{\text{polynomial}}{z^n}\right)$$
$$=\lim_{z\to 1}(\text{polynomial})$$

To summarize:

1. $D_1(z)$ must be of the form

$$D_1(z)=\frac{(z-1)(z-0.368)}{0.0368z+0.0264}\left(\frac{z^n}{\text{polynomial}}-1\right)$$

in order for the overall system to be deadbeat.

2. The unspecified polynomial must have the property

$$\lim_{z\to 1}\,(\text{polynomial})=0$$

for the system's steady state step error to be zero.

The choice of $n=1$ and

$$\text{polynomial}=z-1$$

is simple, has property 2, and results in a simplification of $D_1(z)$:

$$D_1(z)=\frac{z-0.0368}{0.0368z+0.0264}$$

The overall transfer function for this choice is

$$D(z)=\frac{Y(z)}{R(z)}=\frac{1}{z}$$

The digital step response of this system is given by

$$Y(z)=R(z)D(z)=\frac{z}{z-1}\left(\frac{1}{z}\right)$$

$$\frac{Y(z)}{z}=\frac{1}{z(z-1)}=\frac{-1}{z}+\frac{1}{z-1}$$

$$Y(z)=-1+\frac{1}{z-1}$$

$$y(k)=-\delta(k)+u(k)$$

For a step command input, the output reaches its final value in one step, a shown in Fig. 8-24a.

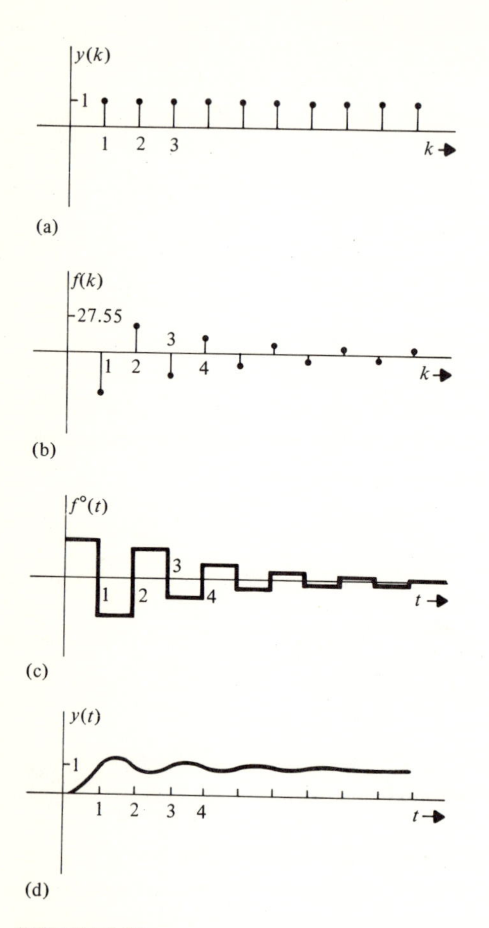

FIGURE 8-24
Tracking system step response. (a) Samples of the output position. (b) Samples of the controller output. (c) Sampled-and-held controller output. (d) Output position as a function of time.

For the step input, the controller's digital signal $F(z)$ is given by

$$F(z) = [R(z) - Y(z)]D_1(z)$$

$$= \left(\frac{z}{z-1} - \frac{1}{z-1}\right)\frac{z-0.368}{0.0368z+0.0264}$$

$$= \frac{z-0.368}{0.0368z+0.0264}$$

$$\frac{F(z)}{z} = \frac{27.17z-10}{z(z+0.717)} = \frac{-13.95}{z} + \frac{+41.1}{z+0.717}$$

$$F(z) = -13.95 + 41.1\,\frac{z}{z+0.717}$$

$$f(k) = -13.95\delta(k) + 41.1(-0.717)^k u(k)$$

This signal, the sampled-and-held waveform $f^0(t)$, and the output $y(t)$ are plotted in Fig. 8-24.

8.11 SUMMARY

The enormous growth of digital technology promises to revolutionize control system design. Both analog and digital concepts are likely to be of equal importance in the future, but system controllers are expected to be predominantly digital. Computer-related topics such as the following are becoming increasingly relevant to the control system designer:

Number systems
Finite-precision arithmetic
Computer organization and architecture
Logic design
Programming languages
Real-time processing
Operating systems

In addition, microprocessors offer remarkable performance at very low cost.

Analog-to-digital and digital-to-analog converters provide the interface between digital processing and a largely analog macroscopic world. A/D converters are used to periodically sample analog waveforms, providing a sequence of numbers, the signal samples, for computer processing. Analog signals are generated from digital output sequences via D/A converters.

For discrete-time sequences, z-transformation plays much the same role as does the Laplace transform for continuous-time functions. Some important sequences and their z-transforms are listed in Table 8-5. Key z-transform properties are given in Table 8-6. The first several terms of a sequence may be found from its z-transform by long division. To find the sequence from the transform as a function of the step index k, a rational transform $F(z)$ is divided by the variable z and expanded into partial fractions:

$$\frac{F(z)}{z} = \frac{K_1}{z - z_1} + \frac{K_2}{z - z_2} + \cdots$$

Then

$$F(z) = \frac{K_1 z}{z - z_1} + \frac{K_2 z}{z - z_2} + \cdots$$

$$f(k) = [K_1 z_1^k + K_2 z_2^k + \cdots]u(k)$$

The relation between an analog signal $f(t)$ and its sample sequence

$$f(k)=f(t=kT)$$

is found by relating the z-transform $F(z)$ of the sequence to the Laplace transform $F(s)$ of the analog signal. Each term in the partial fraction expansion of $F(s)$ is replaced by the corresponding z-transform expansion term from Table 8-5 to form $F(z)$.

Approximate reconstruction of an analog signal $f(t)$ from its samples is commonly performed by a D/A converter with sample-and-hold. The S/H waveform $f^o(t)$ is conveniently represented by the conversion of digital samples to an analog impulse train, followed by a zero-order hold transmittance. More accurate reconstruction may be done with higher-order holds, involving more than a single sample, and low-pass filtering. The sampling theorem states conditions for which a band-limited signal is uniquely represented by its samples.

Discrete-time systems are described by difference equations,

$$\begin{aligned} y(k+n)+a_{n-1}y(k+n-1)+a_{n-2}y(k+n-2)+\cdots \\ +a_1 y(k+1)+a_0 y(k)=b_m r(k+m)+b_{m-1}r(k+m-1) \\ +\cdots+b_1 r(k+1)+b_0 r(k) \end{aligned}$$

where $r(k)$ is the input sequence and $y(k)$ is the output sequence. The z-transfer function of such a system is

$$\begin{aligned} D(z) &= \left.\frac{Y(z)}{R(z)}\right|_{\text{initial conditions}=0} \\ &= \frac{b_m z^m+b_{m-1}z^{m-1}+\cdots+b_1 z+b_0}{z^n+a_{n-1}z^{n-1}+a_{n-2}z^{n-2}+\cdots+a_1 z+a_0} \end{aligned}$$

Block diagrams and signal flow graphs for discrete-time systems, described in terms of z-transforms, are manipulated in the same way as are continuous-time system descriptions in terms of Laplace transforms.

A discrete-time system is termed stable if and only if all poles of its z-transfer function are within the unit circle on the complex plane. The bilinear transformation

$$z=\frac{1+W}{1-W}$$

maps the unit circle to the left half-plane on the complex plane and so is very useful in relating discrete-time situations to equivalent continuous-time ones. With the above substitution for z in a rational z-transfer function, Routh–Hurwitz methods may be applied to determine stability.

State equations offer a systematic and powerful method of system representation. In terms of these,

$$\mathbf{x}(k+1)=\mathbf{F}\mathbf{x}(k)+\mathbf{G}r(k)$$
$$\mathbf{y}(k)=\mathbf{H}\mathbf{x}(k)$$

The z-transfer function matrix of a system is

$$\mathbf{D}(z)=\mathbf{H}(z\mathbf{I}-\mathbf{F})^{-1}\mathbf{G}$$

and the system is stable if and only if all roots of the characteristic equation

$$|z\mathbf{I}-\mathbf{F}|=0$$

are within the unit circle on the complex plane. An nth-order system is completely controllable if and only if

$$M_c=[\mathbf{G} \mid \mathbf{F}\mathbf{G} \mid \mathbf{F}^2\mathbf{G} \mid \cdots \mid \mathbf{F}^{n-1}\mathbf{G}]$$

is of full rank and is completely observable if and only if

$$M_o=\begin{bmatrix} \mathbf{H} \\ \hline \mathbf{H}\mathbf{F} \\ \hline \mathbf{H}\mathbf{F}^2 \\ \hline \vdots \\ \hline \mathbf{H}\mathbf{F}^{n-1} \end{bmatrix}$$

is of full rank.

Change of stable variables

$$\mathbf{x}'(k)=\mathbf{P}\mathbf{x}(k)$$

gives alternative state variable representations of a system and shows available design freedom. If the system characteristic roots are not repeated, the methods of Sec. 7.4 may be used to find a set of state variables for which the state equations are decoupled from one another.

A common analysis problem involves a system containing analog components but with discrete-time input and output. To find the z-transmittance, the system is converted to the form where an impulse train generator drives a continuous-time transmittance $D(s)$, followed by a sampler. The transmittance $D(s)$ is separated into delay terms and rational terms, with the rational terms expanded into partial fractions. Substitutions

$$e^{sT}\rightarrow z$$

into the delay terms and

$$\frac{K}{s+a}\rightarrow\frac{Kz}{z-e^{aT}}$$

for the partial fractions yields the z-transmittance $D(z)$.

One design method for digital control involves replacing a model analog subsystem with a digital subsystem consisting of A/D converter, z-transmittance, and D/A converter. For the step-invariant approximation, the z-transmittance is required to have a step response consisting of samples of the step response of the analog subsystem.

Digital control system design may also be done directly, in terms of overall system performance requirements and objectives.

REFERENCES

Computer Processing

Gothmann, W. H. *Digital Electronics, an Introduction to Theory and Practice.* Englewood Cliffs, N.J.: Prentice-Hall, 1977.

Mano, M. M. *Digital Logic and Computer Design.* Englewood Cliffs, N.J.: Prentice-Hall, 1979.

Osborne, A. *An Introduction to Microcomputers*, 2nd ed. New York: McGraw-Hill, 1980.

Peatman, J. B. *Microcomputer-Based Design.* New York: McGraw-Hill, 1977.

Sampling and Reconstruction

Cadzow, J. A. *Discrete-Time Systems, an Introduction with Interdisciplinary Applications.* Englewood Cliffs, N.J.: Prentice-Hall, 1973.

Hamming, R. W. *Digital Filters.* Englewood Cliffs, N.J.: Prentice-Hall, 1977.

Jury, E. I. *Theory and Application of the z-Transform Method.* New York: Wiley, 1964.

Oliver, B. M.; Pierce, J. R.; and Shannon, C. E. "The Philosophy of PCM." *Proc. IRE* 36 (November 1948): 1324–31.

Stearns, S. D. *Digital Signal Analysis.* Rochelle Park, N.J.: Hayden, 1975.

Tretter, S. A. *Introduction to Discrete-Time Signal Processing.* New York: Wiley, 1976.

Discrete-Time Systems and Filtering

Chen, C.-T. *One-Dimensional Digital Signal Processing.* New York: Marcel Dekker, 1979.

Oppenheim, A. V., and Schafer, R. W. *Digital Signal Processing.* Englewood Cliffs, N.J.: Prentice-Hall, 1975.

Peled, A., and Liu, B. *Digital Signal Processing: Theory, Design and Implementation.* New York: Wiley, 1976.

Rabiner, L. R., and Gold, B. *Theory and Application of Digital Signal Processing.* Englewood Cliffs, N.J.: Prentice-Hall, 1975.

Schwartz, M., and Shaw, L. *Signal Processing.* New York: McGraw-Hill, 1975.

Stanley, W. D. *Digital Signal Processing.* Reston, Va.: Reston, 1975.

Digital Control

Cadzow, J. A., and Martens, H. R. *Discrete-Time and Computer Control Systems.* Englewood Cliffs, N.J.: Prentice-Hall, 1970.

Franklin, G. F., and Powell, J. D. *Digital Control of Dynamic Systems.* Reading, Mass.: Addison-Wesley, 1980.

Freeman, H. *Discrete-Time Systems.* New York: Wiley, 1965.

Kuo, B. C. *Digital Control Systems.* New York: Holt, Rinehart and Winston, 1980.

Monroe, A. J. *Digital Processes for Sampled Data Systems.* New York: Wiley, 1962.
Ragazzini, J. R., and Franklin, G. F. *Sampled Data Control Systems.* New York: McGraw-Hill, 1958.
Schwartz, R. J., and Friedland, B. *Linear Systems.* New York: Wiley, 1965.
Tou, J. T. *Digital and Sampled-data Control Systems.* New York: McGraw-Hill, 1959.

Digital Processing and Control Applications

Allan, R. "Busy Robots Spur Productivity." *IEEE Spec.*, September 1979, pp. 31–36.
______, "The Microcomputer Invades the Production Line." *IEEE Spec.*, January 1979, pp. 53–57.
Andrews, H. C., and Hunt, B. R. *Digital Image Restoration*, Englewood Cliffs, N.J.: Prentice-Hall, 1977.
Kahne, S. "Automatic Control by Distributed Intelligence." *Sci. Amer.*, June 1979, pp. 78–109.
Rabiner, L. R., and Schafer, R. W. *Digital Processing of Speech Signals.* Englewood Cliffs, N.J.: Prentice-Hall, 1978.
Oppenheim, A. V., ed. *Applications of Digital Signal Processing.* Englewood Cliffs, N.J.: Prentice-Hall, 1978.

PROBLEMS

1. List five home appliances for which microprocessor control is today useful and economically feasible. For each, describe the functions performed by the digital system.
2. Carefully describe functions that a microprocessor-based control system might perform in each of the following. Specify the signals to be sensed and the quantities to be controlled.
   (a) An automobile
   (b) A hotel
   (c) Aboard a ship
   (d) An electric power generating plant
   (e) A hospital operating room
3. The analog signal

   $$f(t)=3+4\cos 50t$$

   is sampled at 0.01-sec intervals by an A/D converter preceded by a sample-and-hold (S/H) that freezes the sample while conversion takes place. Then the samples are reconverted to an analog signal with S/H. Sketch your visualization of signal $f(t)$, the first S/H waveform, the A/D samples, and the output waveform.
4. A bank account pays 9% interest per year, compounded monthly. Initially, a deposit of $1000 is made and thereafter $65 per month is deposited into the account each month. Describe the monthly bank balance as a function of the month k after the initial deposit.
5. Sketch the functions $f(t)$ and the samples $f(k)$ with the given sampling interval T:
   (a) $f(t)=3e^{-10t}u(t) \qquad T=0.1$
   (b) $f(t)=\left(3e^{-2t}\cos\dfrac{5\pi t}{4}\right)u(t) \qquad T=0.2$

(c) $f(t) = \left(2 \sin \frac{\pi t}{8}\right) u(t) \qquad T = 1$

(d) $f(t) = (4 + 3e^{0.1t})u(t) \qquad T = 0.5$

6. Find the z-transforms of the following sequences:
(a) $f(k) = u(k) - u(k-4)$
(b) $f(k) = 0.1ku(k) - 0.1ku(k-4)$

ans. $0.1(1 - z^{-4})z/(z-1)^2$

(c) $f(k) = \left[\cos\left(0.1k + \frac{\pi}{4}\right)\right] u(k)$
(d) $f(k) = [(-1)^k \sin 0.5k]u(k)$

ans. $-0.48z/(z^2 + 1.76z + 1)$

7. Find the z-transforms of the sequences consisting of samples of the following functions $f(t)$ with the given sampling interval T:
(a) $f(t) = 5tu(t) \qquad T = 0.5$
(b) $f(t) = 3u(t) - 4u(t) \sin 3t \qquad T = 0.2$

ans. $\frac{3z}{z-1} - \frac{2.26z}{z^2 - 1.65z + 1}$

(c) $15t^2e^{-10t}u(t) \qquad T = 4$
(d) $5u(t)e^{-3t} \cos(\pi t + 45^\circ) \qquad T = 0.1$

ans. $\frac{3.54(z^2 - 0.93z)}{z^2 - 1.4z + 0.55}$

8. Use long division to find $f(0)$, $f(1)$, $f(2)$, and $f(3)$ for each of the discrete-time signals with the following z-transforms:

(a) $F(z) = \frac{4z^2 - 3z + 6}{2z^2 + z - 1}$

(b) $F(z) = \frac{2z - 3}{z^2 - 0.5z + 1}$

ans. 0, 2, -2, -3

(c) $F(z) = \frac{z}{0.3z^3 - 0.1z^2 + 0.2z + 1}$

9. Find the inverse z-transforms:

(a) $F(z) = \frac{4}{z + \frac{1}{3}}$

(b) $F(z) = \frac{3z^2 - 2z}{4z^2 + 5z + 1}$

ans. $-(\frac{11}{3})(-\frac{1}{4})^k u(k) + \frac{20}{3}(-1)^k u(k)$

(c) $F(z) = \frac{z^2}{(2z+1)^2}$

(d) $F(z) = \frac{3z - 2}{z^2 + 1.5z + 0.5}$

ans. $-4\delta(k-1) - 10(-1)^k u(k) + 14(-\frac{1}{2})^k u(k)$

(e) $F(z)=\dfrac{1}{z^3-0.25z}$

10. For the following functions $f(t)$ and sampling periods T, find $f(k)$:
   (a) $f(t)=3e^{-10t}u(t) \qquad T=0.1$
   (b) $f(t)=3e^{-2t}u(t)\cos\dfrac{5\pi t}{4} \qquad T=0.2$

   ans. $3(0.67)^k u(k)\cos(0.785k)$

   (c) $f(t)=2u(t)\sin\dfrac{\pi t}{8} \qquad T=1$
   (d) $f(t)=(t^2e^{-7t}+5e^{-6t}\sin 10t)u(t) \qquad T=0.01$

   ans. $[10^{-4}k^2(0.93)^k+5(0.94)^k\sin 0.1k]u(k)$

11. For each of the following analog signals with given Laplace transform $F(s)$, find the z-transform $F(z)$ of the corresponding sample sequence with the given sampling interval T:

   (a) $F(s)=\dfrac{16}{s^2+4} \qquad T=0.1$

   (b) $F(s)=\dfrac{-2s+20}{s^3+9s^2+20s} \qquad T=0.05$

   ans. $\dfrac{z}{z-1}-\dfrac{7z}{z-0.819}+\dfrac{6z}{z-0.789}$

   (c) $F(s)=\dfrac{4e^{-0.5s}}{s^2+2s+1} \qquad T=0.25$

   (d) $F(s)=\dfrac{3e^{-s}}{s}-\dfrac{4}{s+2}+\dfrac{se^{-2s}}{s^2+9} \qquad T=0.2$

   ans. $\dfrac{3}{z^5-z^4}-\dfrac{4z}{z-0.67}+\dfrac{z-0.825}{z^{11}-1.65z^{10}+z^9}$

   (e) $F(s)=\dfrac{1-e^{-2s}-e^{-3s}}{s^2+2s+10} \qquad T=0.5$

12. For the following sequences $f(k)$, find the corresponding impulse train Laplace transforms $F^*(s)$ for the given sampling interval T:
   (a) $f(k)=(3\cos 2k-4\sin 2k)u(k) \qquad T=0.5$
   (b) $f(k)=[(-1)^k-1]u(k) \qquad T=0.1$

   ans. $-2e^{-0.1s}/(e^{-0.2s}-1)$

   (c) $f(k)=6(-0.5)^k u(k)\cos 3k \qquad T=1$
   (d) $f(k)=10ke^{-0.1k}u(k-3) \qquad T=0.5$

   ans. $9.05/e^{-s}(e^{0.5s}-0.905)^2$

   (e) $f(k)=ku(k)-ku(k-5) \qquad T=0.3$

13. The continuous-time function

$$f(t)=20-15\cos 1000t$$

is sampled at the rate of 100 samples/sec. It is then reconstructed from the impulse train $f^*(t)$, according to the sampling theorem, as if it were a signal band-limited at 50 Hz. Find the signal that results.

14. Find the z-transfer functions of the following discrete-time systems:
   (a) $y(k+2)+0.2y(k+1)-0.5y(k)=r(k)$
   (b) $y(k+3)-y(k)=4r(k+2)-3r(k)$

   ans. $(4z^2-3)/(z^3-1)$

   (c) $y(k+3)=0.75r(k+3)+0.25r(k+2)-0.25r(k+1)-0.75r(k)$
15. For

$$y(k+2)-0.5y(k)=r(k+1)-2r(k)$$

   recursively find $y(2)$, $y(3)$, $y(4)$, and $y(5)$ if

$$r(k)=(\tfrac{1}{2})^k$$
$$y(0)=0$$
$$y(1)=3$$

16. For discrete-time systems with the following z-transfer functions and input sequences, find the output sequences for $k=0$ and thereafter if the initial conditions are zero:

   (a) $D(z)=\dfrac{z}{z+\frac{1}{2}}$
   $r(k)=2\delta(k)$

The equipment in the foreground controls a small manufacturing plant production line. A/D and D/A conversion is done with the rack-mounted equipment to the left. Digital processing is performed by the small computer on the right. (Photo courtesy of Hewlett-Packard.)

(b) $D(z)=\dfrac{1}{z-2}$

$r(k)=3u(k)$

ans. $-3u(k)+3(2)^k u(k)$

(c) $D(z)=\dfrac{z}{z^2+1}$

$r(k)=(-1)^k$

(d) $D(z)=\dfrac{1}{z^2-\frac{1}{2}z}$

$r(k)=(\frac{1}{2})^k$

ans. $4\delta(k)-4(\frac{1}{2})^k u(k)+4k(\frac{1}{2})^k u(k)$

(e) $D(z)=\dfrac{z^2}{8z^2+6z+1}$

$r(k)=3(-\frac{1}{2})^k$

17. Reduce the block diagrams of Fig. P8-1, finding the system z-transfer functions.

ans. (b) $(2z^2+4z-20)/(2z^2+19z-21)$

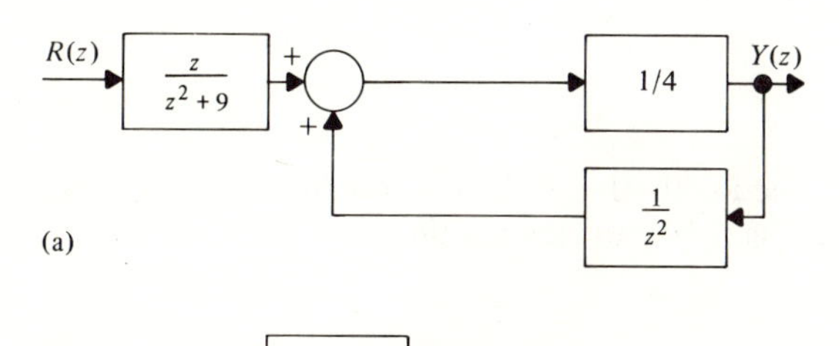

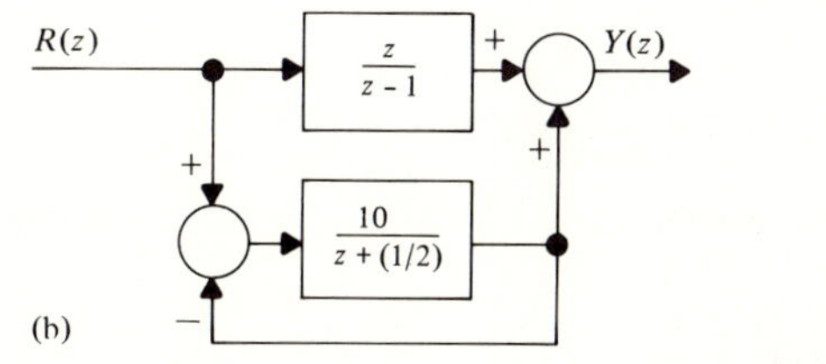

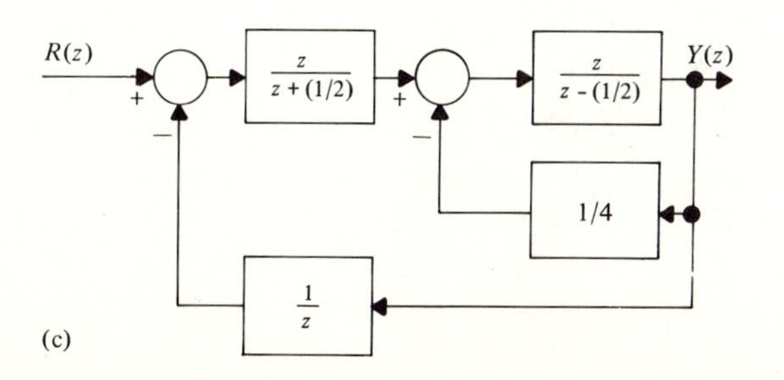

FIGURE P8-1

18. Use Mason's gain rule to find the z-transfer functions of the systems of Fig. P8-2.

ans. (b) $(-6z-13)/(6z^2+z-1)$

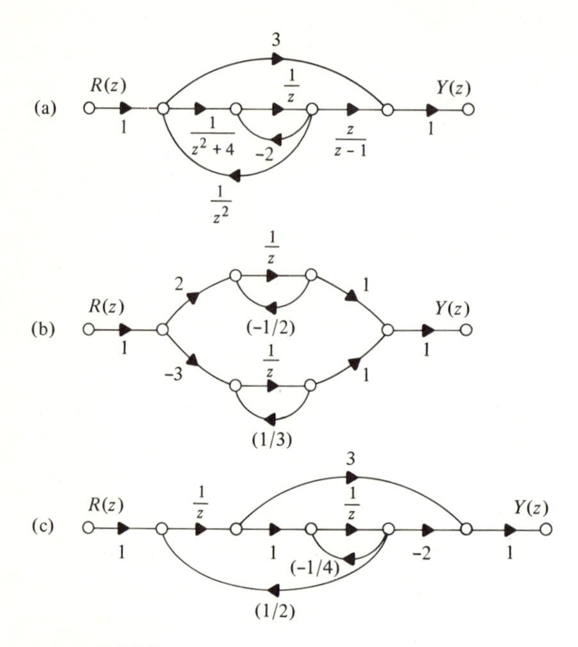

FIGURE P8-2

19. Use the bilinear transformation and Routh–Hurwitz tests to determine whether each of the discrete-time systems with the given z-transfer functions is stable:

(a) $D(z)=\dfrac{4z^3-3z}{z^3-0.8z^2+0.17z-0.01}$

(b) $D(z)=\dfrac{0.63z^2-8.73z+1}{z^3-2.3z^2+0.62z-0.04}$

ans. unstable

(c) $D(z)=\dfrac{0.02}{z^4-2.8z^3+1.77z^2-0.35z+0.02}$

(d) $D(z)=\dfrac{z^4+0.005}{z^4-1.3z^3+0.57z^2-0.095z+0.005}$

ans. stable

20. Draw discrete-time simulation diagrams in phase-variable canonical form for systems with the following z-transfer functions:

(a) $D(z)=\dfrac{z^2-z+2}{z^3-0.5z^2+z}$

(b) $D(z)=\dfrac{10z}{z^4+0.1z^3-0.2z^2+0.3z+0.4}$

(c) $D(z)=\dfrac{4z^2-6z+8}{5z^4+3z^3-2z^2+z+1}$

21. Find state equations in phase-variable matrix form for the discrete-time systems with the following z-transfer functions:

(a) $D(z)=\dfrac{3z+2}{z^2+z+4}$

(b) $D(z)=\dfrac{8}{6z^3+z^2-z}$

ans. $$\begin{bmatrix} x_1(k+1) \\ x_2(k+1) \\ x_3(k+1) \end{bmatrix}=\begin{bmatrix} 0 & 1 & 0 \\ 0 & 0 & 1 \\ 0 & \frac{1}{6} & -\frac{1}{6} \end{bmatrix}\begin{bmatrix} x_1(k) \\ x_2(k) \\ x_3(k) \end{bmatrix}+\begin{bmatrix} 0 \\ 0 \\ 1 \end{bmatrix} r(k)$$

$$y(k)=\begin{bmatrix} \frac{4}{3} & 0 & 0 \end{bmatrix}\begin{bmatrix} x_1(k) \\ x_2(k) \\ x_3(k) \end{bmatrix}$$

(c) $D(z)=\dfrac{4z^2-3}{z^3-0.5z^2+1}$

22. For each of the following discrete-time systems, with the indicated input and initial conditions, recursively find the state vectors $\mathbf{x}(1)$, $\mathbf{x}(2)$, and $\mathbf{x}(3)$ and the outputs $y(0)$, $y(1)$, $y(2)$, $y(3)$:

(a) $$\begin{bmatrix} x_1(k+1) \\ x_2(k+1) \end{bmatrix}=\begin{bmatrix} 1 & -2 \\ -1 & 1 \end{bmatrix}\begin{bmatrix} x_1(k) \\ x_2(k) \end{bmatrix}$$

$$y(k)=\begin{bmatrix} 3 & 0 \end{bmatrix}\begin{bmatrix} x_1(k) \\ x_2(k) \end{bmatrix}$$

$$\begin{bmatrix} x_1(0) \\ x_2(0) \end{bmatrix}=\begin{bmatrix} -4 \\ 0 \end{bmatrix}$$

(b) $$\begin{bmatrix} x_1(k+1) \\ x_2(k+1) \\ x_3(k+1) \end{bmatrix}=\begin{bmatrix} 0 & 1 & 0 \\ -1 & 0 & 1 \\ 2 & 1 & -2 \end{bmatrix}\begin{bmatrix} x_1(k) \\ x_2(k) \\ x_3(k) \end{bmatrix}+\begin{bmatrix} 0 \\ 1 \\ -1 \end{bmatrix}(-1)^k$$

$$y(k)=\begin{bmatrix} 3 & 0 & 4 \end{bmatrix}\begin{bmatrix} x_1(k) \\ x_2(k) \\ x_3(k) \end{bmatrix}$$

$$\begin{bmatrix} x_1(0) \\ x_2(0) \\ x_3(0) \end{bmatrix}=\begin{bmatrix} 0 \\ 0 \\ 0 \end{bmatrix}$$

ans. $$\begin{bmatrix} 0 \\ 1 \\ -1 \end{bmatrix}, \begin{bmatrix} 1 \\ -2 \\ 4 \end{bmatrix}, \begin{bmatrix} -2 \\ 4 \\ -7 \end{bmatrix}, \ 0, -4, 19, -34$$

23. Find step-invariant discrete-time system approximations to the following analog transmittances, using the indicated sampling intervals T:

(a) $G(s)=\dfrac{1}{s}$ $\qquad T=0.01$

(b) $G(s)=\dfrac{1}{s+1} \qquad T=0.5$

ans. $0.393/(z-0.607)$

(c) $G(s)=\dfrac{s}{s^2+1} \qquad T=0.02$

(d) $G(s)=\dfrac{1-e^{-0.5s}}{s+4} \qquad T=0.1$

ans. $(0.0825)(z^5-1)/(z^5-0.67z^4)$

(e) $G(s)=\dfrac{10(1-e^{-0.1s})}{s^2+4s} \qquad T=0.01$

24. An *impulse-invariant* approximation of an analog transmittance by a digital system is the digital system with unit pulse response which consists of samples of the unit impulse response of the analog transmittance. Find the z-transfer function of the impulse-invariant approximation to

$$G(s)=\frac{10}{s+4}$$

with sampling interval $T=0.2$.

25. Find the z-transfer functions of the systems of Fig. P8-3. For each system, the sampling interval is $T=0.3$.

ans. (b) $(0.7z+0.35)/(z^2-0.5z-0.15)$

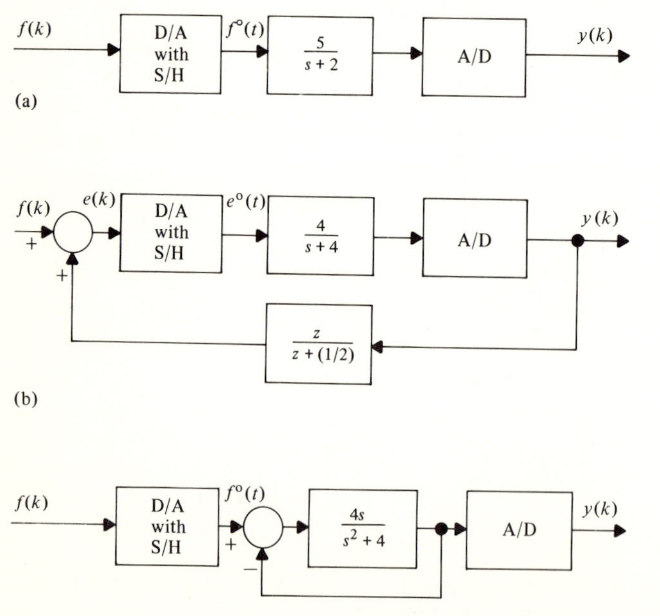

FIGURE P8-3

26. For the system of Fig. P8-4, with sampling interval $T=0.2$, it is desired that the z-transfer function be

$$D(z)=\frac{Y(z)}{F(z)}=\frac{z}{z-1}$$

Find, if possible, the necessary analog transmittance $G(s)$.

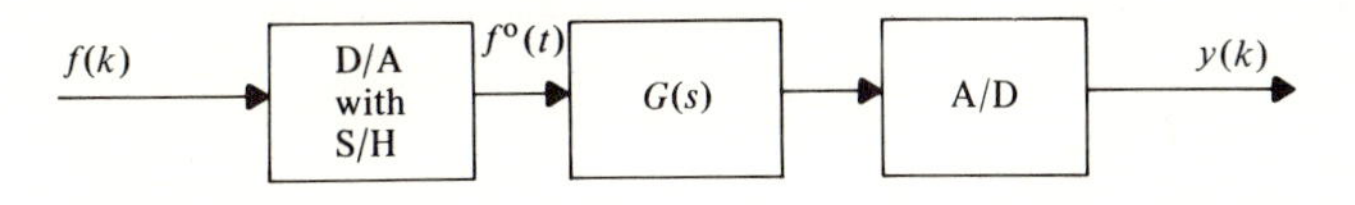

FIGURE P8-4

27. For each of the systems of Fig. P8-5, find the z-transfer function relating $y(k)$ to $r(k)$. The sampling interval is $T=0.2$. In Fig. P8-5 b and c, note that an equivalent system is one in which $Y(s)$ and $R(s)$ are first A/D-converted then summed, as in Fig. P8-5a.

ans. (b) $\dfrac{1.69(z-1)^2}{z(z^2-1.65z+1)+1.69(z-1)^2}$

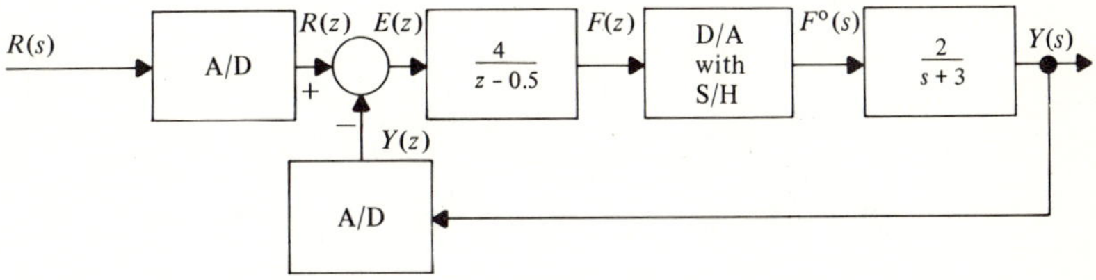

(a)

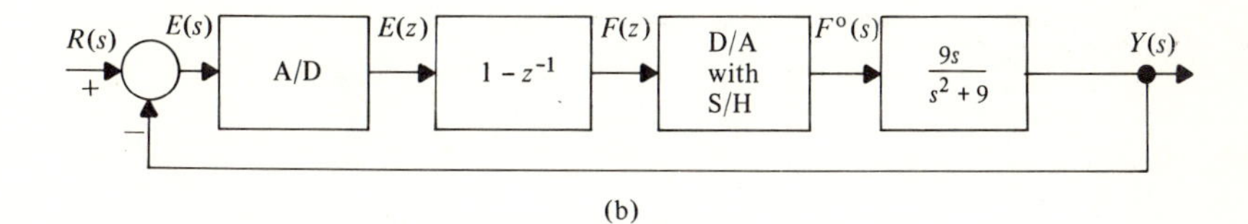

(b)

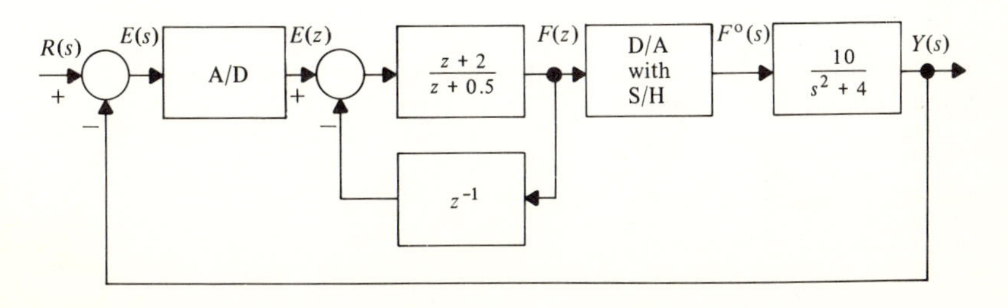

(c)

FIGURE P8-5

28. Determine whether the systems of Fig. P8-6 are stable. For each system, the sampling interval is $T=0.5$.

ans. (b) stable

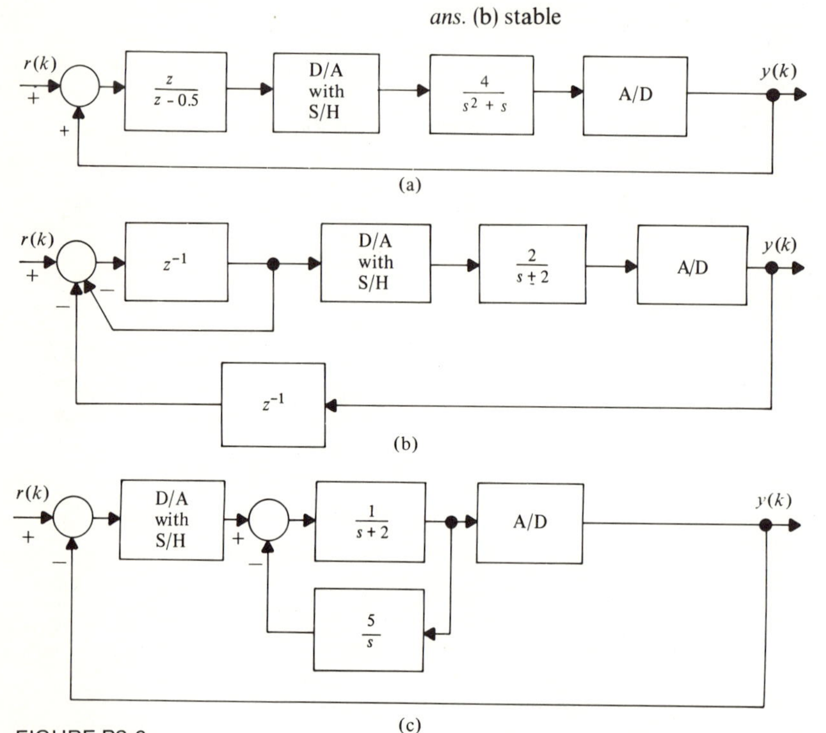

FIGURE P8-6

29. For systems with the following z-transfer functions, find the normalized steady state error, between output and input, for a step input:

(a) $D(z)=\dfrac{20z^2}{12z^2+7z+1}$

(b) $D(z)=\dfrac{0.25z^2-3z}{z^2+1.6z+0.48}$

ans. infinite error; system unstable

(c) $D(z)=\dfrac{2}{2z^3-z^2+z}$

30. For the system of Fig. P8-7, find a transmittance $H(z)$ so that the overall transfer function $D(z)=Y(z)/R(z)$ is deadbeat.

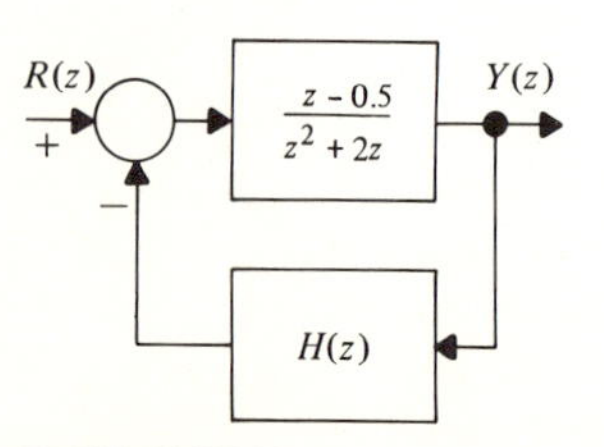

FIGURE P8-7

INDEX

C

D

E

U,V

Z